ADVANCED NANOMATERIALS FOR POLLUTANT SENSING AND ENVIRONMENTAL CATALYSIS

ADVANCED NANOMATERIALS FOR POLLUTANT SENSING AND ENVIRONMENTAL CATALYSIS

Edited by
QIDONG ZHAO

ELSEVIER

Elsevier
Radarweg 29, PO Box 211, 1000 AE Amsterdam, Netherlands
The Boulevard, Langford Lane, Kidlington, Oxford OX5 1GB, United Kingdom
50 Hampshire Street, 5th Floor, Cambridge, MA 02139, United States

British Library Cataloguing-in-Publication Data
A catalogue record for this book is available from the British Library

Library of Congress Cataloging-in-Publication Data
A catalog record for this book is available from the Library of Congress

ISBN: 978-0-12-814796-2

For information on all Elsevier publications visit our website at https://www.elsevier.com/books-and-journals

Publisher: Matthew Deans
Acquisition Editor: Simon Holt
Editorial Project Manager: Leticia Lima
Production Project Manager: Joy Christel Neumarin Honest Thangiah
Cover Designer: Greg Harris

Typeset by MPS Limited, Chennai, India

CONTENTS

List of Contributors ix
Preface xi

1 Hollow Micro- and Nanomaterials: Synthesis and Applications 1
Baojun Liu and Xia Hu

1.1 Introduction 1
1.2 Some Common Features of Hollow Micro- and Nanostructures 7
1.3 Fabrication Methodology of Hollow Structures 9
1.4 Application for Sensing and Catalysis 22
1.5 Utilization in Photocatalysis for Degrading Pollutants 24
1.6 Application for Rechargeable Batteries and Supercapacitors 26
1.7 Conclusion 28
References 29
Further Reading 38

2 Noble Metal–Based Nanosensors for Environmental Detection 39
Wei Xiong, Pancras Ndokoye and Michael K.H. Leung

2.1 Properties of Noble-Metal Nanoparticles 39
2.2 Colorimetric Sensing of Heavy Metal 45
2.3 Fluorescence-Based Sensing Towards Biomolecules 51
2.4 Surface-Enhanced Raman Spectrum-Based Application for Environmental Detection 55
2.5 Electrochemical Sensors 65
2.6 Conclusion and Future Perspectives 71
Acknowledgment 72
References 73

3 Semiconductor Nanocrystal–Based Nanosensors and Metal Ions Sensing 79
Jun Ke

3.1 General Properties of Semiconductor Nanocrystals 79

3.2 Semiconductor Quantum Dot–Based Nanosensors: Principles and Applications 85
3.3 Sensing of Metal Ions in the Environmental Field 99
3.4 In Vivo and In Vitro Sensing of Metal Ions 107
3.5 The Future of Advanced Quantum Dot–Based Nanosensors 109
References 110

4 Semiconductor Nanocrystals for Environmental Catalysis 119
Jun Ke

4.1 Introduction to the Basics of the Catalysis and Flexibility Provided by Nanocrystals 119
4.2 Preparation and Characterization of Nanocrystals for Catalysis 125
4.3 Nanocrystals for CO_2 Catalytic Conversion 140
4.4 The Future of Nanocrystals for Catalysis 154
Acknowledgment 155
References 155

5 Nano-Gold Boosted Environmental Catalysis 165
Wei Xiong, Pancras Ndokoye and Michael K.H. Leung

5.1 Introduction to the Catalytic Properties of Gold 165
5.2 Synthesis and Characterization of Nano-Gold 167
5.3 Environmental Catalysis of Supported Gold Catalysts 177
5.4 Surface Plasmon Resonance–Enhanced Photocatalysis 188
5.5 Conclusion and Future Perspectives 195
Acknowledgments 196
References 197

6 Nanomaterials Developed for Removing Air Pollutants 203
Qidong Zhao, Xinyong Li, Qiang Zhou, Dan Wang and Huixin Xu

6.1 Introduction to the General Principles of Air Pollutants Removal by Nanomaterials 203
6.2 Reactive Nanomaterials With Well-Defined Physical and Chemical Structures 208
6.3 Common Air Pollutants and Challenges in Air Purification 222

6.4 Nanomaterials for Eliminating Air Pollutants Through Adsorption and Separation ... 225
6.5 Converting Air Pollutants Through Catalytic Pathways of Nanomaterials ... 229
6.6 Technical Aspects and Practical Applications ... 235
6.7 Challenges and Perspective ... 239
Acknowledgment ... 241
References ... 241
Further Reading ... 247

7 Advanced Nanomaterials for Degrading Persistent Organic Pollutants ... 249
Qidong Zhao, Xiuming Zhao and Jingjing Cao

7.1 Introduction to the Advantages of Nanomaterials Toward Persistent Organic Pollutants Removal ... 250
7.2 Current Status and Challenges in Degrading Anthropogenic Persistent Organic Pollutants ... 258
7.3 Degrading Persistent Organic Pollutants by Electrochemical and Photocatalytic Techniques Enhanced With Nanomaterials ... 273
7.4 Perspective on Developing Efficient Nanomaterials for Removing Persistent Organic Pollutants ... 290
Acknowledgments ... 295
References ... 295
Further Reading ... 304

8 Power Ready for Driving Catalysis and Sensing: Nanomaterials Designed for Renewable Energy Storage ... 307
Anmin Liu and Xuefeng Ren

8.1 Introduction to the Basics of Renewable Energy Storage and Opportunities From Nanomaterials ... 307
8.2 Nanomaterials for Lithium-Ion Batteries ... 309
8.3 Nanomaterials for Sodium-Ion Batteries ... 327
8.4 Rechargeable Batteries for Driving Catalysis and Sensing ... 337
8.5 Challenges and Future Perspectives ... 338
Acknowledgments ... 339
References ... 339

9 Colloidal Semiconductor Quantum Dot–Based Multicomponent Artificial System for Hydrogen Photogeneration ... 347
Ping Wang

9.1 Introduction to the Basics of Nanomaterials for Hydrogen Photogeneration ... 347
9.2 Structure Design of Colloidal Semiconductor Quantum Dots ... 350
9.3 Surface Treatments ... 359
9.4 Conclusion ... 372
Acknowledgments ... 372
References ... 373

10 Nanocarbon-Based Hybrids as Electrocatalysts for Hydrogen and Oxygen Evolution From Water Splitting ... 379
Muhammad Adnan Younis and Yang Hou

10.1 Introduction to the Principles of Water Splitting Through Electrocatalysis ... 379
10.2 Nanocarbons ... 381
10.3 Synthesis, Structural Characteristics, and Electrochemical Performance of Nanocarbon-Based Hybrids ... 384
10.4 Electrochemical Properties of Electrocatalysts Supported on Graphene/Modified-Graphene ... 394
10.5 Conclusions and Future Perspectives ... 408
Acknowledgments ... 410
References ... 410

Index ... 419

LIST OF CONTRIBUTORS

Jingjing Cao Dalian University of Technology, Dalian, P.R. China

Yang Hou Key Laboratory of Biomass Chemical Engineering of Ministry of Education, College of Chemical and Biological Engineering, Zhejiang University, Hangzhou, P.R. China

Xia Hu Guizhou University, Guiyang, P.R. China

Jun Ke School of Chemistry and Environmental Engineering, Wuhan Institute of Technology, Wuhan, P.R. China

Michael K.H. Leung Ability R&D Energy Research Centre, School of Energy and Environment, City University of Hong Kong, Hong Kong, P.R. China

Xinyong Li Dalian University of Technology, Dalian, P.R. China

Anmin Liu State Key Laboratory of Fine Chemicals, School of Petroleum and Chemical Engineering, Dalian University of Technology, Panjin, P.R. China

Baojun Liu Guizhou University, Guiyang, P.R. China

Pancras Ndokoye Science, Technology and Innovation Unit, Directorate of Education Policy and Planning, Rwanda Ministry of Education, Kigali, Rwanda

Xuefeng Ren School of Food and Environment, Dalian University of Technology, Panjin, P.R. China

Dan Wang Dalian University of Technology, Dalian, P.R. China

Ping Wang State Key Laboratory of Electroanalytical Chemistry, Changchun Institute of Applied Chemistry, Chinese Academy of Sciences, Changchun, P.R. China

Wei Xiong Key Laboratory of Industrial Ecology and Environmental Engineering (Ministry of Education), School of Environmental Science and Technology, Dalian University of Technology, Dalian, P.R. China; Ability R&D Energy Research Centre, School of Energy and Environment, City University of Hong Kong, Hong Kong, P.R. China

Huixin Xu Dalian University of Technology, Dalian, P.R. China

Muhammad Adnan Younis Key Laboratory of Biomass Chemical Engineering of Ministry of Education, College of Chemical and Biological Engineering, Zhejiang University, Hangzhou, P.R. China

Qidong Zhao Dalian University of Technology, Dalian, P.R. China

Xiuming Zhao Dalian University of Technology, Dalian, P.R. China

Qiang Zhou Dalian University of Technology, Dalian, P.R. China

PREFACE

Nanomaterials as building units have brought a unique opportunity for advancing science and technology to create updated products meeting the global demands of environmental remediation and sustainable development. The utilization of nanomaterials in pollution detection and catalysis has been growing rapidly over the past few decades. This book describes current advances in the knowledge and perspectives of nanomaterials for environmental pollutant sensing, removal, and renewable energy through catalytic strategies. It is intended for engineers, researchers, and students engaged in the joint fields of materials, chemistry, environmental science, and engineering as well as nanotechnology.

Advanced Nanomaterials for Pollutant Sensing and Environmental Catalysis focuses on recent insights and scientific discoveries in the joint fields of environmental protection, sensing, and renewable energy with nanotechnology. The chapters in this book present examples of cutting-edge research in three broad categories: (1) emerging applications of nanomaterials in sensing various pollutants, including metal ions and organic molecules; (2) advances in research on pollution control through catalysis methods enforced by nanotechnology; and (3) energy conversion and water splitting based on nanomaterials toward pollutant sensing and environmental catalysis. Throughout, the authors highlight the timely progress with a particular focus on the unique physicochemical properties of nanomaterials and underlying mechanisms which promise their enhanced performances. Topics covered include novel, hollow micro or nanostructures with fantastic functions (Chapter 1: Hollow Micro- and Nanomaterials: Synthesis and Applications), nanosensors (Chapter 2: Noble Metal–Based Nanosensors for Environmental Detection and Chapter 3: Semiconductor Nanocrystal–Based Nanosensors and Metal Ions Sensing), elimination of diverse pollutants (Chapters 4–7), energy storage with lithium-ion and sodium-ion batteries for powering the sensing and catalysis processes (Chapter 8: Power Ready for Driving Catalysis and Sensing: Nanomaterials Designed for Renewable Energy Storage), and advanced nanocatalysts for water splitting (Chapter 9: Colloidal Semiconductor Quantum Dot–Based Multicomponent Artificial System for Hydrogen Photogeneration and Chapter 10: Nanocarbon-Based Hybrids as Electrocatalysts for Hydrogen and Oxygen Evolution From

Water Splitting). The diverse contents demonstrate the basics, rapid developments and challenges in these scientific fields.

It features strong interdisciplinary backgrounds of chemical, material, energy, and environmental science besides nanotechnology. The pollutants concerned include various air pollutants and persistent organic pollutants beyond traditional water pollutants. It also provides the readers a novel view on harvesting renewable energy by water splitting through catalytic techniques based on advanced nanomaterials. We hope the collective information in the book will help engineers, academic researchers, and environmental protection officials and agencies to protect and sustain our environment.

Finally, we would like to express our appreciation for the support from people who helped in preparing and reviewing the manuscript as well as the proposal. We especially thank our families for their patience while we were working on the book. We are also indebted to the literature whose authors have made pioneering contributions toward the rapid progress in this field of science and technology.

1

HOLLOW MICRO- AND NANOMATERIALS: SYNTHESIS AND APPLICATIONS

Baojun Liu and Xia Hu
Guizhou University, Guiyang, P.R. China

1.1 Introduction

Hollow micro and nanomaterials are a special class of functional nanomaterials, defined as architectures with void space surrounded by a shell [1,2]. Over the past two decades, the synthesis of different hollow structures has become a hot research topic and attracted tremendous research efforts.

Before 1998, most hollow materials were of spherical shape and synthesized by using spray-drying or gas-blowing techniques [3]. However, these methods only culminated in the simple sol–gel approaches for coating gold (Au) and (silver) Ag nanoparticles with silica [4,5] before 1998 with Caruso's work on colloidal templating synthesis of hollow spheres [6]. The emerging approaches enabled a more versatile synthesis paradigm for fabricating hollow structures based on hard-templating ways. Currently, various novel self-templated or template-free synthetic approaches based on different mechanisms have been developed such as galvanic replacement, controllable thermal transformation, ion exchange, and inside-out Ostwald ripening to tailor the structural features of hollow nanostructures. Encouragingly, a myriad of hollow nanostructures with hierarchical architectures, polyhedral morphologies, multicompositions, and multishells have been successfully constructed through rationally designed strategies involving these mechanisms (Fig. 1.1).

1.1.1 Single-Shelled Hollow Structures

Owing to their facile preparation, single-shelled spheres composed of numerous nanosized subunits are the most

Advanced Nanomaterials for Pollutant Sensing and Environmental Catalysis. DOI: https://doi.org/10.1016/B978-0-12-814796-2.00001-0

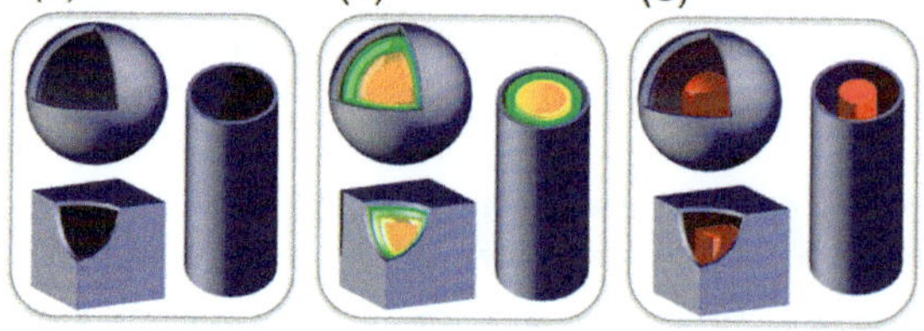

Figure 1.1 Schematic illustration showing various hollow structures: (A) hollow spheres, boxes, and tubes; (B) multishelled hollow spheres, boxes, and tubes; and (C) yolk-shell, cube-in-box, and wire-in-tube structures. Reprinted with permission from X.J. Wang, J. Feng, Y.C. Bai, Q. Zhang, Y.D. Yin, Synthesis, properties, and applications of hollow micro-/nanostructures. Chem. Rev. 116 (2016) 10983–11060. Copyright 2016, American Chemical Society.

common form for hollow micro- and nanostructures. Since the pioneering work by Caruso et al., on the fabrication of single-shelled hollow spheres in 1998 [6], many methods have been reported for the successful preparation of these kinds of hollow structures. The size, thickness, and crystallinity of different shell materials have been well-controlled by adjustments of the preparation parameters. For example, Lou et al., have reported the preparation of TiO_2 hollow spheres constructed from small nanocrystals via a hard-templating method [7]. The released gases from the inner carbonaceous spheres could inhibit crystallite growth in the TiO_2 coating layer during the annealing treatment [8]. One typical example of single-shelled, tubular-like structures is the carbon nanotube (CNT), which has attracted increasing research interest due to the unique functional properties in electrical and thermal transfer [9–12]. Compared to simple spherical and tubular counterparts, great challenges existed in the synthesis of single-shelled hollow structures with well-defined nonspherical morphologies (such as cube-like, prism-like, and bowl-like configurations) by traditional hard-templating methods, which was mainly due to the lack of available templates and less controllable coating around high-curvature surfaces [13]. With the development of self-templated methods through miscellaneous mechanisms and the emergence of metal–organic compounds with variant structures, the creation of single-shelled polyhedral micro and nanostructures has been expanded to a variety of species [14,15]. For example, single-shelled cubic nanocages of Pt with subnanometer thick walls could be prepared using Pd solid nanocubes as the sacrificing templates [16]. In another case, unusual deflated bowl-like polystyrene hollow particles can be derived from normal single-shelled hollow particles through a simple drying treatment [17].

Besides shape control, researchers tried to incorporate different structural features into hollow micro- and nanostructures to infuse new vitality into the utilization of these attractive architectures. Mesoporous channels penetrating the shell enable free diffusion of guest molecules into the hollow interiors [18]. The permeable shells with well-defined porosity could serve as useful platforms in various applications such as biomedicine and catalysis [19]. For instance, Zheng et al., reported the preparation of hollow mesoporous alumino silica spheres with ordered pore arrangements as advanced nanoreactors for catalysis [20]. Yec and Zeng have described the self-assembly of nanobubbles into a large single-shelled manganese silicate hollow sphere [21]. Characterization by transmission electron microscopy (TEM) indicated that the shell of these hierarchical hollow spheres is composed of nanospheres with a diameter range of 7–9 nm.

Fabricating single-shelled hollow micro- and nanostructures from 1D to 3D building blocks would benefit from the structural advantages of primary nanoparticles. For example, Yao et al., reported the synthesis of single-shelled hollow spheres assembled by Co_3O_4 nanosheets [22]. Morphological characterizations reveal that the hierarchical shell is built up from numerous primary building blocks pointing toward the centers. Pang et al., described the assembly of metal organic frame (MOF) nanocubes into single-shelled hollow colloidosomes with tailorable central space and shell thickness [23]. The monolayer of cubic iron-based MOFs (Fe-soc-MOF) in the hollow microspheres could be identified from both scanning electron microscopy and TEM images.

1.1.2 Multishelled Hierarchical Hollow Structures

Hollow structures with multilevel interiors are those hollow materials with multiple shells, chambers, or channels [24]. Multishelled hollow micro- and nanostructures, also known as Matryoshkas-like hollow nanostructures, are the most frequently studied type of complex hollow structures. Compared with single-shelled hollow materials, the synthesis and manipulation of multishelled hollow structures are much more challenging. It appears that the most straightforward method for the creation of hollow structures with multiple shells is the hard-templating strategy based on reduplicative coating and a subsequent etching process. However, sophisticated synthetic procedures bring great difficulties in actual implementation, such as potential agglomeration of nanoparticles and uniformity issues. Also, additional sacrificial layers are needed as barriers between the shells of target materials [25,26]. With the development of synthetic methods, researchers have simplified the preparation steps by exploring new formation concepts or optimizing synthetic routes.

Lou et al., prepared a typical double-shelled SnO_2 hollow nanospheres through a multistep coating and etching process [27]. During the successive coating of target SnO_2 material, barrier layers are not needed due to the structural heterogeneity of different shells. Yolk–shelled or rattle-type structures are another important class of complex hollow structures. Qiao et al., experimentally obtained yolk–shelled SiO_2 spheres with a hierarchical porous structure [28]. Due to the chemical dissimilarity between the yolk and shell, barrier layers are not involved in this case. In recent years, modified formation mechanisms can avoid tedious coating steps for multishelled hollow nanostructures. Wang et al., reported the synthesis of multishelled Co_3O_4 hollow spheres using carbonaceous microspheres as templates [29]. The shell of hollow spheres can be conveniently controlled from 2 to 4 by adjusting synthetic parameters. Aside from hollow spheres with multiple shells, modern synthetic methodologies enable the syntheses of multishelled hollow structures with tailorable morphologies. For example, the tube in-tube nanostructures of various mixed metal oxides could be obtained from 1D electrospun precursor polymers [24].

The design and synthesis of hollow materials with multiple channels or chambers are very intriguing as these structures could mimic the architecture in nature. Unfortunately, due to the complicated microstructures of defined channel configuration, successful achievement of these artificial micro and nanostructures are still rare. The major difficulty lies in the precise control of the number of inner channels or chambers. By great endeavors to the task, several investigators have achieved encouraging results on these unique structures by employing combined methods. Li et al., have synthesized lotus root-like multichannel carbon (LRC) nanofibers through an electrospinning method with a subsequent annealing process [30]. The microemulsion formed by PS in polyacrylonitrile (PAN) solution plays a critical role in controlling the channel number and size in the LRC nanofibers. When the weight ratio of PS to PAN is tuned in a proper range, LRC nanofibers with tunable multichannels could be obtained. Xu et al., developed a "magnetic-guiding" strategy to generate segmented tubes of cobalt chalcogenides [31]. Due to the strong magnetic dipolar interactions between cobalt nanoparticles, a chain-like wire was formed as the precursor. Subject to a sulfidation reaction, the obtained Co_3S_4 hollow nanocrystals preserved the chain-like morphology to form a segmented nanotube.

1.1.3 Hollow Materials with Open Structures

Hollow particles with open structures such as those with large through-holes or hollow polyhedra with frame-like open architectures are a special class of complex hollow micro and nanostructures. These open structures could facilitate fast loading and release of foreign species as well as alter the density of low-coordination atoms on the surface [32]. Various template methods have been developed including sacrificial templates [33,34], galvanic replacement, and preferential etching [35–37] for synthesis of these special hollow structures. For instance, Guan et al., reported an MOF composite-assisted strategy to obtain single-holed cobalt/N-doped carbon hollow particles [33]. Throughout the combustion of the inner template, the released gaseous species help open holes on the surface of hollow spherical particles.

Different structured facets or edges of polyhedra particles usually possess varying densities of defects with different surface energy. Through selective etching, unstable facets or edges might be preferentially cut off to generate frame-like, open structures. Han et al., successfully obtained Ni–Co Prussian-blue-analog (PBA) nanocages with unusual interiors by a facile chemical etching route [37]. During the process, active corners with more defects are dissolved first. The sustaining etching proceeds along the body diagonal direction of the cubes, resulting in cage-like Ni–Co PBA hollow nanoparticles with pyramid-like walls. Kuo and Huang reported that different faces could be exposed with truncated rhombic dodecahedral Cu_2O nanoframes through the sustaining particle growth and etching process [38]. Their experimental observations indicate that the nanoframe possessed different facets under altered reaction durations. Xia et al., reported the selectively etching preparation of Rh cubic nanoframes [39], where Rh atoms only accumulated on the corners and edges of the Pd cube to form Pd@Rh core–frame concave nanoparticles. With removing the Pd core, they got an open frame-like structure.

Besides the above achievements, atomic-level engineering in hierarchical frame-like open nanostructures has proved to be an effective strategy to further increase the atomic utilization [40]. For example, fivefold–twinned PtCu hierarchical nanoframes have been synthesized by a one-pot solution method [41]. High-angle annular dark-field scanning TEM (HAADF-STEM) and TEM images confirm that the framelike products featured dense nanothorns protruding from edges. The length of nanothorns can be controlled by modifying the amount of ethanolamine input to the reactor.

Similar to hollow structures with intact surface, these open structures could also behave as containers to incorporate foreign mass. For example, Neretina et al., synthesized Wulff-shaped nanostructures which were confined within nanoframes in a solution growth mode on substrate-based templates [42]. Wulff-shaped Pt nanoparticles can be incorporated into Wulff-shaped or cube-shaped Ag nanoframes after sequential coating and selective etching processes. Promisingly, such versatile preparation routes can be generalized and adopted for obtaining different hybrid nanomaterials confined in nanoframes with tailored chemical compositions and diverse morphologies.

1.1.4 Hollow-Structured Hybrid Materials

Apart from tailoring morphological structures, compositional adjustment is an important research topic for the establishment and application of hollow micro- and nanostructures [43]. The integration of various chemical species into a desirable hollow-structured particle could effectively enrich the functional properties of a single component for more applications. A variety of hollow-structured hybrid materials between chemical species of the same category or different categories have been realized as functional materials with inspiring results.

For instance, two distinct metals could be integrated into one hollow configuration in various forms (such as core–shell, core–frame, dendritic, and heterostructured structures) [44]. Gonzalez et al., reported the preparation of polymetallic hollow nanoparticles with controlled composition via a strategy of simultaneous or sequential galvanic replacement and the nanoscale Kirkendall growth at room temperature [36]. Hybrid hollow micro- and nanostructures between chemical species of the same category have also attracted much research interest due to their diverse functionality. Yolk–shelled and multishelled structures are two common forms of such hybrid hollow structures [2,45]. Benefitting from the synthetic versatility, hollow nanoboxes with tailorable chemical compositions in different shells could be obtained.

Beyond these studies, chemical species of different categories can also be integrated into a single hollow structured micro- and nanoparticles. Reports on inorganic compounds such as metal oxides, hydroxides, and chalcogenides are emerging more and more, and can be mixed with each other as direct nanoreactors utilized in catalysis, drug delivery, or as precursors for producing more complex structures [19,46,47]. For instance, encapsulating metallic nanocrystals within a protective shell could benefit sustainable catalysis at higher temperatures. Impressively, Chen et al., reported the synthesis of these kinds of hybrid hollow spheres by employing the Stöber method [48].

It is well-known that carbon-based materials in various forms can be applied as conductive additives (such as CNTs, graphene) or protective layers (such as amorphous carbon) for various electrode materials with high-specific capacity/capacitance in secondary batteries and supercapacitors [49–51]. Comprehensive investigations have been conducted for developing carbon-based hybrid hollow structures. Some instances are introduced briefly here. Wang et al., employed CNTs as the support to synthesize hollow Fe_2O_3 nanohorns to further obtain hierarchical nanotubes for constructing electrodes in lithium-ion batteries (LIBs) [52]. Apart from direct coating with carbon mass, the utilization of MOFs helps realize the in situ generation of carbon species with A higher degree of graphenic order around ultrafine nanoparticles composed of metal-based species. In a typical reported work, hollow ball-in-ball nanostructures of NiO/Ni/graphene were achieved through a MOF-engaged method [53]. The obtained composite nanoparticle stably exhibits the same hierarchical structure as its Ni-MOFs precursor throughout the annealing treatment.

1.2 Some Common Features of Hollow Micro- and Nanostructures

Compared with solid counterparts, hollow structures especially those with permeable and porous shells exhibit many unique features such as high-surface area, low density, and high-loading capacity. Thus they show great potential in a wide range of fields such as sensing, chemical catalysis, rechargeable batteries, and supercapacitors [49–59].

Structural parameters of the multishelled hollow structures including the shape, shell number, size, and thickness as well as the intershell spacing and so forth, endow them some special physical or chemical properties, thus, making enormous influence on their performance in practical applications. As a result, geometric manipulation of multishelled hollow structures is of great importance and will be discussed in detail next.

1.2.1 Shell Number

Shell number is one of the most important features for multishelled hollow structures, which distinguishes them from their single or double-shelled counterparts and results in some unique chemical and physical properties. Considerable efforts have been devoted to control the shell number of multishelled hollow structures via various techniques. The shell number of multishelled hollow metal oxide spheres is related to the concentration and penetration depth of metal ions within the carbonaceous templates and can be well adjusted by changing the concentration of the metal salt solution during the adsorption process [56]. And, then, proceeding with an in-depth study, other parameters such as the solvent type and compositions and immersion period also have a remarkable effect on the final shell number of the resulting multishelled hollow structures. For example, in the case of TiO_2 hollow spheres, the shell number could be adjusted from 1 to 3 by regulating the immersion period of Ti coordination ions with carbonaceous templates from 6, 8, to 24 h [59].

1.2.2 Shell Size, Thickness, and Intershell Spacing

Besides the shell number, size (diameter of the shell), and thickness, the intershell spacing also plays an important role in determining the interior structure and have a significant effect on their performances in practical applications [29,60]. With the sequential templating approach, the size of the outer shell is mainly determined by that of templates adopted and could be well-tuned to some extent, while the sizes of inner shells are related to the forming conditions. For hollow SnO_2 microspheres, such special structures with a closed double-shell in the exterior could be produced by simply alkali-treating carbonaceous microspheres and, thus, increasing the absorbed quantity of Sn ions in their outer segment [61]. As a result, an additional shell was formed by this alkali treatment compared with those without it.

1.2.3 Morphology

Until now, multishelled hollow structures are mostly spherical in morphology [62–66]. Well-defined, nonspherical, multishelled hollow structures have been rarely reported due to the difficulty of forming uniform shells with high-curvature surfaces. However, there are some successful examples for constructing nonspherical, multishelled, hollow structures by adopting novel precursors. Wang et al., report a facile "pumpkin-carving" strategy for the preparation of multishelled, hollow, single-crystal $CoSn(OH)_6$ nanoboxes [67]. Perovskite-type $CoSn(OH)_6$ nanocubes were used as precursors and gradually dissolved in a concentrated alkaline solution by coordinating an excess amount of OH^- to form soluble $[Co(OH)_4]^{2-}$ and $[Sn(OH)_6]^{2-}$ at room temperature. An insoluble boundary layer of Co(III) species from the oxidation of $[Co(OH)_4]_2^-$ in air can be readily formed on the surface of $CoSn(OH)_6$ crystals, thereby reducing the alkaline etching. With continuous evacuation of the core materials across the shell, hollow $CoSn(OH)_6$ nanoboxes were eventually formed. More impressively, $CoSn(OH)_6$ hollow architectures with multishelled nanoboxes can be produced by repeating the deposition of $CoSn(OH)_6$ layers onto pregrown $CoSn(OH)_6$ particles (e.g., nanocubes or nanoboxes) and successive alkaline etching.

1.3 Fabrication Methodology of Hollow Structures

In this section, we present a comprehensive overview of synthetic strategies for hollow structures according to their formation mechanisms including well-established hard-templating and soft-templating methods as well as newly emerging self-templated approaches. Apart from these template-based syntheses, template-free methods are also reviewed.

1.3.1 Hard-Templating Methods

The hard templating method for the synthesis of hollow structures is very straightforward [1]. Briefly, hard templates with specific shapes were synthesized first, followed by coating the outer surface with a layer of the desired material. The core materials were then selectively removed to obtain the hollow structure. To achieve successful coating on the surface of the template, a surface modification step that can change the surface functionality such as surface charge or polarity is usually applied. A number of methods, such as sol–gel processes or hydrothermal reactions, could be used to deposit the shell materials on the template surface. The selective removal of the hard template could be achieved through chemical etching, thermal treatment or calcination, or simply by being dissolved in particular solvents. The choice of template removal method is mainly determined by the composition of hard templates. In some cases, posttreatment such as reduction or calcination is required to improve certain properties of the resulting shells. In the following sections, we will introduce several typical hard templates on the basis of their different compositions, followed by detailed discussions on the control over the morphology of the hollow structures.

Preparation of hollow structures by templating against hard particles is conceptually straightforward. In general, it involves the 4 major steps: (1) preparation of hard templates; (2) functionalization/modification of the template surface to achieve favorable surface properties; (3) coating the templates with designed materials or their precursors by various approaches, possibly with posttreatment to form compact shells; and (4) selective removal of the templates to obtain hollow structures. The most commonly employed hard templates include nearly monodispersed silica particles and polymer latex colloids. These templates are advantageous for several reasons including their narrow size distribution, ready availability in relatively large amounts, availability in a wide range of sizes from commercial sources, and simplicity of their synthesis using well-known formulations. Other colloidal systems such as carbon nanospheres and nanoparticles of metals and metal oxides have also been used as templates for the preparation of hollow structures.

1.3.1.1 Polymer-Based Hard Templates

Templating against polymer nanoparticles to synthesize hollow micro and nanostructures might be the most popular method mainly due to the relative ease in selectively removing the templates. The commonly used polymers include polystyrene (PS) and its derivatives, formaldehyde resin, and poly (methyl methacrylate) (PMMA).

Polystyrene templates. The first report on the fabrication of hollow nanostructures was in 1998 [2] when Caruso et al., prepared hollow inorganic silica and inorganic-polymer hybrid spheres through the electrostatic layer-by-layer (LBL) self-assembly of silica nanoparticles and polymer multilayers on colloidal templates. Polystyrene latex particles with a diameter of 640 nm were used as templates. Three-layer cationic poly-(diallyldimethylammonium chloride) (PDADMAC) film was then deposited onto the negatively charged PS particles. The positively charged surface could be utilized for the adsorption of SiO_2 particles with ~25 nm in diameter. Through an additional calcination or tetrahydrofuran (THF) dissolution process, PS templates could be removed, forming hollow silica (HS) or inorganic–polymer hybrid spheres. The shell thickness of the hollow spheres could be adjusted from tens to hundreds of nanometers by simply tuning the number of SiO_2-PDADMAC layer deposition cycles. A similar strategy could be applied to different colloidal cores, such as weakly cross-linked, melamine-formaldehyde (MF) particles. The MF core could be removed by exposing the coated particles to either acidic solutions ($pH < 1.6$) or dimethyl sulfoxide (DMSO).

Formaldehyde resin templates. Because weakly cross-linked MF colloidal particles can decompose in aqueous media with a pH value below 1.6, they have been widely used as hard templates for the synthesis of hollow structures. For instance, Donath et al., reported the synthesis of hollow polymer shells by templating against MF colloidal particles through the LBL assembly of polyelectrolytes [68]. Microsized hollow polyelectrolyte shells of poly(sodium-*p*-styrenesulfonate) (PSS) and poly(allylamine hydrochloride) (PAH) with thicknesses ranging from five to tens of nanometers, were prepared through this method in which polyelectrolytes (negatively charged PSS and positively charged PAH) from dilute aqueous solutions were first deposited onto the surface of monodispersed MF particles through electrostatic forces. After each adsorption step, the polyelectrolyte in solution was removed by repeated centrifugation and washing. Upon exposure to 100 mM HCl, the MF cores decomposed, forming hollow polyelectrolyte shells. The shell thickness could be easily controlled by adjusting the number of the deposited polyelectrolyte layers.

Similarly, monodisperse polyelectrolyte-supported asymmetric lipid-bilayer vesicles were prepared by Katagiri et al. [69]. The only difference from the assembly process is that PDADMAC was used as the cationic polyelectrolyte instead of PAH. When DNA was used as the ionic species, DNA/PAH multilayer microcapsules were synthesized by templating against colloidal MF particles [70].

Other polymer templates. Some other polymers have also been used as templates. For example, Tu et al., reported the synthesis of hollow spheres consisting of alternating titania and graphene nanosheets by templating against PMMA beads via a LBL adsorption technique [71]. PMMA beads were first successively modified with positively charged polyethylenimine

(PEI), negatively charged $Ti_{0.91}O_2$ nanosheets, another layer of PEI, and then a negatively charged graphene oxide (GO) suspension. This process was repeated several times to ensure enough titania and graphene loading. After microwave irradiation in an argon (Ar) atmosphere in the presence of carbon powder, GO was reduced to graphene. The PEI moiety was removed and the PMMA spheres as sacrificial templates were decomposed into exhaust gas. The remaining trifle PMMA residue can be completely removed with THF. Su et al., prepared the polypyrrole hollow nanospheres by in situ polymerization of pyrrole monomer on the PMMA surface and acetone washing that can remove the PMMA core [72].

1.3.1.2 Silica-Based Hard Templates

Silica is one of the most widely used hard templates because of its unique features, including low cost, high uniformity, widely tunable size, and so on. In this section, we will summarize the synthesis of hollow structures using different types of silica as the hard templates, including sol–gel-derived silica, mesoporous silica (MS), and silica shell.

Solid silica templates. Typically, colloidal silica particles are obtained through a classical "Stöber method" in which monodispersed silica spheres with sizes ranging from 50 nm to 2 μm can be prepared by hydrolyzing alkylsilicates in a water–alcohol mixture in the presence of ammonia solution [73]. In 2002 Hyeon et al., reported the fabrication of hollow palladium (Pd) spheres by using silica spheres as the template [74]. Uniform silica spheres were synthesized using the Stöber method. The silica surface was functionalized with modified polystyrene (MPS) by refluxing silica particles with $HS(CH_2)_3\text{-}Si(OCH_3)_3$ in toluene. The palladium precursor, palladium acetylacetonate ($Pd(acac)_2$), was then adsorbed onto the surfaces. The sample was heated at 250°C for 3 h to get metallic Pd. The template was finally removed by using 10 M HF as the etchant, leading to the formation of hollow Pd spheres.

Mesoporous silica templates. In contrast to colloidal silica obtained from the sol–gel process, MS(pore size in the range of 2–50 nm) is another type of material with a high-surface area [75]. By utilizing its porous structure, many hollow spheres or shells could be synthesized. Caruso et al., synthesized nanoporous (pore size <2 nm) polyelectrolyte spheres by sequentially coating the polyelectrolyte on sacrificial MS spheres [76]. First, oppositely charged polyelectrolytes poly(acrylic acid) (PAA) and

PAH were deposited within the bimodal MS spheres. The template used in this synthesis, which is bimodal MS, possesses 10–40 nm pores, making it suitable for infiltration and adsorption of polyelectrolytes. Prior to each deposition process, the MS spheres were heated at 160°C for 2 h to partially cross-link the shells. Finally, the template was removed upon exposure to HF, yielding hollow polyelectrolyte spheres. Yang et al., designed and fabricated a novel type of nanographene-constructed hollow carbon spheres [77].

Silica shell as hard templates. Similarly to silica nanoparticles, whether sol–gel-derived solid silica or mesoporous ones, hollow silica shells (HSs) could also serve as templates. Because HSs have cavities inside, the additional coating layer could grow inside or outside the shell, or even at both sides. Li et al., reported the synthesis of double-layer hollow silica shells (DHSs) by templating against mesoporous HSs [78]. The second hollow shell grew inside the hollow template. The key factor is the surface modification of organosilane onto the template surface. DHSs were synthesized by a simple treatment of HS-TPOAC in tetrapropylammonium hydroxide solution at 150°C. This "immobilized surfactant" strategy could also be used to synthesize other hollow nanostructures with high-level hierarchical complexities.

1.3.1.3 Carbon-Based Templates

Carbon spheres have been widely used as the hard templates because of their low cost and ease of removal. Furthermore some carbon spheres with loose and porous surfaces could absorb many precursors and facilitate the shell formation process.

Mesoporous carbon templates. Ordered mesoporous carbon structure could serve as hard templates to synthesize various mesoporous hollow structures in which the pore structure is an inverse replica of the mesoporous carbon templates. Tang et al., reported the synthesis of mesoporous spheres of metal oxides and phosphates by templating them against mesoporous carbon spheres [79]. The mesopores of carbon templates were first infiltrated with the desired precursors or inorganic salts by wet impregnation. When excess precursor was removed, the hydrolysis of the precursor was initiated upon exposure to the moisture in the air. After crystallization or polymerization of inorganic species at elevated temperature, the carbon templates were finally removed by calcination in air. Following this synthetic strategy, hollow mesoporous ZrO_2 spheres were obtained when the ethanol solution of $ZrCl_4$ was used as the precursor. The formation of hollow structures is due to the incomplete infiltration of the precursors into the inner parts of the carbon spheres.

Solid carbon templates. Titirici et al., synthesized metal oxide hollow spheres via a hydrothermal approach using carbon spheres as the template [80]. It is worth mentioning that the carbon templates were also synthesized during the hydrothermal process and, thus, it is a one-pot synthesis. Various metal salts and D-glucose monohydrates (serving as the carbon precursor) were dissolved in water and then the mixture was transferred into an autoclave. During the hydrothermal treatment, carbon spheres were formed with metal ions incorporated into their hydrophilic shells. The removal of carbon via calcination yielded hollow metal oxide spheres. A wide range of metal oxide hollow spheres such as Fe_2O_3, NiO, Co_3O_4, CeO_2, MgO, and CuO could be produced using this process.

1.3.1.4 Ceramic Templates

Iron oxide is a good hard template for the synthesis of hollow micro and nanostructures because it can be removed easily by acid treatment. Yi et al., prepared HS and rattle-type $Fe_2O_3@SiO_2$ nanoball structures by templating against Fe_2O_3 NPs [81]. First, $Fe_2O_3@SiO_2$ composite was synthesized by a reverse microemulsion technique. A layer of MS was then deposited onto the silica surface. After HCl treatment, the Fe_2O_3 core could be partially or completely removed by varying the concentration of etchant or etching time. Shi et al., also reported the synthesis of hollow MS shells by employing hematite as hard templates [82]. In contrast to Yi's synthesis, they coated MS directly on the hematite surface and used C_{18}TMS as the pore-formation agent. After calcination and HCl treatment, hollow MS spheres could be obtained. When spindle-shaped hematite particles were used as templates, elliptical MS hollow spheres could be obtained. Feyen et al., synthesized Fe_xO_y@PS-SiO_2 mushroom nanostructures and their hollow derivatives, which also employed iron oxide as hard templates [83]. The key point is the partial coating of Fe_xO_y (Fe_3O_4 or Fe_2O_3) with polymer spheres to form the Janus-type Fe_xO_y @PS nanospheres.

1.3.1.5 Hard Templates Based on Inorganic and Complex Salts

Many nano- or microsized inorganic salts, such as carbonates and complex salts, could serve as sacrificial hard templates. For example, nanosized calcium carbonate with controlled size and morphology can act as sacrificial hard templates. Chen et al., prepared porous HS nanoparticles with diameters in the range of 60–70 nm and wall thickness of approximately 10 nm by using $CaCO_3$ nanoparticles as templates [84]. Sodium silicate ($Na_2SiO_3 \cdot 9H_2O$) was added as the silica precursor. After forming a SiO_2 layer on the $CaCO_3$ surface, the template core was removed by HCl etching. Zhao et al., reported the synthesis of silica hollow nanostructures by using needle-like shaped $CaCO_3$ as the template [85]. Besides solid $CaCO_3$, Zhao et al., reported the use of porous amorphous $CaCO_3$ nanospheres prepared by a gas diffusion approach to synthesize hollow MS nanospheres [86]. Both anticancer drug doxorubicin (DOX) and silica precursor tetraethylorthosilicate (TEOS) could be absorbed by the porous templates. In the presence of hexadecyltrimethylammonium bromide (CTAB) surfactant, MS layer was uniformly coated on the $CaCO_3$ surface. Further removal of the templates by mild ethylic acid etching resulted in DOX-loaded hollow MS shells.

1.3.1.6 Natural Materials as Hard Templates

Many naturally existing materials can also be used as the hard templates to prepare hollow micro- and nanostructures. As compared to artificially synthesized micro- and nanomaterials, natural materials could be obtained with large quantity and considerably low cost. Ghadiri et al., reported the synthesis of nanostructured TiO_2 hollow fibers by templating against natural cellulose fibers [87]. Cellulose is an abundant natural polymer and is inexpensive, ecological, and economically favored for practical applications. The cellulose fibers were coated with a TiO_2 layer first, followed by the removal of cellulose templates via calcination. After the samples were heated at 500°C in air for 3 h, hollow TiO_2 fibers were obtained.

1.3.2 Soft-Templating Methods

Templating against hard (solid) templates is arguably the most effective, and certainly the most common, method for synthesizing hollow micro- and nanostructures. However, hard templates have several intrinsic disadvantages, ranging from the inherent difficulty of achieving high-product yields from the multistep synthetic process to the lack of structural robustness of the shells upon template removal. Additionally, key applications of hollow structures such as drug and therapeutic delivery require facile access to the hollow interior space. With the hard templates, refilling the hollow interior with functional species or in situ encapsulation of guest molecules during formation of the shells, though possible, is very challenging. Among these approaches, templating against soft (liquid or gaseous) templates has attracted the greatest attention and significant progress has been made over the past decade. In this section, we will survey commonly used soft templates, including emulsion droplets, surfactant, and other supramolecular micelles, polymer aggregates or vesicles, and gas bubbles and discuss their application to the synthesis of hollow structures.

In this section, we will discuss soft templating–synthetic strategies for preparing hollow nanostructures. The soft templating–synthetic strategies are widely applied for the synthesis of hollow nanostructures composed of polymers, SiO_2, metal oxides, carbons, and composites. The synthetic strategies show advantages on tuning not only the morphology, but also the internal and external structures of the hollow nanoparticles. On the basis of different categories of templates, soft templating synthetic strategies involve 3 methods: (1) emulsion templating method, (2) vesicle/micelle templating method, and (3) gas bubbles templating method. Here, we will first review different methods of soft templating synthetic strategies and then discuss the structural and morphological control of the nanostructures.

1.3.2.1 Emulsion-Based Soft Templates

An emulsion is a mixture of two or more liquids that are usually immiscible [88]. The sizes of emulsion droplets can be varied from 10 nm to 100 μm. On the basis of the sizes of the emulsion droplets, emulsions can be divided into two categories, namely macroemulsions (> 100 nm) and microemulsions (or nanoemulsions <100 nm) [89]. Generally, the formation of emulsions needs an emulsification process, which makes use of mechanical sheer force to break large droplets of a dispersed phase in a continuous phase. The dispersion is usually thermodynamically unstable. Amphiphilic molecules (or surfactants) are employed to reduce the interfacial tension generated at the contact part between the dispersed phase and the continuous phase. On the basis of the different polarities of the dispersed phase and the continuous phase, emulsions can then be divided into direct emulsions and reverse emulsions. Direct emulsions are emulsions that have water as a continuous phase and oil as a dispersed phase. Direct emulsions can also be called oil-in-water (O/W) or water-based emulsions. Unlike direct emulsions, reverse emulsions are emulsions that have oil as a continuous phase and water as a dispersed phase. Reverse emulsions can be W/O or oil-based emulsions. By using these two kinds of emulsions as templates, hollow nanostructures can be obtained via several synthetic methods. We will discuss some typical emulsion templating methods, including deposition of materials on emulsion droplets by sol–gel coating, interfacial precipitation, and hydrothermal reactions. It is worth noting that emulsion polymerization, which uses emulsion droplets to confine the space of the polymerization reaction, is widely used in the synthesis of hollow polymer materials.

1.3.2.2 Micelle/Vesicle-Based Soft Templates

Micelles/vesicles are effective templates for the synthesis of hollow structures with sizes in the nanoscale [90]. Micelles and vesicles are formed by self-assembly of amphiphilic molecules in a single-phase solvent. Amphiphilic molecules are molecules that have both hydrophilic and lipophilic properties. When the concentration of the amphiphilic molecules exceeds the critical micelle concentration, the molecules are packed to form mesostructures. The shape of the mesostructures can be tuned by varying different parameters, such as the concentration of surfactants, addition of additives, ionic strength, temperatures, and pH values. In addition, the geometrical property of the amphiphilic molecule also impacts the packing structure [91]. The synthesis of nanoparticles is realized by coassembly of the templates and precursors of the framework substances. They usually interact with each other through electrostatic interaction and hydrogen bonds. In this section, we will discuss some typical micelles or vesicles templating synthesis routes. Considering the solvent used in synthesis, the micelles or vesicles templates can be divided into 2 parts, namely direct micelle or vesicle templates and reverse micelle templates.

1.3.2.3 Gas Bubble-Based Soft Templates

Gas bubbles dispersed in a liquid host can be used to create stable emulsions and foams, which have recently emerged as promising soft templates for synthesis of an increasing number of hollow particles. In general, the process can be conceptualized in 3 steps: (1) formation of fine nanoparticles and gas bubbles; (2) attachment of fine nanoparticles on the gas/liquid interface; and (3) further aggregation of nanoparticles forming compact shells around the gas bubbles [92]. It is known that the attachment of solid particles to gas bubbles is a complex process that is affected by many factors, such as particle surface properties, particle size, electrostatic interactions, and hydrodynamic conditions [93].

1.3.3 Self-Template Methods

Direct synthesis without the need of additional templates is preferred in practical applications due to significantly reduced production cost and the ease of scaling up [94,95]. A number of self-templating approaches toward hollow structures have been developed on the basis of different principles, including Ostwald ripening, the Kirkendall effect, galvanic replacement, surface-protected etching, and so on. In general, the self-templating methods involve a 2-step synthesis: (1) the synthesis of template nanomaterials, and (2) the transformation of templates into hollow structures. Unlike the conventional templating methods, the templates used in self-templating methods are not only the templates used to create inner hollow structures, but also the integrant composition of the outer shells. The self-templating routes have several advantages, including relatively simple synthesis procedures, high reproducibility, low-production cost, and great control over shell thickness and particle uniformity [94]. In addition, the self-templating approaches eliminate the need for heterogeneous coating, making it more convenient to scale up the synthesis for large quantity production. In this section, we will review the recent progress in the area of self-templated synthesis of hollow nanostructures. The concepts and applications of 4 types of self-templating approaches will be discussed in detail, namely (1) those involving the "surface-protected etching" strategy, (2) Ostwald ripening, (3) the Kirkendall effect, and (4) galvanic replacement.

1.3.3.1 Surface-Protected Etching

Hollow structures can be synthesized by enhancing the relative stability of the surface layer of a single-component nanostructure to make it more stable than the interior. Preferential etching of the core part may allow the formation of hollow structures. Yin et al., introduced the concept of "surface-protected etching" in 2008 that involves the precoating of solid oxide particles with a protecting layer of polymeric ligands and subsequent preferential etching of interior material using an appropriate etching agent, as illustrated in Fig. 1.2 [96–99]. The protection by the polymer enables the oxide particles to retain their original size, while selective etching at the interior produces porous structures and, eventually, hollow spheres. Sol–gel-derived colloidal oxides are typically grown by the condensation of their nanoparticle precursors so that their relative loose structure allows the penetration of the etching agent

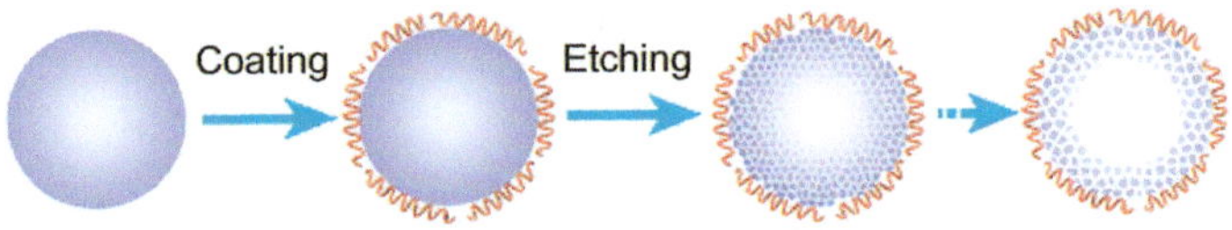

Figure 1.2 Schematic illustration of the concept of "surface-protected etching." Reprinted with permission from Q. Zhang, T.R. Zhang, J.P. Ge, Y.D. Yin, Permeable silica shell through surface-protected etching. Nano Lett. 8 (2008) 2867 – 2871. Copyright 2008, American Chemical Society.

[100]. A polymeric ligand is needed to effectively protect the colloid surface from rapid dissolution by an etching agent. For example, TiO_2 microcapsules with tunable size and wall thickness were synthesized by heating sol–gel-derived TiO_2 microspheres with PAA in a diethylene glycol (DEG) solution [98]. In this case, DEG acted as an etching agent to transform TiO_2 into soluble titanium glycolate, while PAA served as a cross-linker to connect adjacent NPs into a stable shell. By rationally choosing the protecting and etching agents, Yin et al., successfully extended the surface-protected etching process to the preparation of hollow SiO_2 structures [97–99]. Sol–gel-derived solid silica spheres were grown by the condensation of its precursors and are porous in nature [100]. PVP was selected to protect the near surface layer because its carbonyl groups can form strong hydrogen bonds with the hydroxyl groups on the silica surface. NaOH was used to selectively etch the interior of the silica spheres, yielding initially porous structures and eventually HS spheres with porous shells.

1.3.3.2 Ostwald Ripening

Ostwald ripening is a well-known phenomenon that was first systematically investigated by Wilhelm Ostwald around 1900 [101–103]. This phenomenon normally describes the change of an inhomogeneous structure over time in solid solutions or liquid sols and involves matter relocation [104]. The IUPAC in 2007 recommended the definition of Ostwald ripening as the "dissolution of small crystals or sol particles and the re-deposition of the dissolved species on the surfaces of larger crystals or sol particles." [105] The ripening process occurs because larger particles are more energetically favored than smaller particles, giving rise to an apparent higher solubility for the smaller ones [106]. Self-templated synthesis of hollow structures based on the Ostwald ripening mechanism has recently been proposed and extensively studied as research interests toward the preparation of complex nanostructured materials [107–109].

1.3.3.3 The Kirkendall Effect

The Kirkendall effect is a classical phenomenon in metallurgy that describes the motion of the boundary layer between two metals that occurs as a consequence of the difference in diffusion rates of the metal atoms. Experimental observations of unequal matter flows during interfusion, providing the first evidence for vacancy-mediated hopping of atoms being the predominant mechanism for diffusion in crystalline materials [110,111]. The net flow of mass in one direction is balanced by a flux of vacancies. When the vacancy concentration exceeds the saturation value, the vacancies are very likely to coalesce into voids. The formation of Kirkendall voids in alloys and solders is not a desirable process for metallurgical manufacturing, causing the main technological motivation for studying the effect in the past to reduce that negative effect. However, the Kirkendall effect has recently presented its positive aspect for the design and preparation of hollow nanostructures since the first exploitation of the nanoscale Kirkendall effect in 2004 [112]. Fig. 1.3 schematically illustrated the formation of hollow nanocrystals through the nanoscale Kirkendall effect [113]. After solid nanocrystals containing at least one of the elements of the final nanoshell (A) are synthesized, a second element (B) is reacted with A to yield an AB compound. Usually the surface of the solid-core nanocrystals (A) is first reacted with reagents (B) to produce a layer of shell materials (AB). The direct conversion of A to AB is, therefore, hindered by the initially formed AB layer and further reaction will continue by the diffusion of atoms or ions through the interface. If the outward diffusion of the solid core A is much faster than the inward diffusion of B, an inward

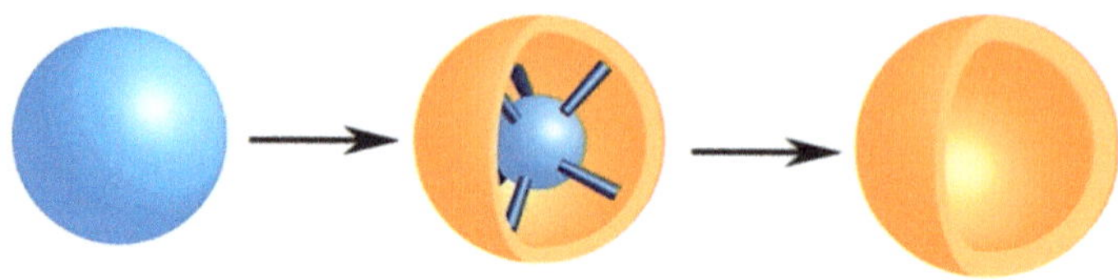

Figure 1.3 Schematic illustration of the nanoscale Kirkendall effect for the formation of hollow nanocrystals. Reprinted with permission from W.S. Wang, M. Dahl, Y.D. Yin, Hollow nanocrystals through the nanoscale Kirkendall effect. Chem. Mater. 25 (2013) 1179 – 1189. Copyright 2013, American Chemical Society.

flux of vacancies accompanies the outward A flux to balance the diffusivity difference. The voids are, thus, formed through coalescence of the vacancies based on the nanoscale Kirkendall effect.

1.3.3.4 Galvanic Replacement

Galvanic replacement reaction has been regarded as an effective self-templating method for producing hollow nanostructures with controlled sizes and shapes, porous walls, and tunable elemental composition, especially noble metals. The driven force comes from the electrochemical potential difference between two metals in which one metal acts as the reducing agent (anode) and the salt of the other metal as the oxidizing agent (cathode). The anode metal nanostructures are synthesized first. Upon contacting the metal ions with higher reduction potential, the anode metal nanoparticles are oxidized and dissolved into the solution, while the metal ions are reduced and plated onto the outer surface of the anode template. Generally, the shape and interior void size of the final hollow structure closely resemble the initial anode metal template with a slight increase in dimensions [114]. With advanced nanotechnology, various metal nanocrystals with controlled shapes and sizes have been prepared [115], which make them suitable for producing different hollow metal nanostructures through the galvanic replacement approach.

1.3.4 Template-Free Methods

It should be clear that, as a group, the templating method has proven very effective and versatile for synthesizing a wide array of hollow structures. Disadvantages related to high cost and tedious synthetic procedures have impeded the scale-up of many of these methods for large-scale applications. For instance, the template-removal step is, in general, indispensable when hard templates are used, which not only significantly complicates the process, but also detrimentally affects the quality (e.g., high impurity levels and inevitable shell collapse) of the as-derived hollow particles. Ideally, a one-step template-free method for controlled preparation of hollow structures in a wide range of sizes is preferred. Therefore it is highly desirable to explore other simpler and more efficient synthetic strategies for hollow structure synthesis.

Before the widespread application of template-involved synthesis, various template-free methods have been developed to produce hollow structures with or without the help of special equipment. During the past decade, the anodization method has been considered as a powerful strategy to obtain TiO_2 nanotube arrays. Usually, anodization of titanium foils or thin films is conducted using an electrochemical cell at a constant potential [116]. The detailed process involves an oxide layer formation under high voltage and an electric field-directed chemical dissolution. By controlling the fabrication variables such as temperature, voltage, pH value, and electrolyte composition, the physical or chemical features (such as roughness, length, crystallinity, and chemical composition) of TiO_2-based nanotubes could be changed for various applications.

1.4 Application for Sensing and Catalysis

In many cases, the particles play the role of fillers or rheological modifiers and their influence can be quantified entirely in terms of gross features such as size, density, volume fraction, and shape. In a growing number of applications, including catalysis, cosmetics, drug and gene delivery, hydrogen production and storage, photonics, photovoltaics, and rechargeable batteries, the chemical make-up and distribution of matter within the particles play important roles in determining function. For example, the large fraction of void space in hollow structures has been successfully used to encapsulate and control release of sensitive materials such as drugs, cosmetics, and DNA. Likewise, the void space in hollow particles has been used to modulate refractive index, lower density, increase the active area for catalysis, improve the particles' ability to withstand cyclic changes in volume, and expand the array of imaging markers suitable for early detection of cancer.

1.4.1 Catalysis

Progress in the synthesis of hollow particles has provided opportunities for their widespread use in catalysis. One of the early works by Kim et al., utilized hollow Pd nanospheres as heterogeneous catalysts for Suzuki coupling reactions [74]. These Pd hollow spheres manifested high catalytic activity (90% yield) even after 7 recycles, and little leaching of Pd was observed. In the past 2 years there has been additional work performed on this theme. Li et al., used PdCo bimetallic hollow nanospheres to catalyze a Sonogashira reaction in aqueous media with high yields [117]. $Ni_{1-x}Pt_x$ hollow spheres were demonstrated to exhibit good catalytic activities for the hydrolysis and thermolysis reactions of NH_3BH_3 to release H_2 [118]. The observed high activities are most readily attributed to the high-surface areas of the materials, compared to dense spheres of similar size, because their open, hollow structure enables both the outer and inner surfaces of the catalyst to come into contact with the reactants, yielding added benefits for the catalytic process.

Rattle-type hollow particles with functional cores are frequently used as catalysts. Ikeda et al., used Pt@carbon nanorattles for heterogeneous hydrogenation of olefins, achieving significantly higher yield and good recyclability compared to other Pt catalysts [119]. The carbon shell is believed to stabilize the nanoparticle core through prevention of coalescence and provide void space for the organic transformation on the Pt nanoparticle, thus, enhancing the catalytic performance. Similarly, Au@ZrO_2 nanorattles have been used as model high-temperature–stable catalysts for CO oxidation [120] where the Au nanoparticles are effectively separated (i.e., catalyst growth is prevented), but still highly accessible to gas molecules.

1.4.2 Sensors

Sensors are widely used in the engineering field. To date, many efforts have been undertaken to develop new devices and materials for different kinds of sensors, such as gas [121,122], chemicals [123], humidity [124,125], optical [126], and biomolecule sensors. Usually, special structures of materials are well designed for higher sensitivity. The sensing material plays an important role in manifesting the sensor's characteristics. Owing to the porous structure, large specific surface areas, low density, good permeation, and less-agglomerated configurations, hollow micro- and nanostructures exhibit potential for accomplishing both high response and fast responding kinetics [127], which allow this kind of material to be successfully applied in gas sensors. Different materials with hollow structures such as ZnO [128], SnO_2 [129], Cu_2O(CuO) [130], and WO_3 [131] have been synthesized though various processes and applied in diverse gas sensing (NH_3, NO, CO, etc.) which shows enhanced sensitivity.

The high surface areas of hollow micro- and nanostructured materials are also advantageous for chemical sensing. Chemical sensors are widely used for industrial process control and are experiencing an increased use in security applications. Current research has mainly focused on metal oxide hollow structures for conductometric sensing of gases. Sensing with these materials is performed through measurement of changes in electrical conductance produced by the adsorption or desorption of a targeted analyte on the oxide surface. This enhancement may be attributed to the hierarchical porous structure, that is, the high porosity and large degree of mesoporosity of the nanoparticle shell could facilitate the diffusion of analytes into the microporous structure.

1.5 Utilization in Photocatalysis for Degrading Pollutants

1.5.1 Photocatalysis

Multishelled hollow micro- and nanostructured materials as photocatalysts have several distinguishing features compared with traditional single-shelled hollow catalysts, including: (1) a special hollow cavity structure and lower density; (2) a tunable surface-to-volume ratio; and (3) short transport lengths for both mass and charge. One of the most prominent characteristics of multishelled hollow micro- and nanostructured materials is that the multishelled hollow structure allows for more multiple reflections of incident light and, thus, enhances the light-harvesting efficiency and further boosts their catalytic performance [45]. These unique features may bring novel collective properties for some catalytic reactions and stimulate further extensive interest in the field, such as degradation of organic compounds, water splitting and water treatment.

1.5.2 Photocatalytic Degradation of Organic Compounds

Along with the development of the global industrialization process, environmental issues have become the most important problem which influences human survival and development. In recent years, the degradation of organic compounds driven by solar energy at room temperature has become an ideal environment pollution–control technology and is considered to be the most promising technology to solve environmental pollution problems and has, therefore, become the research focus in the field of environment sustainability. The process in the synthesis of multishelled hollow micronanostructured materials has provided opportunities for their widespread use in the photocatalytic degradation of organic compounds. One of the early studies by Zeng et al., utilized multishelled titania hollow spheres as photocatalysts for the degradation of rhodamine B [132]. Multishelled hollow spheres exhibit the highest activity of 84% after irradiation for 40 min, while sphere-in-sphere structures and nanoparticles show a lower activity of 54% and 58%, respectively. Since all the samples possess similar crystalline structures, surface areas, and nearly the same residual carbon, the researchers think that the superior activities of multishelled hollow spheres may arise from their unique multishelled hollow structures. The increased number of shells could lead to the multiple reflections of UV light in the interior cavity, resulting in the enhancement of light absorbance and, thus, photocatalytic activity. However, the lower activity of single-shelled hollow spheres is due to negligible multiple reflections because of the too-small size of the inner sphere compared to the outer one [133].

1.5.3 Photocatalytic Energy Conversion

Global energy and environmental crises are important topics [134]. The rate of worldwide energy consumption reached almost 15 TW in 2008 and is expected to double by 2050 due to the dramatic increase of global population and the concomitant demand for an increase of production [135]. Solar energy is widely regarded as a free, abundant, and incessant renewable source of clean energy that could meet the current and future human energy demand in a long term. Forming chemical bonds to collect and store solar energy by photosynthesis is a highly desirable approach to solve the energy challenge.

Solar water splitting directly to hydrogen and oxygen has become one of the most desirable methods for harvesting and conversion of solar energy into chemical energy [135]. At present, a larger number of photocatalysts including metal oxide/hydroxide semiconductors [136], metal sulfide [137], metal nitride [138], and composite semiconductors [139] have been reported. Though these catalysts exhibit improved performance for photocatalytic water splitting, low efficiency still remains, which should originate from low-incident light absorption, fast recombination of electron–hole pairs, and low-photocatalytic stability [140,141]. The design and development of efficient photocatalysts is a critical issue in photocatalytic water splitting, while multishelled, hollow structure, semiconductor microspheres can integrate a series of functions into a microsystem dealing with the subtasks involving enhancing light absorption and speeding up the rate of electron transfer or separation, which can be a new structure photocatalyst for water splitting.

1.6 Application for Rechargeable Batteries and Supercapacitors

Owing to their tunable physical and chemical properties, hollow structures have been demonstrated to be great structural advantages for advanced energy storage and conversion technologies, such as LIBs, hybrid supercapacitors (HSCs), lithium–sulfur batteries, water-splitting devices, and fuel cells [142–147]. Particularly, as electrode materials, sulfur hosts, or electrocatalysts, hollow structures with void space offer abundant electrochemical active sites and large electrolyte-electrode contact area for fast diffusion and reaction [148–150]. Additionally, hollow cavities can effectively prevent encapsulated electroactive nanoparticles from aggregation and accommodate the pronounced volume variation accompanied with repeated charging or discharging processes [151].

Of note is that complex, hollow structures are expected to outperform nanoparticle-assembled hollow architectures with much enhanced electrochemical performance [152,153]. For instance, multishelled hollow nanostructures can generally increase the energy density of energy-storage devices due to their higher weight fraction of active species [45]. Meanwhile hierarchical hollow structures organized by secondary subunits with high-energy facets exhibit enhanced electrocatalytic activity due to highly exposed active sites [154–156]. The rich tailorability of complex hollow nanostructures has stimulated extensive investigations on their energy-related applications.

1.6.1 Lithium-Ion Batteries

Due to the important roles of LIBs in electric energy storage, tremendous effort has been made to develop novel electrode materials for advanced LIBs with improved energy or power densities and lifespans [151–155]. Hollow structures have long been proposed to overcome several major limitations of conventional electrode materials. First, the thin shell and large-exposed surface allow fast intercalation of Li^+ ions and improve the kinetics of electrodes. Second, the volumetric variation and mechanical strain associated with Li^+ insertion could be well buffered by the inner void, thus, improving the cycling stability. In addition, the micro- or submicrosized secondary architectures provide extra mechanical robustness and can prevent unwanted aggregation of electroactive materials.

The self-templated strategy offers the feasibility to tune the features of hollow electrode materials, thus to understand the composition- and/or structure-induced lithium storage capability. Zhang et al., have studied the electrochemical performance of Fe_2O_3/SnO_2 hybrid microboxes derived from MOFs as anode materials for LIBs, which are prepared through a solution-based, ion-exchange reaction [157]. The two integrated active components with different electrochemical characteristics might serve as support and promoter for each other, thus, improving the stability and electrochemical reversibility of the composite electrode. As a result, Fe_2O_3/SnO_2 hybrid microboxes exhibit much-improved cycling performance compared with pure SnO_2 microboxes due to the synergistic effects from the multiple components. In addition, such a micrometer size would render the hollow structure rather robust and minimize the side reaction associated with exterior surface when comparing with conventional nanomaterials.

1.6.2 Supercapacitors

As an emerging type of electrochemical energy storage device, HSCs integrate a battery-type electrode and a capacitive electrode in an electrochemical cell to achieve both high energy and power density [158]. The battery-type electrode, which is typically composed of transition metal-based compounds, should match the high-power characteristics of the capacitive electrode. In this regard, using a hollow structure to improve the electrode kinetics by providing a large exposed surface and shortened ion diffusion length could be a promising solution for HSCs.

Owing to the synthetic versatility, self-templated methods enable the exploration of the relationship between structure or composition-induced physicochemical properties and electrochemical performance of various hollow structures. For example, Hu et al., investigated the structural effects of CoS nanoboxes on electrochemical charge storage [159]. The multistep outward diffusion strategy produces CoS hollow nanoboxes with diverse architectures, ranging from single-shelled nanoboxes assembled by nanoparticles (CoS-NP SSNBs) or nanosheets (CoS-NS SSNBs) to nanoboxes with two different shells (CoS-NP/CoS-NS DSNBs). When evaluated as electrode materials for the battery-type electrodes, single-shelled CoS-NP SSNBs and CoS-NS SSNBs lose their capacitance in 10,000 cycles by 20% and 41%, respectively. Under the same conditions, the double-shelled nanoboxes possess improved capacitance retention of 89%. Such enhanced cycling performance might be attributed to the improved robustness and integrity of the multishelled structure.

1.7 Conclusion

Over the past decade, considerable progress has been made in the synthesis and applications of hollow micro- and nanostructures. The synthetic strategies for hollow structures can be broadly categorized into 4 groups: (1) conventional hard templating; (2) sacrificial templating; (3) soft templating; and (4) template-free approaches. Templating against hard (solid) templates is arguably the most effective and certainly the most common method for synthesizing hollow micro- and nanostructures. Extension of this method to sacrificial template-based syntheses is particularly promising because they generally require no additional surface functionalization and shell formation is guaranteed by chemical reaction. Overall, hard template–based synthetic approaches suffer from several intrinsic disadvantages, ranging from the inherent difficulty of achieving high-product yields from the multistep synthetic process to the lack of structural robustness of the shells upon template removal. These difficulties can be partly overcome by the use of soft (liquid or gaseous) templates. Compared to solid templates, soft templates can be more easily removed or the syntheses are essentially template-free when involving gaseous templates. Additionally, soft templates allow facile and efficient encapsulation of functional species like therapeutic and DNA molecules. More recently, template-free methods based on novel mechanisms (e.g., inside-out Ostwald ripening) have been developed for preparing hollow structures of many materials in a wide range of sizes.

The successes in synthesis of hollow structures have provided opportunities to tune their mechanical, optical, electrical, chemical, and other properties. These advances have, in turn, catalyzed exploration in a growing list of applications. However, it should be noted that high-quality hollow particles will be required in many cases for both fundamental research and practical applications. A survey of the literature in the field shows that methods for producing such high-quality hollow particles are still very limited and that, even if these methods can be found, they are suitable only for a small number of materials. In addition, many of the methods for synthesis of both templates and hollow structures are based on solution synthesis in which the concentration of precursors is usually very low, typically in the millimolar range. Scaling-up these syntheses to produce commercial-scale quantities for applications is expected to introduce significant challenges for size, shape, and shell-thickness control. This aspect of hollow structure synthesis, though still in its infancy, offers exciting opportunities for newcomers to the field.

In this chapter, we summarize some recent developments in the synthesis of hollow-structured, functional materials by self-templated approaches and their exceptional properties for several important electrochemical energy storage and conversion technologies. Based on the formation protocols of internal voids, the self-templated methods for hollow structures are classified as selective etching, outward diffusion, and heterogeneous contraction. Owing to the wide availability of templates and rich physical or chemical processes involved in the synthesis, the self-templated strategy is powerful in building unusual and complex hollow architectures and tuning the chemical compositions. These synthetic merits are further translated into the exceptional structure or composition-induced properties for electrochemical energy applications. Generally speaking, hollow structures offer improved electrochemical reactivity, enhanced kinetics and cyclability as electrode materials for LIBs and HSCs, and high electrocatalytic activity due to the large accessible active surface and improved mass or charge transport for electrochemical reactions.

From the nanomaterial–synthesis point-of-view, exploring novel synthetic protocols for hollow structures is surely one of the most important research directions in the future. At the same time, improving the existing self-templated methods and expanding the established protocols toward novel functional materials are of equal importance. From the perspective of practical applications, designed synthesis according to the needs of specific applications is highly desirable. Overall, we have witnessed the booming development of self-templated formation of functional hollow structures over the past decade, as well as their exceptional potential for electrochemical energy applications. One can expect to see some fast developments in this field and its broader impacts in material science and nanotechnology.

References

[1] X.W. Lou, L.A. Archer, Z.C. Yang, Hollow micro-/nanostructures synthesis and applications, Adv. Mater. 20 (2008) 3987–4019.

[2] X.J. Wang, J. Feng, Y.C. Bai, Q. Zhang, Y.D. Yin, Synthesis, properties, and applications of hollow micro-/nanostructures, Chem. Rev. 116 (2016) 10983–11060.

[3] D.L. Wilcox, M. Berg, T. Bernat, D. Kellerman, J.K. Cochran Jr., Review of low dielectric constant thick film electronic ceramics using hollow microspheres, Res. Soc. Symp. Proc 372 (1995) 239–245.

[4] T. Ung, L.M. Liz-Marzan, P. Mulvaney, Controlled method for silica coating of silver colloids. Influence of coating on the rate of chemical reactions, Langmuir 14 (1998) 3740–3748.

[5] L.M. Liz-Marzan, M. Giersig, P. Mulvaney, Direct observation of chemical reactions in silica-coated gold and silver nanoparticles, Langmuir 12 (1996) 570–575.

[6] F. Caruso, R.A. Caruso, H. Mohwald, Nanoengineering of inorganic and hybrid hollow spheres by colloidal templating, Science 282 (1998) 1111–1114.

[7] G.Q. Zhang, H.B. Wu, T. Song, U. Paik, X.W. Lou, TiO_2 hollow spheres composed of highly crystalline nanocrystals exhibit superior lithium storage properties, Angew. Chem. Int. Ed. 53 (2014) 12590–12593.

[8] H.J. Dai, E.W. Wong, C.M. Lieber, Probing electrical transport in nanomaterials conductivity of individual carbon nanotubes, Science 272 (1996) 523–526.

[9] G.K. Mor, O.K. Varghese, M. Paulose, K. Shankar, C.A. Grimes, Synthesis and photoelectrochemical properties of nanoporous iron (III) oxide by potentiostatic anodization, Sol. Energy Mater. Sol. Cells 90 (2006) 4285–4291.

[10] G.Q. Zhang, B.Y. Xia, C. Xiao, L. Yu, X. Wang, Y. Xie, et al., General formation of complex tubular nanostructures of metal oxides for the oxygen reduction reaction and lithium-ion batteries, Angew. Chem. Int. Ed. 52 (2013) 8754.

[11] Y.M. Chen, L. Yu, X.W. Lou, Hierarchical tubular structures composed of Co_3O_4 hollow nanoparticles and carbon nanotubes for lithium storage, Angew. Chem. Int. Ed. 55 (2016) 5990–5993.

[12] W. Xiao, J. Zhou, L. Yu, D.H. Wang, X.W. Lou, Electrolytic formation of crystalline silicon/germanium alloy nanotubes and hollow particles with enhanced lithium-storage properties, Angew. Chem. Int. Ed. 55 (2016) 7427–7431.

[13] Z.Y. Wang, L. Zhou, X.W. Lou, Metal oxide hollow nanostructures for lithium-ion batteries, Adv. Mater. 24 (2012) 1903–1911.

[14] D. Xie, Q. Su, Z. Dong, J. Zhang, G. Du, L-cysteine-assisted preparation of porous NiO hollow microspheres with enhanced performance for lithium storage, CrystEngComm 15 (2013) 8314–8319.

[15] L. Yu, H.B. Wu, X.W. Lou, Self-templated formation of hollow structures for electrochemical energy applications, Acc. Chem. Res. 50 (2017) 293–301.

[16] L. Zhang, L.T. Roling, X. Wang, M. Vara, M.F. Chi, J.Y. Liu, et al., Platinum-based nanocages with subnanometer-thick walls and well-defined, controllable facets, Science 349 (2015) 412–416.

[17] J. Liang, H. Hu, H. Park, C.H. Xiao, S.J. Ding, U. Paik, et al., Construction of hybrid bowl-like structures by anchoring NiO nanosheets on flat carbon hollow particles with enhanced lithium storage properties, Energy Environ. Sci. 8 (2015) 1707–1711.

[18] W.R. Zhao, H.R. Chen, Y.S. Li, L. Li, M.D. Lang, Uniform rattle-type hollow magnetic mesoporous spheres as drug delivery carriers and their sustained-release property, Adv. Funct. Mater. 18 (2008) 2780–2788.

[19] Y. Chen, H.R. Chen, J.L. Shi, Construction of homogenous/heterogeneous hollow mesoporous silica nanostructures by silica-etching chemistry principles, synthesis, and applications, Acc. Chem. Res. 47 (2014) 125–127.
[20] X.L. Fang, Z.H. Liu, M.F. Hsieh, M. Chen, P.X. Liu, C. Chen, et al., Hollow mesoporous aluminosilica spheres with perpendicular pore channels as catalytic nanoreactors, ACS Nano 6 (2012) 4434–4444.
[21] C.C. Yec, H.C. Zeng, Nanobubbles within a microbubble synthesis and self-assembly of hollow manganese silicate and its metal-doped derivatives, ACS Nano 8 (2014) 6407–6416.
[22] X. Wang, X.L. Wu, Y.G. Guo, Y.T. Zhong, X.Q. Cao, Y. Ma, et al., Synthesis and lithium storage properties of Co_3O_4 nanosheet-assembled multishelled hollow spheres, Adv. Funct. Mater. 20 (2010) 1680–1686.
[23] M.L. Pang, A.J. Cairns, Y.L. Liu, Y. Belmabkhout, H.C. Zeng, M. Eddaoudi, Highly monodisperse M(III)-based soc-MOFs (M = In and Ga) with cubic and truncated cubic morphologies, J. Am. Chem. Soc. 135 (2013) 10234–13179.
[24] Y. Zhao, L. Jiang, Hollow micro/nanomaterials with multilevel interior structures, Adv. Mater. 21 (2009) 3621–3638.
[25] J.B. Joo, Q. Zhang, M. Dahl, I. Lee, J. Goebl, F. Zaera, et al., Control of the nanoscale crystallinity in mesoporous TiO_2 shells for enhanced photocatalytic activity, Energy Environ. Sci. 5 (2012) 6321–6327.
[26] W. Li, Y.H. Deng, Z.X. Wu, X.F. Qian, J.P. Yang, Y. Wang, et al., Hydrothermal etching assisted crystallization: a facile route to functional yolk-shell titanate microspheres with ultrathin nanosheets-assembled double shells, J. Am. Chem. Soc. 133 (2011) 15830–15833.
[27] X.W. Lou, C.L. Yuan, L.A. Archer, Shell-by-shell synthesis of tin oxide hollow colloids with nanoarchitectured walls: cavity size tuning and functionalization, Small 3 (2007) 261–265.
[28] J. Liu, S.Z. Qiao, S. Budi Hartono, G.Q. Lu, Monodisperse yolk – shell nanoparticles with a hierarchical porous structure for delivery vehicles and nanoreactors, Angew. Chem. Int. Ed. 49 (2010) 4981–4985.
[29] J.Y. Wang, N.L. Yang, H.J. Tang, Z.H. Dong, Q. Jin, M. Yang, et al., Accurate control of multishelled Co_3O_4 hollow microspheres as high-performance anode materials in lithium-ion batteries, Angew. Chem. Int. Ed. 52 (2013) 6417–6420.
[30] Z. Li, J.T. Zhang, Y.M. Chen, J. Li, X.W. Lou, Pie-like electrode design for high-energy density lithium-sulfur batteries, Nat. Commun. 6 (2015) 8850.
[31] J.H. Gao, B. Zhang, X.X. Zhang, B. Xu, Magnetic-dipolar-interaction-induced self-assembly affords wires of hollow nanocrystals of cobalt selenide, Angew. Chem. Int. Ed. 45 (2006) 1220–1223.
[32] C. Chen, Y.J. Kang, Z.Y. Huo, Z.W. Zhu, W.Y. Huang, H.L.L. Xin, et al., Highly crystalline multimetallic nanoframes with three-dimensional electrocatalytic surfaces, Science 343 (2014) 1339–1343.
[33] B.Y. Guan, L. Yu, X.W. Lou, Formation of single-holed cobalt/N-doped carbon hollow particles with enhanced electrocatalytic activity toward oxygen reduction reaction in alkaline media, Adv. Sci. 4 (2017) 1700247.
[34] Y. Wang, L. Yu, X.W. Lou, Synthesis of highly uniform molybdenum–glycerate spheres and their conversion into hierarchical MoS_2 hollow nanospheres for lithium-ion batteries, Angew. Chem. Int. Ed. 55 (2016) 7423–7426.
[35] X.H. Xia, Y. Wang, A. Ruditskiy, Y.N. Xia, Galvanic replacement: a simple and versatile route to hollow nanostructures with tunable and well-controlled properties, Adv. Mater. 25 (2013) 6313–6333.

[36] E. González, J. Arbiol, V.F. Puntes, Carving at the nanoscale: sequential galvanic exchange and Kirkendall growth at room temperature, Science 334 (2011) 1377.
[37] L. Han, X.Y. Yu, X.W. Lou, Formation of Prussian-Blue-Analog nanocages via a direct etching method and their conversion into ni-co-mixed oxide for enhanced oxygen evolution, Adv. Mater. 28 (2016) 4601–4605.
[38] C.H. Kuo, M.H. Huang, Fabrication of truncated rhombic dodecahedral Cu_2O nanocages and nanoframes, J. Am. Chem. Soc. 130 (2008) 12815–12820.
[39] S.F. Xie, N. Lu, Z.X. Xie, J.G. Wang, M.J. Kim, Y.N. Xia, Synthesis of Pd-Rh core-frame concave nanocubes and their conversion to Rh cubic nanoframes by selective etching of the Pd cores, Angew. Chem., Int. Ed. 51 (2012) 10266–10270.
[40] S.P. Luo, M. Tang, P.K. Shen, S.Y. Ye, Atomic-scale preparation of octopod nanoframes with high-index facets as highly active and stable catalysts, Adv. Mater. 29 (2017) 1601687.
[41] Z.C. Zhang, Z.M. Luo, B. Chen, C. Wei, L. Zhao, J.Z. Chen, et al., One-pot synthesis of highly anisotropic five-fold-twinned PtCu nanoframes used as a bifunctional electrocatalyst for oxygen reduction and methanol oxidation, Adv. Mater. 28 (2016) 8712–8717.
[42] M. Hajfathalian, K.D. Gilroy, S.D. Golze, A. Yaghoubzade, E. Menumerov, R. A. Hughes, et al., A wulff in a cage: the confinement of substrate-based structures in plasmonic nanoshells, nanocages, and nanoframes using galvanic replacement, ACS Nano 10 (2016) 6354–6362.
[43] X.Y. Yu, L. Yu, X.W. Lou, Metal sulfide hollow nanostructures for electrochemical energy storage, Adv. Energy Mater. 6 (2016) 1501333.
[44] Y.S. Li, J.L. Shi, Hollow-structured mesoporous materials chemical synthesis, functionalization and applications, Adv. Mater. 26 (2014) 3176–3205.
[45] J. Qi, X.Y. Lai, J.Y. Wang, H.J. Tang, H. Ren, Y. Yang, et al., Multi-shelled hollow micro-/nanostructures, Chem. Soc. Rev. 44 (2015) 6749–6773.
[46] K.D. Gilroy, A. Ruditskiy, H.C. Peng, D. Qin, Y.N. Xia, Bimetallic nanocrystals: syntheses, properties, and applications, Chem. Rev. 116 (2016) 10414–10472.
[47] Y. Boyjoo, M. Wang, V.K. Pareek, J. Liu, M. Jaroniec, Synthesis and applications of porous non-silica metal oxide submicrospheres, Chem. Soc. Rev. 45 (2016) 6013–6047.
[48] Y.J. Wong, L.F. Zhu, W.S. Teo, Y.W. Tan, Y.H. Yang, C. Wang, et al., Revisiting the Stöber method: inhomogeneity in silica shells, J. Am. Chem. Soc. 133 (2011) 11422–11425.
[49] J. Jiang, Y.Y. Li, J.P. Liu, X.T. Huang, C.Z. Yuan, X.W. Lou, Recent advances in metal oxide-based electrode architecture design for electrochemical energy storage, Adv. Mater. 24 (2012) 5166–5180.
[50] X.K. Huang, S.M. Cui, J.B. Chang, P.B. Hallac, C.R. Fell, Y.T. Luo, et al., A hierarchical tin/carbon composite as an anode for lithium-ion batteries with a long cycle life, Angew. Chem. Int. Ed. 54 (2015) 1490–1493.
[51] H.B. Wu, G.Q. Zhang, L. Yu, X.W. Lou, One-dimensional metal oxide-carbon hybrid nanostructures for electrochemical energy storage, Nanoscale Horiz. 1 (2016) 27–40.
[52] Z.Y. Wang, D.Y. Luan, S. Madhavi, Y. Hu, X.W. Lou, Assembling carbon-coated alpha-Fe_2O_3 hollow nanohorns on the CNT backbone for superior lithium storage capability, Energy Environ. Sci. 5 (2012) 5252–5256.

[53] F. Zou, Y.M. Chen, K.W. Liu, Z.T. Yu, W.F. Liang, S.M. Bhaway, et al., Metal organic frameworks derived hierarchical hollow NiO/Ni/graphene composites for lithium and sodium storage, ACS Nano 10 (2016) 377–386.
[54] X.Y. Lai, J. Li, B.A. Korgel, Z.H. Dong, Z.M. Li, F.B. Su, et al., General synthesis and gas-sensing properties of multiple-shell metal oxide hollow microspheres, Angew. Chem., Int. Ed. 50 (2011) 2738–2741.
[55] J.W. Nai, Y. Lu, L. Yu, X. Wang, X.W. Lou, Formation of Ni-Fe mixed diselenide nanocages as a superior oxygen evolution electrocatalyst, Adv. Mater. 29 (2017) 1703870.
[56] Z.H. Dong, X.Y. Lai, J.E. Halpert, N.L. Yang, L.X. Yi, J. Zhai, et al., Accurate control of multishelled ZnO hollow microspheres for dye-sensitized solar cells with high efficiency, Adv. Mater. 24 (2012) 1046–1049.
[57] L. Yu, X.Y. Yu, X.W. Lou, The design and synthesis of hollow micro-/ nanostructures present and future trends, Adv. Mater. 30 (2018) 1800939.
[58] Z.M. Li, X.Y. Lai, H. Wang, D. Mao, C.J. Xing, D. Wang, General synthesis of homogeneous hollow core – shell ferrite microspheres, J. Phys. Chem. C 113 (2009) 2792–2797.
[59] H. Ren, R.B. Yu, J.Y. Wang, Q. Jin, M. Yang, D. Mao, et al., Multishelled TiO_2 hollow microspheres as anodes with superior reversible capacity for lithium ion batteries, Nano Lett. 14 (2014) 6679–6684.
[60] M. Wang, J. Han, Y. Hu, R. Guo, Y. Yin, Carbon-incorporated NiO/TiO_2 mesoporous shells with p – n heterojunctions for efficient visible light photocatalysis, ACS Appl. Mater. Interfaces. 8 (2016) 29511–29521.
[61] Z.H. Dong, H. Ren, C.M. Hessel, J.Y. Wang, R.B. Yu, Q. Jin, et al., Quintuple-shelled SnO_2 hollow microspheres with superior light scattering for high-performance dye-sensitized solar cells, Adv. Mater. 26 (2014) 905–909.
[62] J.Y. Wang, H.J. Tang, H. Ren, R.B. Yu, J. Qi, D. Mao, et al., pH-regulated synthesis of multi-shelled manganese oxide hollow microspheres as supercapacitor electrodes using carbonaceous microspheres as templates, Adv. Sci. 1 (2014) 1400011.
[63] J. Qi, K. Zhao, G.D. Li, Y. Gao, H.J. Zhao, R.B. Yu, et al., Multi-shelled CeO_2 hollow microspheres as superior photocatalysts for water oxidation, Nanoscale 6 (2014) 4072–4077.
[64] X.Y. Lai, J.E. Halpert, D. Wang, Multi-shelled CeO_2 hollow microspheres as superior photocatalysts for water oxidation, Energy Environ. Sci. 5 (2012) 5604–5618.
[65] P.F. Xu, R.B. Yu, H. Ren, L.B. Zong, J. Chen, X.R. Xing, Hierarchical nanoscale multi-shell Au/CeO_2 hollow spheres, Chem. Sci. 5 (2014) 4221–4226.
[66] H.Q. Wang, M. Miyauchi, Y. Ishikawa, A. Pyatenko, N. Koshizaki, Y. Li, et al., Single-crystalline rutile TiO_2 hollow spheres room-temperature synthesis, tailored visible-light-extinction, and effective scattering layer for quantum dot-sensitized solar cells, J. Am. Chem. Soc. 133 (2011) 19102–19109.
[67] Z. Wang, Z. Wang, H. Wu, X.W. Lou, Mesoporous single-crystal $CoSn(OH)_6$ hollow structures with multilevel interiors, Sci. Rep. 3 (2013) 1391.
[68] E. Donath, G.B. Sukhorukov, F. Caruso, S.A. Davis, H. Möhwald, Novel hollow polymer shells by colloid-templated assembly of polyelectrolytes, Angew. Chem. Int. Ed. 37 (1998) 2201–2205.
[69] K. Katagiri, F. Caruso, Monodisperse polyelectrolyte-supported asymmetric lipid-bilayer vesicles, Adv. Mater. 17 (2005) 738–743.
[70] O.I. Vinogradova, O.V. Lebedeva, K. Vasilev, H. Gong, J. Garcia-Turiel, B.-S. Kim, Multilayer DNA/poly(allylamine hydro-chloride) microcapsules

assembly and mechanical properties, Biomacromolecules 6 (2005) 1495–1502.
[71] W. Tu, Y. Zhou, Q. Liu, Z. Tian, J. Gao, X. Chen, et al., Robust hollow spheres consisting of alternating titania nanosheets and graphene nanosheets with high photo-catalytic activity for CO_2 conversion into renewable fuels, Adv. Funct. Mater. 22 (2012) 1215–1221.
[72] D. Su, J. Zhang, S. Dou, G. Wang, Polypyrrole hollow nanospheres stable cathode materials for sodium-ion batteries, Chem. Commun. 51 (2015) 16092–16095.
[73] C. Deng, H. Hu, X. Ge, C. Han, D. Zhao, G. Shao, One-pot sonochemical fabrication of hierarchical hollow CuO submicrospheres, Ultrason. Sonochem. 18 (2011) 932–937.
[74] S.W. Kim, M. Kim, W.Y. Lee, T. Hyeon, Fabrication of hollow palladium spheres and their successful application to the recyclable heterogeneous catalyst for suzuki coupling reactions, J. Am. Chem. Soc. 124 (2002) 7642–7643.
[75] C. Kresge, M. Leonowicz, W. Roth, J. Vartuli, J. Beck, Ordered mesoporous molecular sieves synthesized by a liquid-crystal template mechanism, Nature 359 (1992) 710–712.
[76] Y. Wang, A. Yu, F. Caruso, Nanoporous polyelectrolyte spheres prepared by sequentially coating sacrificial mesoporous silica spheres, Angew. Chem. Int. Ed. 44 (2005) 2888–2892.
[77] S. Yang, X. Feng, L. Zhi, Q. Cao, J. Maier, K. Müllen, Nanographene-constructed hollow carbon spheres and their favorable electroactivity with respect to lithium storage, Adv. Mater. 22 (2010) 838–842.
[78] X. Li, X. Liu, Y. Ma, M. Li, J. Zhao, H. Xin, et al., Engineering the formation of secondary building blocks within hollow interiors, Adv. Mater. 24 (2012) 1424–1428.
[79] A. Dong, N. Ren, Y. Tang, Y. Wang, Y. Zhang, W. Hua, et al., General synthesis of mesoporous spheres of metal oxides and phosphates, J. Am. Chem. Soc. 125 (2003) 4976–4977.
[80] M.-M. Titirici, M. Antonietti, A. Thomas, A generalized synthesis of metal oxide hollow spheres using a hydrothermal approach, Chem. Mater. 18 (2006) 3808–3812.
[81] D.K. Yi, S.S. Lee, G.C. Papaefthymiou, J.Y. Ying, Nanoparticle architectures templated by SiO_2/Fe_2O_3 nanocomposites, Chem. Mater. 18 (2006) 614–619.
[82] W. Zhao, M. Lang, Y. Li, L. Li, J. Shi, Fabrication of uniform hollow mesoporous silica spheres and ellipsoids of tunable size through a facile hard-templating route, J. Mater. Chem. 19 (2009) 2778–2783.
[83] M. Feyen, C. Weidenthaler, F. Schüth, A.-H. Lu, Regioselectively controlled synthesis of colloidal mushroom nanostructures and their hollow derivatives, J. Am. Chem. Soc. 132 (2010) 6791–6799.
[84] J.-F. Chen, H.-M. Ding, J.-X. Wang, L. Shao, Preparation and characterization of porous hollow silica nanoparticles for drug delivery application, Biomaterials 25 (2004) 723–727.
[85] H. Zhao, Y. Li, R. Liu, F. Zhao, Y. Hu, Synthesis method for silica needle-shaped nano-hollow structure, Mater. Lett. 62 (2008) 3401–3403.
[86] Y. Zhao, L.-N. Lin, Y. Lu, S.-F. Chen, L. Dong, S.-H. Yu, Templating synthesis of preloaded doxorubicin in hollow mesoporous silica nanospheres for biomedical applications, Adv. Mater. 22 (2010) 5255–5259.
[87] E. Ghadiri, N. Taghavinia, S.M. Zakeeruddin, M. Gratzel, J.-E. Moser, Enhanced electron collection efficiency in dye sensitized solar cells based on nanostructured TiO_2 hollow fibers, Nano Lett. 10 (2010) 1632–1638.

[88] T.G. Mason, J.N. Wilking, K. Meleson, C.B. Chang, S.M. Graves, Nanoemulsions formation, structure, and physical properties, J. Phys. Condens. Matter. 18 (2006) R635–R666.
[89] H. Wennerström, O. Söderman, U. Olsson, B. Lindman, Macroemulsions versus microemulsions, Colloids Surf. A 123-124 (1997) 13–26.
[90] W. Li, P. Xu, H. Zhou, L. Yang, H. Liu, Advanced functional nanomaterials with microemulsion phase, Sci. China Technol. Sci. 55 (2012) 387–416.
[91] J. Liu, A.Y. Kim, L.Q. Wang, B.J. Palmer, Y.L. Chen, P. Bruinsma, et al., Self-assembly in the synthesis of ceramic materials and composites, Adv. Colloid Interface Sci. 69 (1996) 131–180.
[92] Y.S. Han, G. Hadiko, M. Fuji, M. Takahashi, Synthesis of hollow calcium carbonate particles by the bubble templating method, Chem. Lett. 59 (2005) 2519–2522.
[93] X. Fan, Z. Zhang, G. Li, N.A. Rowson, Attachment of solid particles to air bubbles in surfactant-free aqueous solutions, Chem. Eng. Sci. 59 (2004) 2639–2645.
[94] Q. Zhang, W. Wang, J. Goebl, Y. Yin, Self-templated synthesis of hollow nanostructures, Nano Today 4 (2009) 494–507.
[95] X.L. Fang, X.J. Zhao, W.J. Fang, C. Chen, N.F. Zheng, Self-templating synthesis of hollow mesoporous silica and their applications in catalysis and drug delivery, Nanoscale 5 (2013) 2205–2218.
[96] Q. Zhang, J. Ge, J. Goebl, Y. Hu, Z. Lu, Y. Yin, Rattle-type silica colloidal particles prepared by a surface-protected etching process, Nano Res. 2 (2009) 583–591.
[97] Q. Zhang, T.R. Zhang, J.P. Ge, Y.D. Yin, Permeable silica shell through surface-protected etching, Nano Lett. 8 (2008) 2867–2871.
[98] Y.X. Hu, J.P. Ge, Y.G. Sun, T.R. Zhang, Y.D. Yin, A self-templated approach to TiO_2 microcapsules, Nano Lett. 7 (2007) 1832–1836.
[99] Y.X. Hu, Q. Zhang, J. Goebl, T.R. Zhang, Y.D. Yin, Control over the permeation of silica nanoshells by surface-protected etching with water, Phys. Chem. Chem. Phys. 12 (2010) 11836–11842.
[100] L.L. Hench, J.K. West, The sol-gel process, Chem. Rev. 90 (1990) 33–72.
[101] W. Ostwald, L. Allg, Ein Wort der Mathematik an die Energetik, Chem 2 (1896) 39–71.
[102] W. Ostwald, Stoechiometrie und Verwandtschaftslehre. Z. Metastable γ-MnS hierarchical architectures synthesis, characterization, and growth mechanism, Phys. Chem. 22 (1897) 289–330.
[103] W.Z. Ostwald, Biomolecule-assisted synthesis of cobalt sulfide nanowires for application in supercapacitors, Phys. Chem. 34 (1900) 495–503.
[104] L.R.A.P.W. Voorhees, Growth and Coarsening Ostwald Ripening in Material Processing., Springer, New York, 2002.
[105] J. Aleman, A.V. Chadwick, J. He, M. Hess, K. Horie, R.G. Jones, et al., Definitions of terms relating to the structure and processing of sols, gels, networks, and inorganic-organic hybrid materials (IUPAC Recommendations 2007), Pure Appl. Chem 79 (2007) 1801–1827.
[106] H.G. Yang, H.C. Zeng, Preparation of hollow anatase TiO_2 nanospheres via Ostwald ripening, J. Phys. Chem. B 108 (2004) 3492–3495.
[107] C.C. Yec, H.C. Zeng, Synthesis of complex nanomaterials via Ostwald ripening, J. Mater. Chem. A 2 (2014) 4843–4851.
[108] H.C. Zeng, Synthetic architecture of interior space for inorganic nanostructures, J. Mater. Chem. 16 (2006) 649–662.
[109] H.C. Zeng, Ostwald ripening: a synthetic approach for hollow nanomaterials, Curr. Nanosci. 3 (2007) 177–181.

[110] A.D. Smigelskas, E.O.K. Zinc, Zinc diffusion in alpha brass, Am. Inst. Min. Metall. Eng. 171 (1947) 130–142.
[111] E.O. Kirkendall, Diffusion of zinc in alpha brass, Trans. Am. Inst. Min. Metall. Eng. 147 (1942) 104–110.
[112] Y. Yin, R.M. Rioux, C.K. Erdonmez, S. Hughes, G.A. Somorjai, A.P. Alivisatos, Formation of hollow nanocrystals through the nanoscale Kirkendall Effect, Science 304 (2004) 711–714.
[113] W.S. Wang, M. Dahl, Y.D. Yin, Hollow nanocrystals through the nanoscale Kirkendall effect, Chem. Mater. 25 (2013) 1179–1189.
[114] Y. Sun, B.T. Mayers, Y. Xia, Template-engaged replacement reaction: a one-step approach to the large-scale synthesis of metal nanostructures with hollow interiors, Nano Lett. 2 (2002) 481–485.
[115] Y. Xia, Y. Xiong, B. Lim, et al., Shape-controlled synthesis of metal nanocrystals simple chemistry meets complex physics, Angew. Chem. Int. Ed. 48 (2009) 60–103.
[116] G.K. Mor, O.K. Varghese, M. Paulose, C.A.K. Shankar, A review on highly ordered, vertically oriented TiO_2 nanotube arrays: fabrication, material properties, and solar energy applications, Sol. Energy Mater. Sol. Cells 90 (2006) 744–750.
[117] Y.G. Li, P. Zhou, Z.H. Dai, Z.X. Hu, P.P. Sun, J.C. Bao, A facile synthesis of PdCo bimetallic hollow nanospheres and their application to Sonogashira reaction in aqueous media, New J. Chem. 30 (2006) 832–837.
[118] F.Y. Cheng, H. Ma, Y.M. Li, J. Chen, $Ni_{1-x}Pt_x$ (x = 0-0.12) hollow spheres as catalysts for hydrogen generation from ammonia borane, Inorg. Chem. 46 (2007) 788–794.
[119] S. Ikeda, S. Ishino, T. Harada, N. Okamoto, T. Sakata, H. Mori, et al., Preparation of magnetite core-titania shell and hollow titania nanoparticles via layer-by-layer (LbL) assembly method, Angew. Chem. Int. Ed. 45 (2006) 223–226.
[120] P.M. Arnal, M. Comotti, F. Schuth, High-temperature-stable catalysts by hollow sphere encapsulation, Angew. Chem. Int. Ed. 45 (2006) 8224–8227.
[121] S.H. Wei, M.H. Zhou, W.P. Du, Improved acetone sensing properties of ZnO hollow nanofibers by single capillary electrospinning, Sens. Actuators B 160 (2011) 753–759.
[122] J. Zhang, X. Liu, S. Wu, M. Xu, X. Guo, S.J. Wang, Au nanoparticle-decorated porous SnO_2 hollow spheres a new model for a chemical sensor, J. Mater. Chem. 20 (2010) 6453–6459.
[123] D. Zang, M. Yan, P. Zhao, L. Ge, S. Liu, J. Yu, A novel high selectivity chemiluminescence sensor for fenvalerate based on double-sided hollow molecularly imprinted materials, Analyst 137 (2012) 4247–4253.
[124] C.N. Xu, K. Miyazaki, T. Watanabe, Humidity sensors using manganese oxides, Sens. Actuators B 46 (1998) 87–96.
[125] Z. Chen, C. Lu, Humidity sensors: a review of materials and mechanisms, Sens. Lett. 3 (2005) 274–295.
[126] A.P. Goodey, J.T. McDevitt, Multishell microspheres with integrated chromatographic and detection layers for use in array sensors, J. Am. Chem. Soc. 125 (2003) 2870–2871.
[127] L.L. Wang, T. Fei, J.N. Deng, Z. Lou, R. Wang, T. Zhang, Synthesis of rattle-type SnO_2 structures with porous shells, J. Mater. Chem. 22 (2012) 18111–18114.
[128] J. Zhang, S.R. Wang, Y. Wang, M.J. Xu, H.J. Xia, S.M. Zhang, et al., ZnO hollow spheres preparation, characterization, and gas sensing properties, Sens. Actuators B Chem. 139 (2009) 411–417.

[129] F. Gyger, M. Hubner, C. Feldmann, N. Barsan, U. Weimar, Reactive and photocatalytic degradation of various water contaminants by laser ablation-derived SnO_x nanoparticles in liquid, Chem. Mater. 22 (2010) 4821–4827.
[130] J. Zhang, J. Liu, Q. Peng, X. Wang, Y. Li, Nearly monodisperse Cu_2O and CuO nanospheres: preparation and applications for sensitive gas sensors, Chem. Mater. 18 (2006) 867–871.
[131] X.L. Li, T.J. Lou, X.M. Sun, Y.D. Li, Highly sensitive WO_3 hollow-sphere gas sensors, Inorg. Chem. 43 (2004) 5442–5449.
[132] Y. Zeng, X. Wang, H. Wang, Y. Dong, Y. Ma, J.N. Yao, Multi-shelled titania hollow spheres fabricated by a hard template strategy enhanced photocatalytic activity, Chem. Commun 46 (2010) 4312–4314.
[133] H.X. Li, Z.F. Bian, J. Zhu, D.Q. Zhang, G.S. Li, Y.N. Huo, et al., Mesoporous titania spheres with tunable chamber stucture and enhanced photocatalytic activity, J. Am. Chem. Soc. 129 (2007) 8406–8407.
[134] A. Kudo, Y. Miseki, Heterogeneous photocatalyst materials for water splitting, Chem. Soc. Rev. 38 (2009) 253–278.
[135] J. Ran, J. Zhang, J. Yu, M. Jaroniec, S.Z. Qiao, Earth-abundant cocatalysts for semiconductor-based photocatalytic water splitting, Chem. Soc. Rev. 43 (2014) 7787–7812.
[136] C.G. Silva, Y. Bouizi, V. Fornés, H. García, Layered double hydroxides as highly efficient photocatalysts for visible light oxygen generation from water, J. Am. Chem. Soc. 131 (2009) 13833–13839.
[137] X. Zong, H. Yan, G. Wu, G. Ma, F. Wen, L. Wang, et al., Enhancement of photocatalytic H_2 evolution on CdS by loading MoS_2 as cocatalyst under visible light irradiation, J. Am. Chem. Soc. 130 (2008) 7176–7177.
[138] E. Nurlaela, S. Ould-Chikh, M. Harb, S. del Gobbo, M. Aouine, E. Puzenat, et al., Critical role of the semiconductor–electrolyte interface in photocatalytic performance for water-splitting reactions using Ta_3N_5 particles, Chem. Mater. 26 (2014) 4812–4825.
[139] S.J. Hong, S. Lee, J.S. Jang, J.S. Lee, Heterojunction $BiVO_4/WO_3$ electrodes for enhanced photoactivity of water oxidation, Energy Environ. Sci. 4 (2011) 1781–1787.
[140] M. Ni, M.K. Leung, D.Y. Leung, K. Sumathy, A review and recent developments in photocatalytic water-splitting using TiO_2 for hydrogen production, Renew. Sustain. Energy Rev. 11 (2007) 401–425.
[141] D. Barreca, G. Carraro, V. Gombac, A. Gasparotto, C. Maccato, P. Fornasiero, et al., Supported metal oxide nanosystems for hydrogen photogeneration quo vadis, Adv. Funct. Mater. 21 (2011) 2611–2623.
[142] D.P. Dubal, O. Ayyad, V. Ruiz, P. Gomez-Romero, Hybrid energy storage the merging of battery and supercapacitor chemistries, Chem. Soc. Rev. 44 (2015) 1777–1790.
[143] Y. Jiao, Y. Zheng, M.T. Jaroniec, S.Z. Qiao, Design of electrocatalysts for oxygen-and hydrogen-involving energy conversion reactions, Chem. Soc. Rev. 44 (2015) 2060–2086.
[144] T.Y. Ma, S. Dai, S.Z. Qiao, Self-supported electrocatalysts for advanced energy conversion processes, Mater. Today 19 (2016) 265–273.
[145] A. Manthiram, Y.Z. Fu, S.H. Chung, C.X. Zu, Y.S. Su, Rechargeable lithium-sulfur batteries, Chem. Rev. 114 (2014) 11751–11787.
[146] M.V. Reddy, G.V.S. Rao, B.V.R. Chowdari, Metal oxides and oxysalts as anode materials for Li ion batteries, Chem. Rev. 113 (2013) 5364–5457.
[147] P. Simon, Y. Gogotsi, Materials for electrochemical capacitors, Nat. Mater. 7 (2008) 320–329.

[148] B.Y. Xia, H.B. Wu, X. Wang, X.W. Lou, Highly concave platinum nanoframes with high-index facets and enhanced electrocatalytic properties, Angew. Chem., Int. Ed. 52 (2013) 12337–12340.
[149] Q.F. Zhang, E. Uchaker, S.L. Candelaria, G.Z. Cao, Nanomaterials for energy conversion and storage, Chem. Soc. Rev. 42 (2013) 3127–3171.
[150] C.Z. Yuan, H.B. Wu, Y. Xie, X.W. Lou, Mixed transition-metal oxides design, synthesis, and energy-related applications, Angew. Chem. Int. Ed. 53 (2014) 1488–1504.
[151] C. Liu, F. Li, L.P. Ma, H.M. Cheng, Advanced materials for energy storage, Adv. Mater. 22 (2010) E28–E62.
[152] S.M. Xu, C.M. Hessel, H. Ren, R.B. Yu, Q. Jin, M. Yang, et al., α-Fe_2O_3 multi-shelled hollow microspheres for lithium ion battery anodes with superior capacity and charge retention, Energy Environ. Sci. 7 (2014) 6679–6684.
[153] J.Y. Wang, N.L. Yang, H.J. Tang, Z.H. Dong, Q. Jin, M. Yang, et al., Accurate control of multishelled Co_3O_4 hollow microspheres as high-performance anode materials in lithium-ion batteries, Angew. Chem. Int. Ed. 52 (2013) 6545–6548.
[154] L. Yu, B.Y. Guan, W. Xiao, X.W. Lou, Formation of yolk-shelled Ni-Co mixed oxide nanoprisms with enhanced electrochemical performance for hybrid supercapacitors and lithium ion batteries, Adv. Energy Mater. 5 (2015) 1500981.
[155] L. Zhang, H.B. Wu, Y. Yan, X. Wang, X.W. Lou, Hierarchical MoS_2 microboxes constructed by nanosheets with enhanced electrochemical properties for lithium storage and water splitting, Energy Environ. Sci. 7 (2014) 3302–3306.
[156] F.X. Ma, H.B. Wu, B.Y. Xia, C.Y. Xu, X.W. Lou, Hierarchical β-Mo_2C nanotubes organized by ultrathin nanosheets as a highly efficient electrocatalyst for hydrogen production, Angew. Chem. Int. Ed. 54 (2015) 15615–15619.
[157] L. Zhang, H.B. Wu, X.W. Lou, Metal-organic-frameworks derived general formation of hollow structures with high complexity, J. Am. Chem. Soc. 135 (2013) 10664–10672.
[158] P. Simon, Y. Gogotsi, Materials for electrochemical capacitors, Nat. Mater. 7 (2008) 845–854.
[159] H. Hu, B.Y. Guan, X.W. Lou, Construction of complex CoS hollow structures with enhanced electrochemical properties for hybrid supercapacitors, Chem 1 (2016) 102–113.

Further Reading

J. Gao, G. Liang, J.S. Cheung, Y. Pan, Y. Kuang, F. Zhao, et al., Multifunctional yolk-shell nanoparticles a potential MRI contrast and anticancer agent, J. Am. Chem. Soc. 130 (2008) 11828–11833.

2

NOBLE METAL–BASED NANOSENSORS FOR ENVIRONMENTAL DETECTION

Wei Xiong[1,2], Pancras Ndokoye[3] and Michael K.H. Leung[2]

[1]Key Laboratory of Industrial Ecology and Environmental Engineering (Ministry of Education), School of Environmental Sciences and Technology, Dalian University of Technology, Dalian, P.R. China [2]Ability R&D Energy Research Centre, School of Energy and Environment, City University of Hong Kong, Hong Kong, P.R. China [3]Science, Technology and Innovation Unit, Directorate of Education Policy and Planning, Rwanda Ministry of Education, Kigali, Rwanda

2.1 Properties of Noble-Metal Nanoparticles

The detection of chemical and biological pollutants plays a fundamental role in environmental analyses [1,2]. The development of highly sensitive and stable sensors requires advanced technology coupled with fundamental knowledge in chemistry, biology, and material sciences. Noble-metal nanoparticles (NMPs) have attracted much attention in environmental sensing due to their unique physical and chemical properties, including good conductivity, easy functionalization with a range of ligands, large electronic field enhancement, fluorescence quenching, and catalytic behavior (Fig. 2.1).

2.1.1 Surface Plasmon Resonance

Surface plasmon resonance (SPR), which relies on the surface electromagnetic mode originating from collective coherent oscillation of conduction-band electrons induced by incident light, is the most exceptional property of metal nanostructures, especially gold and silver [4]. The various colors of colloidal gold nanomaterials with different shapes in the solution are mainly due to the SPR effect of gold nanoparticles.

Advanced Nanomaterials for Pollutant Sensing and Environmental Catalysis. DOI: https://doi.org/10.1016/B978-0-12-814796-2.00002-2

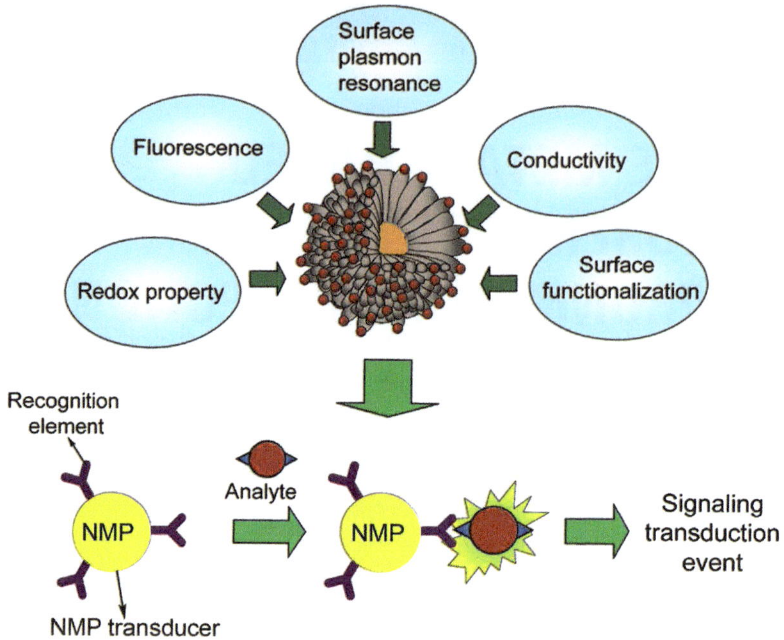

Figure 2.1 Physical properties of noble-metal nanoparticles (NMPs) and a schematic illustration of the NMP-based sensing platform. Reprinted with permission from K. Saha, S.S. Agasti, C. Kim, X. Li, V.M. Rotello, Gold nanoparticles in chemical and biological sensing, Chem. Rev. 112 (2012) 2739–2779 [3]. Copyright (2012) American Chemical Society.

Since the Mie theory can give a simple and accurate solution to the Maxwell equation, it has become a fundamental theoretical tool for predicting the plasmon resonance effect of metal nanomaterials [5]. According to the Mie theory, the resonance absorption peak and bandwidth of metal nanoparticles can be affected by many factors such as the material, shape, and size of the nanoparticles and the dielectric constant of the surrounding environment [6]. The SPR of metal nanostructures is highly sensitive to these factors. The SPR properties of the metal nanoparticles can be predicted accurately through the simulation by discrete dipole approximation, finite–difference time domain, and finite-element modeling. The effects of the metal's composition, shape, size, and surrounding environmental media of the nanoparticles on the SPR properties can be summarized as:

1. *Metal composition*. Different metal materials have different plasmon resonance frequencies and, therefore show different SPR properties. Gold and silver nanostructures are the most widely used materials in sensing. Silver nanoparticles (AgNPs) with a diameter of 40 nm show an SPR peak located at 400 nm [7], while the plasmonic peak for gold nanoparticles (AuNPs) with the same size locates at 530 nm [8]. The absorption peak of AuNPs shows more obvious redshift in comparison with AgNPs with the same size.

2. *Morphology of materials.* The morphology of materials can be precisely controlled by different fabrication methods and reaction conditions, and it shows an important effect on tuning SPR properties [9]. For instance, gold nanospheres (AuNSs) and gold nanorods (AuNRs) exhibit completely different SPR properties. The SPR peak of AuNSs with a diameter of about 13 nm is mainly distributed around 521 nm, while the plasmon resonance absorption of AuNRs divides into two peaks, namely the strong absorption peak attributed to the longitudinal dipolar resonance mode and the weak absorption peak at around 515 nm corresponding to transverse dipolar resonance modes, respectively.
3. *Particle size.* Nanoparticles with different sizes have different SPR properties. The most typical examples are also AuNSs and AuNRs. With the increase of the diameter of AuNSs, the dipole resonance mode of SPR continuously redshifts. The plasma absorption peak for the AuNSs with a diameter of 13 nm is around 521 nm. However, it redshifts to around 575 nm when increasing the diameter to 99 nm [8]. In the case of AuNRs, the longitudinal and transverse dipole resonances show different changing trends with increasing aspect ratios [10]. In more detail, the high-energy transverse resonance peaks hardly shift with the increase of the aspect ratio of the nanorods, while the peak of the low-energy longitudinal dipole resonance shows a significant redshift. The absorption peak for the longitudinal SPR can even be tuned to the near-infrared region by further increasing the aspect ratio.
4. *Surrounding environmental media.* The surrounding media can also cause changes in the SPR properties of the metal nanoparticles. After transferring the AuNRs from the aqueous solution to toluene, the longitudinal dipole plasma resonance peaks of AuNRs turn widely and redshift to 765 from 720 nm. This change of the SPR performance can be attributed to the refractive index of toluene (1.50) which is remarkably higher than that of water (1.33) [9].

Benefiting from the significant difference of plasmonic wavelengths rising from the geometrical or surrounding environment change, SPR has been employed as an advanced, label-free, and real-time optical analytical sensing technique for environmental monitoring. However, the sensitivity of the SPR sensors exactly depends on the change of the plasmonic bands of metal nanostructures and the effect of the local environment for the SPR is quite important in terms of environmental sensing.

Plasmon-coupling effect. When the colloidal AuNPs in the solution aggregate, its color will change from red to blue [11]. This phenomenon is regarded as a typical optical property of gold nanomaterials. The color change induced by aggregation can be attributed to the interaction between each other when getting close as the AuNPs are aggregated, therefore lead to a redshift of SPR peak and the color of the solution changes correspondingly. This phenomenon is called the plasmon-coupling effect [6].

In order to facilitate the principle of this interaction between AuNPs, the coupling effect of adjacent oscillators was investigated by using two closed dipoles as a model. The energy of interaction between two dipoles can be described by Eq. (2.1):

$$E \propto \frac{p_1 p_2}{r^3} \tag{2.1}$$

where p_1, p_2 represent the sizes of the dipoles and r is the distance between them. It can be seen from Eq. (2.1) that the energy of the interaction becomes significantly stronger while reducing the separation between the dipoles, and the resonance frequency also shifts, resulting in the formation of a new plasma resonance peak compared to the single nanoparticle. For two adjacent AuNPs, the low-energy resonant elements corresponding to the longitudinally aligned dipoles produce a significant redshift in the spectrum, while the coupled dipoles counteract each other for high-energy resonant elements, resulting in a net dipole moment of zero and hardly any resonance effect by the incident light. High-order multipoles also produce this interaction, and the effect depends primarily on the size of the nanoparticles and their mutual distance.

Metal nanoparticles with coupling plasmon resonance can produce three distinct optical signals for sensing: (1) color change due to distance variation; (2) significant redshift of SPR peak; and (3) the enhancement of the surface-enhanced Raman spectrum (SERS) performance benefitting from the enhanced electromagnetic field induced in the gap between the nanoparticles. The three changes caused by the plasma-coupling effect have a wide range of applications. For example, the color change caused by the effect of plasmon-resonance coupling depends on the distance between the nanoparticles. Based on this characteristic, the coupling effect can be used to investigate changes in the distance between the nanoparticles induced by chemical or biological forces. Sonnichsen et al. [11] used double-stranded DNA to tune the distance of NMP assemblies. Since the DNA as a linker is flexible, the distance between the nanoparticle dimers can be precisely regulated. A decrease in the distance between the nanoparticles leads to their color change under dark-field conditions. The color of the AuNP solution turns from green to red, while enhanced plasmonic coupling can also lead to a significant redshift of the SPR peaks of AgNPs and AuNPs. In addition, the plasmon-coupling effect between the adjacent nanoparticles causes great convergence of the electromagnetic field energy to the gaps between the nanoparticles. For instance, when the distance between two AuNRs are decreased to only 1 nm, the electric-field strength between them can be magnified to 10^3 or more [12]. This ultrahigh-intensity of the local electromagnetic field due to the plasmon resonance coupling effect can greatly improve the SERS performance.

2.1.2 Fluorescence

NMPs sized 3 nm or smaller usually possess strong fluorescent properties. In this size regime, they commensurate to the Fermi wavelength of electrons and exhibit molecule-like properties due to the discrete electronic states caused by the strong-quantum confinement of free electrons. As their size increases, NMPs can also lead to a shift of fluorescence emission and fluorescence quenching. NMPs with adsorbed fluorescent ligands show size-tunable fluorescence emission that shifts to higher wavelengths with increasing size. Due to the overlap between the emission spectrum and the metal surface–plasmon band, fluorescence quenching can be caused by the SPR effect of NMPs serving as quenchers for various systems based on Förster resonance energy transfer (FRET).

2.1.3 Catalytic Activity

In the 1980s, Haruta et al. [13,14] unexpectedly discovered that highly dispersed AuNPs showed good activity for CO catalytic oxidation with good stability and poison resistance at the low temperature of 77K. This research enables us to realize the excellent catalytic activity of AuNPs, which can be attributed to the quantum-size effect of AuNPs and the charge-transfer effect on the support. In recent years, studies have shown that the catalytic properties of gold nanomaterials may come from the gold particles themselves by using unsupported AuNPs in the efficient catalytic process of the aerobic oxidation of glucose [15]. The outstanding catalytic activity of gold nanomaterials is assumed to originate from the nanostructured gold itself, which is much different from the mechanism of the effective charge transfer, and quantum-size effect described above. The catalytic activity of NMPs leads to their molecule-like redox property and outstanding activity in sensing and catalysis.

2.1.4 Surface Functionalization

NMPs are effortless for surface functionalization, benefiting from their high specific surface area. Conjugation with specific ligand molecules that facilitate specific binding to the target analyte is also a prerequisite for efficient NMP sensing platform [16]. Three main approaches have been carried out for the functionalization of NMPs by specific molecules: (1) physical adsorption by electrostatic or hydrophobic interaction; (2) covalent coupling between metal and S group; and (3) specific recognition based on the principle of explicit specificity of ligand molecules to the analyte molecule. The electrostatic or hydrophobic interaction induced physical adsorption avoids the complex synthesis process, nevertheless the adsorption is sensitive to the change of environmental parameter. Any changes in pH or ionic strength will result in the desorption of specific ligands. Small molecules within the nitro, amino or carboxyl groups have been widely employed to functionalize NMPs using this approach.

With the advantages in SPR effect, fluorescence changing or quenching, catalytic activity, good conductivity, and easy surface functionalization with a range of ligands, NMPs have been employed in applications in environmental sensing as colorimetric (or SPR) sensors, fluorescent sensors, SERS-based sensors, electrochemical sensors and others, as shown in Table 2.1 [16].

Table 2.1 Different Types of Sensing Platform-Based NMPs

Type	Principle	Property of NMPs
Colorimetric sensor	Optical property	Sensitivity to change in size, aggregation state, and refractive index
Fluorescent sensor	Fluorescence changes or quenching	High-molar extinction coefficient and broad energy bandwidth
SERS-based sensor	Local electromagnetic field enhancement	Sensitivity to change in size, shape, orientation, and aggregation of particles
Electrochemical sensor	Electrical change	High surface area and conductive and catalytic properties

Source: Reprinted from E. Priyadarshini, N. Pradhan, Gold nanoparticles as efficient sensors in colorimetric detection of toxic metal ions: a review, Sens. Actuators B 238 (2017) 888–902, Copyright (2017), with permission from Elsevier.

2.2 Colorimetric Sensing of Heavy Metal

The word "heavy" in the "heavy metal" (HM) means the metals have relatively "high" densities. HM usually refers to metals having a specific density above 5 g cm^{-3} and the most common metals are iron, copper, and tin and precious metals such as silver, gold, and platinum. However, the various HMs have very different effects on humans and the environment. Some HMs are either essential nutrients that are required for various biochemical and physiological functions (typically iron, cobalt, and zinc). Other metals such as ruthenium, silver, gold, and indium have no established biological functions and are relatively harmless. However, it needs to be pointed out that all metals can be toxic at high concentrations. Other HMs such as cadmium, mercury, arsenic, and lead have been reported to be highly poisonous to the cellular organelles and components involved in metabolism, detoxification, and damage repair even at low levels of exposure [17,18]. In addition, these kinds of HMs can also result in the reduction of soil quality, crop failure, poor quality of agricultural products, and the endangerment of ecosystems. Due to their acute toxicity, safe limits or maximum contaminant levels of the HM ions in drinking water (especially cadmium, mercury, arsenic, and lead) have been defined that are of great significance for human and environment health [19]. The World Health Organization (WHO) and Environmental Protection Agency (EPA) have recommended standards for hazardous cadmium (0.003, 0.005 mg L^{-1}), mercury (0.001, 0.002 mg L^{-1}), arsenic (0.010, 0.010 mg L^{-1}), and lead (0.010, 0.015 mg L^{-1}) in drinking water.

Currently, the development of sensing platforms for HM ions at low concentrations in environmental samples is receiving considerable attention. Several effective methods have been developed for the determination of HM ions, including atomic absorption spectroscopy, atomic fluorescence spectrometry, inductively coupled plasma–mass spectrometry, vapor generation methods, and electrochemical sensing platforms [20]. They show efficient applications in the detection of HM ions within a short time, at low cost, with short preconcentration steps and without requiring sophisticated equipment. A simple, rapid, inexpensive, and real-time sensing system remains desirable for the detection of HM ions and colorimetric sensing technology based on NMPs (such as Ag and Au) has become one of the most efficient approaches to meet this demand. In addition, the most important advantage of colorimetric sensing technology is monitoring HM ions by the

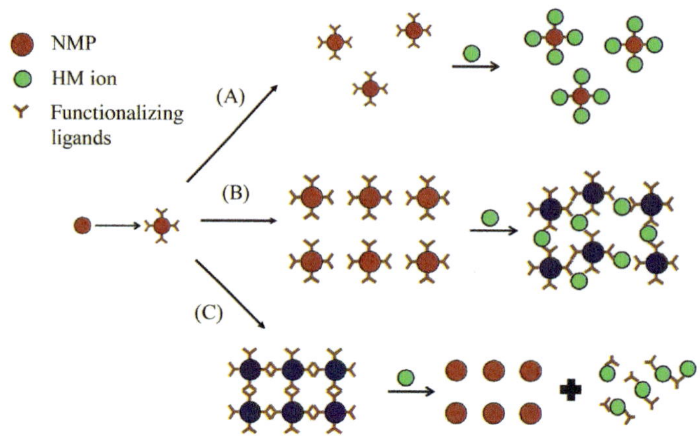

Figure 2.2 Mechanisms behind colorimetric detection of HM ions: (A) control of SPR property; (B) aggregation of dispersed NMPs; and (C) disassembly of aggregated NMPs. Reprinted from E. Priyadarshini, N. Pradhan, Gold nanoparticles as efficient sensors in colorimetric detection of toxic metal ions: a review, Sens. Actuators B 238 (2017) 888–902, Copyright (2017), with permission from Elsevier.

color change of the solution even by the naked eye, thereby avoiding complex instrumentation.

Colorimetric sensing technology for the detection of HM ions arises from the SPR of NMPs. Any change in the size, shape, composition, geometry or the surrounding environment of NMPs alters the local electron confinement, which is reflected in the tunable change in the SPR properties of the colloidal nanoparticles in the solution. Colorimetric sensors based on noble metals for the detection of HM ions fundamentally relies on the HM ions induced SPR absorption peak shift and color change of colloidal solution, which is attributed to the change in the interparticle interactions or dielectric constant of the surrounding environment of the NMPs in the colloidal solution (Table 2.1). There are three optional sensing mechanisms in the design of the interactions in the colorimetric sensing progress: HM ions induced control of SPR property, aggregation of dispersed NMPs, and disassembly of aggregated NMPs (Fig. 2.2).

2.2.1 Control of Surface Plasmon Resonance Properties

As emphasized, NMPs are highly sensitive to the change of dielectric constant of the local surrounding environment. Tiny changes lead to obvious broadening or shifting of SPR peaks. The sensing mechanism is based on the investigation of the change of the surface-plasmon properties during the binding of external HM ions at, or near, the surface of NMPs. This strategy has been successfully applied in the sensing of Hg^{2+} by using individual plasmonic AuNPs–based substrate with a detection limit at the picomolar level [21]. The AuNPs were functionalized with 1,6-hexanedithiol, a kind of thiolated molecule with the specific adsorption of Hg^{2+}. The scattering property of single 1,6-hexanedithiol functionalized AuNPs was altered with the binding of Hg^{2+}. The SPR peak redshifts after the adsorption of Hg^{2+} onto the AuNP surface, resulting from the change of the dielectric constant of the surrounding environmental media. It indicated that the control of the dielectric constant of the interface between NMPs and surrounding media was a crucial point in the detection of HM ions with low concentration.

Different surrounding environments not only change the dielectric constant around NMPs, but also tunes their sizes, shapes or morphologies. Nonmodified AuNRs have been reported to detect Hg^{2+} by monitoring the SPR peak shift caused by size change [22]. The selective etching along the tips of AuNRs when reacting with Hg atoms shortened the length of AuNRs, which results in the blueshift of the absorption peak, even in the presence of ultralow traces of Hg^{2+}. This interesting approach provides a detection limit of 6.6×10^{-13} g L^{-1} of Hg^{2+} in solution.

2.2.2 Aggregation of Dispersed Noble-Metal Nanoparticles

In case of the aggregation of dispersed metal particles, the inducing of HM ion leads to the binding of the dispersed NMPs. Their aggregation causes a visible change in color of the solution as well as the shift and broadening of SPR peak [16]. For example, the 14.2 nm spherical AuNPs prepared by citrate-mediated reduction had been reported for the selective colorimetric detection of aqueous Hg^{2+}, Ag^{+} and Pb^{2+} [23]. The sensing system was based on the AuNPs aggregation induced by the HM ions, with detection limits up to nanomolar metal concentration. Alkanethiols have been employed as strong electron-releasing ligands of high polarizability in the sensing progress. The competition between strong Hg^{2+}–S bonds and Au–S linkages induces the aggregation of nanoparticles, resulting in the color change from red to blue and a declining ratio of the extinction coefficients from 650 to 520 nm of the AuNP solution. The aggregation of aqueous AuNPs induced by HM ions and alkanethiols enabled us to develop label-free assays for the sensitive and selective detection of the HM ions.

Appropriately functionalized NMPs are more common in the aggregation-based colorimetric sensing of HM ions, wherein NMPs are conjugated with target specific ligand molecules, such as small molecules, polymers and biomolecules, which not only provide sufficient steric hindrance ensuring stability of the colloid, but also facilitate specific binding to the target HM ions of interest [16]. During this process, the NMP's surface is chemically anchored by the functionalizing molecules. The other end of the molecules is exposed with functional groups freely, thereby binding to the target HM ions. Therefore the selectivity and sensitivity of the sensing platform are improved. By adding a naturally occurring bifunctional molecule such as gallic acid during the synthesis of NMPs, AuNPs, and AgNPs can be employed as Pb^{2+} selective colorimetric sensors [24]. The unique coordination behavior between Pb^{2+} ions and carboxylate groups on the surface of NMPs results in the formation of a stable supramolecular complex. The visual color change with the aggregation of gallic acid–capped NMPs enables the detection of micromolar quantities (ppm level) of Pb^{2+} ions in the presence of other metal ions. The sensitivity of gallic acid–capped NMPs could be enhanced to nanomolar level by adding $NaClO_4$ to minimize electrostatic repulsion between gallic acid and NMPs, which reduces the energy barrier to overcome for Pb^{2+} [25].

In addition, polymers, with the major advantage of their good biocompatibility, can also be employed to functionalize the NMPs by physical adsorption [26]. Chitosan, a heteropolymer with the free-amine groups, in repeat units can enable its application of HM ion detection. With the presence of the reactive amino group, chitosan-capped AuNPs would get protonated in dilute acidic media, which makes it useful in chelating HM ions by forming the multiple bonding sites. The protonated amines promote the chitosan-capped AuNPs to be agglomerated in the presence of HM ions even at low concentration. The comparison of the SPR peak between 600 and 700 nm of chitosan-capped AuNPs suspension before and after exposure to metal ions make it a good strategy of colorimetric sensing toward HM ions. The sensitivity of chitosan-capped AuNPs in the detection of Hg^{2+} was as low as 5 pM, which is one of the lowest values achieved using the colorimetric method.

The covalent coupling between metal and S group can induce the strong bonding interaction, which is the most popular and widely used approach for the generation of functionalized NMPs. A large number of thiol-containing molecules and sulfur-containing compounds such as alkanethiolates, glutathione, disulfide, and thioethers have been reported attach on the surface of NMPs. By this approach, NMPs can be designed as the simple colorimetric platform for the detection of aqueous HM ions. After the capture of 11-mercaptoundecanoic acid with an Au–S bond, AuNPs were aggregated in the presence of divalent HM ions by an ion-templated chelation process, resulting in the naked-eye distinguishable color change of the solution [27]. Benefitting from the strong Au–S band interaction, biomolecules such as protein and DNA have also been widely used for the functionalization of NMPs for HM ion detection. Papain, an enzyme with abundant thiol groups, can adhere directly onto the NMPs for the detection of Hg^{2+}, Pb^{2+}, and Cu^{2+} [28]. Papain-functionalized AuNPs are highly stable with no significant aggregation at $pH \geq 6$. However, in the presence of mercury, a color change from red to blue can be observed by the naked eye, which is attributed to the aggregation of papain-functionalized AuNPs. After adding a mercury solution, DNA-functionalized AuNPs also show a purple-to-red color change resulting from the DNA-linked aggregates of two cDNA–AuNPs, which can be employed for colorimetric Hg^{2+} ion detection [29]. The major advantage of DNA used for HM ion detection is that the ions' concentration can be quantified from the change of the solution color at the melting temperature because the mercury coordinates selectively to the bases that make up a T–T mismatch. The concentration of Hg^{2+} ions usually shows a linear relation with the melting temperature, allowing one to quantify the concentration of Hg^{2+} ions in this sensing platform using DNA for the functionalization.

2.2.3 Disassembly of Aggregated Noble-Metal Nanoparticles

Disassembly can be induced by the interaction between the aggregated particles and the target HM ions, which leads to a change in color of the solution to blue from red accompanied by blueshift of the SPR band. In this case, the disassembly of the aggregated NMPs can be employed as the probes in the colorimetric sensing of HM ions. By employed unmodified AuNPs and a phytochelatin-like peptide such as $(g\text{-}Glu\text{-}Cys)_3$-Gly-Arg ($PC_3R$), the strategy of the disaggregation-based approach can be used in the detection of As^{3+} ions [30]. PC3R is an oligomer of glutathione and can react with As^{3+} to form a trigonal-pyramidal complex, for this thiol contains molecules that have a strong binding affinity with As^{3+}. In the absence of As^{3+}, PC3R can bind to the surface of AuNPs, thus resulting in the aggregation of AuNPs and a change in color from red to blue. However, the aggregation of AuNPs would be disassembled with the addition of As^{3+} due to the formation of As^{3+} phytochelatin peptide complex, thereby averting its aggregation and retaining the original red color of GNPs. The coordination chemistry between Hg and thymine can also be used for Hg^{2+} detection based on the disassembly of citrate-capped AuNPs [31]. The AuNPs can bind to thymine via Au–N bonds, thus inducing the aggregation. However, with the presence of Hg^{2+}, thymine would prefer to bind with Hg^{2+} by forming T-Hg complex rather than with AuNPs, which leads to detection of Hg^{2+}. The detection limit of the sensing platform was 2 nM. AuNPs conjugated with carboxylate and 15-crown-5 can be assembled into the monolayer due to the hydrogen bond between carboxylic acid residues in the methanol–water interface [32]. However, the AuNPs monolayer can be disrupted in the presence of Pb^{2+}. The electronic repulsion between AuNPs results in a change in color from blue to red, thus providing the detection of Pb^{2+}.

2.2.4 The Sensitivity of Noble-Metal Nanoparticles-Based Colorimetric Sensing

In order to improve the sensitivity of the NMPs-based colorimetric sensing platform, a lot of effort has been applied toward the fabrication of the system with a larger SPR shift for a given change in the refractive index. Though AuNPs have been widely used in many fields, their SPR absorption ranges in a small wavelength area (520–580 nm) with the diameters increasing from 2 to 100 nm. AuNRs have attracted increasing interest due to the unique properties arising from their anisotropy. Attributed to the oscillation of electrons along the transverse and longitudinal axes, the absorption spectra of AuNRs are characterized by two distinct plasmon bands, with a very weak transverse SPR band at 512 nm and a dominant longitudinal SPR band [10]. The position of the longitudinal SPR peak can be tuned as a function of the AuNRs aspect ratio from 550 to 1400 nm, which is extremely sensitive to any change in dielectric properties of the surrounding environment. This sensing response mode makes them exhibit much higher sensitivity than that of AuNPs.

2.3 Fluorescence-Based Sensing Towards Biomolecules

Fluorescence is a physical phenomenon of photoluminescence. When irradiated with light or other electromagnetic radiation, a fluorescent molecule such as a molecule, atom, or nanostructure absorbs the photon energy and its orbital electron enters an excited singlet state (S_1). Fluorescence occurs when the electron relaxes to its ground state (S_0) by emitting a photon:

$$\text{Excitation: } S_0 + h\nu_{ex} \rightarrow S_1 \tag{2.2}$$

$$\text{Fluorescence (emission): } S_1 \rightarrow S_0 + h\nu_{em} \tag{2.3}$$

Here h is Planck's constant, ν is the frequency of light and $h\nu_{ex}$ and $h\nu_{em}$ are generic terms for photon energy of exciting and emitted light, respectively. In most cases, the emitted light has a smaller frequency and longer wavelength and, therefore lower energy than the excited light. Benefiting from the excited light with longer wavelength, fluorescence can give the material a bright visible color when exposed to ultraviolet light which is invisible to the naked eye. Due to this unique property, fluorescence has been widely used in the areas of sensing, imaging, and others.

2.3.1 Fluorescence of Ultrasmall Gold Nanoparticles

The fluorescence of NMPs has been observed some 40 years ago, but less attention has not been paid to them due to their extremely low quantum yield (QY) of 10^{-10} [33]. Recently, researchers have developed various approaches to synthesize water-soluble fluorescent NMPs with much-enhanced QY in the range of 10^{-3}–10^{-1}, thus sufficiently bright for the detection of metal ions and proteins, using aggregation-induced quenching or enhanced fluorescence of NMPs [34]. In general, the AuNPs with strong fluorescent properties are usually smaller than 3 nm. In this size regime, the AuNPs commensurate to the Fermi wavelength of electrons and the discrete electronic states caused by the strong-quantum confinement of free electrons invests them with the molecule-like properties. A series of systems for sensing have been developed. A fluorescence quenching-based glucose sensing platform can be built by using glucose oxidase-functionalized AuNPs. In the presence of glucose, the enzymatic product of H_2O_2 can lead to the oxidation of the Au core to form Au^+ and the aggregation of nanoparticles. As a result, the fluorescence of AuNPs is quenched. Fluorescence quenching can be used for the detection of glucose in human urine and serum samples, with a detection limit of 0.7 μM (Fig. 2.3A) [35].

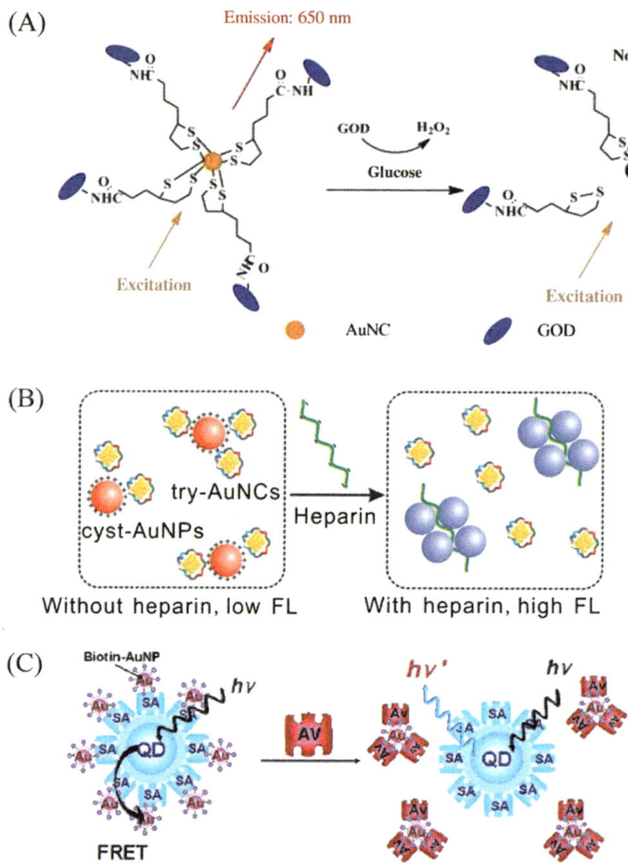

Figure 2.3 (A) Fluorescence quenching of glucose oxidase-functionalized AuNPs in presence of glucose; (B) schematic illustration for selective detection of heparin based on SPR enhanced energy transfer between cyst-AuNPs and try-AuNPs; and (C) schematic illustration of the inhibition assay method based on the fluorescence quenching of streptavidin-conjugated-QDs by biotinylated AuNPs. (A) Reprinted from X. Xia, Y. Long, J. Wang, Glucose oxidase-functionalized fluorescent gold nanoclusters as probes for glucose, Anal. Chim. Acta 772 (2013) 81–86, Copyright (2017), with permission from Elsevier. (B) Reprinted with permission from J.-M. Liu, J.-T. Chen, X.-P. Yan, Near infrared fluorescent trypsin stabilized gold nanoclusters as surface plasmon enhanced energy transfer biosensor and in vivo cancer imaging bioprobe, Anal. Chem. 85 (2013) 3238–3245. Copyright (2013) American Chemical Society. (C) Reprinted with permission from E. Oh, M.-Y. Hong, D. Lee, S.-H. Nam, H.C. Yoon, H.-S. Kim, Inhibition assay of biomolecules based on fluorescence resonance energy transfer (FRET) between quantum dots and gold nanoparticles, J. Am. Chem. Soc. 127 (2005) 3270–3271. Copyright (2005) American Chemical Society.

Synthetic strategies for fluorescent NMPs include "bottom–up" and "top–down" approaches, which entail the reduction of metal ion precursors and the etching of NMPs. Stable, ultrasmall NMPs can be commonly synthesized by the chemical reduction of Au^{3+} or Ag^{+} to Au or Ag in the presence of reducing and capping agents [36]. Thiol compounds are often employed as the capping agent because of the formation of strong metal–S bonding with the atoms or ions. Sodium borohydride ($NaBH_4$) is a common reducing agent in the presence of thiol compounds. A series of thiols such as glutathione, tiopronin, phenylethyl thiolate, polyethylene glycol appended–lipoic acid and thiolate cyclodextrin have been used to prepare ligand-stabilized NMPs in the presence of $NaBH_4$ as a reducing agent [35]. However, the ligand can affect the QYs of thiol-stabilized NMPs. When using glutathione as the capping reagent, the thiolate-Au complex can acquire a QY of 15% through the in situ generation of the fluorescent Au cores. Photoreduction approaches have also been applied in the preparation of NMPs instead of using an inorganic reducing agent such $NaBH_4$, which is hazardous. Tridentate thioether terminated polymers are usually employed to synthesize fluorescent NMPs through photoreduction. The nature and the amount of the polymers used can decide the size and QYs of the NMPs [37]. In addition, chemical etching can be used for the synthesis of ultrasmall NMPs from larger sizes by adding excess ligands. For example, the glutathione-stabilized AuNPs ($Au_{25}SG_{18}$) can be etched with the presence of octane thiol, leading to the formation of $Au_{23}SG_{18}$, which shows a bright red emitting [38]. In addition, glutathione can be used as an etching reagent to synthesize fluorescent AuNPs with a QY of 5.4% from nonfluorescent AuNPs.

2.3.2 Fluorescence Quenching by Surface Plasmon Resonance

Fluorescence quenching, or the decrease of the intensity of the fluorescence emission, may occur by several mechanisms, including static and dynamic quenching. Static quenching usually occurs when a small molecule makes a ground state complex with the fluorophore so that it becomes nonfluorescent. In the case of dynamic quenching, FRET is most common, which is caused by the energy transfer when the random, noninteractive collision of a small molecule deactivates the excited state of the fluorophore. FRET depends on the overlap of the spectra and the relative orientation of the transition dipole moments of the donor and acceptor [3].

Due to the overlap between the emission spectrum of fluorescence probe and the surface plasmon band of NMPs, fluorescence quenching is commonly observed when fluorophores are attached onto NMPs. This feature makes NMPs serve as excellent fluorescence quenchers for various FRET-based systems. In this case of fluorescence quenching in the presence of NMPs, it can be employed for the sensing of small organic molecules. For example, after the functionalization of AuNPs with fluorescent trypsin and cysteamine, the heparin can be detected via this FRET-based platform of the mixture solution of try-AuNPs and cyst-AuNPs (Fig. 2.3B) [39]. In this system, the SPR absorption band peak of cyst-AuNPs locates around at 524 nm and the emission spectrum of try-AuNPs shows an obvious peak at 690 nm. The positively charged cyst-AuNPs and negatively charged try-AuNPs get close due to the electrostatic interaction in the buffer solution. The aggregation of AuNPs leads to the redshift of the SPR band, thus enlarging the overlap of the SPR spectrum and emission spectrum. The energy transfer is attracted to the fluorescence quenching. In addition, the enhanced SPR caused by the plasmon coupling is also responsible for the great energy transfer efficiency. However, the addition of negatively charged heparin will induce the aggregation of cyst-AuNPs, thus leading to a weaker interaction between cyst-AuNPs and try-AuNPs. The presence of heparin induces an increase of fluorescence, even with concentrations down to 0.05 $\mu g\, mL^{-1}$. In addition, the fluorescence intensity is further enhanced as the concentration of heparin increases. In this sensing case, the quenched fluorescence of labeled fluorophores on the NMPs can be enhanced with the presence of target molecules. By employing this mechanism of FRET-induced fluorescence quenching, NMPs have been widely used in the sensing the other target objects, especially DNA [40]. The NMPs are usually labeled with organic dye through the conjugation of a nucleic acid probe. The dye usually lies very close to the NMPs by forming the hairpin structure on the surface, which leads to effective fluorescence quenching. However, the distance between the dye and the NMP will increase through the complementary hybridization after adding the target DNA, therefore resulting in the enhancement of the quenched fluorescence.

Besides detecting biomolecules, the FRET-based sensing system can also be utilized for the detection of HM ions by designing a suitable fluorescence quenching approach, such as the Cu^{2+} ion sensor of the bispyridyl perylene–bridged AuNPs

[41]. The fluorescence of the bispyridyl perylene on AuNPs is usually quenched. However, the fluorescence can be restored by the replacement of bispyridyl perylene by Cu^{2+} ion, leading to the sensing of copper ions. This specific interaction between the NMPs and the target molecules has been designed based on the fluorescence quenching approaches for the sensing of HM ions. As a comparison, the specific interaction is unnecessary for the detection of some analytes with large sizes, such as the proteins, pathogens, and mammalian cells. The electrostatic interaction between the NMPs and those analytes is usually utilized to achieve the efficient detection.

2.3.3 Assembly With Quantum Dots

Quantum dots (QDs) are semiconductor particles of only several nanometers in size, which have unique optical electronic properties compared to those of larger sizes [42]. QDs usually exhibit a strong fluorescence effect that can be tuned by the size, shape, and material [43,44]. QDs can be utilized for the labeling of NMPs as the inorganic fluorescent agent, with high efficiency and stability for the detection of proteins and DNA [45]. In this sensing platform, the functional AuNPs and QDs are employed as FRET donor–acceptor couples. Similarly, the SPR property of AuNPs causes the fluorescence quenching of the QDs. However, due to competitive binding with the biotinylated AuNPs, avidin can release them from QDs, thus allowing the successful detection of avidin by this FRET-based system (Fig. 2.3C) [46].

2.4 Surface-Enhanced Raman Spectrum-Based Application for Environmental Detection

Raman spectroscopy is a spectroscopic technique to study vibrational, rotational, and other low-frequency modes of molecules based on Raman scattering discovered by the Indian physicist C.V. Raman in 1928 [47]. In general, most scattered photons from an atom have the same energy and wavelength as the incident photons, while a small fraction (10^{-7}) of them are different. These phenomena are Rayleigh scattering and Raman scattering, respectively. The shift in the wavelength of the Raman-scattered photons can be used to identify the chemical and structural information of the molecules [4,48]. However, the signal of Raman spectrum is very week, due to the poor scattering compared to the Rayleigh scattering. Raman spectrum intensity I_R shows a relationship with the frequency and intensity of the incident light and the concentration of the molecule as:

$$I_R \propto \nu^4 I \, \sigma \, e^{-Ei/kT} C \tag{2.4}$$

where ν and I are the frequency and intensity of incident light, σ is the Raman cross-section, k is the Boltzmann constant, E_i is the energy for state i at the temperature of T, and C is the concentration of the molecule in the system. The Raman cross-section σ is usually 10^{-31}–10^{-26} per molecule, which limits the sensitivity in the application of analyzing the composition of solids, liquids, and gases.

In 1973 Martin Fleischmann, Patrick J. Hendra, and A. James McQuillan of the University of Southampton accidentally observed the enhanced Raman scattering from pyridine adsorbed on electrochemically roughened silver [49]. Since then, it is well-known that the enhancement of Raman spectra can be acquired on electrochemically roughened, coinage metal surfaces. SERS has grown dramatically, demonstrating its power as a surface-sensitive tool for analyzing molecules adsorbed on rough metal surfaces or NMPs, which provides the same signal with normal Raman spectroscopy, but with a greatly enhanced intensity.

The enhancement of Raman spectra has been understood as the product of two contributions, namely chemical enhancement and electromagnetic enhancement [4,50]. The enhancement factor (E_F) attributed to the enhanced local electromagnetic field induced by the SPR of NMPs can be as much as 10^{14}. Chemical enhancement relies on the charge-transfer effect of the adsorbed molecule with the E_F up to 10^6. It's very difficult to distinguish these two effects independently by experiments. However, SERS only occurs when target molecules are located within a few nanometers of the surface of substrates in both mechanisms. Silver (Ag), gold (Au), and copper (Cu) are classic metals for SERS application. Recently, other metals such as iron (Fe), cobalt (Co), nickel (Ni), platinum (Pt), palladium (Pd), and ruthenium (Ru) also demonstrated the SERS effect with surface enhancements from one to three orders of magnitude, which are much lower than those of Au and Ag because of their poor SPR properties in the visible light region [4].

Benefiting from the sensitivity of different vibrational modes, NMPs-based SERS can provide a "fingerprint" of the target molecules, which has made it a powerful approach for environmental sensing. Normally, E_F demonstrates the sensitivity of a SERS system and can be evaluated in practical use by:

$$E_F = \frac{(I_{SERS} \times N_{surf})}{(I_{surf} \times N_{SERS})} \tag{2.5}$$

where I_{SERS} and N_{SERS} are the intensity of enhanced-Raman intensity and the number of molecules absorbed on the metallic substrate, while the I_{surf} and N_{surf} are the intensity of normal Raman intensity and the number of molecules in the excitation volume, respectively [4].

Since then, SERS-based sensing has attracted increasing interest with studies usually focusing on the fabrication of SERS substrates, which entails NMPs used in SERS. The SERS substrates are often chemically stable, easy to prepare in a reproducible manner and exhibit a spatially uniform, high, enhanced factor. The substrates used in SERS can be divided into two classes, namely NMPs and their assemblies, and nanostructured metal arrays and films [51].

2.4.1 Noble-Metal Nanoparticles and Assemblies as Substrates

Metal nanoparticles employed as SERS substrates are usually chemically stable, reproducibly prepared, and exhibit excellent uniform enhancement for SERS [4]. As shown in Fig. 2.4A, metal nanoparticles such as those from Ag, Au, and Cu are deposited on a solid support such as silicon, glass or metal oxide to act as the SERS substrates and the molecules to be detected are in direct contact with the surface of metal nanoparticles. However, these bare metal nanoparticles may suffer from the unrepeatable SERS results, which limits their breadth of practical applications. The stability of metal nanoparticles can be enhanced by being coated with transition metals. In addition, the coating of transition metals can also extend the SERS activity of single silver and gold nanospheres to other wavelength regions, while they usually show the activity in the blue and green regions. Most transition metals show very good catalytic activity in many chemical reactions. This means that the SERS of metal-coated substrates can be extended to monitor the transition metal catalyzed reactions (Fig. 2.4B). Fig. 2.4C depicts tip-enhanced Raman scattering (TERS), which operates in a noncontact mode. In this modality, the probed molecules on the surface are separated

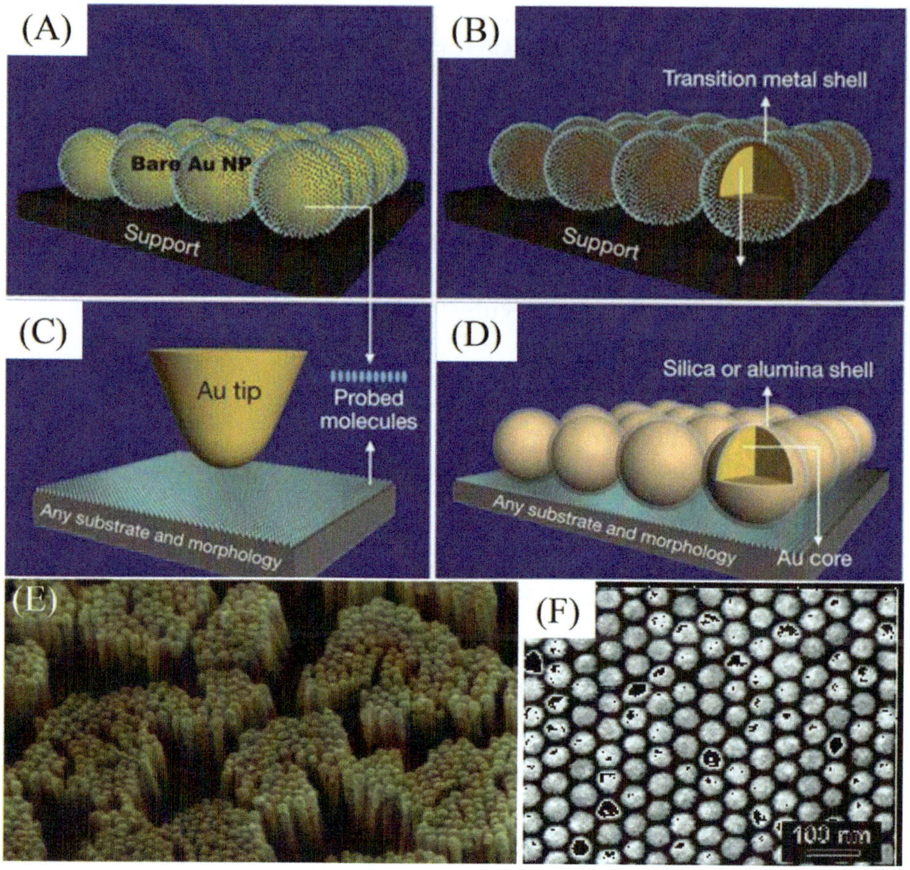

Figure 2.4 Schematic of the contact mode. (A) Bare Au nanoparticles: contact modes. (B) Au core–transition metal shell nanoparticles adsorbed by probed molecules: contact mode. (C) Tip-enhanced Raman spectroscopy: noncontact mode, and (D) SHINERS: shell-isolated mode. (E) SEM images of silver nanowires fabricated in porous aluminum oxide as the template. (F) SEM image of sub-10-nm gap AuNP arrays. (A–D) Reprinted by permission from Springer: J.F. Li, Y.F. Huang, Y. Ding, Z.L. Yang, S.B. Li, X.S. Zhou, et al., Shell-isolated nanoparticle-enhanced Raman spectroscopy, Nature 464 (2010) 392, Copyright 2010. (E) Reprinted with permission from S.J. Lee, A.R. Morrill, M. Moskovits, Hot spots in silver nanowire bundles for surface-enhanced Raman spectroscopy, J. Am. Chem. Soc. 128 (2006) 2200–2201. Copyright (2006) American Chemical Society. (F) Reprinted with permission from H. Wang, C.S. Levin, N.J. Halas, Nanosphere arrays with controlled sub-10-nm gaps as surface-enhanced Raman spectroscopy substrates, J. Am. Chem. Soc. 127 (2005) 14992–14993. Copyright (2005) American Chemical Society.

from the Raman signal amplifier of the Au tip. This isolated mode prevents a potentially disturbing interaction between Au and the molecules. TERS can act as a powerful tool to image any substrate at nanometer scale without additional constraints on the material composition and surface topography. Tian et al., designed another noncontact mode, shell-isolated, nanoparticle-enhanced, Raman spectroscopy (SHINERS) by using AuNSs coated by a 2 nm ultrathin silica shell (Fig. 2.4D) [52]. The coated-silica shell can isolate the gold surface from the molecules to be probed. This isolated mode cannot only prevent the disturbing interaction, but can also lead to a significantly enhanced Raman signal with a separation of 2 nm compared to the bare ones. By using these silica coated Au nanoparticles as SERS substrates, a very large number of "tips" could simultaneously sense the probed molecules on various surfaces, such as metal films, living cell walls, or fruits.

Since single metal nanoparticles only exhibit moderate E_F controlled synthesis of metal nanoparticles with sharp corners and edges is a promising approach to improve electromagnetic field enhancement and, therefore obtain a highly active sensing performance. For example, in order to improve the SERS activity of Ag nanoparticles, anisotropic etching by NH_4OH/H_2O_2 mixture solution is performed to selectively etch the (100) faces [53]. A star-like octopod structure is obtained after the etching. Benefiting from the plasmon coupling effect, a lot of "hot spots" occur in the gaps between the octopod silver nanoparticles, leading to the outstanding SERS activity for these anisotropic nanostructures. The design of multiple structures is another feasible approach to enhance the SERS activity. By using gold nanorods as the template, the caged gold nanorods can be successfully synthesized through the silver layer coating and the subsequent galvanic replacement reaction [54,55]. The hot spots occur in the gap between gold nanorod in the center and the cage outside. Benefiting from the plasmon-coupling effect between them, the caged gold nanorods show enhanced SERS activity.

Assembly of NMPs provides another effective approach to obtain highly active SERS substrate attracted to the hot spot occurred in the gap between the nanoparticles [51]. A range of methods and tools have been proposed to direct the assembly of NMPs as building blocks [6]. By functionalization with inorganic layer, small organic molecules, polymer, DNA, the NMPs can be assembled into dimers, trimers, core-satellite structures, and 1D, 2D, 3D superlattice structure. The assembly is usually induced by electrostatic attraction, covalent forces, hydrogen bonding, metal ion–organic ligand complexation, and DNA hybridization. Xiong et al. [56] indicated the core-satellite structure induced by electrostatic attraction. The caged AuNRs are coated by a silica layer and functionalized with small organic molecules with a positive charge. Attributing to the electrostatic self-assembly, the negatively charged AuNPs are anchored to their surface to form core–satellite structures. The gap between the AuNP and the cage can be tuned by changing the thickness of the silica coat, enabling to control the SERS performance.

2.4.2 Nanostructured Metal Arrays and Films as Substrates

Silver or gold nanorod arrays offer a simple yet powerful substrate for obtaining hot spots between the neighboring nanorods. By using highly ordered porous aluminum oxide as hard templates, the gold or silver nanorod can be synthesized through the electrochemical deposition and subsequent acid etching of alumina matrix (Fig. 2.4E) [57]. Polystyrene spheres are another widely used type of hard template. A monolayer of polystyrene spheres is self-assembled together by capillary forces onto the surface of a substrate through solvent evaporation. This assembled monolayer of polystyrene spheres can be served as the template for the fabrication of porous metal nanostructures by the physical vapor deposition or electrochemical deposition. Removal of the polystyrene spheres by sonication or calcination leads to the formation of ordered metal arrays [58]. Self-assembly also offers a useful approach to fabricate the SERS substrate. The cetyltrimethylammoniumbromide-functionalized AuNSs can be self-assembled into a 2D monolayer on the surface of indium-doped tin oxide through solvent evaporation (Fig. 2.4F) [59]. In addition, with the functionalization of polymers or DNA, free-standing NMPs film can also be fabricated as the SERS substrates [5].

The development of SERS substrates makes it a powerful tool for the multiplex detection of analytes, because of the enhancement in Raman signal with unique molecular fingerprints. The large enhancement in detectable signal enables the sensing of single molecule. After the development of decades, the SERS-based sensing is not only successfully performed in the laboratory, but also widely used in the quantitative detection of environmental, biomedical, food hygiene and safety. The SERS-based sensing has promoted the development of different subject areas and socioeconomic development.

2.4.3 Detection of Environmental Pollutants

Silver-silicone nanocomposites are employed as SERS substrates to improve the qualitative and quantitative measurement of typical environmental contaminants in water, including aromatics, chlorides and sulfides [60]. The protecting of silicone not only reduces the influence of the inherent light and heat during the detection, but also improves the sensitivity and reproducibility of SERS sensing. The results show a linear dynamic range of at least two orders of magnitude with a relative standard deviation of less than 10%. Polycyclic aromatic hydrocarbons (PAHs) are a group of persistent organic chemical pollutants with low concentration, but high toxicity in the environment. A hydrophobic SERS substrate has been designed to solve the problem of the detection of PAHs with low solubility in aqueous media. Firstly, the nonpolar polystyrene microspheres and AuNPs are banded cleverly into core–shell structures through the Debye adsorption between the permanent dipole and the induced dipole. The SERS-active substrates are then fabricated by depositing the AuNPs coated polystyrene microspheres onto the quartz substrate and they could be utilized for the quantitative detection of naphthalene in a concentration range of 1–20 $mg\,L^{-1}$ [61]. The substrate can suppress the increase of organic terminal functional groups effectively, therefore avoiding the signal disturbance by this functional group to the sensing performance. The thiol-functionalized Fe_3O_4@silver core–shell magnetic nanoparticles were also employed as the SERS substrate for the quantitative sensing of PAHs [62]. The resulting SERS signal showed a linear relationship to PAHs concentration in the range of 1–50 $mg\,L^{-1}$.

The combination of droplet microfluidic chips and SERS technology offers a new method for the rapid and sensitive analysis of trace HM ions in water. In this strategy, the AuNPs are served as the SERS substrate, which shows a strong interaction with the detected mercury ions [63]. The SERS signal intensity of the probe molecule rhodamine B changes with different concentrations of mercury ions in the solution. The quantitative analysis of mercury ions was performed by calculating the peak area of rhodamine B at 1647 cm^{-1}, which shows a good linear relationship in the range of 0.1–2.0 $\mu g\,L^{-1}$. The detection limit is 100–500 $ng\,L^{-1}$. Compared with the fluorescence method for the analysis of trace mercury ions, the sensitivity of SERS is increased by one order of magnitude.

Presently, SERS has also made some progress in the real-time analysis of environmental monitoring. For the detection of environmental pollutants, the traditional standard procedure is to send the collected samples to a designated laboratory for analysis. This process is usually very expensive, requires a large amount of the sample and complex chemical pretreatments to get available. The result usually takes days to weeks. Hatab et al.

[64] have employed SERS technology to achieve real-time quantitative detection of perchlorate and trinitrotoluene (TNT) in groundwater, which shows extremely high reproducibility and sensitivity. By using the highly ordered, bow-shaped gold nanoprotuberances as SERS substrates, perchlorate, and TNT in water samples can be quantified at concentrations as low as 0.66 and 0.20 mg L^{-1}, respectively.

2.4.4 Detection of Food Residual Pesticides

Food security has become one of the most important livelihood issues of enormous public concern. Unsafe foods will lead to a series of acute and lifelong diseases ranging from diarrhea to various cancers. SERS offers high-efficiency, low-cost, and portable quantitative detection for harmful substances in foods. By employing a portable Raman spectrometer, the content of Sudan Red 1 in a complex food matrix has been quantitatively analyzed through the high sensitivity of SERS. The detectable concentration ranges from 10^{-3} to 10^{-4} mol L^{-1} [65]. SHINERS can also be used for inspecting food safety [52]. Fig. 2.5A shows the Raman spectra recorded on fresh orange

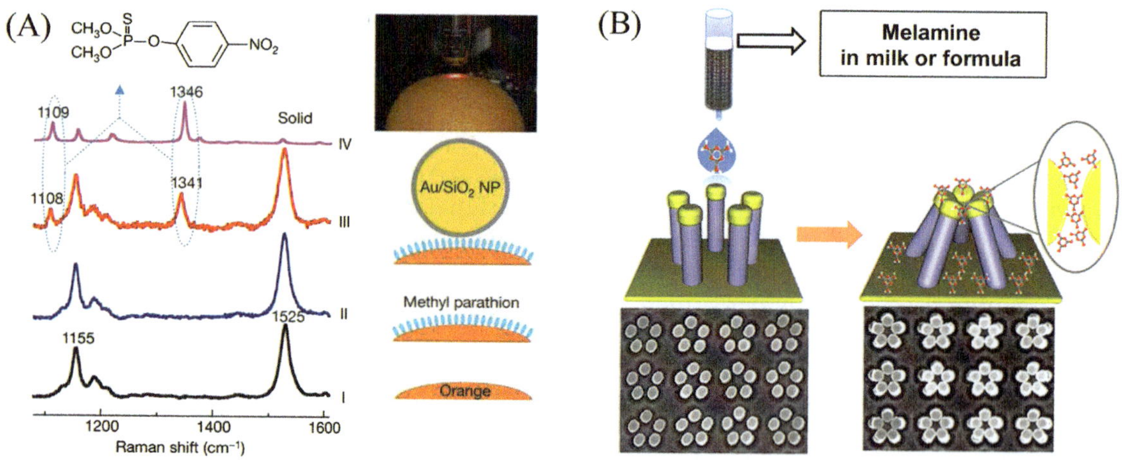

Figure 2.5 (A) Normal Raman spectra on fresh citrus fruits. Curve I, with clean pericarps; curve II, contaminated by parathion, curve III, SHINERS spectrum of contaminated orange modified by Au/SiO_2 nanoparticles, and curve IV, Raman spectrum of solid methyl parathion. (B) The representative SEM images of open (left) and closed (right) pentamer nanofinger-based sensing platforms before and after treatment with the filtered milk. (A) Reprinted by permission from Springer: J.F. Li, Y.F. Huang, Y. Ding, Z.L. Yang, S.B. Li, X.S. Zhou, et al., Shell-isolated nanoparticle-enhanced Raman spectroscopy, Nature 464 (2010) 392, Copyright 2010. (B) Reprinted with permission from A. Kim, S.J. Barcelo, R.S. Williams, Z. Li, Melamine sensing in milk products by using surface enhanced Raman scattering, Anal. Chem. 84 (2012) 9303–9309. Copyright (2012) American Chemical Society.

without, and with, shell-isolated nanoparticles on the surface. It can be seen that two peaks at 1108 and 1341 cm^{-1} (curve III) are widened with the nanoparticles, which are characteristic bands of parathion residues.

The detection of melamine in food has also attracted wide attention. SERS shows the advantages of compact sample preparation procedure and rapid signal response in the detection of melamine in wheat bran, chicken feed, cake and noodle, compared to high performance liquid chromatography. Many SERS substrates have been fabricated for the detection of melamine, such as AgNR matrices and colloid AgNPs. For example, by using silver/carbon nanospheres as the SERS substrate, the detection limit of melamine molecules is as low as 5.0×10^{-8} M [66]. Quantitative detection in the concentration range of 1.0×10^{-4} to 5.0×10^{-8} M can be achieved, while using the peak intensity of 682 cm^{-1} as the normalization standard with E_F of 10^7. A SERS substrate of gold nanofinger chips has been designed for ultrasensitive detection of melamine in milk. The gold nanofingers collapse into well-defined groups after exposed to the filtered milk samples (Fig. 2.5B) [67]. The gaps between the gold nanofinger tips offer hot spots for sensing. The detection limit of the melamine is 120 ppt in water and 100 ppb in infant formula.

2.4.5 Selectivity of Surface-Enhanced Raman Spectrum-Based Sensing

The environmental pollutants usually have no functional groups that can specifically interact with SERS substrates, resulting in low selectivity of the substrate. The selective adsorption of pollutants on the surface can be realized by the functionalization of the SERS substrates, resulting in higher selectivity.

The physical interaction such as hydrophobic interaction, van der Waals' force, $\pi-\pi$ stacking action can be used to induce the adsorption of the target molecules onto the surface of SERS substrates for sensing. Alkanethiols are often used to modify the NMPs. After being functionalized by alkanethiol and perfluoroalkanethiol on the surface, the silver layer supported on the silicon nanospheres is employed as the SERS substrate and shows the signal of polychlorinated biphenyls (PCBs) such as PCB-47 and PCB-77, with the detection limit of 5×10^{-11} mol L^{-1} [68]. This can be attributed to the hydrophobic interaction between the thiol molecules and the PCB molecules.

Carbon nanotubes, humic acids, and polystyrene molecules can produce $\pi-\pi$ stacking interactions with target molecules, therefore they have also commonly been used as substrate modifiers. With modified with humic acid, AgNPs reveal excellent SERS-based sensing performance of PAHs [69]. Theoretical simulation proves that the adsorption orientation of PAH molecules on the surface of this substrate is parallel to the aromatic ring in humic acid, leading to the $\pi-\pi$ stacking effect, which not only greatly enhances the selectivity for PAH molecules, but also increases the SERS-sensing sensitivity [70].

The chemical reaction between the target molecules and the NMPs can also improve the selectivity of NMPs-based SERS sensing. Dasary et al. [71] first proposed the use of cysteine-modified AuNPs as a SERS substrate to trap TNT molecules. The sulfhydryl group in cysteine are easy to bind with AuNPs, while the amino group can react with TNT molecules to form a complex via the Meisenheimer reaction. A large number of hot spots have been formed in the aggregation of AuNPs by this chemical reaction. The SERS substrate shows specific detection of TNT molecules with a detection limit of 2×10^{-12} mol L^{-1}. Hao et al. [72] modified the silver film with thiolethylamine containing ammonium groups for the purpose of the detection of hypochlorite ions. Hypochlorite ions in solution can be combined with ammonium groups and, therefore are adsorbed on the surface of SERS substrates. The minimum detectable concentration of hypochlorite ions by this SERS substrate is 5 $\mu g\ L^{-1}$.

The reaction site in this physical and chemical interaction is constituted by one functional group of target molecules, while the reaction site of molecular recognition is the entire molecule. It provides much higher selectivity for SERS-based sensing. The molecular recognition effect in SERS-based sensing can also avoid the interference of the environmental matrix during detection. Molecular recognition can be divided into two categories. First, molecularly imprinted polymers are modified on the surface of the substrate. Targeted molecules are used as templates to form a hollow structure on the surface, enabling the SERS substrate to actively select and adsorb contaminant molecules. Second, modifiers on the surface of the SERS substrate can self-assemble into a cavity structure, such as amethyst dication, calixarenes, and cyclodextrin, which can capture specific environmental contaminants to achieve SERS detection. Silver nanoparticles on silver molybdate nanowires are employed as the SERS substrate. The π–electron interaction between TNT and dinonylazobenzene forms an imprint on the surface. This SERS substrate with a TNT molecularly imprinted template has a large number of hot spots between the AgNPs and shows a detection limit of 10^{-12} mol L^{-1} for TNT [73]. The amethyst dicationic luster is also used to modify the AgNPs for the SERS detection of PAH molecules. The amethyst dications can form molecular cavities at the hot spots of the nanoparticles. When the PAH molecules enter the cavity, they can increase the Raman scattering cross-section and enhance the Raman signals. The detection limit of the quinone can reach 10^{-9} mol L^{-1} [74].

2.5 Electrochemical Sensors

NMPs, especially AuNPs, feature excellent conductivity, high surface area, good biocompatibility, and catalytic properties. These advantages make them excellent materials for the electrochemical detection of a wide range of analytes, mainly as an electroactive label, or active interface for constructing electrochemical sensor and investigating the direct electron transfer [3].

2.5.1 Noble-Metal Nanoparticles as Electroactive Labels

NMPs usually possess high surface areas and good electrical conductivity compared to nonmetallic nanoparticles. Benefiting from their large specific surface area and high surface free energy, NMPs can strongly adsorb organics, polymers, and DNA through special functional groups such as thiols and others, which can interact strongly with AuNPs [3]. In addition, for the AuNPs fabricated with citrate as the surfactant, the negative charges can enhance the electrostatic adsorption between them. Benefiting from the strong covalent bond to sulfhydryl groups, the colloidal NMPs can combine with the sulfhydryl-labeled molecules to form probes, which are easy to use for the environmental electrochemical detection of HM ions and biological systems.

A very successful approach for using NMPs as the electroactive labels is to combine nanotechnology, nucleic acid hybridization technology, and electrochemical analysis technology on the surface of the electrode to achieve highly sensitive detection of DNA. For example, with AuNPs attached to the oligonucleotide as an electroactive probe, the hybridization of a target oligonucleotide to magnetic bead-linked oligonucleotide probes is followed by dissolving the AuNPs into aqueous metal ions via a hydrobromic acid/bromine (HBr/Br_2) solution. The potentiometric stripping measurements of the dissolved metal tag at single-use, thick-film carbon electrodes realize the goal of indirect determination of the DNA [75]. However, The HBr/Br_2 solution is highly toxic, which limits the application of indirect electrochemical detection by AuNPs labels.

Electrochemical detection for the Factor V Leiden mutation from polymerase chain reaction (PCR) amplicons can also be acquired by investigating the oxidation signal of colloidal gold (Fig. 2.6A) [76]. Factor V Leiden mutation was immobilized onto the cysteamine modified AuNPs with from their amino groups with EDC and NHS as coupling agents to perform as the electroactive probes. The complementary DNA hybridization immobilized on the graphite electrode was used to prepare an electrochemical DNA sensor due to the appearance of the Au oxide peak at +1.2 V, with good reproducibility and stability and a detection limit of 0.78 fmol L^{-1}.

The sensitivity for this electrochemical detection by AuNPs as labels can be improved by employing silver, hence obtaining a better detection limit [77]. With the precipitation of silver on AuNPs tags and subsequent dissolution in HNO_3, the electrochemical potentiometric stripping detection of Ag can achieve highly sensitive detection, which represents an attractive alternative to indirect optical affinity assays for electrochemical sensing. The high sensitivity can also be acquired by using Cu@Au core–shell nanoparticles as the labels [78]. Au thin layer coated Cu cores are successfully prepared, functionalized with oligonucleotides and labeled to a 5′-alkanethiol capped oligonucleotides probe. During the sensing process, the target oligonucleotides are immobilized on the surface of polypyrrole/glassy carbon electrode with the electrostatic adsorption, followed by hybridizing with the Cu@Au DNA probe. The release of the copper metal ions anchored on the hybrids by oxidative metal dissolution could be monitored by sensitive anodic stripping voltammetry for the indirect determination of target oligonucleotides. The detection limit of target oligonucleotides is 5 pM.

2.5.2 Noble-Metal Nanoparticles as the Active Interface for Constructing Electrochemical Sensing

In the construction of electrochemical sensing devices, the immobilization of probes is directly related to the sensor's reproducibility, sensitivity, and stability. Using NMPs as the supports

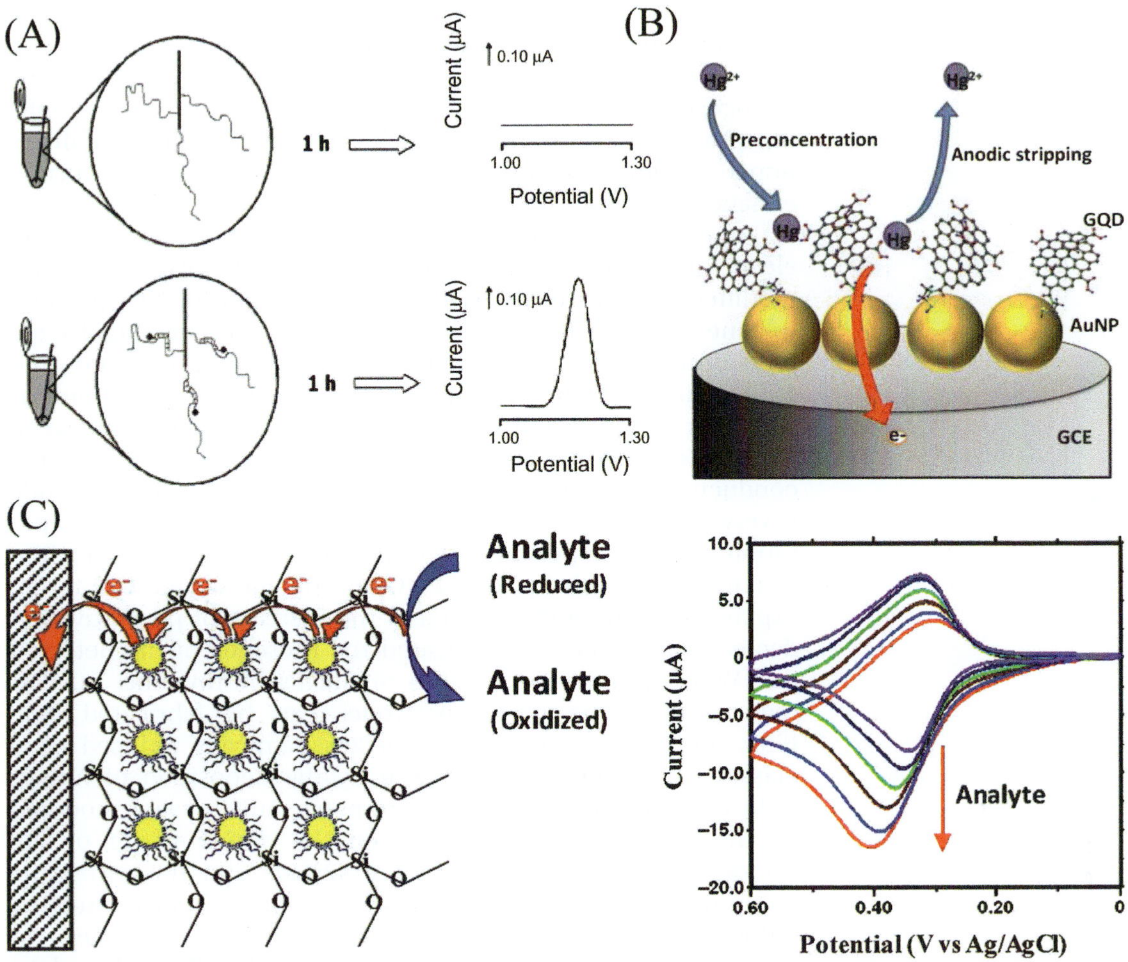

Figure 2.6 (A) Detection of hybridization using AuNP-tagged capture probes: polymerase chain reaction amplicon modification and hybridization with AuNPs labeled DNA probe. (B) Schematic illustration of Hg^{2+} detection based on GQD-AuNPs system. (C) Mechanism depicting the mediated electrocatalytic oxidation and ensuing electron transport across the entrapped Au25 in sol–gel electrode. (A) Reprinted with permission from M. Ozsoz, A. Erdem, K. Kerman, D. Ozkan, B. Tugrul, N. Topcuoglu, et al., Electrochemical genosensor based on colloidal gold nanoparticles for the detection of Factor V Leiden mutation using disposable pencil graphite electrodes, Anal. Chem. 75 (2003) 2181–2187. Copyright (2003) American Chemical Society. (B) Reprinted from S.L. Ting, S.J. Ee, A. Ananthanarayanan, K.C. Leong, P. Chen, Graphene quantum dots functionalized gold nanoparticles for sensitive electrochemical detection of heavy metal ions, Electrochim. Acta 172 (2015) 7–11, Copyright (2015), with permission from Elsevier. (C) Reprinted with permission from S.S. Kumar, K. Kwak, D. Lee, Electrochemical sensing using quantum-sized gold nanoparticles, Anal. Chem. 83 (2011) 3244–3247. Copyright (2011) American Chemical Society.

for immobilizing the probe provides a potential approach to constructing the active interface of sensing platforms with good performance due to NMPs' good biocompatibility and high-surface adsorption capacity. The good biocompatibility of NMPs can provide a microenvironment that can maintain the activity of biological materials effectively. The high-surface adsorption capacity makes them a suitable medium for immobilizing biological materials. In addition, NMPs modified with specific functional groups can be easily achieved for the purpose of directional alignment and orientation regulation, therefore further enhancing their sensing activity.

Functionalized NMPs have been widely employed in electrochemical sensing as active interfaces. For example, thiol-modified AuNPs can be used for the detection of NO_2 gas and toluene vapor [79]. In this sensing platform, the AuNPs are functionalized with 4-methylbenzenethiol, 1-hexanethiol, and 1-dodecanethiol and then fabricated into films by Langmuir-Schaeffer deposition. Any change of thiol/air medium between adjacent conductive AuNPs could lead to a rise or decline of the conductivity of the electrode. The conductivity of the electrode is very sensitive to small changes in the inter-nanoparticle separation and/or the permittivity of the environment of AuNPs film electrodes. The adsorption of NO_2 or toluene would lead to the expansion of the AuNPs film and the increase of permittivity. A decrease in the current indicated the presence of the detected gas. With the functionalization of AuNPs by thymine, AuNPs/reduced graphene oxide–electrodes can be fabricated [80]. First, the graphene oxide was electrochemically reduced on a glassy carbon substrate, followed by the deposition of AuNPs onto the surface by cyclic voltammetry. AuNPs were modified by the covalent coupling between the amine group of the cysteamine self-assembled on the surface and the carboxylic group of the thymine-1-acetic acid. This functionalized AuNPs/reduced graphene oxide–electrode can be employed in the sensing of mercury ions due to its specific affinity based on thymine–mercury–thymine coordination chemistry. The proposed sensor shows high sensitivity to Hg^{2+} in the range of $10\ ng\ L^{-1}$–$1.0\ \mu g\ L^{-1}$, and good stability during the regeneration. Graphene quantum dots (GQDs) can also be employed for constructing the active interface in the sensing platform (Fig. 2.6B) [81]. As a kind of zero-dimensional material, GQDs show the probability in various novel applications due to their extraordinary physicochemical properties. GQDs can be easily conjugated on the surface of cysteamine-capped AuNPs for the propose of electrochemical detection of heavy metal ions (Hg^{2+} and Cu^{2+}). This electrochemical sensing platform shows an ultralow detection limit (0.02 nM with $S/N = 6.25$ for Hg^{2+} and 0.05 nM with $S/N = 4.81$ for Cu^{2+}) and high sensitivity ($2.47\ \mu A\ nM^{-1}$ for Hg^{2+} and $3.69\ \mu A\ nM^{-1}$ for Cu^{2+}).

2.5.3 Electron Transferring–Based Sensing Platform

The electron transfer between the detected objects and the NMPs electrode due to the redox reaction can also be used in electrochemical sensing. This sensing platform can be divided into direct NMP sensing, protein-NMPs sensing, and protein-mediator NMPs sensing.

A type A, direct NMPs sensing platform is usually based on the direct electron transfer between the electrode and the detected objects due to the catalytic oxidation or reduction of NMPs. In the detection of CO by this strategy, Au/Co_3O_4 was fabricated and deposited on the anode of a galvanic cell. Attracted to the strong catalytic oxidation of CO, Au/Co_3O_4 displayed a good response time and high sensitivity for electrochemical sensing [82]. By employing DNA as the template, Ag–DNA nanoparticles with controlled narrow size distribution were electrodeposited on a glassy carbon electrode for H_2O_2 detection [83]. Benefiting from the favorable catalytic ability to the reduction of H_2O_2, the Ag–DNA nanoparticles modified electrode showed a limit of detection of 0.6 μM for H_2O_2 and a sensitivity of $773\ \mu A\ mM^{-1}\ cm^{-2}$. SnO_2-modified Au also showed the reversible electrochemical response for cytochrome C due to the facile electron transfer between them [84].

In addition, bimetallic nanoparticles have also attracted significant attention in electrochemical sensing due to their high catalytic activity, good resistance to deactivation and high catalytic selectivity with the addition of a second metal [85]. Au–Ag bimetallic nanoparticles electrodeposited on a glassy carbon electrode was successfully employed for the detection of H_2O_2 in the linear range of 1–250 μM in lab samples, and 1×10^{-3}–2×10^{-2} M in real samples [86]. PtPd nanoparticles have also been employed in sensitive sensors for the electrochemical detection of H_2O_2 with the support of multiwalled carbon nanotubes [85]. Attracted to the electron transfer rate constants of $\sim 1.23 \times 10^{-3}\ cm\ s^{-1}$, the PtPd/MWCNTs/GC electrode exhibited a low detection limit (1.2 μM) and high sensitivity ($414.8\ \mu A\ mM^{-1}\ cm^{-2}$) with the linear range of 2.5–125 μM.

In order to enhance the sensitivity of the detection of ascorbic acid and uric acid, the 3D assembly of AuNPs on the electrodes has been designed by entrapped into the sol–gel network through the hydrolysis of ethyl trimethoxysilane. In this sensing platform, the AuNP acts as an electronic conductor and a redox mediator (Fig. 2.6C) [87]. During detection, AuNPs with the size of 3 nm are oxidized first and then electrocatalytically oxidize the analyte while it is reduced to Au^- (Eq. (2.7)).

$$Au^- \rightarrow Au^0 \tag{2.6}$$

$$Au^0 + analyte_{red} \rightarrow Au^- + analyte_{ox} \tag{2.7}$$

Taking advantage of the structural regulation can also improve sensing activity. By designing microelectrodes with a seamless solid/nanoporous gold/cobalt oxide hybrid structure, synergistic electrocatalytic activity of the gold skeleton and cobalt oxide nanoparticles toward glucose oxidation lead to a multilinear detection ranges with ultrahigh sensitivities [88]. This direct NMPs sensing platform shows an ultralow detection limit of 5 nM for glucose. The nanostructured Au particles are especially attractive to increase the immobilization number of enzymes owing to their porous film architectures which provide increased surface area and offer unlimited mass transport.

The type B, protein-NMPs sensing platform is based on the electron transfer through the redox proteins immobilized on the surface of NMPs. In this sensing platform, the NMPs not only provide a friendly microenvironment for immobilizing proteins, but also act as the conducting tunnel to enhance the electron transfer between redox centers in proteins and electrode surfaces due to their excellent conductivity. The AuNPs are usually attached on the surface of electrodes through physical adsorption, chemical bonds, electrodeposition, and entrapment for facilitating protein and detecting H_2O_2. For example, by immobilizing the AuNPs onto the surface of 3-mercaptopropyl trimethoxysilane functionalized ITO glass electrode due to the strong binding interactions, and followed by coupling with cyt c, the electron transfer between cyt c and the electrode during the decomposition of H_2O_2 can be observed [89]. This cyt c-AuNPs/ITO glass electrode displayed an excellent electrocatalytic response for the detection of H_2O_2. The assembly of AuNPs into a 3D structure can also increase the immobilization amount of protein, therefore enhancing the sensitivity. Combining interfacial assembly and layer-by-layer assembly without the assistance of organic linker molecules can be employed in the fabrication of multilayer AuNPs film [90].

The type C, protein-mediator NMPs sensing, the most original approach for an electrochemical platform, is acquired by adding a mediator into the test solution to realize the electron transfer between the protein and electrode. In this sensing platform various electron-shuttling mediators such as catechol, ferrocene, hydroquinone, methylene blue, toluidine blue, thiamine, and hexacyanoferrates have been introduced. For example, horseradish peroxidase is successfully immobilized on the nanometer-sized Au colloids with thiol-tailed groups of cysteamine functionalized. The horseradish peroxidase-labeled Au colloids show excellent electrocatalytic response to the reduction of H_2O_2. The detection limit of H_2O_2 is 0.15 μM [91].

2.6 Conclusion and Future Perspectives

In summary, this chapter introduced the application of noble-metal particles as colorimetric sensing, fluorescence-based sensing, SERS, and electrochemical sensing platforms in environmental detection due to their unique physical and chemical properties, including good conductivity, easy functionalization with a range of ligands, large electronic field enhancement, fluorescence quenching, and catalytic behavior.

Advanced environmental sensing techniques need to meet the demands of easy operation, rapid response, low cost, and multiplexed identification of pollutants. However, the optimization of the sensing technology based on NMPs is still essential for their commercial application in environmental detection. In particular, the development of efficient sensors to detect analytes in complex biological fluids such as human urine, serum, and blood remains a challenge. These issues can be addressed using three parallel methods: (1) Further functionalization of the NMPs for specific capture of an analyte with a more powerful combination. It's necessary to meet various functional requirements for NMPs in order to meet different detection platforms. Inorganic materials such as silica, surfactants such as polyethylene glycol, and organisms such as DNA and protein have been widely used for the functionalization of NMPs by taking advantage of their high biocompatibility, hydrophilicity, and good chemical and colloidal stability. Functionalization can prevent the particles from aggregation, and the coating layer after functionalizing helps to bind a wide variety of biochemical ligands to the surface of the NMPs, therefore enhancing the sensitivity and stability of the sensing platform. Many attempts have already been extensively researched and explored. However, a more precise and controllable fabrication process would be very helpful to improve the repeatability and stability of the sensing platform. Biocompatible organic polymer materials are also widely used as stabilizers for NMPs. Polymers can also form single or double-layered structures on the surface of

NMPs by chemical bonding or physical adsorption. The development of synthetic polymers with high performance and the modification of NMPs using them are of great significance for enhancing the specific adsorption of the analyte. In addition, by selecting natural biomolecules with specific recognition with the molecules to be detected can also be used for the functionalization of NMPs after comprehensively evaluating the physicochemical properties of the analyte. (2) Further tuning the structure, morphology of NMPs or designing the composite with other materials for superior electric, optical properties to produce more resonant response for the analyte. Tuning NMPs in size and shape is beneficial for obtaining various desired physical or chemical properties. It's necessary to control the monodispersity of the NMPs in order to explore the mechanisms of the NMPs-based sensing platform. In the synthesis of NMPs, the chemical properties of the reducing agent, the precursor, the surfactant or the reaction conditions have an unpredictable influence on the monodispersity of the synthesized nanoparticles. The pH of the reaction or the redox and hydrophilicity of the reagent need to be comprehensively considered in the design of recipes for the synthesis of NMPs. In addition, the self-assembly of the NMPs contributes to the composite with metal oxides or other substrates and facilitates the discovery of the interaction between the NMPs and other materials. The combination of different synthetic routes is the future trend for the controllable synthesis of particle size and morphology and excavating its mechanism. (3) Developing more accurate signal conversion and amplification systems. With the development of science and technology, the conversion and amplification of chemical signals to electrical and optical signals have made great progress and play a huge role in the application of environmental detection. The design of these high-efficiency sensing systems is inseparable from the development of the biology, engineering, and electronic informatics. The fundamental advantages of NMPs have generated an exponential increase in their applications in environmental detection that will continue to revolutionize practical applications.

Acknowledgment

The authors are grateful for support from the Hong Kong Scholars Program (No. XJ2017051).

References

[1] D.-W. Li, W.-L. Zhai, Y.-T. Li, Y.-T. Long, Recent progress in surface enhanced raman spectroscopy for the detection of environmental pollutants, Microchim. Acta 181 (2014) 23–43.
[2] H. Wei, S.M. Hossein Abtahi, P.J. Vikesland, Plasmonic colorimetric and SERS sensors for environmental analysis, Environ. Sci. Nano 2 (2015) 120–135.
[3] K. Saha, S.S. Agasti, C. Kim, X. Li, V.M. Rotello, Gold nanoparticles in chemical and biological sensing, Chem. Rev. 112 (2012) 2739–2779.
[4] P.L. Stiles, J.A. Dieringer, N.C. Shah, R.P. Van Duyne, Surface-enhanced Raman spectroscopy, Annu. Rev. Anal. Chem. 1 (2008) 601–626.
[5] S.J. Tan, M.J. Campolongo, D. Luo, W. Cheng, Building plasmonic nanostructures with DNA, Nat. Nanotechnol. 6 (2011) 268.
[6] S.K. Ghosh, T. Pal, Interparticle coupling effect on the surface plasmon resonance of gold nanoparticles: from theory to applications, Chem. Rev. 107 (2007) 4797–4862.
[7] M. Rycenga, C.M. Cobley, J. Zeng, W. Li, C.H. Moran, Q. Zhang, et al., Controlling the synthesis and assembly of silver nanostructures for plasmonic applications, Chem. Rev. 111 (2011) 3669–3712.
[8] Q. Ruan, L. Shao, Y. Shu, J. Wang, H. Wu, Growth of monodisperse gold nanospheres with diameters from 20 nm to 220 nm and their core/satellite nanostructures, Adv. Opt. Mater. 2 (2014) 65–73.
[9] J. Pérez-Juste, I. Pastoriza-Santos, L.M. Liz-Marzán, P. Mulvaney, Gold nanorods: synthesis, characterization and applications, Coord. Chem. Rev. 249 (2005) 1870–1901.
[10] X. Huang, M.A. El-Sayed, Gold nanoparticles: optical properties and implementations in cancer diagnosis and photothermal therapy, J. Adv. Res. 1 (2010) 13–28.
[11] C. Sönnichsen, B.M. Reinhard, J. Liphardt, A.P. Alivisatos, A molecular ruler based on plasmon coupling of single gold and silver nanoparticles, Nat. Biotechnol. 23 (2005) 741.
[12] L. Shao, K.C. Woo, H. Chen, Z. Jin, J. Wang, H.-Q. Lin, Angle-and energy-resolved plasmon coupling in gold nanorod dimers, ACS Nano 4 (2010) 3053–3062.
[13] M. Haruta, T. Kobayashi, H. Sano, N. Yamada, *Novel gold catalysts for the oxidation of carbon monoxide at a temperature far below 0°C*, Chem. Lett. 16 (1987) 405–408.
[14] M. Haruta, N. Yamada, T. Kobayashi, S. Iijima, Gold catalysts prepared by coprecipitation for low-temperature oxidation of hydrogen and of carbon monoxide, J. Catal. 115 (1989) 301–309.
[15] M. Comotti, C. Della Pina, R. Matarrese, M. Rossi, The catalytic activity of "naked" gold particles, Angew. Chem. Int. Ed. 43 (2004) 5812–5815.
[16] E. Priyadarshini, N. Pradhan, Gold nanoparticles as efficient sensors in colorimetric detection of toxic metal ions: a review, Sens. Actuators B 238 (2017) 888–902.
[17] P.B. Tchounwou, C.G. Yedjou, A.K. Patlolla, D.J. Sutton, Heavy metal toxicity and the environment, in: A. Luch (Ed.), Molecular, Clinical and Environmental Toxicology: Volume 3: Environmental Toxicology, Springer Basel, Basel, 2012, pp. 133–164.
[18] R. Singh, N. Gautam, A. Mishra, R. Gupta, Heavy metals and living systems: an overview, Indian J. Pharmacol. 43 (2011) 246–253.

[19] G. Aragay, J. Pons, A. Merkoçi, Recent trends in macro-, micro-, and nanomaterial-based tools and strategies for heavy-metal detection, Chem. Rev. 111 (2011) 3433–3458.
[20] O.T. Butler, J.M. Cook, C.F. Harrington, S.J. Hill, J. Rieuwerts, D.L. Miles, Atomic spectrometry update. environmental analysis, J. Anal. At. Spectrom. 22 (2007) 187–221.
[21] S. Hyeon Don, C. Inhee, Y. Young In, H. Surin, L. Suseung, K. Taewook, et al., *Picomolar selective detection of mercuric ion (Hg^{2+}) using a functionalized single plasmonic gold nanoparticle*, Nanotechnology 21 (2010) 145501.
[22] M. Rex, F.E. Hernandez, A.D. Campiglia, Pushing the limits of mercury sensors with gold nanorods, Anal. Chem. 78 (2006) 445–451.
[23] Y.-L. Hung, T.-M. Hsiung, Y.-Y. Chen, Y.-F. Huang, C.-C. Huang, Colorimetric detection of heavy metal ions using label-free gold nanoparticles and alkanethiols, J. Phys. Chem. C 114 (2010) 16329–16334.
[24] K. Yoosaf, B.I. Ipe, C.H. Suresh, K.G. Thomas, In situ synthesis of metal nanoparticles and selective naked-eye detection of lead ions from aqueous media, J. Phys. Chem. C 111 (2007) 12839–12847.
[25] K.-W. Huang, C.-J. Yu, W.-L. Tseng, Sensitivity enhancement in the colorimetric detection of lead(II) ion using gallic acid–capped gold nanoparticles: improving size distribution and minimizing interparticle repulsion, Biosens. Bioelectron. 25 (2010) 984–989.
[26] A. Sugunan, C. Thanachayanont, J. Dutta, J.G. Hilborn, Heavy-metal ion sensors using chitosan-capped gold nanoparticles, Sci. Technol. Adv. Mater. 6 (2005) 335–340.
[27] Y. Kim, R.C. Johnson, J.T. Hupp, Gold nanoparticle-based sensing of "spectroscopically silent" heavy metal ions, Nano. Lett. 1 (2001) 165–167.
[28] Y. Guo, Z. Wang, W. Qu, H. Shao, X. Jiang, Colorimetric detection of mercury, lead and copper ions simultaneously using protein-functionalized gold nanoparticles, Biosens. Bioelectron. 26 (2011) 4064–4069.
[29] J.-S. Lee, M.S. Han, C.A. Mirkin, *Colorimetric detection of mercuric ion (Hg^{2+}) in aqueous media using DNA-functionalized gold nanoparticles*, Angew. Chem. Int. Ed. 46 (2007) 4093–4096.
[30] N. Xia, Y. Shi, R. Zhang, F. Zhao, F. Liu, L. Liu, Simple, rapid and label-free colorimetric assay for arsenic based on unmodified gold nanoparticles and a phytochelatin-like peptide, Anal. Methods 4 (2012) 3937–3941.
[31] T. Lou, L. Chen, C. Zhang, Q. Kang, H. You, D. Shen, et al., A simple and sensitive colorimetric method for detection of mercury ions based on anti-aggregation of gold nanoparticles, Anal. Methods 4 (2012) 488–491.
[32] S.-Y. Lin, S.-H. Wu, C.-H. Chen, A simple strategy for prompt visual sensing by gold nanoparticles: general applications of interparticle hydrogen bonds, Angew. Chem. Int. Ed. 45 (2006) 4948–4951.
[33] A. Mooradian, Photoluminescence of metals, Phys. Rev. Lett. 22 (1969) 185–187.
[34] Z. Wu, R. Jin, On the ligand's role in the fluorescence of gold nanoclusters, Nano. Lett. 10 (2010) 2568–2573.
[35] X. Xia, Y. Long, J. Wang, Glucose oxidase-functionalized fluorescent gold nanoclusters as probes for glucose, Anal. Chim. Acta 772 (2013) 81–86.
[36] L.-Y. Chen, C.-W. Wang, Z. Yuan, H.-T. Chang, Fluorescent gold nanoclusters: recent advances in sensing and imaging, Anal. Chem. 87 (2015) 216–229.

[37] L. Li, Z. Li, H. Zhang, S. Zhang, I. Majeed, B. Tan, Effect of polymer ligand structures on fluorescence of gold clusters prepared by photoreduction, Nanoscale 5 (2013) 1986–1992.
[38] M.M.A. Habeeb, V.P. Kumar, P.S. Kumar, K.R.C. Arun, P. Soumya, O.R. Vyomakesannair, et al., Bright, NIR-emitting Au23 from Au25: characterization and applications including biolabeling, Chem. Eur. J. 15 (2009) 10110–10120.
[39] J.-M. Liu, J.-T. Chen, X.-P. Yan, Near infrared fluorescent trypsin stabilized gold nanoclusters as surface plasmon enhanced energy transfer biosensor and in vivo cancer imaging bioprobe, Anal. Chem. 85 (2013) 3238–3245.
[40] B. Dubertret, M. Calame, A.J. Libchaber, Single-mismatch detection using gold-quenched fluorescent oligonucleotides, Nat. Biotechnol. 19 (2001) 365.
[41] X. He, H. Liu, Y. Li, S. Wang, Y. Li, N. Wang, et al., Gold nanoparticle-based fluorometric and colorimetric sensing of copper(II) ions, Adv. Mater. 17 (2005) 2811–2815.
[42] A.P. Alivisatos, Semiconductor clusters, nanocrystals, and quantum dots, Science 271 (1996) 933–937.
[43] B.O. Dabbousi, J. Rodriguez-Viejo, F.V. Mikulec, J.R. Heine, H. Mattoussi, R. Ober, et al., *(CdSe)ZnS core – shell quantum dots: synthesis and characterization of a size series of highly luminescent nanocrystallites*, J. Phys. Chem. B 101 (1997) 9463–9475.
[44] S. Kim, B. Fisher, H.-J. Eisler, M. Bawendi, Type-II quantum dots: CdTe/CdSe(Core/Shell) and CdSe/ZnTe(core/shell) heterostructures, J. Am. Chem. Soc. 125 (2003) 11466–11467.
[45] X. Michalet, F. Pinaud, L. Bentolila, J. Tsay, S. Doose, J. Li, et al., Quantum dots for live cells, in vivo imaging, and diagnostics, Science 307 (2005) 538–544.
[46] E. Oh, M.-Y. Hong, D. Lee, S.-H. Nam, H.C. Yoon, H.-S. Kim, Inhibition assay of biomolecules based on fluorescence resonance energy transfer (FRET) between quantum dots and gold nanoparticles, J. Am. Chem. Soc. 127 (2005) 3270–3271.
[47] D.A. Long, D. Long, Raman Spectroscopy, vol. 276, McGraw-Hill, New York, 1977.
[48] M. Moskovits, Surface-enhanced Raman spectroscopy: a brief retrospective, J. Raman Spectrosc. 36 (2005) 485–496.
[49] M. Fleischmann, P.J. Hendra, A.J. Mcquillan, Raman spectra of pyridine adsorbed at a silver electrode, Chem. Phys. Lett. 26 (1974) 163–166.
[50] A. Campion, P. Kambhampati, Surface-enhanced Raman scattering, Chem. Soc. Rev. 27 (1998) 241–250.
[51] S. Schlücker, Surface-enhanced Raman spectroscopy: concepts and chemical applications, Angew. Chem. Int. Ed. 53 (2014) 4756–4795.
[52] J.F. Li, Y.F. Huang, Y. Ding, Z.L. Yang, S.B. Li, X.S. Zhou, et al., Shell-isolated nanoparticle-enhanced Raman spectroscopy, Nature 464 (2010) 392.
[53] M.J. Mulvihill, X.Y. Ling, J. Henzie, P. Yang, Anisotropic etching of silver nanoparticles for plasmonic structures capable of single-particle SERS, J. Am. Chem. Soc. 132 (2010) 268–274.
[54] W. Xiong, D. Sikdar, M. Walsh, K.J. Si, Y. Tang, Y. Chen, et al., Single-crystal caged gold nanorods with tunable broadband plasmon resonances, Chem. Commun. 49 (2013) 9630–9632.
[55] W. Xiong, R. Mazid, L.W. Yap, X. Li, W. Cheng, Plasmonic caged gold nanorods for near-infrared light controlled drug delivery, Nanoscale 6 (2014) 14388–14393.

[56] W. Xiong, D. Sikdar, L.W. Yap, M. Premaratne, X. Li, W. Cheng, Multilayered core–satellite nanoassemblies with fine-tunable broadband plasmon resonances, Nanoscale 7 (2015) 3445–3452.
[57] S.J. Lee, A.R. Morrill, M. Moskovits, Hot spots in silver nanowire bundles for surface-enhanced Raman spectroscopy, J. Am. Chem. Soc. 128 (2006) 2200–2201.
[58] D.-Y. Wu, J.-F. Li, B. Ren, Z.-Q. Tian, Electrochemical surface-enhanced Raman spectroscopy of nanostructures, Chem. Soc. Rev. 37 (2008) 1025–1041.
[59] H. Wang, C.S. Levin, N.J. Halas, Nanosphere arrays with controlled sub-10-nm gaps as surface-enhanced Raman spectroscopy substrates, J. Am. Chem. Soc. 127 (2005) 14992–14993.
[60] D.J. Masiello, G.C. Schatz, Many-body theory of surface-enhanced Raman scattering, Phys. Rev. A 78 (2008) 042505.
[61] O. Péron, E. Rinnert, T. Toury, M.L. De La Chapelle, C. Compere, Quantitative SERS sensors for environmental analysis of naphthalene, Analyst 136 (2011) 1018–1022.
[62] J. Du, C. Jing, *Preparation of thiol modified Fe_3O_4@Ag magnetic SERS probe for PAHs detection and identification*, J. Phys. Chem. C 115 (2011) 17829–17835.
[63] G. Wang, C. Lim, L. Chen, H. Chon, J. Choo, J. Hong, Surface-enhanced Raman scattering in nanoliter droplets: towards high-sensitivity detection of mercury (II) ions, Anal. Bioanal. Chem. 394 (2009) 1827–1832.
[64] N.A. Hatab, C. Rouleau, S.T. Retterer, G. Eres, P.B. Hatzinger, B. Gu, An integrated portable Raman sensor with nanofabricated gold bowtie array substrates for energetics detection, Analyst 136 (2011) 1697–1702.
[65] W. Cheung, I.T. Shadi, Y. Xu, R. Goodacre, Quantitative analysis of the banned food dye sudan-1 using surface enhanced raman scattering with multivariate chemometrics, J. Phys. Chem. C 114 (2010) 7285–7290.
[66] L.-M. Chen, Y.-N. Liu, Surface-enhanced Raman detection of melamine on silver-nanoparticle-decorated silver/carbon nanospheres: effect of metal Ions, ACS Appl. Mater. Interfaces 3 (2011) 3091–3096.
[67] A. Kim, S.J. Barcelo, R.S. Williams, Z. Li, Melamine sensing in milk products by using surface enhanced Raman scattering, Anal. Chem. 84 (2012) 9303–9309.
[68] K.C. Bantz, C.L. Haynes, Surface-enhanced Raman scattering detection and discrimination of polychlorinated biphenyls, Vib. Spectrosc. 50 (2009) 29–35.
[69] R.A. Alvarez-Puebla, D.S. Dos Santos Jr, R.F. Aroca, SERS detection of environmental pollutants in humic acid–gold nanoparticle composite materials, Analyst 132 (2007) 1210–1214.
[70] P. Leyton, I. Córdova, P. Lizama-Vergara, J.S. Gómez-Jeria, A. Aliaga, M. Campos-Vallette, et al., Humic acids as molecular assemblers in the surface-enhanced Raman scattering detection of polycyclic aromatic hydrocarbons, Vib. Spectrosc. 46 (2008) 77–81.
[71] S.S. Dasary, A.K. Singh, D. Senapati, H. Yu, P.C. Ray, Gold nanoparticle based label-free SERS probe for ultrasensitive and selective detection of trinitrotoluene, J. Am. Chem. Soc. 131 (2009) 13806–13812.
[72] J. Hao, M.-J. Han, J. Li, X. Meng, Surface modification of silver nanofilms for improved perchlorate detection by surface-enhanced Raman scattering, J. Colloid. Interface. Sci. 377 (2012) 51–57.

[73] L. Yang, L. Ma, G. Chen, J. Liu, Z.Q. Tian, Ultrasensitive Sers detection of TNT by imprinting molecular recognition using a new type of stable substrate, Chem. Eur. J. 16 (2010) 12683–12693.
[74] L. Guerrini, J.V. Garcia-Ramos, C. Domingo, S. Sanchez-Cortes, Nanosensors based on viologen functionalized silver nanoparticles: few molecules surface-enhanced Raman spectroscopy detection of polycyclic aromatic hydrocarbons in interparticle hot spots, Anal. Chem. 81 (2009) 1418–1425.
[75] J. Wang, D. Xu, A.-N. Kawde, R. Polsky, Metal nanoparticle-based electrochemical stripping potentiometric detection of DNA hybridization, Anal. Chem. 73 (2001) 5576–5581.
[76] M. Ozsoz, A. Erdem, K. Kerman, D. Ozkan, B. Tugrul, N. Topcuoglu, et al., Electrochemical genosensor based on colloidal gold nanoparticles for the detection of Factor V Leiden mutation using disposable pencil graphite electrodes, Anal. Chem. 75 (2003) 2181–2187.
[77] J. Wang, R. Polsky, D. Xu, Silver-enhanced colloidal gold electrochemical stripping detection of DNA hybridization, Langmuir 17 (2001) 5739–5741.
[78] H. Cai, N. Zhu, Y. Jiang, P. He, Y. Fang, Cu@Au alloy nanoparticle as oligonucleotides labels for electrochemical stripping detection of DNA hybridization, Biosens. Bioelectron. 18 (2003) 1311–1319.
[79] M.D. Hanwell, S.Y. Heriot, T.H. Richardson, N. Cowlam, I.M. Ross, Gas and vapour sensing characteristics of Langmuir-Schaeffer thiol encapsulated gold nanoparticle thin films, Colloids Surf. A 284-285 (2006) 379–383.
[80] N. Wang, M. Lin, H. Dai, H. Ma, Functionalized gold nanoparticles/reduced graphene oxide nanocomposites for ultrasensitive electrochemical sensing of mercury ions based on thymine–mercury–thymine structure, Biosens. Bioelectron. 79 (2016) 320–326.
[81] S.L. Ting, S.J. Ee, A. Ananthanarayanan, K.C. Leong, P. Chen, Graphene quantum dots functionalized gold nanoparticles for sensitive electrochemical detection of heavy metal ions, Electrochim. Acta 172 (2015) 7–11.
[82] R.-J. Wu, C.-H. Hu, C.-T. Yeh, P.-G. Su, Nanogold on powdered cobalt oxide for carbon monoxide sensor, Sens. Actuators B 96 (2003) 596–601.
[83] S. Wu, H. Zhao, H. Ju, C. Shi, J. Zhao, Electrodeposition of Silver–DNA hybrid nanoparticles for electrochemical sensing of hydrogen peroxide and glucose, Electrochem. Commun. 8 (2006) 1197–1203.
[84] K.R. Brown, A.P. Fox, M.J. Natan, *Morphology-dependent electrochemistry of cytochrome C at Au colloid-modified* SnO_2 *electrodes*, J. Am. Chem. Soc. 118 (1996) 1154–1157.
[85] K.-J. Chen, K. Chandrasekara Pillai, J. Rick, C.-J. Pan, S.-H. Wang, C.-C. Liu, et al., *Bimetallic PtM (M = Pd, Ir) nanoparticle decorated multi-walled carbon nanotube enzyme-free, mediator-less amperometric sensor for* H_2O_2, Biosens. Bioelectron. 33 (2012) 120–127.
[86] T.-H. Tsai, S. Thiagarajan, S.-M. Chen, Green synthesized Au–Ag bimetallic nanoparticles modified electrodes for the amperometric detection of hydrogen peroxide, J. Appl. Electrochem. 40 (2010) 2071–2076.
[87] S.S. Kumar, K. Kwak, D. Lee, Electrochemical sensing using quantum-sized gold nanoparticles, Anal. Chem. 83 (2011) 3244–3247.
[88] X.-Y. Lang, H.-Y. Fu, C. Hou, G.-F. Han, P. Yang, Y.-B. Liu, et al., Nanoporous gold supported cobalt oxide microelectrodes as high-performance electrochemical biosensors, Nat. Commun. 4 (2013) 2169.

[89] A.K. Yagati, T. Lee, J. Min, J.-W. Choi, *Electrochemical performance of gold nanoparticle–cytochrome C hybrid interface for H_2O_2 detection*, Colloids Surf. B 92 (2012) 161–167.

[90] W.-T. Li, M.-H. Wang, Y.-J. Li, Y. Sun, J.-C. Li, *Linker-free layer-by-layer self-assembly of gold nanoparticle multilayer films for direct electron transfer of horseradish peroxidase and H_2O_2 detection*, Electrochim. Acta 56 (2011) 6919–6924.

[91] Y. Xiao, H.-X. Ju, H.-Y. Chen, Hydrogen peroxide sensor based on horseradish peroxidase-labeled Au colloids immobilized on gold electrode surface by cysteamine monolayer, Anal. Chim. Acta 391 (1999) 73–82.

3

SEMICONDUCTOR NANOCRYSTAL–BASED NANOSENSORS AND METAL IONS SENSING

Jun Ke

School of Chemistry and Environmental Engineering, Wuhan Institute of Technology, Wuhan, P.R. China

3.1 General Properties of Semiconductor Nanocrystals

3.1.1 Electronic Structure of Semiconductor Nanocrystals

The unique optical properties of semiconductor nanocrystals (NCs) or QDs have been paid more attention over the past few decades. Specifically, the attractive photoluminescence feature of QDs derives from the irradiative recombination of excited trapped carriers, which offers significant advantages in optical labels for chemo/biosensing, imaging, and light emitting diodes (LED). Despite the significant advances achieved by the pioneering efforts of many groups in synthetic procedures, structural characterizations, electronic, and optical properties of semiconductor QDs, studies on QDs are still on going. Why are QDs so charming?

It is well known that for a semiconductor material, one of the most important parameters is the energy gap width, that is, the energy difference between conduction bands and valence bands constituted by various energy levels. Typically, the width of this bandgap for a bulk semiconductor is a fixed and intrinsic value, depending on the material itself. However, intensive research has led to the discovery that once the size of the semiconductor decreases to less $\sim$10 nm, the situation of the fixed bandgap will change. In a bulk semiconductor, these energy

Advanced Nanomaterials for Pollutant Sensing and Environmental Catalysis. DOI: https://doi.org/10.1016/B978-0-12-814796-2.00003-4

levels can be considered as continuous state because the difference in energy is negligible, where the excitation spectrum corresponds to the bandgap width. As the semiconductor is miniaturized to nanosized scale, the effect of dimensional confining becomes more apparent. The averaged characteristics in bulk are no longer continuous and tend to be discrete and noncontinuous, corresponding to the quantized energy spectrum, which results in small and finite separation between energy levels [1]. This phenomenon is known as the quantum size effect, where NCs or QDs are referred to as quantized nanoscale particles. As a result, it is unquestionable that when the size of QDs diminishes, the corresponding bandgap increases, thus resulting in a blueshift of the characteristic absorption and emission wavelength [2], as shown in Fig. 3.1. Hence, many synthetic procedures have been developed to prepare high-quality QDs with different emission wavelengths and sizes for adapting to different application conditions.

To adjust the electronic structure and optical properties of bare QDs, introducing a new energy level by doping has been demonstrated to be an effective approach. The doped impurities can add one more valence electron into the host system to form n-type doping. In contrast, a substitutional impurity with one less valence electron can bring about an extra hole, leading to p-type doping [3]. Consequently, more carriers can be provided in comparison to undoped QDs and the characteristic

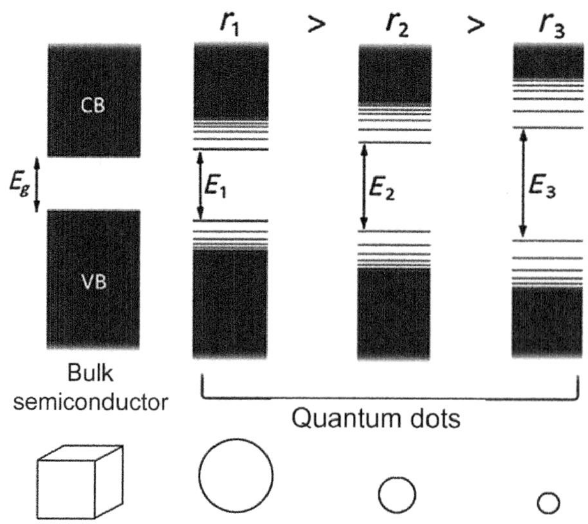

Figure 3.1 Illustration of the relationship between size of semiconductor and its bandgap [7].

fluorescence peak and intensity can also be enhanced or reduced, depending on the method used and impurity density [4]. The most studied and best dopant is manganese (specifically Mn^{2+}) due to its unique half-filled 3D valence shell, which is doped into wider-bandgap QDs such as ZnS, ZnSe, offering a new system that differs from conventional Cd-based QDs. Fainblat et al. [5] summarized the advance of individual magnetic impurity in colloidal 0D semiconductor QDs and described the effect of Mn doping on the electronic structure of the host QDs. Because of the special electron distribution of Mn^{2+} ions, high electron spin and first internal (d-d) excited state are extended into the host by doping, resulting in hybridization of Mn orbitals with the host orbitals. As a result, a new energy level and corresponding emission occurred together in the host QDs. Apart from the introduction of impurity, nanostructure is another path to vary the electronic distribution in semiconductor QDs. Borys et al. demonstrated that the electronic distribution of NCs can be tailored by exquisitely controlling the shapes of NCs, including wire, rod, and tetrapods. The different shapes mean the different exposures to specific facet, resulting in significant differences in charge density distribution. For example, various nanostructured CdSe/CdS NCs prepared exhibited distinguished differences in physical shape, photoluminescence spectra, and quantum confinement parameters [6].

3.1.2 Optimizing the Photoluminescence of Semiconductor Nanocrystals

Unlike molecular fluorophores which have narrow excitation ranges due to their unique spatial structures, semiconductor QDs usually possess broad absorbance bands because higher energetic photons excite the QDs to emit fluorescence. In addition, although the theoretical band is certain for specific QDs with fixed diameters, the synthetic procedure is hardly controlled finely to obtain highly uniform QDs, leading to a broad excitation range. Therefore, pursuing a novel and optimal synthesis route to obtain a high quality of QDs with high quantum yield, strong photoluminescence and uniform morphology is a goal in sensing, imaging and LED area. To achieve this goal, different synthetic strategies have been studied widely in the past decades, such as the strategies for controlling size and morphology [8], epitaxial growth of inorganic shell [9], surface passivation [10–12] and doping.

As previously mentioned, the fluorescence peak and quantum yield can be tuned by changing the size of QDs due to the discrete energy levels found in nanosized QDs, thus affecting the detection performance of QD-based nanosensors. For example, Zhu et al. [13] fabricated CdSe QDs with different diameters in the range of 1.98–3.68 nm to assemble mercury nanosensors by linking bovine serum albumin (BSA) to QDs, which exhibited different sensing performances in aqueous solutions. Experimental results revealed that both the sensitivity and the selectivity of nanosensors to Hg^{2+} increase with decreasing the size of CdSe QDs, where the smaller-sized CdSe QDs can provide more surface electrons and high reactivity, thus allowing more convenient coordination with Hg^{2+} ions. Apart from photoluminescence, the photoresponse feature of QDs is also closely related to the size of QDs. The reason has been mentioned previously, that is, inside QDs, the energy levels emerge in noncontinuous and discrete states due to the quantum confinement effect resulting from the decreasing of the size to ~10 nm. Kongkanand et al. [14] investigated the influences of sizes of CdSe QDs on the photocurrent intensity of CdSe QDs/TiO_2 photoelectrochemical solar cell, in which the photoexcited hot electrons from CdSe QDs were injected into TiO_2 to form photocurrent. The decreased size of the CdSe QDs could apparently enhance the photocurrent intensity ascribed to the shifting of the conduction band to a more-negative position, thus increasing the driving force of the photogenerated charge injection into the conduction band of TiO_2. However, it was also observed that the decreasing size did not always lead to an enhancement of photocurrent due to the enlarged bandgap and low harvesting efficiency of visible light.

Apart from size, the shape of QDs also has significant influences on optical properties due to the unique dimension confinement; therefore shape control has become another interesting and challenging study area in the synthesis of NCs [15,16]. Peng et al. [17] reported the controllable synthesis of CdSe nanorods and discussed the growth kinetics of II–VI CdSe from nearly spherical to rod-like (Fig. 3.2A–D). The study revealed that the CdSe rod-like NCs exhibited greater splitting between the absorbance peak and emission peak than that of CdSe spherical dots (Fig. 3.2E and G), which means the reabsorption issue can be efficiently overcome in many applications, for instance, in LEDs. Meanwhile, the quantum yield of the prepared rods was about 1%, increasing by a factor of 5 compared with the core–shell QDs by growth of the wide-bandgap materials. Although the quantum yield is lower than the developed spherical dots, it can also meet the demand of biolabeling and is expected to further improve by optimizing the synthetic

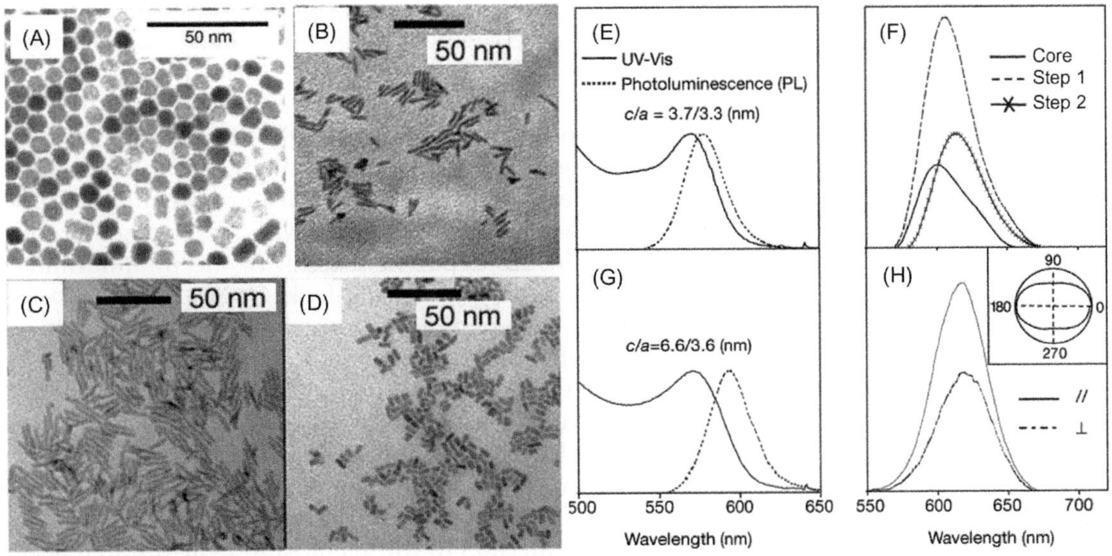

Figure 3.2 (A–D) TEM images of the CdSe dots and CdSe rods of different sizes; (E) absorption spectrum and emission of CdSe dots; (F) photoluminescence spectra of CdSe dot–based core–shell with different shell thicknesses; (G) absorption spectrum and emission of CdSe rods; and (H) polarization resolved photoluminescence spectra of prepared CdSe rods at 4.7K [17]. Copyright 2000, Springer Nature.

procedure and parameters in future study. Based on this study, various morphologies of semiconductor-based NCs have been developed [18,19] and the growth mechanism and emission ways have been further perfected [20].

Forming a protected shell on the nanosized core is another approach to optimize photoluminescence for achieving practical applications. It has been proven that the growth of wide-bandgap shell materials could efficiently enhance the photostability toward resisting oxidation and increase the fluorescence quantum yield by separating excited electrons from the outside oxidative environment. To date, many approaches have been well developed for preparing core–shell QDs such as microemulsion, the sequential injection method, and the reverse micelle route. Kumar et al. [21] demonstrated that the introduction of a ZnS shell significantly improved photoluminescence when comparing CdS with CdS/ZnS through a low-temperature wet chemical route. The enhancement of photoluminescence for CdS/ZnS was assigned to the interband connection between the ZnS shell and the CdS core, where the wide-bandgap ZnS shell could confine the photogenerated electron and holes to the interface of the CdS core because of the quantum confinement effect and thus the nonradiative transitions were suppressed efficiently. Zhu et al. [22] prepared CdTe–CdS core–shell QDs in aqueous phase through a facile route and studied the effects of the CdTe core size on the fluorescence properties of the core–shell QDs. The study revealed that the smaller the initial CdTe core is, the higher the enhanced fluorescence of the CdTe–CdS QDs is. Furthermore, it was found that the shell became gradually thicker with increasing amounts of CdS, resulting in the apparent redshift of the photoluminescence peak attributed to the partial leakage of exciton into the CdS shell. These studies demonstrate that epitaxial growth of the shell could efficiently enhance the fluorescence intensity because the growth of the shell could smooth the core surface and reduce the surface traps that largely diminish the fluorescence as a nonradiative recombination center. Besides, indium phosphide (InP) with a band gap of 1.35 eV, being one of the most promising QDs, has also been paid more attention because of its size-tunable emission in the visible and near-infrared spectral ranges and its low intrinsic toxicity [23,24]. Like Cd-based QDs, the enhancement of photoluminescence of InP QDs was achieved by means of forming large-bandgap Zn-based (ZnS [25,26] or ZnSe [27,28]) and silica [29] shells in the past. Nevertheless, the lattice matching should be considered when choosing shell materials because mismatching between the core and the shell could possibly lead to severe quenching of photoluminescence. A large lattice difference could lead to lattice distortion at the contact interface where the excited electrons and holes recombine in a nonradiative way.

In the aspect of optimizing photoluminescence, doping impurities into QDs can significantly alter the electronic structure and physical properties of semiconductor-based QDs, which is beneficial for enhancing the performance of QDs in various fields. Introducing Mn^{2+} ions into semiconductor QDs is well studied due to the unique electronic structure of Mn^{2+}. The Mn^{2+} ion has a half-filled 3D orbital so it possesses a high spin and large energy gap between the ground state and the excited state (d-d transition) [5]. Meanwhile, it is observed that the characteristic emission for diluted Mn-doped QDs is always located at about 580 nm deriving from the $_4T^1$–$_6A^1$ transition. Apart from Mn, Ag [30] and Cu [31] have also been widely used to dope semiconductor QDs, which exhibit PL enhancements of varying degrees in comparison to undoped QDs. Therefore doped QDs have been explored in the assembly of various nanosensors for probing metal ions in simulated or real aqueous solutions [32,33]. In other words, for nanosized QDs, their large exposed surface is another factor to consider for optimizing photoluminescence feature. The excited hot electrons can thermodynamically move to anywhere on the whole QD including the shallow surface, while a wealth of dangling bonds and surface defects exist on the shallow surface, which often act as recombination centers in radiative or nonradiative ways. Therefore the composition, structure, and defects of the surface atom layer are correlated with the emission of QDs, and further efforts are required in the investigation of the surface engineering of QDs [34].

To eliminate the potential toxicity of Cd-based QDs in bioimaging, biosensing, and biolabeling, Cd-free QDs such as ternary semiconductors including $CuInS_2$, $CuInSe_2$, and $AgInS_2$, are viewed as good candidates to substitute Cd-based QDs for sensing, labeling, and LEDs [35–38]. Nevertheless, although until now ternary QDs have been intensively investigated and their growth mechanism has also been discussed, the quantum yield, photostability, and functionalization are not comparable with binary QDs, especially Cd-based QDs. In addition to the mentioned weaknesses, the emission path of ternary QDs is unclear, though surface defect emission is deemed as the primary recombination path [39,40], requiring an aimless optimizing preparation procedure. Therefore unveiling the emission mechanism is a key factor in improving the quality of ternary QDs and boosting their practical applications.

3.2 Semiconductor Quantum Dot–Based Nanosensors: Principles and Applications

3.2.1 Composition of Nanosensors

"Nano" refers to objects measured in nanometers or billionths of a meter. Similarly, nanosensors are sensors whose active elements include nanomaterials (NMs). In general, nanosensors consist of a signal source (QDs, noble metals [41,42], fluorophores [43,44], etc.,), surface-functionalize dentities that are capable of interacting with specific compounds, and supports (if necessary) including electrodes. The signal source can exhibit a specific signal such as excited photoluminescence, a solution color that is visible to the naked eye, or electronic signals, which is used to build direct correlation with the concentration of the target substances, as described in Fig. 3.3. On the

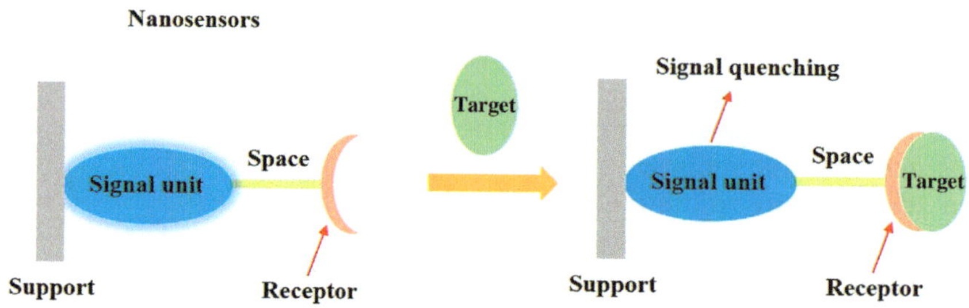

Figure 3.3 Illustration of the functional components in nanosensors and sensing processes.

other hand, the surface entity functions as a receptor that directly and selectively connects to the target substances, which should result in an apparent change of the signal to satisfy the ultimate target detection. For the support, it usually turns up in electronic analysis, where the electronic signal must be delivered through an electrode as the support.

To design a nanosensor, choosing a signal model is the first step. Second, confirming a signal emitter is a key point. Owing to their high reactive surface area and small particle size, NMs as signal emitters offer significant advantages including a superior surface area to volume ratio, which allows for higher catalysis and sensing response as well as better optical, magnetic, and electrical properties and are suitable for environmental and biological applications. It has been demonstrated that when reducing the size of a sensor, there are many benefits such as faster response, better signal-to-noise, more accurate data, increased data density, less impact on the phenomenon being measured. Therefore nanosensors consisting of NMs are powerful analytical tools due to their portability, self-contained nature, and low cost. For example, commercial pregnancy test kits are comprised of Au NCs and antibody in order to determine human chorionic gonadotropin through the double antibody sandwich method and they indicate whether women are pregnant or not via a simple color change. Although the accuracy of these test kits only reaches 90%, they can provide a prejudgment in a rapid, simple, and cheap way.

Third, based on the purpose target, a suitable receptor should be chosen, which not only is highly sensitive to the target, but also affects significantly the signal intensity. The affecting mechanism and analysis models will be further discussed in the next section. When the target is added into the system including nanosensors, the corresponding signal intensity rapidly decreases or increases, and then the linear range and detection limit could be obtained by building a relationship between the analyst concentration and signal intensity. Once the signal model is ensured, the main study is the assembly of nanosensors. Although the fabrication process of nanosensors is constituted by two steps, namely synthesis and functionalization of NMs, the two steps are often integrated to finish so that the surface receptor can tightly anchor onto the surface of the NMs and insignificantly undermine the signal intensity. If the two steps are performed separately, the initial signal intensity is unavoidably weakened with the conjugation of the surficial entity because the signal, especially photoluminescence, derives from the surface transmission and movement of excited charges. The conjugation of the functionalized entity must anchor at the surface and affect the behavior of the excited charges in NMs.

3.2.2 Principles of Nanosensor Design and Applications

Generally, nanosensors can be categorized into three kinds of models according to the signal emitted by the subunit. First, colorimetric sensors are based on changes of the probe solution, which originate from significant differences in the electronic properties of the signal sources in the absence or presence of the target by the intra/intermolecular charge transfer mechanism. Second, fluorogenic sensors take effect upon the introduction of an analyst, which influences fluorescent emission by photoinduced electron transfer, excited state intramolecular proton transfer, and fluorescence resonance energy transfer (FRET). Third, electrochemical sensors are related to the measurement of changes in the redox potential with changes in analyst concentration.

3.2.2.1 Colorimetrics

Colorimetric sensing is a promising technique that is beneficial for qualitatively and quantitatively detecting ionic or neutral species without the aid of any instrumentation; therefore it is known as a "naked eye" analytical method [45]. Often, it is applied in the form of a test kit for real-time and rapid detection, simplicity, and high selectivity and sensitivity, which makes it a desirable technique among researchers. Taking cation detection as an example, noble nanoparticles such as Au and Ag are combined with binding subunits that display certain colors stemming from the specific light absorption. When the cation is added into the probe solution, it can selectively bind with the receptor whilst the color of the probe solution becomes heavier or lighter that can be qualitatively distinguished by naked eyes or colorimetric cards. If the probe solution is visible light, the characteristic absorption peak should be located in the range of visible light. Although gold and silver nanoparticles have strong merits and are well developed in comparison with other NMs, their precious properties limit, to some degree, their wide application. Therefore some semiconductor QDs including Cd-based QDs, InP, and ZnTe, can still satisfy the requirement of visible light and low cost-efficiency.

In the past few decades, a growing number of organic fluorophores such as pyrene, bodipy, rhodamine, 1,8-naphthalimide, and near-infrared dyes have been developed to detect various substances, especially metal ions, in aqueous or organic solvents because these fluorescent compounds can selectively chelate with metal ions, resulting in a variation of the spatial molecular structure and absorption spectra that correspond to the color of the probe solution. For example, photoactive moieties, coumarin and nitrobenzoxazole, are often used to achieve the colorimetric determination of cations through Schiff base and amide linkages. Moreover, the rich functional groups of organic fluorophores are beneficial for conjugating with proteins and biomacromolecules through specific reactions, which can extremely enlarge target ranges and improve sensing performance. Therefore choosing a suitable surficial ligand is vital in the design of high-quality semiconductor QD–based nanosensors, which directly determines whether a nanosensor is sensitive to the target metal ion and has a high antiinterference to other metal ions in one sample. To achieve an efficient colorimetric detection of metal ions, two sensing mechanisms have been reported in the past. The one is the D-π-A system, where both an electron donator (D) and an electron acceptor (A) are introduced into the chemosensor. When metal ions bind with the electron donor in priority, the conjugation between the D and A is undermined, resulting in a blueshift in the absorption spectrum and an enhancement in the charge transfer from metal to ligand, while when metal ions preferentially bind with the electron receptor, it leads to the strengthening of the charge transfer from the ligand to the metal and the occurrence of a redshift in the absorption spectrum, as described in Fig. 3.4 [45]. The other way is to introduce Rhodamine dyes into the sensor system because the present spirolactam is able to magnify the colorimetric changes upon the addition of metal ions, in particular, mercury ions, which is beneficial for improving detection quality of nanosensors. However, there is a challenge in the utilization of Rhodamine derivatives because of their

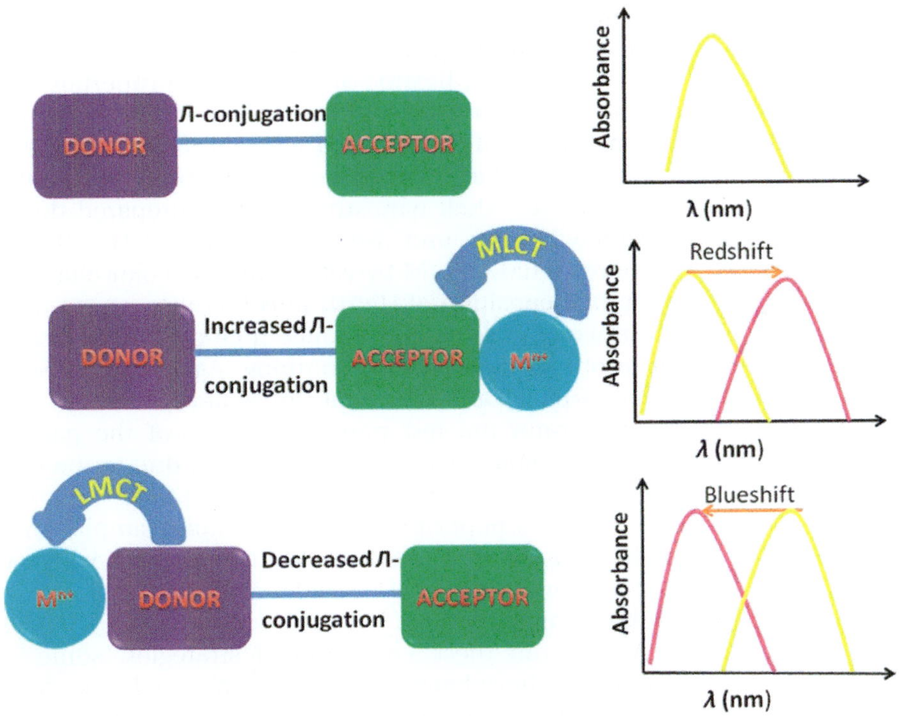

Figure 3.4 Illustrated effects of the interaction between cation and acceptor on the absorption spectrum [45]. Copyright 2017, Elsevier.

water insolubility; in contrast, QDs are generally used in aqueous solutions [46]. Hence, combining QDs with Rhodamine derivatives is not a useful approach to determine the presence of metal ions in aqueous solutions unless transferring metal ions into organic solvents. For the former strategy, specific ligands and nanostructures are vital factors and should be selected carefully in the light of different metal ions. For instance, Gore et al., prepared thioglycolic acid (TGA)-capped CdS QDs via a developed classic route in an aqueous solution for the colorimetric probing of cobalt (Co (II)) ions. When the Co (II) ion was introduced, the aggregation of TGA–CdS QDs was induced, leading to an apparent enhancement in the absorption peak at 360 nm, accompanied by a visible color variation from its original colorlessness to a yellowish brown. The phenomenon was ascribed to the charge transfer from the QDs to the metal ions through the ligand via the chelating interaction between the carboxyl group in the TGA and the Co (II) ions [47]. On the other hand, TGA–CdS QDs can be used to determine mercury ions through a fluorescence model instead of a colorimetric process [48]. The results indicate that the effects of surficial ligands on the optical properties of semiconductor QDs stem from various aspects. In additional to bare QD systems, coupling noble NPs with QDs is an efficient path to assemble excellent colorimetric detection of metal ions. An Au–HgS core–shell nanostructure was prepared through an in situ growth route and used to determined Hg (II) ions in the range of 10 nM–80 μM by visualizing the color changes deriving from the deposition of Hg (II) ions [49].

Butwong et al. [50] developed a colorimetric paper test strip for probing mercury by combing Ag-doped CdS QDs with chitosan-encapsulated cellulose. When the Hg^{2+} ions were dropped onto the test paper, the color of the paper changed from its original yellow to a deep brown due to the capturing of the Hg^{2+} by mercaptoacetic acid (MA)-coated Ag-doped CdS QDs. The test paper proved to be a rapid, simple, and low-cost analysis tool with a detection limit of 124 μM, which to a large degree promotes QD-based nanosensors for practical applications.

Apart from these conventional strategies, some innovative approaches have been explored which have hardly been categorized. In the first section, it is discussed that QDs are easily excited by the incident of light and produce hot electrons that can reduce some organic compounds, leading to the aggregation and dispersion of QDs. Tang et al. [51] took advantage of this strategy to prepare glutathione (GSH)-capped CdS QDs for probing trace Cu (II) ions in the presence of 3,3′,5,5′-tetramethylbenzidine (TMB). On the one hand, the photoexcited CdS QDs can oxidize the TMB and produce oxidized TMB with a blue color. On the other hand, Cu (II) ions can efficiently oxidize GSH, resulting in the aggregation of CdS QDs. As a result, the CdS QDs cannot further oxidize the TMB and the color of the probe solution changes from blue to colorless. According to this strategy, under optimal conditions, the linear range of Cu (II) concentration is from 10 nM to 2.0 μM with a detection limit of 5.3 nM. At the same time, the nanosensor displayed a high selectivity to Cu (II) ions.

3.2.2.2 Fluorescence Mode

Fluorescence emission is another unique optical signal of QDs that is used in many applications such as sensing, imaging, and LED. The formation of fluorescence can be described by the diagram in Fig. 3.5.

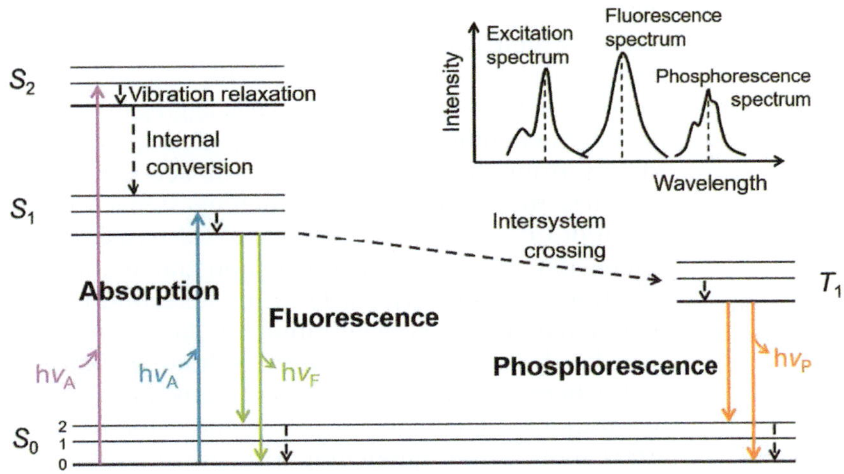

Figure 3.5 Diagram of fluorescence and phosphorescence phenomenon [52]. Copyright 2018, Elsevier.

Commonly, an electron at the ground state (S_0) can rapidly jump to the excited state (S_1 or S_2) by absorbing sufficient energy such as excitation light or applied voltage, while the excited electron is unstable and tends to come back to the ground state because of the minimum energy principle. Instead when transitioning from the excited S_2 state to the ground state (S_0), the hot electron firstly moves to another excited state S_1 through nonradiative vibration relaxation or internal conversion and then returns to the ground state (S_0) accompanied with fluorescence emission. Owing to the existence of a nonradiative transition, the characteristic peak of fluorescence emission shifts to a larger wavelength in comparison with the excitation spectrum, known as Stokes' shift. During the procedure of target substance determination, when the nanosensor selectively interacts with analysts, the characteristic fluorescence signals will be influenced such as fluorescence quenching or enhancement, blueshift or redshift, and lifetime prolonging or shortening, depending on the different signal transduction mechanisms. Based on this phenomenon, the quantitative relationship between signal changes and the target concentration can be built in order to realize rapid determination. Specifically, the fluorescence intensity is easily varied in QDs-based nanosensors upon the addition of metal ions, in particular, the quenching of fluorescence resulting from the blocking of the emitting procedure [53]. For instance, when two kinds of glutathione-capped CdTe and CdZnSe QDs are applied for ultrasensitive determination of Pb (II) ions, the fluorescent emission could be significantly quenched, named by "turn-off" mode, which is attributed to the competitive binding to glutathione between Pb^{2+} and QDs. When the glutathione preferentially bind with Pb^{2+} ions, the QDs aggregate together, resulting in the quenching of emission [54]. This quenching mechanism differs from the one mentioned in Section 3.2.2.1 on colorimetrics, the fluorescence quenching stems from the replacement of Cd^{2+} ions at the surface of Cd-based QDs by the target metal ion [55]. It has been demonstrated that the change of fluorescence intensity (F_0/F) in the detection process is linearly correlated to the concentration of the target material, which can be described by the Stern–Volmer equation: $F_0/F = 1 + K[Q]$, where K is the Stern–Volmer constant and $[Q]$ is the concentration of quencher [33,56]. If the value of K is higher than 100 L mol^{-1}, the quenching process is ascribed to a static quenching process; in contrast, if the K value is less than 100 L mol^{-1} it corresponds to a dynamic quenching process. In static quenching, the quencher rapidly reacts with the emitter to form a complex, which has no influence on the fluorescence lifetime regardless of the quencher; while in the case of dynamic quenching, the quencher needs a diffusive movement to encounter the emitter, which causes the fluorescence lifetime to decrease. Therefore this equation can facilitate the exploration of the kinetics of the quenching process and unveil which quenching mechanism(static, dynamic, or combined) is the primary quenching process in the determination of metal ions [57].

In contrast to fluorescence quenching, fluorescence enhancement is another mode, defined as "turn-on," where the recognized target can enhance the fluorescence signal. Under this mode, there is another more complicated fluorescence enhancement mechanism called as fluorescence resonance energy transfer (FRET) [58], resulting in blue/redshift of characteristic peak with an intensity enhancement, which will be discussed separately in the next section. Here, only the enhancement of fluorescence is analyzed excluding the FRET mode. According to this discussion, there is no doubt that enhanced fluorescence is beneficial for enforcing the sensitivity of a nanosensor owing to its lower background interference and higher signal-to-noise ratio. After all, a bright signal is more easily observed than a dimming signal in the dark. For example, Mn dopant in the lattice is usually recognized as intrinsic radiation recombination center while the residual Mn ions at the surface of QDs is often an effective quencher, which can severely undermine the intrinsic fluorescence emission. Zhou et al. [59] discovered that mercury ions efficiently displaced residual Mn ions on the surface of ZnSe QDs and thus resulted in the enhancement of the characteristic fluorescence emission. Based on this strategy, synthesized Mn-doped ZnSe QDs were used to selectively determine Hg^{2+} ions with an limitation of detection (LOD) of 7 nM in a real aqueous sample. Departing from unintentional quenching, many groups intend to quench the fluorescence of QDs before metal ion detection. Ethylene diamine tetraacetic acid (EDTA) was used to etch mercaptopropionic acid (MPA)-capped CdTe QDs, resulting in an effective quenching of emission, while the quenched emission was recovered with the addition of Zn^{2+} ions due to the interaction between Zn^{2+} with EDTA [60].

3.2.2.3 Fluorescence Resonance Energy Transfer

In general, FRET is viewed as a nonradiative process, in which a donor D in excited state can transfer energy to an acceptor A in ground state through a long-distance dipole–dipole interaction. Theoretically, the donor D is excited by the incident of light or current and emits certain wavelengths of light that can be absorbed by the acceptor A. In turn, the acceptor A not only emits light with different wavelengths, but also quenches the emission, both of which can be considered as efficient energy transferring between donor D and acceptor A (Fig. 3.6). The rate of energy transfer between donor D and acceptor A highly depends on the distance between the two components, the relative orientation of the dipoles, and the extent of spectral overlay, which could be described using the equation [61]:

$$R_0 = 9.78 \times 10^3 \left[k^2 \times n^{-4} \times Q_D \times J(\lambda)\right]^{1/6}$$

R_0 is used to denote the Förster distance where 50% of the excited D molecules decay due to energy transfer, while the other 50% decay through the radiative or nonradiative process. k^2 denotes the dipole orientation between donor D and acceptor A and the range of k^2 is from 0 (perpendicular) to 4 (collinear). n is the refractive index of the medium where the system is used such as 1.4 for aqueous solutions. Q_D is the quantum yield of the donor and $J(\lambda)$ is the overlap integral between the donor emission and the acceptor absorption. From the equation, it is observable that the value of R_0 increases with a higher acceptor extinction coefficient and greater overlap between the donor emission spectrum and acceptor absorption spectrum. Therefore the particular r can be estimated to determine whether the FRET is effective when choosing a specific donor–acceptor pair to assemble nanosensors.

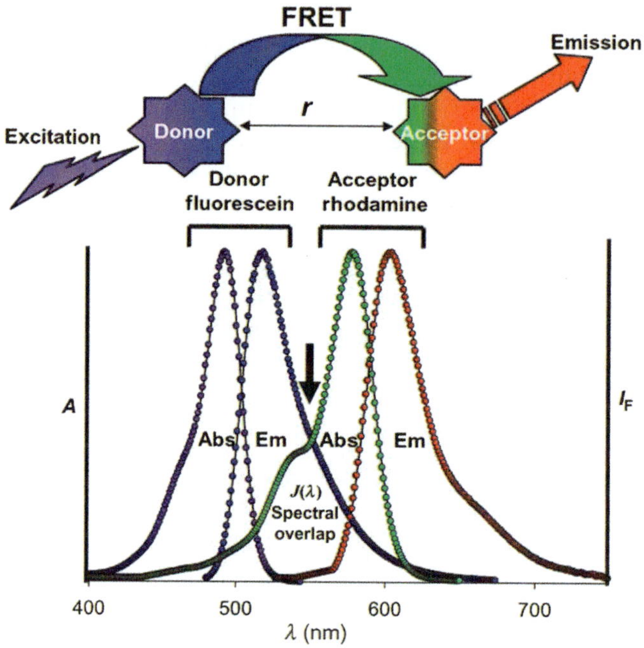

Figure 3.6 Illustration of the FRET process: upon excitation, the excited state donor molecule transfers energy to the acceptor molecule located at distance *r* from the donor [61]. Copyright 2006, John Wiley and Sons.

According to the mentioned principle, it is expected that FRET is a very useful strategy to design chemo/biosensors due to its intrinsic sensitivity to the distance between the donor and the acceptor. In the past few decades, FRET techniques have been extensively utilized in real-time imaging and biosensing in vivo and for investigating the interaction between antigens and antibodies in vitro. Simultaneously, pioneering investigations have demonstrated that colloidal semiconductor QDs could be utilized to determine various metal ions, proteins, and DNA through versatile, smart design and synthetic routes. Owing to the unique, size-tunable emission, broad absorption spectra, and large Stokes' shifts, QDs can be excited by a wavelength far from the emission wavelengths. This means that QDs can be finely tuned to adapt the acceptor to obtain satisfying spectra overlap in FRET applications. Accompanied by the high quantum yield of QDs, the increasing of spectral overlap could result in the proportional increasing of the Förster distance R_0, which refers to longer separation distances in FRET systems. The best available QDs for FRET applications are CdSe/ZnS QDs, which are constituted by a CdSe core and a ZnS shell [62].

The proper shell not only passivates the core, thus protecting the core from oxidation, but also apparently increases the quantum yield and photoluminescence intensity, which has been demonstrated in the optimal synthetic process. In general, CdSe/ZnS QDs with diameters of 5 nm emit fluorescence of 510-nm wavelength and 8 nm for 610-nm. If CdSe/ZnS QDs were encapsulated by organic ligands or proteins, the overall size of the colloidal nanoparticle would reach 20 nm. For instance, for many bioconjugates in which QDs act as the donor and dye-labeled protein as the acceptor, the value of R_0 often falls in the range of the QDs diameter, thus leading to a low FRET efficiency for a single QD and a single acceptor [63]. Nevertheless, it has been demonstrated that conjugating a central QD donor with multiple acceptors could enhance the FRET efficiency in proportion to the ratio of the FRET acceptor to the QDs due to the enhancement of the fluorescence spectrum overlap ratio [64]. In addition, semiconductor QDs have often been combined with noble NPs to obtain better assaying performance or give insight into the FRET mechanism between QDs and noble NPs. For example, *n*-alkanethiol stabilized Au NPs were utilized to conjugate with streptavidin-coated QDs, where the fluorescence emission from the QDs was totally quenched due to the efficient energy transfer from the QDs to the Au NPs. From TEM imagery, it was observed that streptavidin-coated QDs were surrounded by the *n*-alkanethiol stabilized Au NPs to form large-sized clusters, causing the distance between the QDs and the Au NPs to fall within the range of R_0. On the other hand, when high density biotin was added to encapsulate the Au NPs, the distance between the QDs and the Au NPs was enlarged, thus blocking the energy transfer and recovering the strong photoluminescence [65].

Departing from the CdSe–ZnS core–shell QDs, various other binary and ternary semiconductors including ZnSe [66], CdS [67], PbS [68], and CdHgTe [69] with emissions ranging from the UV to the IR spectra have also been synthesized in the past. Nowadays, versatile strategies have been developed to attach acceptors to QDs such as covalent coupling, electrostatic self-assembly, and biotin–avidin chemistry. However, before conjugating various functional entities, another obstacle needs to be overcome to assemble high-quality chemo/biosensors with effective FRET. Typically, high-quality semiconductor QDs are prepared by means of insoluble salts so they are often not water soluble and have to be kept in organic solvent such as chloroform and toluene. Therefore the original organic ligands utilized in the synthetic route must be exchanged using a bifunctional ligand to simultaneously offer solubility and possible bioconjugation sites. Although a great number of ligands have been studied, each one has its own advantages and disadvantages [70]. For example, some ligands can hamper QDs dispersion into the basic pH range and some obviously increase the overall size, thereby severely undermining photoluminescence intensity [71]. Many groups are dedicated to pioneering the exploration of QD-based FRET applied for bioassays and contaminant determination through elaborate design and controlling the energy transfer process. For ultrasensitive and selective bioanalysis, QDs and dye-labeled proteins were typically paired to assemble biosensors for the in vivo or in vitro detection of proteins, antibodies, antibiotics, and even intracellular pH [72]. In addition, other applications of QD-based FRET nanosensors involve the detection of Cr (III), Cu (II), Hg (II), and organic contaminants such as TNT and oleic acid. When foreign substances are introduced, the initial FRET are cut off or new FRET are created, thus achieving quantitative recognition of the foreign target.

3.2.2.4 Surface-Enhanced Raman Scattering

Surface-enhanced Raman scattering (SERS) has attracted intense interest in the past several decades. It has been demonstrated that compared with the conventional Raman signal, the signal from SERS can be dramatically enhanced by a factor of up to 10^{14} due to the strong light-induced electric fields at locations on nanostructured metal NMs (Au, Ag, Pt, etc.). Similarly to fluorescence and colorimetric tools, the intensity of SERS is closely related to the target substance structure, composition, and even size, which easily distinguishes different analysts in accordance with its characteristic fingerprints [73]. The characteristic peak intensity as a function of the concentration of a given analyst is directly shown, thus realizing the purpose of the analysis. In addition, SERS has other attractive advantages such as its narrow Raman band and high antiinterference ability to environment factors including humidity, oxygen, and temperature. Consequently, SERS is considered as one of the most promising techniques for ultrasensitive determination on a single-molecule level. For instance, although Au nanoparticles were reported to be synthesized with a range colors due to the localized surface plasmon resonance (LSPR) effect that is often utilized to determine metal ions or DNA molecules via the mentioned colorimetric models, resulting from the aggregation of the Au nanoparticles, when it was functionalized by 4-mercaptobenzoic acid the selective capture of mercury ions was achieved in the presence of 2.6-pyridinedicarboxylic acid. From the generated fingerprint of the SERS peak at 374 cm^{-1}a linear correlation between the peak intensity and the concentration of trace mercury ions was built, ranging from 10 ppt to 500 ppb with a detection limit of 5 ppt [74]. Zhou et al. presented a novel SERS nanosensor consisting of an urchin-like TiO_2–Ag nanostructure for the highly sensitive detection of Cr (VI) ions in a linear range up to 2 μM through the strong interaction between GSH and Cr (VI) ions, which led to the severe aggregation of TiO_2–Ag nanoparticles [75].

Although the phenomenon of SERS is often observed in noble metal nanoparticles (MNPs) and widely used to determine various targets such as proteins, organics, DNA, and environmental contaminants, nonnoble metal systems have been demonstrated to possess the ability to utilize a SERS strategy to realize ultrasensitive determination [76]. Wang et al. [77] demonstrated that CdTe QDs combined with 4-mercaptopyridine showed an enhancement in the Raman spectrum by a factor 10^4 in comparison with bare 4-mercaptopyridine due to the charge transfer between the CdTe QDs and the 4-mercaptopyridine, which indicates that the use of semiconductor QDs is likely to achieve ultratrace detection in some assay and bioimaging systems. In addition to organic compounds, the energy transfer between semiconductor QDs and Au nanowires was discovered by hybridizing CdSe/ZnS QDs with Au nanowires. Owing to the overlap of the QD absorption bands with the LSPR absorption bands, the Raman spectra of QDs was enhanced, supporting the possibility of SERS in QD systems [78]. However, commonly, interpreting SERS spectra can be incredibly challenging and time-consuming because a variety of instabilities and modifications contribute to the enhancement, as well as the conditions and NMs used to measure SERS spectra [79]. Therefore except designing novel systems, emphasis on the interpretation of SERS spectra and mechanisms are urgent in realizing the practical applications of the SERS tool in future studies.

3.2.2.5 Voltammetry

The voltammetric technique is well developed in analytical chemistry because of its sensitivity, versatility, and low cost. Commonly, upon the application of a potential sweep to a working electrode, the reduction or oxidation of a target analyte is triggered, resulting in the generation of a current that is measured and linked with the concentration of the target analyte. Actually, there are several types of voltammetric techniques such as anodic stripping voltammetry (ASV), cyclic voltammetry, and linear sweep voltammetry (LSV) according to the applied potential models [80]. During the assembly of voltammetric nanosensors, to improve the detection performance, NMs including carbon NMs, noble MNPs, and QDs, have been introduced to facilitate electron transfer between the analysts and the transducer and to provide an anchor point for the functionalized entity (acceptors). In addition, the introduced NMs could offer a more quantifiable signal, which could result in the enhancement of sensitivity and selectivity of the voltammetric nanosensors to trace levels of analysts [81].

For Cd-based QDs, it is a superior strategy to construct QD-modified voltammetric nanosensors because the Cd-based QDs could be dissolved and release Cd^{2+} ions that are accumulated on the functionalized electrode and reduced to Cd in the presence of diluted HNO_3, which would form a significant current peak associated directly with the amount of target immobilized on the surface of the electrode [82,83]. For example, CdS QD–modified sandwich-type nanosensors have been prepared through an in situ growth route for sensitive bioanalysis trough an ASV model, as described in Fig. 3.7 [84]. In comparison with other probing methods, CdS-modified bioassay has exhibited enhanced performance for detecting human immunoglobulin G, human carcinoembryonic antigen, human fetoprotein, and thrombin, corresponding to detection limits of $4.5\ fg\ mL^{-1}$, $3.0\ fg\ mL^{-1}$, $4.9\ fg\ mL^{-1}$, and 0.9 fM, respectively. In addition, owing to the high density of active sites on the surface of graphene, combining graphene with QDs is used as a strategy to amplify the signal and sensitivity of electrochemical nanosensors. Based on this strategy, Shiddiky et al. [85] designed a novel biosensor for detecting epithelial cell adhesion molecule antigen by immobilizing amine-functionalized CdSe QDs on carboxylated graphene nanosheets based on a carbodiimide coupling reaction, which exhibited a detection limit of $100\ fg\ mL^{-1}$.

In addition, the voltammetric technique has expanded to determine metal ions in recent years, where the electrode is functionalized by various metal sensitive acceptors. When specific metal ions are introduced, the electronic signal will increase or decrease depending on the added concentration of metal ions. For example, Ahour et al. [86] constructed a graphene QD-capped graphite electrode for highly selective and sensitive detection of nanomolar level Cu (II) ions using the ASV method. When trace Cu (II) ions were added into the solution, they rapidly reacted with the graphene QDs via the surficial oxygen-containing groups, which led to an increasing in square wave voltammetry. The graphene QD-modified

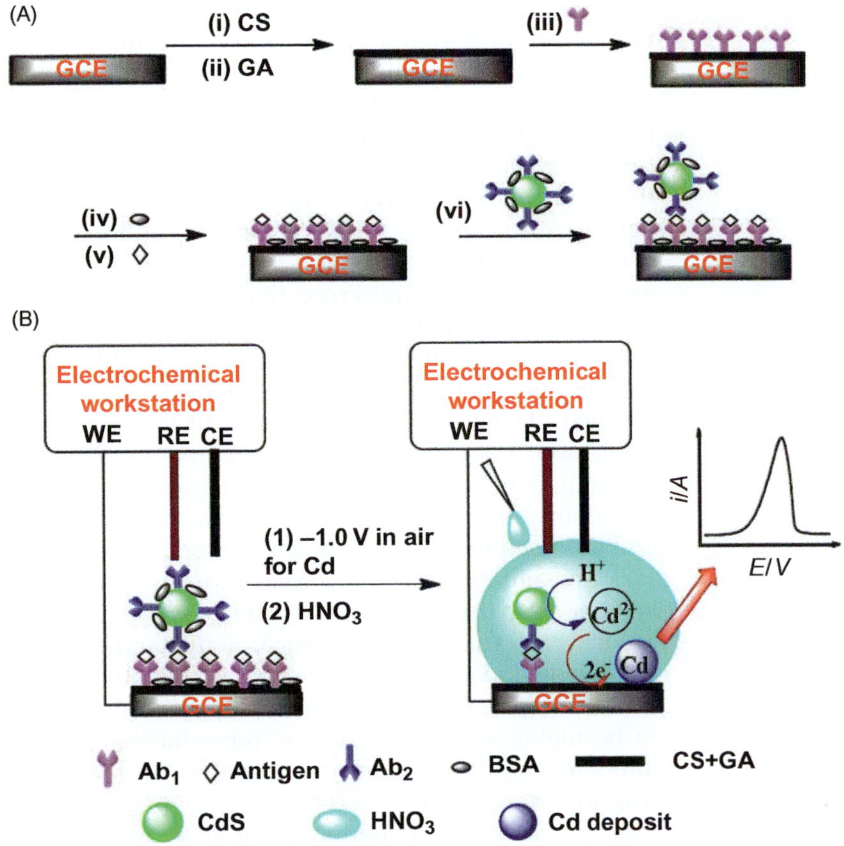

Figure 3.7 (A) Synthetic steps for voltammetric nanosensors and (B) the sensing process [84].

electrosensor exhibited great determination performance of Cu (II) ions in the linear range of 50 pM to 4 nM with a lower detection limit of 12 pM. By means of polyaniline/graphene QDs, a screen-printed carbon electrode was modified to rapidly detect Cr(VI) ions in aqueous solutions through the LSV method. The optimized electrosensor displayed a determination ability of Cr(VI) in the linear range of 0.1–10 mg L^{-1} with a detection limit of 0.097 mg L^{-1} [87].

3.3 Sensing of Metal Ions in the Environmental Field

The rapid increase in toxic heavy metal ions (e.g., As, Cr, Cu, Cd, Pb, Hg, Ni, and Zn) being released into the environment and the impact thereof on human health are now considered as major environmental problems throughout the world. Furthermore, it is well known that exposure to highly toxic heavy metal ions can cause a series of diseases including neurological disorders, cancer, kidney failure, DNA, liver, etc. Considering these dangers, many strategies have been established to control the levels of toxic heavy metal ions in various systems in accordance with scientific toxicity standards and exposure guidelines such as those provided by the World Health Organization (WHO) and the Environmental Protection Agency (EPA). To date, methods to detect heavy metal ions at trace levels (even in the ppt range) have been established using sophisticated analytical instruments such as atomic absorption spectroscopy, inductively coupled plasma-mass spectrometry (ICPMS), mass spectroscopy (MS), and X-ray fluorescence spectroscopy. Nevertheless, these techniques have drawbacks including their high expense, lack of portability, low throughput, time-consuming pretreatment steps, and the need for highly skilled operators.

Many efforts have been made to develop improved sensing systems that use NMs and a wide range of detection principles (optical, electrical, ion-exchange, semiconducting metal oxides, conductive polymers, quartz crystal microbalance sensors, and electronic nose devices). These sensing systems have the potential to be highly sensitive, selective, reproducible, less susceptible to interference (from humidity), and stable. To develop such systems, two important aspects need to be considered, namely the receptor (e.g., a heavy metal ionophore or biological receptor) and the immobilizing/transducing platform, which can be based on optical, electrochemical, magnetic, or miscellaneous principles.

Chemically recognizing heavy metal ions by detecting specific substances/species, called receptors, are key to developing diverse sensing techniques. Generally, these receptors specifically interact with heavy metal ions through noncovalent bonds, hydrogen bonds, metal coordination, hydrophobic forces, van der Waals forces, p–p interactions, or electrostatic or electromagnetic effects. In addition, NM properties, when bound to biological receptors (such as enzymes, DNA, RNA, and antibodies) allow for a highly specific, sensitive, and cost-efficient detection of heavy metal ions. Many advances have been achieved in the development of NM-based techniques to monitor heavy metal ions in various samples including (1) biological, (2) optical, (3) electrochemical, and (4) miscellaneous sensing strategies (Fig. 3.1). These strategies have been established with diverse classes of NMs such as MNPs, QDs, nanometal organic frameworks, magnetic nanoparticles, carbon nanotubes, and nanocomposites.

In comparison with organic fluorophores, semiconductor QDs are brighter, exhibit lower photobleaching, and are non-blinking, making them suitable as a signal source in the assembly of nanosensors to determine metal ions. In addition, the majority of organic fluorophores are hydrophobic, which hampers in the extreme their application in contaminated water analysis. In contrast, the surface properties of QDs can be tuned by anchoring different polar ligands according to the determination environment. In Section 3.3.1, QDs with different compositions and structures will be summarized and discussed for probing heavy metal ions in the form of a liquid or film. The effects of fabrication procedures and the functionalized entity on the detection performance of nanosensors are also overviewed in detail.

3.3.1 II–VI Nanocrystals for Nanosensors

In general, II–VI NCs mainly denote a compound consisting of I and II transition metal and chalcogenide elements such as CdX, ZnX, Ag_2X, and Cu_2X (where X = S, Se, or Te, etc.). From a maturity standpoint, Cd-based QDs have been widely developed over the past few decades [88]. Owing to their high quantum yield, stability, and nonblinking property, QDs often are employed in imaging [89], sensing, and labeling [90]. Wang et al. [91] prepared a series of thiol-coated CdTe QDs for probing arsenic (As) ions in aqueous solution. To endow sensing abilities to CdTe QDs, surficial ligands, for example, MA or GSH, are necessary during the preparation process. When As(III) was added into the probe solution, the emission of GSH-capped CdTe QDs was apparently quenched, which could efficiently determine the concentration of As(III) ranging from 5 to 250 μM with a detection limit of 20 nM. Meanwhile, the selective experiments testified that the GSH-coated CdTe QDs had the strongest response to As (III) in comparison with others metal ions, for example, Fe^{3+}, Cu^{2+}, Zn^{2+}, Mn^{2+}, Cr^{3+}, and Ni^{2+}. In the past few decades, it has been demonstrated that the introduction of surficial ligands, especially thiol organic ligands, could not only passivate QDs, preventing them from aggregation, but also strongly interact with target analysts, in particular, heavy metal ions. Among these, thiol ligand includes a sulfur atom which has a high affinity to heavy metal ions (Hg, Pb, As) [92]; and this is usually used as a strategy to design QD-based chemosensors through choosing different thiol ligands and QDs. Xia et al. [93] fabricated TGA-coated CdTe QDs through a developed injection process in aqueous medium, which has been widely applied in the preparation of various nanocrystals. The obtained TGA-coated CdTe QDs present strong quenching responses to Hg^{2+} and Cu^{2+} ions, which indicates that TGA-coated CdTe QDs could be utilized to determine Hg^{2+} and Cu^{2+} in certain concentration ranges. To improve the fluorescent emission, the TGA-capped CdTe QDs were further functionalized by means of BSA. The experimental data showed that the BSA-functionalized QDs had a linear detection range for $Hg^{2+}$from 12 to 1500 nM with a detection limit of 4.0 nM. Meanwhile, the probe solution was successfully applied in real natural samples. For functionalized ligands, except the thiol groups, carboxyl and amine can also be applied to modify QDs in order to achieve the ability to detect metal ions. Li et al. [94] prepared L-carnitine-coated CdSe–ZnS core–shell QDs in ethanol. It was reported that the L-carnitine has a good sensitivity to Cd^{2+} in comparison with other metal ions, for instance, Ca^{2+}, Ni^{2+}, Fe^{2+}, Ag^{+}, Pb^{2+}, and Zn^{2+}. Therefore the L-carnitine-capped QDs possessed better selectivity to Cd^{2+} at the interference of other metal ions. Unlike the former model herein, the addition of Cd^{2+} resulted in the enhancement of the fluorescent emission, which determined the Cd^{2+} concentration to be in the range of 0–50 μM with a limit of detection of 0.15 μM calculated according to the international union of pure and applied chemistry (IUPAC) criteria.

By summarizing the colorimetric nanosensors consisting of noble MNPs such as Au, it is well known that DNA has a high affinity to heavy metal ions, for instance, Hg^{2+} and Pb^{2+}, because these heavy metal ions cause apparent distortion of DNA molecules, thus leading to the aggregation of Au nanoparticles. Based on this strategy, a DNA fragment could be utilized to functionalize QDs for detecting heavy meal ions. Long et al. [95] combined DNA fragments with CdS QDs for the assembly of an ultrasensitive mercury chemosensor through a facile one-step route. The Cd precursor was directly added into the solution containing the DNA pieces, where Cd^{2+} ions chelated with the organic groups in the DNA molecules and then the CdS QDs formed rapidly on the surface of the DNA in the presence of Na_2S. The obtained CdS-encapsulated DNA nanocomposite presented strong fluorescent emission and displayed high sensitivity and selectivity to trace Hg^{2+} ions in a wide concentration range from 0.04 to 13 μM by quenching the fluorescent emission. A detection limit of 4.3 nM was achieved. Freeman et al. [96] also took advantage of DNA pieces to functionalize CdSe–ZnS core–shell QDs for the simultaneous detection of Hg^{2+} and Ag^{+} in the form of a QD array.

Moreover, the emission of QDs originated from the surficial recombination of excited electrons, thus the blocking of this process also could result in the severe quenching of QDs. Liang et al. [97] reported functionalized MA-coated CdSe QDs for determining Ag (I) ions in aqueous solutions. Owing to the ultra-small diameter of QDs, aggregation is the main problem in the preparation and utilization processes, which could lead to the total quenching of the fluorescent emission. BSA as a protein could make CdSe QDs stable in water and improve the fluorescent emission. Functionalized MA-coated CdSe QDs could efficiently and rapidly detect Ag (I) ions at pH 5.0 in a linear range of 0.4–15 μM with a detection limit of 70 nM. When the Ag (I) ion was introduced into the probe solution, the surface Cd (II) ion on the CdSe QDs was displaced by the Ag (I) owing to the low solubility of Ag_2Se, which also triggered the apparent quenching of the fluorescent emission. Based on this model, Chen et al. [56] also developed L-cysteine-capped CdS QDs for detecting heavy and transition metal ions in water through a facile seed-assisted route. The obtained QDs exhibited different response abilities to five metal ions, where the added Ag (I) efficiently enhanced the emission intensity of the CdS QDs, and Hg^{2+}, Cu^{2+}, Co^{2+}, and Ni^{2+} apparently quenched the fluorescence. Other metal ions have slight influences on the emission of CdS QDs. The phenomenon might stem from the stronger affinity between metal ions and sulfur in comparison to CdS. The solubility product constants (K_{sp}) of relative metallic sulfide, $K_{sp}(Ag_2S) = 6.0 \times 10^{-50}$, $K_{sp}(CoS) = 2.0 \times 10^{-25}$, $K_{sp}(Cu_2S) = 3.0 \times 10^{-48}$, $K_{sp}(CuS) = 6.0 \times 10^{-36}$, $K_{sp}(NiS) = 1.4 \times 10^{-24}$, and $K_{sp}(HgS) = 1.6 \times 10^{-52}$, are much lower than that of CdS, K_{sp} $(CdS) = 7.1 \times 10^{-28}$. The introduction of Hg^{2+} and Cu^{2+} ions results in the formation of ultra-small clusters on the surface of CdS QDs by replacing the Cd^{2+} ions, thus undermining the surface emission of CdS QDs and the redshift of the luminescent emission. In contrast, the addition of Ag (I) led to the formation of an Ag_2S phase on the surface, which could efficiently suppress the nonradiative recombination at the ligands, thus improving the emission.

To further improve the emission intensity and anchor more functional ligands, hybridized systems have been intensively developed. Han et al. [98] fabricated a Fe_2O_3-wrapped CdTe QDs system for the detection of Cu^{2+} and Ag^{+} ions in water. The CdTe QDs were first prepared through a developed route, and then added into the precursor solution of Fe_2O_3 using a microemulsion method. The fluorescent CdTe QDs were wrapped into Fe_2O_3 superparticles (SPs) and continued to emit strong luminescence despite the apparent redshifting. The prepared hybridized system exhibited great response abilities and selectivity to Cu^{2+} and Ag^{+} in comparison with other metal ions. Simultaneously, this assembled model and strategy was further expanded to two emissions for bioimaging. This work could enlighten the development of multifunctional luminescent materials for sensing and imaging by combing various QDs and substances based on the intended purposes. Li et al. [99] reported a QD/DNA/Au chemosensor to detect mercury ions via the energy transfer mechanism.

Apart from the solution form, solid membrane nanosensors have also been studied in the past. For these QD-based film nanosensors, the key point is to balance the fabrication of the film and the emission of the QDs. During the preparation process, it was revealed that the emission of QDs can be severely quenched when the distance between two QDs is too small due to the nonradiative recombination of excited electrons. This is the reason that QDs are usually kept in the form of a solution and are passivated by organic ligands. Nevertheless, the fabrication of film is to transform liquid to solid state, which can immobilize QDs in the membrane. It inevitably undermines the emission of QDs. Therefore balancing the film fabrication and the QDs is important for assembling QD-based film nanosensors. Wang et al. [100] synthesized a multilayer film sensor consisting of CdTe QDs and poly(dimethyl-diallyl ammonium chloride) (PDDA) for the detection of mercury ions. The water-soluble, stable, monodisperse MA-coated CdTe QDs were fabricated by the developed route, and then deposited on the quartz slides accompanied by the addition of PDDA. The PDDA/QD film presents a strong and stable photoluminescence emission that is higher than that of bare QDs due to the fine controlling of the distance between each QD. The obtained film exhibited a considerable linear determination ability to Hg^{2+} ions in the range of 0.01–1 μM despite other metal ions also weakening the emission intensity of the fluorescent film. It can be envisaged that the attempt of combining QDs with membrane is to extremely promote the practical application of QD-based nanosensors in real aqueous samples. Moreover, it is clearly observed that the QD-embedded film sensors were easily affected by variations in pH values, which should be overcome in future studies.

3.3.2 III–V Nanocrystals for Nanosensors

It can be observed that for II–VI NCs, the Cd-included QDs account for a major proportion of QD-based nanosensors due to their stable luminescence, high quantum yield, and developed synthesis route. However, the toxicity of Cd is always a concerning issue, especially in biosensing and the biomedical field. Although some literatures have reported that passivated Cd-based QDs by organic ligand [101] or ZnS inorganic shell only release trace Cd ions into liquid phase, the worries deriving from uncertain complex environment in real water or living thing still limit their practical usage widely. Therefore developing low toxic even nontoxic element QDs is a promising strategy for QD-based nanosensors. From this viewpoint, III–V QDs including InP [102,103], InN [104,105], carbon QDs [106–108], and silicon QDs [109–111] have been paid more attention and are widely studied. Herein, carbon and silicon are not categorized as semiconductors so systematical studies on nanosensors constituted by carbon or silicon QDs are ruled out.

Like II–VI QDs, the assembly process of nanosensors using III–V QDs focuses on the preparation and specific functionalization of QDs. Zhu et al. [112] prepared a novel near-infrared chemosensor consisting of water-soluble MA-coated InP NCs for probing mercury ions in aqueous solutions. Li et al. [113] designed a novel colorimetric sensor for detecting Cu (II) through the photoreductive ability of QDs under visible light irradiation. Colorless TMB can be oxidized to blue TMB as an oxidation product under UV illumination, while the introduction of InP/ZnS QDs could release photoinduced electrons that will reduce oxidized blue TMB to colorless TMB. Therefore conjugating TMB with InP–ZnS core–shell QDs is done in order to produce a colorless probe solution. On the other hand, interestingly, the presence of Cu (II) efficiently deposited on the surface of InP/ZnS QDs formed Cu-doped InP/ZnS QDs in which the photogenerated electrons are consumed, thus resulting in a decrease of the photoreductive ability. Finally, the probe solution became blue due to the oxidation of the colorless TMB. The color is closely related to the concentration of Cu (II) ions in the aqueous solution. Furthermore, the designed nanosensor exhibited a high selective to Cu (II) ions and a partial response to Hg (II) ions. The absorbance increased with increases in the Cu (II) concentration in the linear range of 0.025–2.5 μM with a detection limit of 10 nM, which is comparable with ICP spectrometry.

3.3.3 Ternary Nanocrystals

In comparison with binary QDs, ternary QDs consisting of low toxic elements in the sensing area alone have been paid attention in the past few decades due to the low quantum yield and immature synthetic procedures. However, preparation strategies and tools have been developed with a deep understanding of the NC growth theory, which gradually substitutes the traditional binary as promising emitters for LED, sensing, and imaging. Based on the improvements in synthetic techniques, high-quality ternary NCs have been developed to assemble chemosensors for probing metal ions [114] and organic contaminants [115–117] in aqueous solutions in the past. Liu et al. [118] prepared water-soluble MPA-capped $CuInS_2$ ternary QDs in an aqueous solution for detecting Cu (II) and Cd (II) ions. Owing to the passivation of the MPA, the ternary $CuInS_2$ QDs proved to be stable in the aqueous solution, accompanied by a strong luminescent emission. In the presence of Cu (II) ions, the emission of the $CuInS_2$ QDs was apparently quenched, while the photoluminescence intensity of the $CuInS_2$ QDs was enhanced by the addition of Cd (II) ions. It was demonstrated that the introduced Cu (II) was reduced to Cu (I), leading to the formation of Cu_2S on the surface of the ternary $CuInS_2$ QDs and blocking the trap emission which is the primary emission path of ternary $CuInS_2$ QDs. In contrast, the addition of Cd (II) ions resulted in the activation of surface states, which was ascribed to the adsorption interaction between the Cd (II) and the $CuInS_2$ QDs by analyzing the adsorption model. The adsorbed Cd (II) ions were anchored at the surface binding sites of the QDs, thus enhancing the emission. The MPA-coated $CuInS_2$ QDs exhibited a high sensitivity to Cu (II) and Cd (II) with good linear ranges of 0.2–10 and 0.8–72 μM, respectively. The limit of detection for Cu (II) and Cd (II) were 0.1 and 0.19 μM, respectively.

For ternary NCs, the photoluminescence results from the surface recombination of photoexcited electrons and holes at the surface defects at ground state, while the surface defects are not controllable in comparison with intrinsic emission, which is easily affected by the preparation procedure. Meanwhile, the functionalization of QDs involving in phase transfer also influences the emission property. Therefore the deposition of inorganic shell is an effective approach to overcome the fluctuation of the emission. Zi et al. [119] synthesized TGA-coated $CuInS_2$–ZnS core–shell QDs for the assembly of a cobalt ion nanosensor through the cation exchange route. Compared with bare $CuInS_2$ QDs, the characteristic emission peak of the $CuInS_2$/ZnS QDs shifted to low wavelength and the intensity was significantly enhanced due to the formation of core/shell structure. It was demonstrated that because the partial Cu^+

ions in the $CuInS_2$ QDs were replaced by Zn^{2+}ions, a ZnS shell grew on the surface of the $CuInS_2$ core, resulting in a smaller core size with blueshift and enhanced emission [120]. With an increase in Co (II) concentration, the fluorescent emission of the TA-coated $CuInS_2$–ZnS core–shell QDs apparently decreased. The metal nanosensor displayed a wide linear range from 0.3 to 90 μM with a detection limit ($S/N = 3$) of 0.16 μM. It was concluded that the quenching mechanism stems from the interaction between the TA and the Co (II), where the Co (II) ions preferentially coordinated with the TA molecules stripping from the QDs, thus leading to the aggregation of the QDs.

Apart from introducing multifunctional ligands to meet the demands, combining ternary NCs with other NMs not only integrates the unique emission properties of the NCs, but also suppresses negative effects from the functionalization procedure. It was reported that by means of inorganic NCs as "artificial atoms" to fabricate SPs, one can realize the enhancement of physical properties of single NCs and integration of the unique properties of the subunits, which has become a novel and efficient way to employ advanced functional materials in sensing and catalysis fields. Sun et al. [121] successfully attempted to combine $CuInS_2$/ZnS QD SPs with amphibious Fe_3O_4 nanoparticles and ultimately achieved the coupling of strong photoluminescence and permanent magnetism in one system. It was observed that the fluorescence intensity was decreased significantly when the ternary NCs were transferred from nonpolar cyclohexane to polar ethanol, which is in agreement to the binary NCs. Nevertheless, the purpose of this study was to explore the combination of multiple components with unique properties in one system, which could endow the novel system with more functions such as bioimaging, orientational movement, and sensing.

3.4 In Vivo and In Vitro Sensing of Metal Ions

In Section 3.3, a host of applications in metal ion detection in natural or simulated aqueous solutions are mainly discussed. However, sensing metal ions in vivo or in vitro is another challenging field because more factors need to be taken into consideration in the design and fabrication of biosensors such as biocompatibility, overall sizes, cellular environment, and interferents. Therefore it is not feasible to fabricate biosensors according to the design principle of metal ion nanosensor in bare aqueous solution. In in vivo and in vitro nanosensors, the FRET tool has the highest usage frequency because its sensitivity and assembly procedure are suitable for determining metal ions inside cells or tissues. Based on the FRET strategy, Ge et al. [122] reported an upconversion nanoparticle (UCNPs)–$HmSiO_2$–polyetherimide (PEI) nanosensor for probing Hg^{2+} ions in vitro by taking advantage of the upconversion property of rare earth–doped UCNPs. The upconversion process could convert long-wavelength incident light into short-wavelength emission based on a large anti-Stokes' shift. In the original system, the photons emitting from the upconversion of the UCNPs under excitation at 980 nm could be absorbed by the Ru complex, displaying a low emission intensity, while in the presence of Hg^{2+}, the absorption peak of the Ru complex moves to the low wavelength due to the strong interaction between Ru complex and Hg^{2+} ion, leading to the enhancement of the emission intensity of the UCNPs–$HmSiO_2$–PEI in the aqueous solution. This core–shell nanosensor displays sensitive probing for Hg^{2+} with a detection limit of 0.16 μM as well as good biocompatibility and low cytotoxicity, and is further used for in vivo MRI in Kunming mouse as an MRI agent. Gao et al. [123] prepared a carbon-based QD sensor that has a great organ penetration depth and low tissue photodamage because the carbon QDs possess a unique and suitable red emission. In addition, the carbon QDs showed better biocompatibility in comparison with the mentioned semiconductor NCs. The prepared carbon QDs exhibited a 12.1% quantum yield, good stability against photobleaching that dye and QDs often encounter, and low cytotoxicity. When Pd^{2+}, Pt^{2+}, and Au^{3+} were introduced into the carbon QD solution, the fluorescent emission of the carbon QDs was quenched apparently, where the detection of limit for Pt^{2+}, Au^{3+}, and Pd^{2+} were estimated to be 0.886, 3.03, and 3.29 μM, respectively. Nevertheless, by means of time-correlated single photon counting experiments, it was found that the quenching mechanism between the QDs and $Pt^{2+}$was different from that of Au^{3+} and Pd^{2+}. For the Pt^{2+}, the obvious reduction of the fluorescence lifetime demonstrates that the quenching of the emission can be ascribed to the fast electron transfer between the carbon QDs and the Pt^{2+}; in turn, no significant changes of the fluorescence lifetime was observed for Au^{3+} and Pd^{2+}, which is mainly assigned to the occurrence of a strong inner filter effect after the addition of Au^{3+} or Pd^{2+} into the carbon QDs solution. In addition, the red emissive carbon QDs were fabricated to probe noble-metal ions in vivo through incubating PC12 cells with carbon QDs. After incubation, the fluorescent carbon QDs were observed in the PC12 cells. When Pt^{2+} was added into the PC12 cell solution, it was clearly found that the red fluorescence was gradually quenched with increases in the Pt^{2+} concentration, which indicates that the carbon QDs could successfully detect Pt^{2+} in the cells in vivo.

3.5 The Future of Advanced Quantum Dot–Based Nanosensors

From the summary provided, it is clearly observed that the preparation of QDs and the exploration of functionalizing applications for metal determination have been well studied in the past few decades due to the significant advantages such as high sensitivity and selectivity, low dosage and consumption, versatile functions and categories. In fact, until now, some QD-based biosensors for blood sugar probing and immune screening have been commercialized by biotechnology companies and enterprises. However, although QD-based metal nanosensors have promising potential applications, their practical and wide commercialization is still a long way off. On the one hand, their reliable and stable lab performance should be further improved by designing novel and feasible approaches to functionalization. Owing to the variety of target substances, the functionalized procedures have to be confirmed repeatedly for gaining a stable sensing performance even in the presence of various interferent effects such as pH value, suspended particles, organics, etc., which have to be investigated systematically during the performance evaluation procedure. Therefore expected interference experiments must be included in the evaluation system to better judge the efficacy of QD-based nanosensors.

Above all, there is a prerequisite that signal source QDs must have a high quantum yield, strong photoluminescence, photostability, and low toxicity. It is undoubted that QD-based biosensors are often injected into living things, and thereby must have low side effects on acceptors. Otherwise, they will hardly get approval from supervision departments. It is well known that toxicity experiments often need to be conducted over a long period of time. Therefore it required that QDs should contain low toxicity elements, which in turn has pushed the development of Cd-free QDs forward over the past few decades, though the optical performances of Cd-free QDs are still inferior at this stage. Among Cd-free QDs, ternary even quaternary QDs are a promising choice due to their various compositions in comparison to binary QDs. Meanwhile, ternary or quaternary

QDs often exclude high toxic metal elements such as Cd, Pb, Hg, which can easily be accepted. Hopefully, in the coming years, these requested QDs will be expanded upon.

References

[1] C.C. Yang, Y.W. Mai, Size-dependent absorption properties of CdX (X = S, Se, Te) quantum dots, Chem. Phys. Lett. 535 (2012) 91–93.

[2] X. Peng, An essay on synthetic chemistry of colloidal nanocrystals, Nano Res. 2 (2010) 425–447.

[3] D. Mocatta, G. Cohen, J. Schattner, O. Millo, E. Rabani, U. Banin, Heavily doped semiconductor nanocrystal quantum dots, Science 332 (2011) 77–81.

[4] D.J. Norris, A.L. Efros, S.C. Erwin, Doped nanocrystals, Science 319 (2008) 1776–1779.

[5] R. Fainblat, C.J. Barrows, D.R. Gamelin, Single magnetic impurities in colloidal quantum dots and magic-size clusters, Chem. Mater. 29 (2017) 8023–8036.

[6] N.J. Borys, M.J. Walter, J. Huang, D.V. Talapin, J.M. Lupton, The role of particle morphology in interfacial energy transfer in CdSe/CdS heterostructure nanocrystals, Science 330 (2010) 1371–1374.

[7] G. Dong, H. Wang, G. Chen, Q. Pan, J. Qiu, Quantum dot-doped glasses and fibers: fabrication and optical properties, Front. Mater. 2 (2015) 13.

[8] P. Ramasamy, N. Kim, Y.S. Kang, O. Ramirez, J.S. Lee, Tunable, bright, and narrow-band luminescence from colloidal indium phosphide quantum dots, Chem. Mater. 29 (2017) 6893–6899.

[9] S. Kim, B. Fisher, H.J. Eisler, M. Bawendi, Type-II quantum dots: CdTe/CdSe (core/shell) and CdSe/ZnTe (core/shell) heterostructures, J. Am. Chem. Soc. 125 (2003) 11466–11467.

[10] R. Jose, Z. Zhelev, T. Nagase, R. Bakalova, Y. Baba, M. Ishikawa, Self-surface passivation of CdX (X = Se, Te) quantum dots, J. Nanosci. Nanotechnol. 6 (2006) 618–623.

[11] P. Singhal, P. Maity, S.K. Jha, H.N. Ghosh, Metal-ligand complex-induced ultrafast charge-carrier relaxation and charge-transfer dynamics in CdX (X = S, Se, Te) quantum dots sensitized with nitrocatechol, Chem. Eur. J. 23 (2017) 10590–10596.

[12] Y. Chen, J. Vela, H. Htoon, J.L. Casson, D.J. Werder, D.A. Bussian, et al., Giant multishell CdSe nanocrystal quantum dots with suppressed blinking, J. Am. Chem. Soc. 130 (2008) 5026–5027.

[13] J. Zhu, Z.J. Zhao, J.J. Li, J.W. Zhao, CdTe quantum dot-based fluorescent probes for selective detection of Hg (II): the effect of particle size, Spectrochim. Acta A Mol. Biomol. Spectrosc. 177 (2017) 140–146.

[14] A. Kongkanand, K. Tvrdy, K. Takechi, M. Kuno, P.V. Kamat, Quantum dot solar cells. tuning photoresponse through size and shape control of CdSe-TiO_2 architecture, J. Am. Chem. Soc. 130 (2008) 4007–4015.

[15] J. Yang, J.S. Son, J.H. Yu, J. Joo, T. Hyeon, Advances in the colloidal synthesis of two-dimensional semiconductor nanoribbons, Chem. Mater. 25 (2013) 1190–1198.

[16] A.C. Bartnik, A.L. Efros, W.K. Koh, C.B. Murray, F.W. Wise, Electronic states and optical properties of PbSe nanorods and nanowires, Phys. Rev. B 82 (2010) 4706–4712.

[17] X. Peng, L. Manna, W. Yang, J. Wickham, E. Scher, A. Kadavanich, et al., Shape control of CdSe nanocrystals, Nature 404 (2000) 59–61.
[18] L. Manna, E. Scher, A.P. Alivisatos, Synthesis of soluble and processable rod-, arrow-, teardrop-, and tetrapod-shaped CdSe nanocrystals, J. Am. Chem. Soc. 122 (2000) 12700–12706.
[19] L. Manna, D.J. Milliron, A. Meisel, E.C. Scher, A.P. Alivisatos, Controlled growth of tetrapod-branched inorganic nanocrystals, Nat. Mater. 2 (2003) 382–385.
[20] Z.A. Peng, X. Peng, Mechanisms of the shape evolution of CdSe nanocrystals, J. Am. Chem. Soc. 123 (2001) 1389–1395.
[21] H. Kumar, P.B. Barman, R.R. Singh, Effect of size and shell: enhanced optical and surface properties of CdS, ZnS and CdS/ZnS quantum dots, Phys. E 67 (2015) 168–177.
[22] J. Zhu, S.N. Wang, J.J. Li, J.W. Zhao, The effect of core size on the fluorescence emission properties of CdTe@CdS core@shell quantum dots, J. Lumin. 199 (2018) 216–224.
[23] C.K. De, T. Routh, D. Roy, S. Mandal, P.K. Mandal, Highly photoluminescent InP based core alloy shell QDs from air-stable precursors: excitation wavelength dependent photoluminescence quantum yield, photoluminescence decay dynamics, and single particle blinking dynamics, J. Phys. Chem. C 122 (2018) 964–973.
[24] E.A. Baquero, H. Virieux, R.A. Swain, A. Gillet, A. Cros-Gagneux, Y. Coppel, et al., Synthesis of cxide-free InP quantum dots: surface control and H_2-assisted growth, Chem. Mater. 29 (2017) 9623–9627.
[25] L. Li, P. Reiss, One-pot synthesis of highly luminescent InP/ZnS nanocrystals without precursor injection, J. Am. Chem. Soc. 130 (2008) 11588–11589.
[26] L. Zhang, X.Q. Yang, J. An, S.D. Zhao, T.Y. Zhao, F. Tan, et al., In vivo tumor active cancer targeting and CT-fluorescence dual-modal imaging with nanoprobe based on gold nanorods and InP/ZnS quantum dots, J. Mater. Chem. B 6 (2018) 2574–2583.
[27] V. Chandrasekaran, M.D. Tessier, D. Dupont, P. Geiregat, Z. Hens, E. Brainis, Nearly blinking-free, high-purity single-photon emission by colloidal InP/ZnSe quantum dots, Nano Lett. 17 (2017) 6104–6109.
[28] K.R. Reid, J.R. McBride, N.J. Freymeyer, L.B. Thal, S.J. Rosenthal, Chemical structure, ensemble and single-particle spectroscopy of thick-shell InP-ZnSe quantum dots, Nano Lett. 18 (2018) 709–716.
[29] E.P. Jang, J.H. Jo, M.S. Kim, S.Y. Yoon, S.W. Lim, J. Kim, et al., Near-complete photoluminescence retention and improved stability of InP quantum dots after silica embedding for their application to on-chip-packaged light-emitting diodes, RSC Adv. 8 (2018) 10057–10063.
[30] G. Huang, C. Wang, S. Xu, Z. Qi, C. Lu, Y. Cui, Ag- and Mn-doped ZnInS/ZnS dual-emission quantum dots with zone tunability in the color coordinate, Nanotechnology 27 (2016) 185602.
[31] S. Bhowal, A. Ghosh, S.P. Chowdhuri, R. Mondal, B.B. Das, A novel metallogel based approach to synthesize (Mn, Cu) doped ZnS quantum dots and labeling of MCF-7 cancer cells, Dalton Trans. 47 (2018) 6557–6569.
[32] L.Q. Lu, P. Deng, Y.H. Ma, L.L. Wang, F. Peng, X.K. Tian, Room-temperature phosphorescence detection of silver ions based on glutathione capped Mn-doped ZnS quantum dots, J. Nanosci. Nanotechnol. 17 (2017) 6759–6764.
[33] T. Gan, N. Zhao, G. Yin, M. Tu, J. Liu, W. Liu, Mercaptopropionic acid-capped Mn-doped ZnS quantum dots as a probe for selective room-

temperature phosphorescence detection of Pb^{2+} in water, New J. Chem. 41 (2017) 13425–13434.

[34] S.J. Lim, L. Ma, A. Schleife, A.M. Smith, Quantum dot surface engineering: toward inert fluorophores with compact size and bright, stable emission, Coordin. Chem. Rev. 320–321 (2016) 216–237.

[35] F. Zhu, J. Zhu, Z. Zhang, Selective detection of glufosinate using $CuInS_2$ quantum dots as a fluorescence probe, RSC Adv. 7 (2017) 48077–48082.

[36] Z. Lin, X. Fei, Q. Ma, X. Gao, X. Su, $CuInS_2$ quantum dots@silica near-infrared fluorescent nanoprobe for cell imaging, New J. Chem. 38 (2014) 90–96.

[37] S.H. Park, A. Hong, J.H. Kim, H. Yang, K. Lee, H.S. Jang, Highly bright yellow-green-emitting $CuInS_2$ colloidal quantum dots with core/shell/shell architecture for white light-emitting diodes, ACS Appl. Mater. Interfaces 7 (2015) 6764–6771.

[38] C. Zhao, Z. Bai, X. Liu, Y. Zhang, B. Zou, H. Zhong, Small GSH-capped $CuInS_2$ quantum dots: MPA assisted aqueous phase transfer and bio-imaging applications, ACS Appl. Mater. Interfaces 7 (2015) 17623–17629.

[39] G. Wang, H. Wei, J. Shi, Y. Xu, H. Wu, Y. Luo, et al., Significantly enhanced energy conversion efficiency of $CuInS_2$ quantum dot sensitized solar cells by controlling surface defects, Nano Energy 35 (2017) 17–25.

[40] C. Cai, L. Zhai, Q. Wu, Y. Ma, L. Zhang, Y. Yang, et al., Tailoring defects of $CuInS_2$ quantum dots for sensitized solar cells, J. Alloy. Compd 719 (2017) 227–235.

[41] G. Yue, S. Su, N. Li, M. Shuai, X. Lai, D. Astruc, et al., Gold nanoparticles as sensors in the colorimetric and fluorescence detection of chemical warfare agents, Coordin. Chem. Rev. 311 (2016) 75–84.

[42] H. Wei, S.M. Hossein Abtahi, P.J. Vikesland, Plasmonic colorimetric and SERS sensors for environmental analysis, Environ. Sci. Nano 2 (2015) 120–135.

[43] R.M. Duke, E.B. Veale, F.M. Pfeffer, P.E. Kruger, T. Gunnlaugsson, Colorimetric and fluorescent anion sensors: an overview of recent developments in the use of 1,8-naphthalimide-based chemosensors, Chem. Soc. Rev. 39 (2010) 3936–3953.

[44] H. Zhu, J. Fan, B. Wang, X. Peng, Fluorescent, MRI, and colorimetric chemical sensors for the first-row d-block metal ions, Chem. Soc. Rev. 44 (2015) 4337–4366.

[45] B. Kaur, N. Kaur, S. Kumar, Colorimetric metal ion sensors - a comprehensive review of the years 2011–2016, Coordin. Chem. Rev. 358 (2018) 13–69.

[46] H.N. Kim, S.W. Nam, K.M. Swamy, Y. Jin, X. Chen, Y. Kim, et al., Rhodamine hydrazone derivatives as Hg^{2+} selective fluorescent and colorimetric chemosensors and their applications to bioimaging and microfluidic system, Analyst 136 (2011) 1339–1343.

[47] A.H. Gore, D.B. Gunjal, M.R. Kokate, V. Sudarsan, P.V. Anbhule, S.R. Patil, et al., Highly selective and sensitive recognition of cobalt(II) ions directly in aqueous solution using carboxyl-functionalized CdS quantum dots as a naked eye colorimetric probe: applications to environmental analysis, ACS Appl. Mater. Interfaces 4 (2012) 5217–5226.

[48] M. Koneswaran, R. Narayanaswamy, Mercaptoacetic acid capped CdS quantum dots as fluorescence single shot probe for mercury (II), Sens. Actuat. BChem. 139 (2009) 91–96.

[49] F. Zhang, L. Zeng, C. Yang, J. Xin, H. Wang, A. Wu, A one-step colorimetric method of analysis detection of Hg^{2+} based on an in situ formation of Au@HgS core-shell structures, Analyst 136 (2011) 2825–2830.
[50] N. Butwong, P. Kunthadong, P. Soisungnoen, C. Chotichayapong, S. Srijaranai, J.H.T. Luong, Silver-doped CdS quantum dots incorporated into chitosan-coated cellulose as a colorimetric paper test stripe for mercury, Microchim. Acta 185 (2018) 126.
[51] S. Tang, M. Wang, Z. Li, P. Tong, Q. Chen, G. Li, et al., A novel sensitive colorimetric sensor for Cu^{2+} based on in situ formation of fluorescent quantum dots with photocatalytic activity, Biosens. Bioelectron. 89 (2017) 866–870.
[52] Q. Yang, J. Li, X. Wang, H. Peng, H. Xiong, L. Chen, Strategies of molecular imprinting-based fluorescence sensors for chemical and biological analysis, Biosens. Bioelectron. 112 (2018) 54–71.
[53] J. Ke, X. Li, Y. Shi, Q. Zhao, X. Jiang, A facile and highly sensitive probe for Hg (II) based on metal-induced aggregation of ZnSe/ZnS quantum dots, Nanoscale 4 (2012) 4996–5001.
[54] E.M. Ali, Y. Zheng, H.-h Yu, J.Y. Ying, Ultrasensitive Pb^{2+} detection by glutathione-capped quantum dots, Anal. Chem. 79 (2007) 9452–9458.
[55] C. Dong, H. Qian, N. Fang, J. Ren, Study of fluorescence quenching and dialysis process of CdTe quantum dots, using ensemble techniques and fluorescence correlation spectroscopy, J. Phys. Chem. B 110 (2006) 11069–11075.
[56] J. Chen, A. Zheng, Y. Gao, C. He, G. Wu, Y. Chen, et al., Functionalized CdS quantum dots-based luminescence probe for detection of heavy and transition metal ions in aqueous solution, Spectrochim. Acta A Mol. Biomol. Spectrosc. 69 (2008) 1044–1052.
[57] H. Wang, L. Sun, Y. Li, X. Fei, M. Sun, C. Zhang, et al., Layer-by-layer assembled Fe_3O_4@C@CdTe core/shell microspheres as separable luminescent probe for sensitive sensing of Cu^{2+} ions, Langmuir 27 (2011) 11609–11615.
[58] L. Chang, X. He, L. Chen, Y. Zhang, A novel fluorescent turn-on biosensor based on QDs@GSH-GO fluorescence resonance energy transfer for sensitive glutathione S-transferase sensing and cellular imaging, Nanoscale 9 (2017) 3881–3888.
[59] Z.Q. Zhou, R. Yan, J. Zhao, L.Y. Yang, J.L. Chen, Y.J. Hu, et al., Highly selective and sensitive detection of Hg^{2+} based on fluorescence enhancement of Mn-doped ZnSe QDs by Hg^{2+}-Mn^{2+} replacement, Sensor. Actuat. B Chem. 254 (2018) 8–15.
[60] W. Liu, F. Wei, G. Xu, Y. Wu, C. Hu, Q. Song, et al., Ethylene diamine tetraacetic acid etched quantum dots as a "turn-on" fluorescence probe for detection of trace Zinc in food, J. Nanosci. Nanotechnol. 16 (2016) 6511–6519.
[61] K.E. Sapsford, L. Berti, I.L. Medintz, Materials for fluorescence resonance energy transfer analysis: beyond traditional donor-acceptor combinations, Angew. Chem. Int. Ed. 45 (2006) 4562–4589.
[62] T. Skajaa, Y. Zhao, D.J. van den Heuvel, H.C. Gerritsen, D.P. Cormode, R. Koole, et al., Quantum dot and Cy5.5 labeled nanoparticles to investigate lipoprotein biointeractions via Forster resonance energy transfer, Nano Lett. 10 (2010) 5131–5138.
[63] A. Clapp, I. Medintz, J.M. Mauro, B. Fisher, M. Bawendi, H. Mattoussi, Fluorescence resonance energy transfer between quantum dot donors and dye-labeled protein acceptors, J. Am. Chem. Soc. 126 (2004) 301–310.

[64] I.L. Medintz, A.R. Clapp, H. Mattoussi, E.R. Goldman, B. Fisher, J.M. Mauro, Self-assembled nanoscale biosensors based on quantum dot FRET donors, Nat. Mater. 2 (2003) 630–638.

[65] E. Oh, M.Y. Hong, D. Lee, S.H. Nam, H.C. Yoon, H.S. Kim, Inhibition assay of biomolecules based on fluorescence resonance energy transfer (FRET) between quantum dots and gold nanoparticles, J. Am. Chem. Soc. 127 (2005) 3270–3271.

[66] V.R. Hering, T.E.S. Faulin, E.R. Triboni, S.D. Rodriguez, D.L. Bernik, R.I. Schumacher, et al., Violet ZnSe/ZnS as an alternative to green CdSe/ZnS in nanocrystal-fluorescent protein FRET systems, Bioconj. Chem. 20 (2009) 1237–1241.

[67] A.H. Gore, M.B. Kale, P.V. Anbhule, S.R. Patil, G.B. Kolekar, A novel FRET probe for selective and sensitive determination of vitamin B12 by functionalized CdS QDs in aqueous media: applications to pharmaceutical and biomedical analysis, RSC Adv. 4 (2014) 683–692.

[68] Z. Lingley, S. Lu, A. Madhukar, A high quantum efficiency preserving approach to ligand exchange on lead sulfide quantum dots and interdot resonant energy transfer, Nano Lett. 11 (2011) 2887–2891.

[69] H. Sun, J. Zhang, Y. Tian, Y. Ning, H. Zhang, J. Ju, et al., Multifunctional composites obtained by incorporating nanocrystals into decorated PVK polymers, J. Nanomater. 2007 (2007) 1–7.

[70] F. Dubois, B. Mahler, B. Dubertret, E. Doris, C. Mioskowski, A versatile strategy for quantum dot ligand exchange, J. Am. Chem. Soc. 129 (2007) 482–483.

[71] W.R. Algar, U.J. Krull, Quantum dots as donors in fluorescence resonance energy transfer for the bioanalysis of nucleic acids, proteins, and other biological molecules, Anal. Bioanal. Chem. 391 (2008) 1609–1618.

[72] A.M. Dennis, W.J. Rhee, D. Sotto, S.N. Dublin, G. Bao, Quantum dot–fluorescent protein FRET probes for sensing intracellular pH, ACS Nano 6 (2012) 2917–2924.

[73] L. Rodriguez-Lorenzo, L. Fabris, R.A. Alvarez-Puebla, Multiplex optical sensing with surface-enhanced Raman scattering: a critical review, Anal. Chim. Acta 745 (2012) 10–23.

[74] P. Zheng, X. Shi, K. Curtin, F. Yang, N. Wu, Detection of mercury (II) with a surface-enhanced Raman scattering sensor based on functionalized gold nanoparticles, Mater. Res. Express 4 (2017) 055017.

[75] W. Zhou, B.C. Yin, B.C. Ye, Highly sensitive surface-enhanced Raman scattering detection of hexavalent chromium based on hollow sea urchin-like TiO_2@Ag nanoparticle substrate, Biosens. Bioelectron. 87 (2017) 187–194.

[76] L.G. Quagliano, Observation of molecules adsorbed on III-V semiconductor quantum dots by surface-enhanced raman scattering, J. Am. Chem. Soc. 126 (2004) 7393–7398.

[77] Y. Wang, J. Zhang, H. Jia, M. Li, J. Zeng, B. Yang, et al., Mercaptopyridine surface-functionalized CdTe quantum dots with enhanced Raman scattering properties, J. Phys. Chem. C 112 (2008) 996–1000.

[78] Yb Lee, S. Ho Lee, S. Lee, H. Lee, J. Kim, J. Joo, Surface enhanced Raman scattering effect of CdSe/ZnS quantum dots hybridized with Au nanowire, Appl. Phys. Lett. 102 (2013) 033109.

[79] N.P.W. Pieczonka, R.F. Aroca, Inherent complexities of trace detection by surface-enhanced Raman scattering, Chemphyschem. 6 (2005) 2473–2484.

[80] C. Kokkinos, A. Economou, Emerging trends in biosensing using stripping voltammetric detection of metal-containing nanolabels - a review, Anal. Chim. Acta 961 (2017) 12–32.
[81] S. Song, Y. Qin, Y. He, Q. Huang, C. Fan, H.Y. Chen, Functional nanoprobes for ultrasensitive detection of biomolecules, Chem. Soc. Rev. 39 (2010) 4234–4243.
[82] C. Kokkinos, A. Economou, P.S. Petrou, S.E. Kakabakos, Microfabricated tin-film electrodes for protein and DNA sensing based on stripping voltammetric detection of Cd (II) released from quantum dots labels, Anal. Chem. 85 (2013) 10686–10691.
[83] A.L. Sun, Y.F. Zhang, X.N. Wang, Sensitive voltammetric determination of DNA via a target-induced strand-displacement reaction using quantum dot-labeled probe DNA, Microchim. Acta 182 (2015) 1403–1410.
[84] X. Qin, L. Wang, Q. Xie, Sensitive bioanalysis based on in-situ droplet anodic stripping voltammetric detection of CdS quantum dots label after enhanced cathodic preconcentration, Sensors 16 (2016) 1342.
[85] M.J. Shiddiky, S. Rauf, P.H. Kithva, M. Trau, Graphene/quantum dot bionanoconjugates as signal amplifiers in stripping voltammetric detection of EpCAM biomarkers, Biosens. Bioelectron. 35 (2012) 251–257.
[86] F. Ahour, M. Taheri, Anodic stripping voltammetric determination of copper (II) ions at a graphene quantum dot-modified pencil graphite electrode, J. Iran. Chem. Soc. 15 (2017) 343–350.
[87] E. Punrat, C. Maksuk, S. Chuanuwatanakul, W. Wonsawat, O. Chailapakul, Polyaniline/graphene quantum dot-modified screen-printed carbon electrode for the rapid determination of Cr (VI) using stopped-flow analysis coupled with voltammetric technique, Talanta 150 (2016) 198–205.
[88] S. Mussa Farkhani, A. Valizadeh, Review: three synthesis methods of CdX (X = Se, S or Te) quantum dots, IET. Nanobiotechnol. 8 (2014) 59–76.
[89] D.V. Freitas, S.G.B. Passos, J.M.M. Dias, A. Mansur, S.M. Carvalho, H. Mansur, et al., Toward greener electrochemical synthesis of composition-tunable luminescent CdX-based (X = Te, Se, S) quantum dots for bioimaging cancer cells, Sensors Actuat. B Chem. 250 (2017) 233–243.
[90] J.K. Oh, Surface modification of colloidal CdX-based quantum dots for biomedical applications, J. Mater. Chem. 20 (2010) 8433–8445.
[91] X. Wang, Y. Lv, X. Hou, A potential visual fluorescence probe for ultratrace arsenic (III) detection by using glutathione-capped CdTe quantum dots, Talanta 84 (2011) 382–386.
[92] G. Chen, Z. Guo, G. Zeng, L. Tang, Fluorescent and colorimetric sensors for environmental mercury detection, Analyst 140 (2015) 5400–5443.
[93] Y.S. Xia, C.Q. Zhu, Use of surface-modified CdTe quantum dots as fluorescent probes in sensing mercury (II), Talanta 75 (2008) 215–221.
[94] H. Li, Y. Zhang, X. Wang, L-Carnitine capped quantum dots as luminescent probes for cadmium ions, Sensors Actuat. B Chem. 127 (2007) 593–597.
[95] Y. Long, D. Jiang, X. Zhu, J. Wang, F. Zhou, Trace Hg^{2+} analysis via quenching of the fluorescence of a CdS-encapsulated DNA nanocomposite, Anal. Chem. 81 (2009) 2652–2657.
[96] R. Freeman, T. Finder, I. Willner, Multiplexed analysis of Hg^{2+} and Ag^{+} ions by nucleic acid functionalized CdSe/ZnS quantum dots and their use for logic gate operations, Angew. Chem. Int. Ed. 48 (2009) 7818–7821.
[97] J.G. Liang, X.P. Ai, Z.K. He, D.W. Pang, Functionalized CdSe quantum dots as selective silver ion chemodosimeter, Analyst 129 (2004) 619–622.
[98] J. Han, X. Zhang, Y. Zhou, Y. Ning, J. Wu, S. Liang, et al., Fabrication of CdTe nanoparticles-based superparticles for an improved detection of Cu^{2+} and Ag^{+}, J. Mater. Chem. 22 (2012) 2679–2686.

[99] M. Li, Q. Wang, X. Shi, L.A. Hornak, N. Wu, Detection of mercury (II) by quantum dot/DNA/gold nanoparticle ensemble based nanosensor via nanometal surface energy transfer, Anal. Chem. 83 (2011) 7061–7065.
[100] C. Wang, J. Zhao, Y. Wang, N. Lou, Q. Ma, X. Su, Sensitive Hg (II) ion detection by fluorescent multilayer films fabricated with quantum dots, Sensor. Actuat. B Chem. 139 (2009) 476–482.
[101] N. Ma, J. Yang, K.M. Stewart, S.O. Kelley, DNA-passivated CdS nanocrystals luminescence, bioimaging, and toxicity profiles, Langmuir 23 (2007) 12783–12787.
[102] V. Brunetti, H. Chibli, R. Fiammengo, A. Galeone, M.A. Malvindi, G. Vecchio, et al., InP/ZnS as a safer alternative to CdSe/ZnS core/shell quantum dots: in vitro and in vivo toxicity assessment, Nanoscale 5 (2013) 307–317.
[103] A.L. Moreau, R. Janissen, C.A. Santos, L.A. Peroni, D.R. Stach-Machado, A. A. de Souza, et al., Highly-sensitive and label-free indium phosphide biosensor for early phytopathogen diagnosis, Biosens. Bioelectron. 36 (2012) 62–68.
[104] N.H. Alvi, P.E.D. Soto Rodriguez, V.J. Gómez, P. Kumar, G. Amin, O. Nur, et al., Highly efficient potentiometric glucose biosensor based on functionalized InN quantum dots, Appl. Phys. Lett. 101 (2012) 153110.
[105] Nu.H. Alvi, P.E.D.S. Rodriguez, V.J. Gómez, P. Kumar, M. Willander, R. Nötzel, Highly sensitive and fast anion-selective InN quantum dot electrochemical sensors, Appl. Phys. Express 6 (2013) 115201.
[106] Y. Dong, R. Wang, G. Li, C. Chen, Y. Chi, G. Chen, Polyamine-functionalized carbon quantum dots as fluorescent probes for selective and sensitive detection of copper ions, Anal. Chem. 84 (2012) 6220–6224.
[107] S. Liu, J. Tian, L. Wang, Y. Zhang, X. Qin, Y. Luo, et al., Hydrothermal treatment of grass: a low-cost, green route to nitrogen-doped, carbon-rich, photoluminescent polymer nanodots as an effective fluorescent sensing platform for label-free detection of Cu (II) ions, Adv. Mater. 24 (2012) 2037–2041.
[108] X.T. Zheng, A. Ananthanarayanan, K.Q. Luo, P. Chen, Glowing graphene quantum dots and carbon dots: properties, syntheses, and biological applications, Small 11 (2015) 1620–1636.
[109] S. Chinnathambi, S. Chen, S. Ganesan, N. Hanagata, Silicon quantum dots for biological applications, Adv. Healthc. Mater. 3 (2014) 10–29.
[110] N. Dhenadhayalan, H.L. Lee, K. Yadav, K.C. Lin, Y.T. Lin, A.H. Chang, Silicon quantum dot-based fluorescence turn-on metal ion sensors in live cells, ACS Appl. Mater. Interfaces 8 (2016) 23953–23962.
[111] J. Zhao, J. Deng, Y. Yi, H. Li, Y. Zhang, S. Yao, Label-free silicon quantum dots as fluorescent probe for selective and sensitive detection of copper ions, Talanta 125 (2014) 372–377.
[112] C. Zhu, L. Li, F. Fang, J. Chen, Y. Wu, Functional InP nanocrystals as novel near-infrared fluorescent sensors for mercury ions, Chem. Lett. 34 (2005) 898–899.
[113] H. Li, H. Bai, Q. Lv, W. Wang, Z. Wang, H. Wei, et al., A new colorimetric sensor for visible detection of Cu (II) based on photoreductive ability of quantum dots, Anal. Chim. Acta 1021 (2018) 140–146.
[114] Z. Liu, G. Li, Q. Ma, L. Liu, X. Su, A near-infrared turn-on fluorescent nanosensor for zinc (II) based on $CuInS_2$ quantum dots modified with 8-aminoquinoline, Microchim. Acta 181 (2014) 1385–1391.
[115] S. Liu, F. Shi, X. Zhao, L. Chen, X. Su, 3-Aminophenyl boronic acid-functionalized $CuInS_2$ quantum dots as a near-infrared fluorescence

probe for the determination of dopamine, Biosens. Bioelectron. 47 (2013) 379–384.
[116] S. Liu, J. Hu, H. Zhang, X. Su, $CuInS_2$ quantum dots-based fluorescence turn off/on probe for detection of melamine, Talanta 101 (2012) 368–373.
[117] S. Liu, F. Shi, L. Chen, X. Su, Bovine serum albumin coated $CuInS_2$ quantum dots as a near-infrared fluorescence probe for 2,4,6-trinitrophenol detection, Talanta 116 (2013) 870–875.
[118] S. Liu, Y. Li, X. Su, Determination of copper (II) and cadmium (II) based on ternary $CuInS_2$ quantum dots, Anal. Methods 4 (2012) 1365–1370.
[119] L. Zi, Y. Huang, Z. Yan, S. Liao, Thioglycolic acid-capped $CuInS_2$/ZnS quantum dots as fluorescent probe for cobalt ion detection, J. Lumin. 148 (2014) 359–363.
[120] J. Park, S.W. Kim, $CuInS_2$/ZnS core/shell quantum dots by cation exchange and their blue-shifted photoluminescence, J. Mater. Chem. 21 (2011) 3745–3750.
[121] X. Sun, K. Ding, Y. Hou, Z. Gao, W. Yang, L. Jing, et al., Bifunctional superparticles achieved by assembling fluorescent $CuInS_2$@ZnS quantum dots and amphibious Fe_3O_4 nanocrystals, J. Phys. Chem. C 117 (2013) 21014–21020.
[122] X. Ge, L. Sun, B. Ma, D. Jin, L. Dong, L. Shi, et al., Simultaneous realization of Hg^{2+} sensing, magnetic resonance imaging and upconversion luminescence in vitro and in vivo bioimaging based on hollow mesoporous silica coated UCNPs and ruthenium complex, Nanoscale 7 (2015) 13877–13887.
[123] W. Gao, H. Song, X. Wang, X. Liu, X. Pang, Y. Zhou, et al., Carbon dots with red emission for sensing of Pt^{2+}, Au^{3+}, and Pd^{2+} and their bioapplications in vitro and in vivo, ACS Appl. Mater. Interfaces 10 (2018) 1147–1154.

4

SEMICONDUCTOR NANOCRYSTALS FOR ENVIRONMENTAL CATALYSIS

Jun Ke

School of Chemistry and Environmental Engineering, Wuhan Institute of Technology, Wuhan, P.R. China

4.1 Introduction to the Basics of the Catalysis and Flexibility Provided by Nanocrystals

Catalysis refers to an increase in the rate of a chemical reaction through the introduction of an additional substance called a catalyst, which is not consumed in the catalyzed reaction and can continue to act repeatedly. The name catalysis is mainly involved in changes in the route to equilibrium. It is about reaction kinetics, not thermodynamics. It should be known that catalysts do not and cannot change the thermodynamic equilibrium. Imagine that you are crossing a mountain range on your way from A to B. You start at point A, and after walking for some hours you reach point B. The reaction thermodynamics describe your elevation at the starting point and at your destination. Conversely, the reaction kinetics depend on which pathway you have taken.

Catalysts may be classified as either homogeneous or heterogeneous. A homogeneous catalyst is one whose molecules are dispersed in the same phase (usually gaseous or liquid) as the reactant molecules such as organometallic compounds and enzymes. In contrast, in heterogeneous catalysis, the catalyst and reactant are separately located at different phases, where the reactants are firstly adsorbed onto the surface of solid catalyst and then finish the specific reaction, for example, V_2O_5 catalyst in sulfuric acid industry, V-W-Ti catalyst in automotive catalytic converter. At the bridge of these two categories, there is another type of catalyst called a semiheterogeneous catalyst,

Advanced Nanomaterials for Pollutant Sensing and Environmental Catalysis. DOI: https://doi.org/10.1016/B978-0-12-814796-2.00004-6

colloid, or nanocluster, whose synthetic path and formation mechanism are ascribed to colloidal chemistry. Meanwhile, it is noteworthy that colloidal catalysts are often used in a solution manner. That is why they are called "artificial atom" nanoparticles (NPs). In fact, they are still categorized as heterogeneous catalysts because they have an independent crystallinity and shape in reactive systems [1].

For the majority of commercial catalysts, an active component possessing very small diameter, for example, of approximately nanometer scale, is often highly dispersed on solid supports that have high surface area. Small particles with a diameter ranging from 1 to 20 nm are referred to as NPs. In heterogeneous catalysis systems, the active component plays a key role in achieving excellent performance, which requires extensive efforts to investigate and characterize the properties of various components using characterization tools. This is the study of nanoscience which combines physics, chemistry, materials, and chemical engineering. Based on nanoscience, the relationship between the local size, composition, structure, and catalyst performance should be discussed and understood, which is useful to provide principles for obtaining high catalytic activity and selectivity. Until now, it has been demonstrated that the size of active component significantly affects the catalytic activities because of the high sensitivity of surface properties and electronic structure to the size, named by quantum confinement effects. For example, an Au/TiO_2 composite was verified to be effective for oxidizing CO to CO_2 at ambient conditions, which can be utilized to mitigate the CO concentration in buildings. The study revealed that the catalytic activity of Au/TiO_2 was sensitive to the size of deposited Au NPs, which was attributed to the high oxidation ability of exposed Au atoms, as shown by the scanning transmission electron microscopy (STEM) image in Fig. 4.1 [2]. The ratio of exposed Au atoms increased with decreases in the Au particle size, resulting in an appropriate enhancement of catalytic performance.

In the clean energy industry, the photocatalytic H_2 evolution performance has been demonstrated to be closely correlated with the size of the photocatalyst used. Taking Pt nanocrystal (NCs) as an example, Kotani et al. [3] investigated how the sizes and shapes of Pt NPs affected the photocatalytic H_2 generation in the presence of 9-mesityl-10-methylacridinium (Acr^+-Mes) and dihydronicotinamide adenine dinucleotide. They found that the electron transfer (ET) from Acr^+-Mes to Pt was the key step to controlling the evolution of H_2, whereas the size and shape of Pt NPs had an important influence on the ET rate due to the

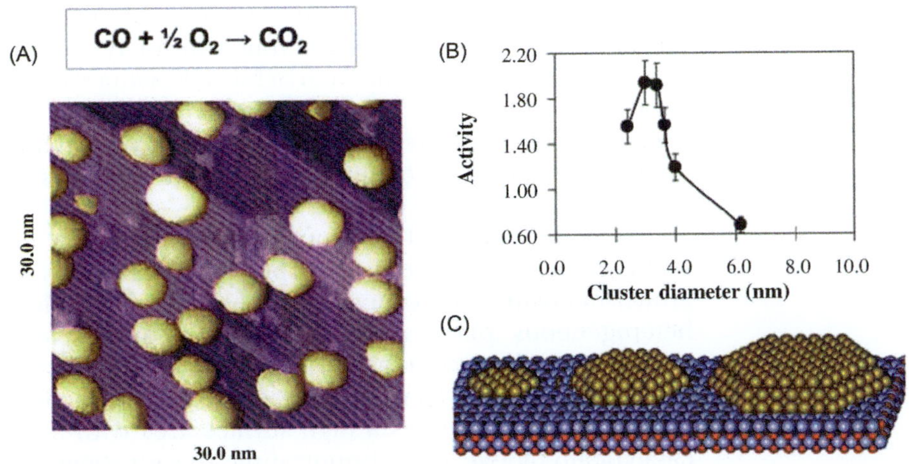

Figure 4.1 (A) STEM image of Au/TiO_2 catalysts, (B) activities versus diameters of catalysts, and (C) atomic illustration of Au/TiO_2 [2].

quantum size effect. It is well known that the quantum size effect, deriving from the decrease of size, is closely related to the electronic structure of Pt NPs, thus affecting the catalytic activity.

For a semiconductor material, it is well-known that one of the most important parameters is the energy gap width, that is, the energy difference between conduction bands (CBs) and valence bands (VBs) constituted by various energy levels. Typically, the width of this bandgap for a bulk semiconductor is a fixed and intrinsic value, depending on the material itself. However, intensive research has led to the discovery that once the size of the semiconductor decreases to less than ~20 nm, the situation of the fixed bandgap will change. In bulk semiconductors, these energy levels can be considered as continuous state because the difference in energy is negligible, where the excitation spectrum corresponds to the bandgap width. As the semiconductor is miniaturized to nanosize, the effect of dimension confining becomes more apparent. The averaged characteristics in bulk are no longer continuous and tend to be discrete and noncontinuous, corresponding to the quantized energy spectrum, which results in small and finite separation between energy levels. This phenomenon is known as the quantum size effect, where NCs are referred to as quantized nanoscale particles. Nowadays, developed semiconductor-based NCs include traditional binary metal chalcogenides such as CdX, ZnX, and PbX (where X = S, Se, or Te), ternary Cu-based NCs such as $CuInS_2$ and $AgInS_2$, and innovative metal oxides such as

TiO_2, Co_3O_4, and Cu_2O. From the occurrence of landmark NCs preparation in 1993, the breakthrough in finely controlling NPs and their size distributions was achieved owing to the usage of many phosphine-chalcogen precursors and coordinating solvent [4]. Over the past few decades, the essential components for synthesizing high-quality NCs have remained significantly unchanged.

Based on the well-developed synthetic route and skills, semiconductor NCs, either dispersed in aqueous media or grown/deposited on supports, have been at the forefront of heterogeneous photocatalysis because many have favorable band gap and band edge positions, a high extinction coefficient, efficient free charge generation, a short carrier pathway to the surface as well as a high surface area with plentitude of catalytically active sites. Importantly, these properties are further tunable through modification of the size, shape, doping level, aggregation state, and surface functionalization of the NCs, allowing for several degrees of freedom in optimizing the material. Zhao et al., demonstrated that the sizes of CdSe NCs are closely related with the photocatalytic generation of H_2 from water splitting, as presented in Fig. 4.2A and B [5], which provided the first quantitative correlation between the size of NCs and photocatalytic activities and gives deep insight into the size-controlled interface reaction between semiconductor and solution. Owing to the sensitive dependence on the size of NCs, the bandgap and positions of the VB and CB shift to the corresponding directions. For example, the decreasing of sizes results in the enlarging of the bandgap, an upshift of CB and a downshift of VB, which affects the wavelength of effective excitation photons and the transfer path of photoexcited electrons and holes that generally diffuse to the interface and react with the substance adsorbed on the surface of NCs. Therefore the changes in the positions of the CB and VB directly influence the degree of difficulty of surface reactions. Taking H_2 production into consideration, as shown in Fig. 4.1C, the photoinduced electrons transfer to the surface of the CdSe NCs and reduce H^+ to H_2, and the holes in turn oxidize Na_2SO_3 to Na_2SO_4. When the CdSe NCs decrease to 2.6 nm, the position of the CB is more negative than that of the surface Cd^{2+} reduction and the VB position is more positive than that of SO_3^{2-}/SO_4^{2-}, which ensure that the H_2 generation reaction occurs continuously under visible light irradiation. Therefore for NC-based catalysts, the size has an important influence on photocatalytic performance due to the close relationship between surface properties and sizes, which unveils that the selection of NCs

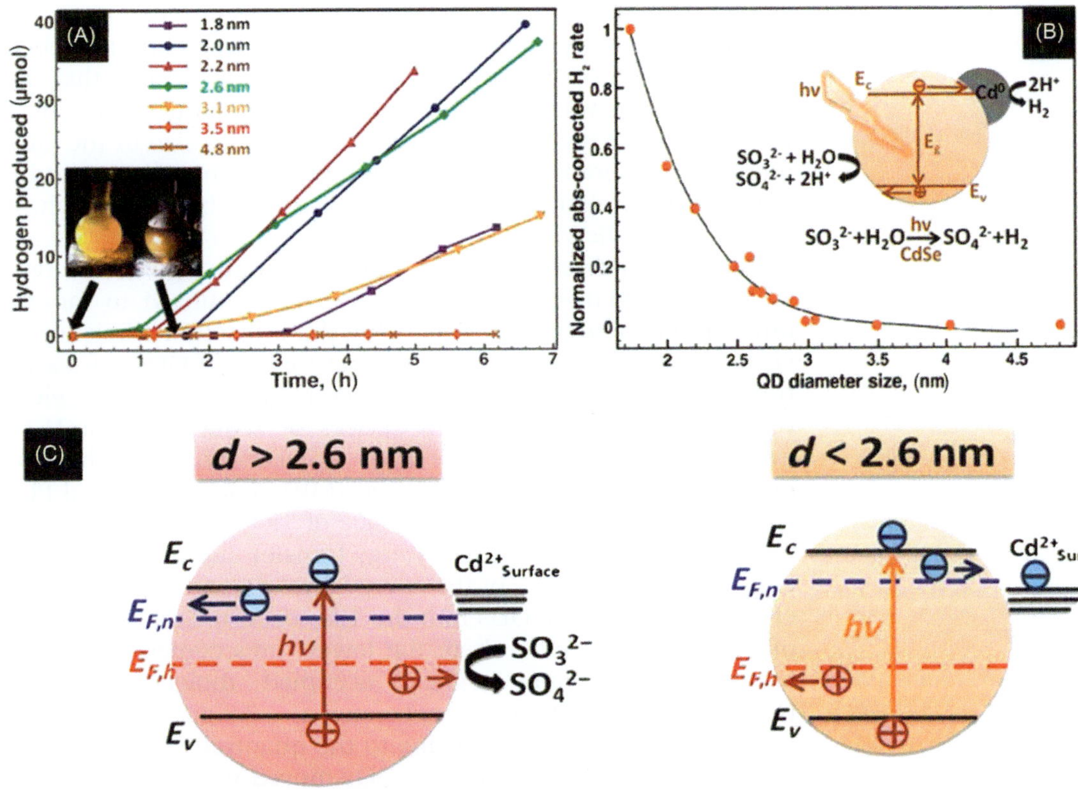

Figure 4.2 (A) Photocatalytic H_2 generation from CdSe quantum dots (QDs) in Na_2SO_3 (0.1 M) solution. Inset: visible color change of CdSe QD solution before (left) and during (right) irradiation; (B) relative H_2 production rates versus size of CdSe QDs; (C) schematic diagrams of the surface catalytic reaction switch when the size of QDs decreases from large to small [5].

should well match with the reaction system for achieving the highest catalytic conversion, especially in photocatalysis and photoelectrocatalysis.

NCs act as light harvesters and are excited by incident photons with an appropriate wavelength. The excited charge carriers are separated spatially and delocalized inside the NCs. To take part in a catalytic reaction, the ET process must be proceeded between photoexcited NCs and electron acceptors. Scholz et al. [6] revealed that the ET process from CdSe quantum dots (QDs) to the electron acceptor methylviologen (MV^{2+}) in a high concentration occurred on the femtosecond scale using ultrafast transient absorption spectroscopy, while the ET rate significantly reduced in a low concentration of electron acceptor. Meanwhile, they demonstrated that the ET rate was closely related with the size of the QDs, which depends on the contact probability between delocalized electrons and adsorbed MV^{2+} molecules. On the other hand, it was observable that the size of the QDs was correlated with the free energy of ET, where the free energy increased with decreases in the size of the QDs due to the difference in 1*S* transition energy. Therefore optimizing QDs in catalytic systems is necessary to obtain the highest possible catalytic yield.

Apart from directly injecting electrons into a given reaction system, as electron donors, QDs are often utilized to modify other semiconductors for enhancing the utilization efficiency of visible light. Photoexcited electrons at the CB of QDs can easily transfer to the CB of the semiconductor, which can output electronic energy in photovoltaic devices or reduce CO_2 or H_2O, even if irradiated by visible light or near-infrared light. For example, 3-mercaptopropionic acid (MPA)-coated colloidal PbS QDs were anchored onto the surface of TiO_2 NPs, which exhibited a strong size-dependent energy transfer and photoresponse current [7]. An analysis of the energy level demonstrated that an efficient ET from the QDs into the TiO_2 occurred for a diameter of QDs below 4.3 nm, where the lowest unoccupied molecular orbital (LUMO) and highest occupied molecular orbital (HOMO) levels of PbS QDs are located at -3.9 and -5.1 eV versus vacuum, respectively. Owing to the quantum size effect, the LUMO and HOMO level of PbS QDs moved to a more negative and positive potential, respectively, which facilitated the free energy of ET to TiO_2 NPs. When the diameter of the PbS QDs was 4.8 nm, ET from the QDs to the TiO_2 was not observed using time-integrated and steady-state fluorescence spectra, where no obvious changes were found between the colloidal MPA-capped PbS QDs and the PbS QDs/TiO_2. That is why photoluminescence is often used to indicate the recombination efficiency of photoinduced electrons and holes in catalytic systems. Based on the discussion above, it can be concluded that the size of QDs plays a key role in the intentional use of QDs to enhance catalytic performance. Therefore highly efficient, repeatable, and low-cost synthetic routes for high-quality QDs are highly desirable.

Combining with metal NPs is another intentional form of constructing NC-based systems for efficient catalytic activity. It is well known that many reactions are involved in multielectron procedures in nature, which inevitably need multiple sequential separations and the transfer of photoexcited charge carriers in one system, as described in Fig. 4.3 [8]. In addition, when the excited charge is not scavenged by other substances

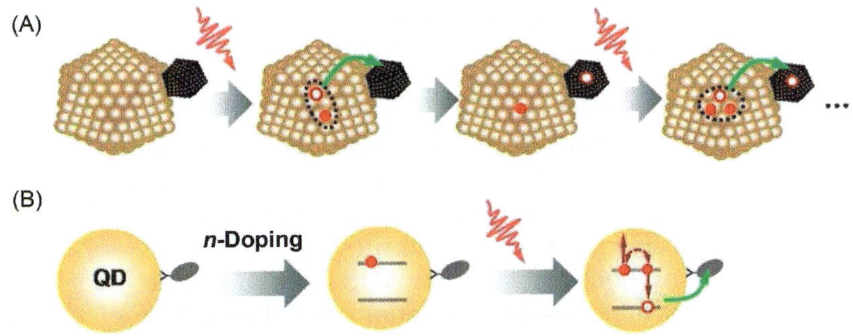

Figure 4.3 (A) Production mechanism of photoinduced excitons in multielectron photocatalysis; (B) simulating the charge transfer event in an *n*-doped quantum dot with an extra electron [8].

before absorbing the following photons, the NCs are charged and not neutral, which leads to sequent photoexcitation difficulty and limits multielectron photocatalysis. Therefore the absorption and excitation processes should be explored using time-resolved spectroscopies such as time-resolved transient absorption and fluorescence spectroscopy in order to understand the charge transfer from NCs to the capturing center including adsorbed substances, traps, and metallic NPs. Wang et al. [8] took advantage of time-resolved transient absorption spectroscopy to probe the charge transfer dynamics from *n*-doped CdS to phenothiazine as an electron acceptor. The results revealed that the Auger decay of excitons lowered the charge separation efficiency in comparison to neutral NCs without charge, which demonstrated that an efficient consumption of photoinduced charges is beneficial for enhancing the multielectron reaction.

4.2 Preparation and Characterization of Nanocrystals for Catalysis

Following on from the discussion above, NCs have an important effect on the catalytic activity of heterogeneous catalysts and thus the quality of prepared NCs directly affects catalyst performance. In this section, the preparation routes and characterization tools for obtaining controllable, high-quality NCs are further discussed along with enhanced catalytic activities in various catalytic systems, for example, water splitting and CO_2 reduction.

4.2.1 Compositions

4.2.1.1 Metal Chalcogenides Binary Nanocrystals

Chalcogen elements are located at the VIIA column of periodic table, including three elements, sulfur (S), selenium (Se), and tellurium (Te). Metal elements mainly contain transition metals, for example, Cd, Zn, Cu, In, Hg, Pb, and others. In 1993, Murray et al. [9] finished a pioneering work on synthesis of monodisperse binary CdE NCs (E = S, Se, or Te). After that, the synthesis and applications of Cd-based NCs for desired bandgap, good quantum yield, and strong fluorescent emission have been extensively explored in the past decades. Previous literatures have demonstrated that high quality CdE should be prepared using tri-*n*-octylphosphine (TOP) as chalcogen precursors and trioctylphosphine oxide (TOPO) as the coordinating solvent. Peng et al., reported a series of synthetic routes to prepare high-quality CdE NCs under elevated temperatures (~300°C) through adjusting the Cd-precursor and chalcogen-precursors [10,11]. When one precursor is rapidly injected into the other precursor solution, Cd and chalcogen quickly bind together to form high concentration CdE nuclei, which is known as the nucleation phase. Then the remaining precursor is continuously deposited onto the nuclei, resulting in an increase in the nuclei size, which is known as the growth phase. If the precursor reacts completely at the nucleation phase, partial nucleation would discompose and deposit on the other nuclei to allow bigger NCs to be formed. The nucleation and growth processes of colloidal NCs are described in

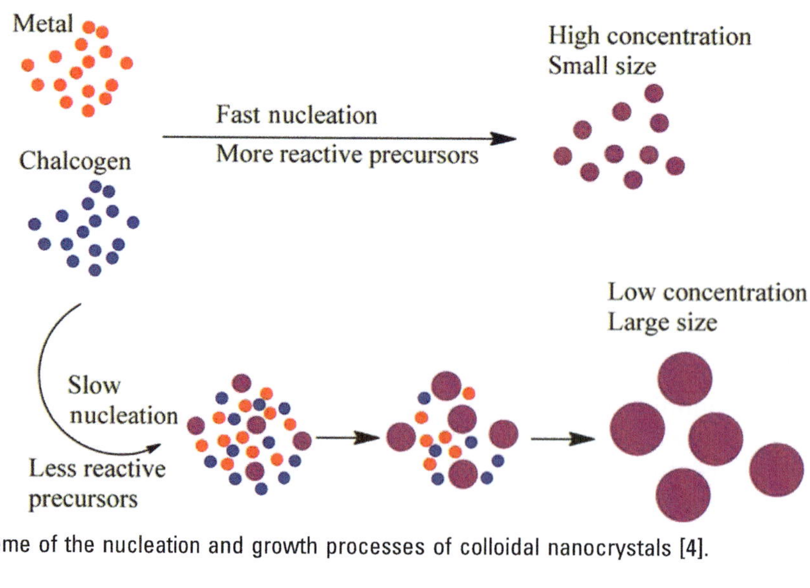

Figure 4.4 Scheme of the nucleation and growth processes of colloidal nanocrystals [4].

Fig. 4.4 [4]. It is observed that the size of NCs can be easily controlled by choosing different precursors and coordination solvents. When the precursors are highly reactive, they are consumed completely in a short time so that the size of the NCs is small and a focused distribution is achieved. In contrast, if the precursors are less reactive, the rest of the precursors are gradually deposited on the surface of NCs so the size is large. Therefore according to the purpose, corresponding precursors are selected to prepare high-quality NCs. Based on this principle, other non-Cd NCs have been widely prepared by nucleation and growth procedures such as ZnSe [12], InP [13], PbSe [14], and InAs [15]. After the well-established method, more recent studies have proved that colloidal QDs and their hybrid heterostructures can be applied as novel visible-light photocatalysts for energy conversion, CO_2 reduction, and environment treatment. Zhang et al. [16] took advantage of CdSe QDs to sensitize TiO_2 for efficient photocatalytic H_2 evolution under visible light illumination. The high-quality CdSe QDs were firstly prepared through the hot injection method using TOP and TOPO as coordinating solvents, while ligand exchange was conducted to transfer CdSe QDs from the organic solvent into the aqueous phase, which benefited the anchoring of the QDs on the surface of TiO_2 and the photoinduced charge transfer from the QDs to the TiO_2. The H_2 yield of CdSe QDs/TiO_2 was the highest compared with three other sensitizers used in TiO_2 systems including phenol-formaldehyde resin, poly(4-vinylphenol), Ru(4,40-$(CO_2H)_2$bpy$)_3$. Meanwhile, as a similar system to PbS/TiO_2, CdSe/TiO_2 heterojunction exhibited a remarkable size-dependent feature in H_2 evolution activity, where the tunable CdSe size resulted in the movement of LUMO and HOMO of CdSe and enhanced the charge transfer between CdSe and TiO_2 [7]. On the other hand, when the association between QDs and semiconductor is improper, the charge transfer does not result in enhancement of photocatalytic performance in QD-sensitized photocatalytic system. For example, Sardar et al. [17] reported a PbS/ZnO system in which TOPO-capped PbS QDs were deposited on the surface of ZnO NPs, which exhibited enhanced photovoltaic performance and undermined the photocatalytic activity of organic dye despite the photoluminescence (PL) spectra of PbS QDs at 820 nm being quenched apparently upon the attachment of the PbS QDs onto the ZnO NPs. The phenomenon was mainly attributed to the unsuitable matching of band energy levels between PbS and ZnO. For 3.2 nm PbS QDs, the energy levels of the CB and VB of the PbS QDs were situated at -3.7 and -5.1 eV versus vacuum, respectively, and the CB bottom of the ZnO NPs was located at −4.3 eV versus vacuum. Under visible light excitation, the photogenerated electrons at the CB of the PbS QDs was expected to transfer to the CB of the ZnO NPs, while the electrons at the CB of the ZnO NPs disabled to produce reactive oxygen species thus damaging the methyl blue (MB).

On the other hand, it was reported that phosphine-containing organic compounds are highly toxic and easily flammable which become a dilemma in synthetic procedures. It has been verified that phosphine-containing organic compounds not only function as coordinating organics to dissolve metal salts and chalcogens, for instance, Cd-TOP and Se-TOP, but also act as mediators of reaction rate, where the breakage of Cd-O and P-Se bonds is sensitive to the reaction temperature, thus easily controlling the formation of Cd-Se. Therefore some groups have attempted to prepare semiconductor NCs using "green" phosphine-free precursors and solvents, for example, oleic acid (OA) and 1-octadecene (ODE), to substitute the phosphine-containing organic compounds. Deng et al. [18] utilized OA and ODE to prepare high-quality CdSe NCs on a large scale, where 2.8 nm CdSe NCs had a maximum quantum yield of 60%, which is comparative with the CdSe NCs quantum yield of 90% prepared using TOP. Shen et al. [19] systematically investigated synthetic processes of high quality CdTe NCs using different Cd and Te stock precursors with phosphine-containing and phosphine-free compounds. Meanwhile, it was revealed that the photoluminescence range, size, and shape could be controlled by simply integrating the injected amount and growth temperature.

As mentioned previously, the preparation of high-quality NCs was conducted at elevated temperatures due to the decomposition requirement of the precursors, which results in some disadvantages such as long-chain organic compounds, high energy consumption, and complicated posttreatment. Henglein et al., has done pioneering research to develop an aqueous method to prepare water-soluble semiconductor NCs. Until now, the aqueous route has been well established to fabricate comparative NCs through the high-temperature injection method. In comparison with the latter, the former has some advantages, for example, being environmentally friendly, making use of biocompatible solvents, and its easy scalability and functionalization. In the aqueous synthetic strategy, chalcogens in binary NCs are introduced in the form of gaseous or ionic states as a starting material such as H_2Te, H_2S, and H_2Se, or Te^{2-}, S^{2-}, and Se^{2-}. On the other hand, the corresponding metals are added in the form of water-soluble salt. First, metal–thiol complexes are formed by dissolving a metal salt and thiol compounds in water with subsequent pH adjustment, accompanied by complete purging by inert gas (Ar, N_2). Second, a chalcogen source is injected into the mentioned deaerated solution, forming an ME precursor (where M = Cd, Zn, Pb, etc., and E = S, Se, Te). Third, the ME precursor proceeds with the nucleation and growth stages to the desired size and shape under a reflux process [20]. Extensive researches have proven that the aqueous synthetic route can also fabricate satisfying QDs for constructing QD-hybridized photocatalysts [21,22]. For example, Lv et al. [23] coupled PbS QDs with $BiVO_4$ using a molecular linker, polydopamine (PDA), to form a heterojunction for enhanced photocatalytic performance. The PbS QDs were directly prepared on the surface of PDA/$BiVO_4$ in an aqueous solution, where $Pb(NO_3)_2$ and Na_2S were used. The PbS/PDA/$BiVO_4$ exhibited enhanced photocatalytic activity to glyphosate in comparison with bare $BiVO_4$, P25, PbS/$BiVO_4$, and PDA/$BiVO_4$ under visible light irradiation. The PDA served as a stabilizer to prevent the PbS QDs from aggregation ascribed to the lack of passivating ligands, thus proving useful for exerting multiple exciton generation of PbS QDs sufficiently, where multiple electron holes could be produced following the absorption of a single photon beyond the Schockley–Queisser limiting efficiency. In addition, a green and environment-friendly strategy has been developed, where the preparation procedure proceeds under room temperature in an N_2 atmosphere. Recently, ultrasonic energy has become viewed as an efficient and clean energy donator to trigger the formation of QDs under benign conditions. For example, Shi et al., reported a "green" route for preparing water-soluble CdTe QDs using a sonoelectrochemical procedure in the absence of N_2 [24]. The quality of CdTe QDs is comparable with that of QDs synthesized by the hot-injection route, which is a promising method in practical applications.

Doping binary NCs is another approach to mediate the composition of NCs, where the introduced elements bring a new energy level into the original one, thus modifying the electronic, optical, and magnetic properties of semiconductor NCs. For instance, Mn is widely utilized to dope into wide-bandgap hosts such as ZnS, ZnSe, and even CdS. Besides, combing two or more semiconductor NCs together, specifically core–shell, dot-in-rods, or tetrapod nanostructures is a new way to regulate the electronic behavior of NC systems, thus controlling the charge transfer paths in photocatalytic systems. For instance, Saunders et al. [25] investigated an interesting linear CdTe/CdSe/CdTe nanorod (NR) heterostructure that has been successfully tested using position-resolved EDS mapping. Talapin et al. [26] reported CdSe/CdS NRs and tetrapods using CdSe NCs as the seed. The morphology and quantum yield can be tuned by varying the crystalline structure of the CdSe seed. The strategy provides a highly convenient approach to construct heterostructures with different compositions and shapes and to explore their unique electronic structures and photophysical properties.

4.2.1.2 Ternary Nanocrystals

I-III-VI_2-type chalcogenides, as ternary semiconductors, have attracted much research attention owing to their excellent optical and electrical properties, which can offer alternatives for desired bandgap energies and be widely applied in linear and nonlinear optical devices and photovoltaic solar cells such as the $CuInS_2$ bandgap of 1.45 eV, the $AgInS_2$ bandgap of 1.19 eV, and the $CuGaS_2$ bandgap of 2.4 eV. With years of development, various methods have been proposed to efficiently prepare ternary NCs including hot-injection, noninjection, thermal decomposition, and solvothermal routes. Li et al. [27] summarized extensive studies on the synthesis of I-III-VI_2 NCs and their applications in solar cells, light-emitting diodes, bioimaging, and photocatalysis. The most typical and successful method is the direct thermal decomposition route, where salt and chalcogens are synthesized in advance and then mixed together to finish the preparation in one-pot at a certain temperature. For instance, Wang and Li et al., reported a general synthesis route for ternary NCs including $AgInS_2$, $CuInS_2$, and $AgInSe_2$ [28]. According to this strategy, the shape (NPs, NRs, and worm-like particles) and size of NCs can be tuned by varying the preparation parameters. Besides, Tang et al. [29] studied the synthesis of Zn-doped $AgInS_2$ NCs using the hot-injection method. By varying the reaction temperature from 120°C to 210°C, the amount of Zn in the $AgInS_2$ could be well controlled and the emission wavelength of the obtained Zn-doped $AgInS_2$ NCs could be tuned from 680 to 520 nm.

Based on these optical properties and surface electronic structures, many studies have proven that ternary semiconductor NCs can be applied in photo/electrocatalytic water splitting, pollutant abatement, and CO_2 reduction. For example, owing to the desired bandgap of 2.4 eV, Tabata et al. [30] applied chalcopyrite-type $CuGa_3S_5$ for photocatalytic H_2 evolution in the presence of a cocatalyst. The prepared $CuGa_3S_5$ exhibited a strong visible light response and the proper position of the CB, where the excited electrons were transfer to the cocatalyst, thus reducing H^+ to H_2. Meanwhile, the results showed that Rh, Ir, NiS, and FeS acted as efficient cocatalysts in comparison to conventionally and widely used Pt. $CuGa_3S_5/NiS$ exhibited the maximum H_2 evolution performance among these systems due to the release of sulfur from the NiS. Recently, a solid solution has been developed to control the bandgap of ternary NCs, which could efficiently improve the absorption and band position. Tsuji et al. [31] prepared a $(CuIn)_xZn_{2(1-x)}S_2$ solid solution from ZnS and $CuInS_2$ for visible-light driven H_2 evolution from water splitting based on a controlled band structure. It was demonstrated that the composition of NCs significantly affects the photophysical properties, band structures, and photocatalytic activities. Therefore in Tsuji's work, the band structure was adjusted by a combination of wide-bandgap ZnS and narrow $CuInS_2$, which resulted in an apparent redshift in the absorption spectrum with an increase in the amount of $CuInS_2$ in the ZnS host.

4.2.1.3 Metal Oxide Nanocrystals

On the basis of the quantum size effect, the use of intentional synthesis to control the facet exposure, size, and electronic behavior of semiconductor metal oxide NCs is the heart of heterogeneous nanocatalysis. Similarly to traditional QDs' synthesis mechanism, metal oxide NCs have also been intensively prepared in the past. Some published reviews have summarized the preparation processes and conditions. To date, based on solution-phase colloidal chemistry, various chemical synthesis routes have been established, especially three representative methods, namely reduction, thermal decomposition, and nonhydrolytic sol–gel process, which have been widely used to synthesize a series of metal oxide NCs including Fe_2O_3, Co_3O_4, TiO_2, ZrO_2, Cu_2O, and ZnO [32]. Alivisatos et al. successfully synthesized moderately monodisperse Fe_2O_3, Cu_2O, and Mn_3O_4 NCs through the thermal decomposition of metal-*N*-nitrosophenylhydroxylamine complex. The size of the metal oxide NCs could be manipulated by varying the reaction temperature and the amount of precursor used [33]. In addition, Yin et al. reported the synthesis of monodisperse Cu_2O NCs based on the thermal decomposition method, where copper (I) acetate [$Cu(Ac)_2$] and a mixture of OA and trioctylamine were used as the Cu source, surfactant, and solvent, respectively. The sizes of the prepared Cu_2O NCs were in the range of 3.6–10.7 nm, which was achieved by varying the molar ratio of copper acetate and OA that functioned as a surfactant to control the growth rate of the Cu_2O NCs [34]. Zhou et al. prepared monodispersed hematite nanocubes and nanoplates with well-defined facets through the hydrothermal route [35]. The degradation results indicated that the Fe_2O_3 nanocubes showed higher photocatalytic activity in comparison with the nanoplates. By means of density functional theory calculation it was found that the dominant facet of the nanocubes was the (012) facet which had a higher surface energy than that of the (001) facet in the Fe_2O_3 nanoplates, which demonstrated that the (012) facet was more reactive than the (001) facet [36]. Xu et al., investigated the transformation process of cobalt hydroxide and hydroxide nitrate into Co_3O_4 nanocubes with a particle size of 47 nm through a low-temperature decomposition method in an aqueous solution [37]. Following that, He et al. reported a thermal decomposition method for synthesizing Co_3O_4 NCs with a diameter of 5 nm [38]. Shi et al., took advantage of benzyl alcohol, ammonium hydroxide, and cobalt acetate to form 4.5 nm Co_3O_4 QDs, which were used in photocatalytic water oxidation in a $[Ru(bpy)_3]^{2+}$-persulfate system [39]. In addition, Yehezkeli et al. [40] reported Co_3O_4 NCs/CdS NRs nanostructure by coupling 6 nm Co_3O_4 NCs and CdS NRs based on an electrostatic assembling route for photocatalytic oxidation of water and MB degradation under visible light irradiation.

Based on thermal decomposition, In_2O_3 NCs with sizes ranging from 4 to 8 nm were prepared through the dissociation of indium acetylacetonate in oleylamine by either varying the molar ratio of precursor to surfactant or by a multiple injection of the precursor [41]. Larger-sized In_2O_3 NCs in the range of 11–20 nm also were fabricated through a modified dynamic injection route with subsequent aging. Chen et al. [42] prepared rod-like In_2O_3 NCs through the decomposition of InOOH at 400°C, whose band absorption had an apparent blueshift in comparison with bulk In_2O_3 due to the remarkable quantum size effect. Compared with the In_2O_3 nanocubes, the prepared rod-like In_2O_3 NCs possessed enhanced photocatalytic activity under UV irradiation. Apart from the regular single-crystalline In_2O_3 NCs, other hierarchical nanostructures of In_2O_3 NCs have also been fabricated in the past. For example, Gondal et al. [43] prepared ordered mesoporous In_2O_3 NCs for the enhanced photocatalytic conversion of CO_2 into methanol using SBA-15 as a hard template. Yin et al. reported the synthesis of hollow In_2O_3 NCs through a similar thermal decomposition to that used for InOOH but which were fabricated under different conditions, and they exhibited an excellent photocatalytic degradation of rhodamine B (RhB) and MB [44].

4.2.2 Nanostructures for Catalysis

4.2.2.1 Zero-Dimension Nanocrystals-Based Catalysts

In general, nanosized spherical-like NCs are categorized as Zero-dimension (0D) because of the finite spatial regime.

According to their growth mechanism, 0D NCs are easily prepared by simply controlling the ratio of meta to chalcogen precursor and the reaction temperature. To obtain the desired size, extra precursor is added into the solution by a microinjection system, which can suppress the formation of new nuclei in the solvent and facilitate the added precursor being deposited on the surface of nuclei NPs formed in the first step. In most of cases, the size of NCs is limited to within 5–6 nm due to the low growth rate in the second stage. It has been demonstrated that the first nucleation stage maintains for ~1 min and the second growth stage spans serval hours. In addition, the following growth process easily leads to the defocusing of the size distribution. Therefore the synthesis route should be optimized to fabricate larger-sized NCs. Jana et al. [45] design a novel multistep injection strategy to prolong the growth duration with an appropriate sulfur precursor, which ensured a rapid decomposition rate of the precursor. As a result, thiourea as the sulfur source was used to retain extra precursor and delay the ripening process. Finally, the size of NCs grew larger than 10 nm. By means of this strategy, larger-size ZnS, CdS, even Mn-doped ZnS have been successfully prepared. Now, imagine when the same metal and chalcogen precursor are used, the size of NCs can be increased; however, if the precursors are changed, what is obtained? Right, core–shell NCs can be obtained.

The synthetic idea of core–shell NCs stems from the multistep injection method with variation of precursors, which could be utilized to produce different crystal phases on the external surface of core NCs, such as CdSe/ZnS, CdSe/CdS, InP/ZnS and ZnSe/ZnS. In 1996 Hines et al. [46], for the first time, described the preparation of monodispersed CdSe–ZnS core–shell NCs through a two-step hot injection method, where the diameter of the CdSe core was 2.7–3.0 nm and the thickness of ZnS shell was 0.3–0.9 nm. The capping of the ZnS not only did similar work to that of long-chain organic compounds, for example, TOP, but also reduced the surface traps of the CdSe core, thus enhancing the quantum yield to 50%. Following successful synthesis, the wide tentative applications of core–shell NCs have been explored in the photocatalysis and photovoltaic fields. In addition, Zhu et al., prepared type-I CdS–ZnS core–shell NCs and investigated the effects of the thickness of the ZnS shell on the charge separation and recombination rates using time-resolved transient absorption and anthraquinone as an electron acceptor. The experimental data demonstrated that a thickness increase in the ZnS shell resulted in an exponential decrease in charge separation and recombination rates with exponential factors of 0.35 ± 0.03 and $0.91 \pm 0.14\ Å^{-1}$, respectively [47]. This inferred that the thicker ZnS shell suppressed the recombination, thus relatively enhancing the charge separation. Based on the analysis, core–shell nanostructures have been successfully applied in photo/electrocatalysis or photovoltaics in the past. For example, Huang et al. [48] reported type-I CdS–ZnS core–shell NCs for effective and stable photocatalytic H_2 generation in the presence of dual cocatalysts. The ZnS shell was passivated on the surface of the CdS core, which ensured that the photocatalysts worked efficiently and stably under visible light irradiation. Pt and Ni metallic NPs captured electrons as reduction cocatalysts and PdS or PbS collected holes as oxidation cocatalysts, thus efficiently enhancing the charge separation and transfer even though the ZnS shell was formed externally of the CdS core.

4.2.2.2 One-Dimensional Nanocrsytals-Based Catalysts

Unlike 0D QDs, in which the excited charges are localized inside a finite regime in three dimensions, the exciton movements in One-dimensional (1D) NRs are confined in the radial direction but are bulk-like along the axial direction with diameters of below ~10 nm and lengths of below ~100 nm. On the other hand, according to the analysis discussed previously, it is well known that the spatial distribution of excited electrons at the CB and holes at the VB can easily be adjusted by varying the sizes and composition to tune the photophysical properties including photoluminescence, exciton lifetime, and charge separation and recombination rate. In the case of 1D NRs, the wave function engineering deriving from the controlling of electronic structures can be achieve by means of the formation of heterostructures such as NRs, dot-in-rods, and tetrapods.

In 2002 Nair et al. [49] reported the synthesis of CdS NRs via a single source route through the thermal decomposition of an organometal complex in a TOPO solution. Meanwhile, Peng et al. [50] synthesized CdSe NRs through a hydrothermal route in an aqueous solution. Nevertheless, uniformity and quality are not enough to satisfy the requirements for NC-based applications. With the promotion of synthesis techniques and instruments, more and more high-quality NRs with single or heterostructures are being fabricated, though the yield is still low in single synthesis operations. Besides, Li et al. [51] fabricated uniform CdTe NRs through a mixture of two short-chain ligands in an aqueous solution, which exhibited strong photoluminescence and a quantum yield. Except for the direct preparation method, the cation exchange strategy was established to synthesize NCs of various morphologies, where the original NCs with a specific morphology are used as a starting frame material and then the intended cation is introduced to replace the original cation to form a new phase associated with no change of morphology. Alivisatos et al. reported the synthesis of PbS NRs using easily synthesized CdS NRs as a starting material under benign reaction conditions [52,53]. The cation exchange reaction was driven by completely different solubilities of cations in organic solvents and ligands such as tributylphosphine. The CdS NRs were prepared first and then converted into Cu_2S and PbS NRs using different solvents and ligands. This strategy can extremely enrich the compositions and heterostructures of NCs which are difficultly synthesized through the traditional multistep hot injection or seeded routes. The formation of multiple components and heterostructures can promote deep study of electronic structures and behaviors in NCs, and can thus be intentionally applied in the photo/electrocatalysis field.

In aspect of catalytic applications, various rod-like NCs have been successfully applied in photocatalytic H_2 evolution and environment treatment. For example, Chen et al. [54] reported the facile synthesis of thioacetamide-coated CdS NRs with enhanced photocatalytic degradation activity toward methyl orange under UV irradiation. Besides, Ran et al. [55] coupled Ni $(OH)_2$ with CdS NRs for enhanced visible light driven photocatalytic H_2 production. The introduction of $Ni(OH)_2$ was helpful for water splitting due to the formation of Ni^0 ascribed to a reduction of photoinduced electrons at the CB of the CdS NRs. In other literatures, it has been demonstrated that Ni^0 metallic NPs are able to facilitate a charge transfer from CdS NRs to Ni^0, thus enhancing the photocatalytic performance. In addition to combining them with other semiconductors, these NRs are further assembled into considerably special heterostructures such as dot-in-rod and dot-in-tetrapod, for exploring their exciting electronic structures and enhanced catalytic activities. For example, Wu et al. [56] designed a CdSe/CdS dot-in-rod system and investigated its light-driven H_2 generation in the presence of Pt at the tip of the NRs. In the dot-in-rod system, a large type-II VB offset existed, driving the ultrafast localization of the holes in the CdSe core, which was competitive with the ultrafast hole trapping in the CdS NRs. The competitive processes resulted in three types of exciton species in the system with different spatial distributions. Owing to the strong quantum confinement in the radial direction, the ET and hole removal were significantly enhanced. On the other hand, in the axial direction, the fast carrier mobility enabled long distance charge separation and slow recombination, thus ensuring an efficient light-driven H_2 evolution with MV^{2+} as an electron acceptor and Pt NPs in a dispersed manner. Although the heterostructure can offer a unique electronic structure and band level distribution, it is not always useful for increasing photoactivity. For instance, Qiu et al., systematically compared the photocatalytic H_2 activity of four different QD systems including CdSe QDs, CdSe–CdS core–shell, CdSe NRs, and CdSe/CdS dot-in-rods [57]. The experimental data inferred that the photocatalytic H_2 evolution of CdSe/CdS dot-in-rods was the lowest among these prepared samples, which was mainly ascribed to the reaction system, that is, the electron acceptor, Ni-dihydrolipoic acid. In the first section, it was discussed that the photocatalytic reaction is closely related with the redox potential of the surficial intermediate reaction regardless of the final production. Therefore, it is observed that catalysts are similar, but the conversion efficiency is distinctly different.

Departing from binary rod-like NCs, ternary rod-like NCs have also been proved to have strong photoactivity. For example, Ng et al. [58] synthesized 1D orthorhombic $AgInSe_2$ NRs with dimensions of 50 nm × 14.5 nm through the thermal decomposition of a single-source precursor in a mixed solution of oleylamine and dodecanethiol, while the synthesis duration spanned 17 h in an Ar atmosphere. The oleylamine functioned as both an activating agent and a capping ligand and the dodecanethiol guided the growth of the $AgInSe_2$ into the form of 1D NRs.

4.2.2.3 Two-Dimensional Nanocrystal-Based Catalysts

Over the past few decades, two-dimensional (2D) crystals have been paid a tremendous amount of attention owing to their exceptional optical, electronic, and physiochemical properties. Triggered by the growing research interest and well-established synthesis and characterization techniques in 2D nanomaterials, single- and few-layer metal dichalcogenide nanosheets have been extensively investigated, which exhibit many promising properties for a wide range of applications [59]. Based on extensive studies, it has been concluded that 2D semiconductor NCs have many exciting properties that offer chances to enhance the photocatalytic performance of NC-based photocatalysts including their high surface area, strong excitonic resonances, absorption cross section, and large charge carrier mobilities. Consequently, using 2D semiconductor NCs to assemble efficient catalytic systems attracts more and more attention in photocatalysis, photovoltaics, and solar-to-fuel conversion. Ithurria et al. reported a series of synthetic methods to prepare colloidal semiconductor nanoplatelets such as CdS, CdSe, and CdTe through a modified hot-injection method by varying the amount of metal and chalcogen precursor, ligand, and solvent. The thicknesses of the nanoplatelets were tuned by adjusting the reaction parameters and these were measured using high resolution transmission electron microscopy (HRTEM) and atomic force microscopy (AFM) [60–62]. Meanwhile, the steady-state and time-resolved fluorescence spectra were used to study the recombination rate and exciton lifetimes. These revealed that the 2D NCs exhibited shorter exciton lifetimes in comparison with other spherical-like NCs due to the fast exciton transfer rate at the surface, similar to that of ultrathin graphene nanosheets.

Owing to the high surface area of nanosheets, Xu et al., prepared ultrathin CdS nanosheets with a thickness of 4 nm through a hydrothermal route using L-cysteine and diethylenetriamine as shaper and stabilizer agents [63]. By combining them with a molecular nickel complex, the ultrathin CdS nanosheets were used as a noble-metal-free system to produce H_2 from water splitting under visible light irradiation in the presence of triethylamine as a sacrificial electron donor under an alkaline environment [64]. The large surface area of ultrathin nanosheets could provide more surficial active sites where the excited electrons could react with the molecular nickel complex and generate H_2 with a turnover number (TON) of ~28,000. Wu et al. [65] integrated CdSe nanosheets with Pt NPs to investigate the charge transfer between CdSe nanosheets and metallic NPs. Owing to the strong quantum confinement in the thickness direction and high exciton mobility, 85% of the photoinduced excitons with short lifetime transferred from the CdSe nanosheets to Pt NPs through the rapid energy transfer way. In contrast, the rest excitons possessing long lifetime were localized by holes and transferred to Pt NPs in a form of charging. This metal/semiconductor nanosheets model can be used to catalyze water to produce H_2. However, Zhukovskyi et al. [66] reported a Ni NPs/CdS nanosheets system and its photocatalytic H_2 evolution ability. The ultrathin CdS nanosheets were controllably synthesized through the traditional thermal decomposition of cadmium diethyldithiocarbamate in an ODE solvent, where the thickness of the nanosheets could be tuned by varying the molar ratio of Cd to the surfactant (stearic acid). By means of transient differential absorption (TDA), the recombination and relaxation processes of photoinduced charges were investigated because the TDA spectrum mainly reflects the movement of electrons rather than holes owing to the lighter effective mass of electrons. From the results of TDA, as compared with CdS nanosheets, the Ni/CdS nanosheets exhibited a higher photocatalytic H_2 generation due to the larger absorption efficiency, stronger exciton binding energy, and faster ET from the CdS to the Ni.

4.2.2.4 Three-Dimensional Nanocrystals-Based Catalysts

Three-dimensional (3D) catalysts refer to catalysts that have three-dimensional, multiscale, and porous structures. Hence, 3D NC-based catalysts are categorized into two kinds, namely mesoporous NC catalysts and NC-decorated mesoporous catalysts, which usually consist of semiconductor NCs and 3D host materials where the NCs are anchored at the surface of the 3D host materials which have many families such as metal oxide, silica, and carbon materials. For example, Fe_2O_3 NCs were used to modify mesoporous carbon, thus resulting in an enhancement of the electrocatalytic activity of a four-electron oxygen reduction reaction. The 3D mesoporous carbon was impregnated into an Fe^{3+} solution, adsorbing Fe ions on the surface. And then the mixed solid was calcined in an inert atmosphere. This method ensured that the Fe_2O_3 was dispersed uniformly on the surface of the mesoporous carbon [67]. In addition, other metal oxide NCs can also be deposited onto mesoporous carbon nanostructures following a similar synthetic route such as SnO_2 NCs/C [68], TiO_2 NCs/SiO_2 [69], CdS QDs/SiO_2 [70].

Interestingly, some research groups have tried to couple NCs with mesoporous metal organic frameworks (MOFs) for enhancing the photocatalytic performance in water purification [71–73]. On the one hand, the NCs could provide strong sunlight absorption and photoexcited charge; while on the other hand, the rich pore structure from the MOFs could benefit the dispersal of the NCs and avoid excess aggregation leading to the disappearance of unique optoelectronic properties. For example, CdSe QDs were embedded in UIO-66 by successively adding Cd and Se precursors into the UIO-66 suspension for preparing CdSe QDs at room temperature, where the high content of small-size CdSe QDs was highly dispersed throughout the UIO-66 [74]. The CdSe QDs–UIO-66 indicated obvious adsorption and photoactivity toward RhB as compared with the bare CdSe QDs and UIO-66. The synthetic route in this paper is called "ship in the bottle" [75]. The precursor to QDs is added in the proper sequence and then the precursor molecules penetrate through the pore windows of MOFs. Finally, the precursors react each other and produce QDs inside the pore of MOFs.

This method is feasible for pursuing desired unique structures and sizes even bigger than the inner cavity size of MOFs. Nevertheless, some requirements should be considered: (1) precursors should be compatible with MOFs, so that they can move smoothly into the pores of the MOFs; (2) the framework must be stable under the synthetic conditions of the QDs. Upon choosing this method, some problems often confront researchers. MOFs consist of metal ions and organics to form frameworks under certain conditions, while other metal ions from the precursors of QDs might replace the original metal ions, resulting in the collapse of the initial framework. Besides, precursors of QDs disperse in whole solutions including the exterior of MOFs. Therefore a major part of prepared QDs distribute at the surface of MOFs and in the solvent phase. In contrast, another synthetic strategy has also been developed to achieve the same purpose, which is called "bottle around the ship." The desired QDs are fabricated first and then stabilized by organic capping agents, which are added into the MOF precursors and then the framework is assembled around the QDs [76,77]. This method could completely clear up the presence of QDs at the external surface of MOFs. Instead, the added QDs are enclosed inside the MOF matrix and do not occupy the pore space. Moreover, the shape of the QDs can be tuned according to the requirements (Fig. 4.5).

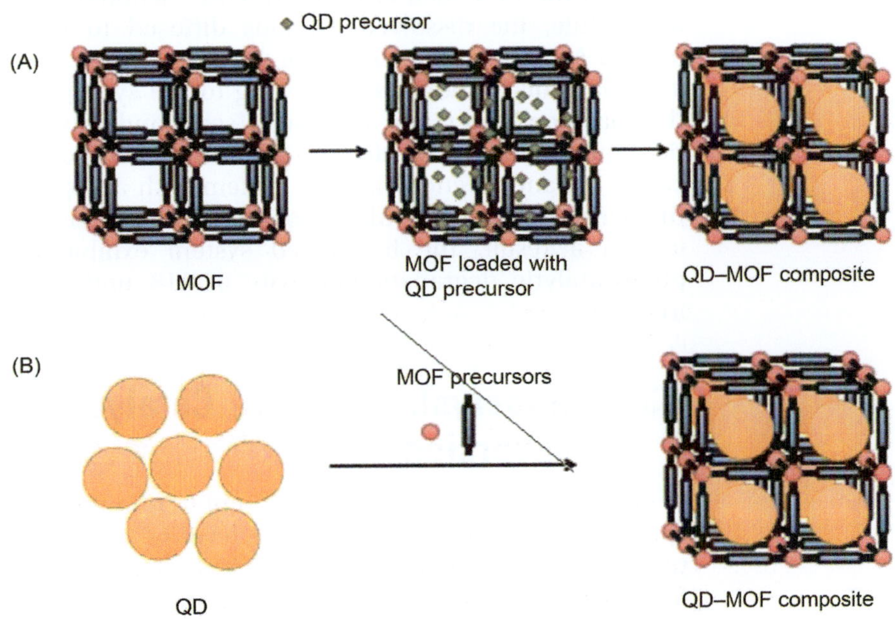

Figure 4.5 Two methodologies used to prepare QD–metal organic framework (MOF) composites. (A) Ship in the bottle; (B) bottle around the ship [75].

On the other hand, the intrinsic 3D nanostructure of NCs has also been explored to improve photocatalytic activities in the past. To prepare 3D nanostructured NCs such as porous NCs, a hard or soft template is required, which not only functions as a platform to support the formation of semiconductor NCs, but also acts as a pore space formed as a result of the etching removal operation [78]. Vamvasakis et al., used a polymer as a template to directly assemble ordered mesoporous CdS and ZnS NCs through a polymer-templated oxidative polymerization process. The synthesized mesoporous NC network had a 3D open-pore structure, a high surface area, and uniform pores, which exhibited high photocatalytic activity and stability [79]. In addition, a 3D nanostructure can also be formed without the use of a template method. Feng et al. [80] prepared 3D Au–ZnS–AgAuS yolk–shell NCs through a cation exchange route under hydrothermal conditions. The cavity-free Au–AgAu core–shell NCs were first synthesized, followed by the addition of a sulfur precursor, which resulted in the formation of Au–AgAuS. Second, the Zn precursor was introduced into the Au–AgAuS suspension to form a mixed solution and then transferred to the autoclave for producing Au–ZnS–AgAuS yolk–shell NCs. Owing to the cation replacement of Au with Zn, the size of the Au core was gradually reduced, thus forming a space between the ZnS shell and the Au core, denoted as yolk–shell. Meanwhile, the dissolved Au ions diffused to the shell and formed Au NPs on the surface of the ZnS, which could function as a plasmon-exciton center. Owing to the synergistic effect of the plasmon effect and the metal/semiconductor heterojunction, the photoinduced electron and hole separation was significantly enhanced, endowing the system with an optimization of the performance of the 3D nanostructure and electronic behavior. As a result, the hybridized system exhibited enhanced photocatalytic degradation activity of MB under visible light irradiation.

4.3 Nanocrystals for CO_2 Catalytic Conversion

In recent years, the concentration of CO_2 in the Earth's atmosphere has exceeded the alarming level of 400-ppm, as compared to 280-ppm, which is the level that had been maintained for hundreds of thousands of years until the beginning of the first industrial revolution. This clearly confirms that the increased level of CO_2 in the atmosphere is due to heavy industrialization, deforestation, and other reckless human activities. Greenhouse gases like CO_2 are key players in the natural greenhouse effect that makes life possible on Earth; however, the excessive emission of greenhouse gases (mainly CO_2) increases the global temperature and thereby leads to global warming and harmful climatic effects. It was reported that over the past 100 years the surface temperature of Earth has increased by 0.75°C due to the heat trapped in CO_2 and other greenhouse gases. In addition to various international strategies and regulations to control CO_2 emissions, the development of appropriate technologies to reduce the level of CO_2 in the atmosphere is imperative. Among many technologies adopted for controlling the level of CO_2 in the atmosphere, carbon capture and sequestration (CCS) is generally considered as a competent and viable method. In CCS, CO_2 emissions from sites of power generation and industrial processes are captured, safely transported to storage facilities using pipelines, tankers, and ships, and finally stored in a secured place. However, this technology becomes commercially and practically not viable, particularly in power plants, as it requires an extra 25% of power generation, a vast amount of space, and complex instrumentation for the processes of the capture, compression, and storage of CO_2.

4.3.1 CO_2 Catalytic Reduction

Apart from the semicommercial CCS technique, the catalytic technique of converting CO_2 into other higher value-added chemicals and fuels is an attractive and promising approach to solve the CO_2 concentration increase. Up to now, various electrochemical, photocatalytic, thermochemical, and biological processes for directly converting CO_2 into value-added hydrocarbons, for example, methane, methanol, formic acid, and ethanol have been explored and reported [81–84]. Most of these methods of conversion require huge amounts of energy and lengthy procedures, owing to the inherently stable and inert property of the CO_2 molecule. However, simple, benign, and cost-effective methods based on photocatalysis to realize CO_2 conversion have become attractive [85].

In photocatalytic CO_2 reduction, the electron–hole pairs generated on the surface of a semiconducting photocatalyst mediate the photooxidation and photoreduction reactions that result in the desired end product, as shown in Fig. 4.6A. Reducing CO_2 to other organics need to consume electrons that can derive from photoexcitation of semiconductor under solar light irradiation. This method does not require complex

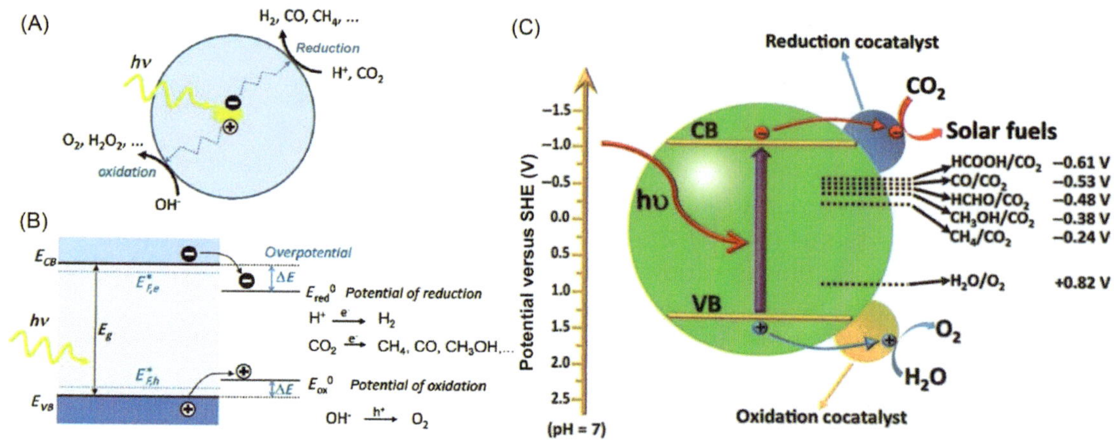

Figure 4.6 (A) Illustrations of the photocatalytic process; (B) CO_2 reduction mechanism; and (C) potentials of various reduction reactions [82].

instrumentations as the reactions take place in ambient conditions, and also photocatalysts are quite inexpensive and are capable of working under the naturally abundant and inexhaustible solar radiation without any external circuitry or bias. Instead of a complex device wired together, the process relies on multiple very simple reactors placed in a reaction medium. The economic analysis of the NP-based photocatalytic approach indicates that in comparison to other options, it has the potential to facilitate a substantial reduction in fabrication costs due to the simplicity of its design, provided that the yield of the reactions improves. However the challenge is that as both half reactions proceed in the same reaction space, the prevention of various parasitic back reactions and product extractions becomes critical for achieving a high efficiency. In Fig. 4.6B and C, it can be observed that CO_2 can capture electrons and produce many organic productants at different potentials. This complex process gives rise to a bigger barrier in implementing photocatalytic CO_2 reduction application, for high selectivity and production purification.

4.3.2 Metal Chalcogenides Nanocrystals for CO_2 Reduction

In the past, there have been many semiconductor photocatalysts synthesized and applied for the process of the photocatalytic reduction of CO_2 to methane, methanol, and other high-value chemicals including pure TiO_2 and TiO_2-based composite catalysts, silicon carbide, and WO_3 using UV radiation. Taking into consideration the electronic structure and optical properties of NCs, the pioneering contributions of semiconductor NCs to CO_2 photoreduction have also been made to promote the effective conversion and utilization of CO_2 as fuels and chemicals. Wang et al., demonstrated that CdSe NCs were able to fix CO_2 using first principle calculations, where the role of facet, photoexcitation, doping, and electron confinement have been further discussed [86]. It was uncovered that the CO_2 molecule was directly activated by NCs from neutrally uncharged to negatively charged through remotely polarized interacton between the NCs and CO_2. The results suggested that the CO_2 molecule does not have to undergo the chemisorption process. This result infers that NCs have the ability to activate and even catalyze CO_2 molecules, but this result must be verified by experimental data. Cho et al. [87] explored the application of CdSe NCs in the photoelectrochemical reduction of CO_2 by depositing previously synthesized CdSe NCs onto an NiO electrode (Fig. 4.7). They revealed that the size of CdSe NCs significantly affected the conversion efficiency of CO_2 due to the close relationship between size and the CB/VB positions. It has been demonstrated that the CB and VB energy levels of NCs or QDs could present a discrete and quantized state when the size of the NCs deceases to a specific diameter, known as the Bohr diameter. Therefore the CB position moves to a more-negative

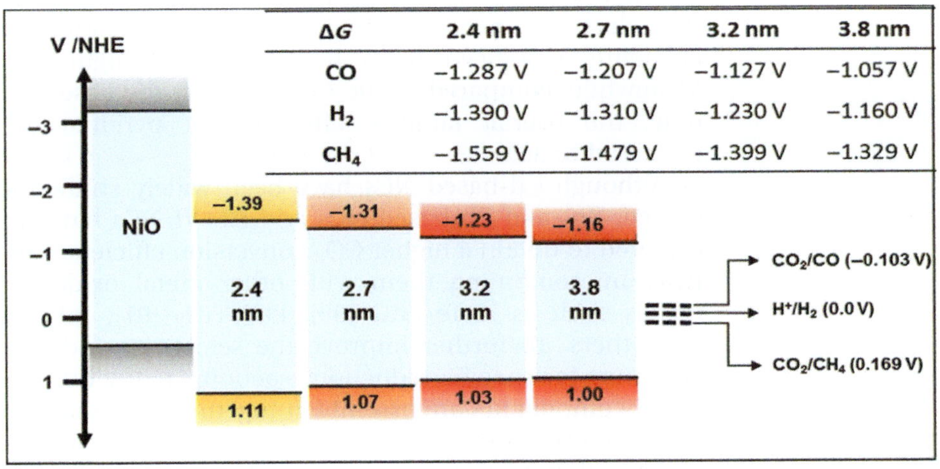

Figure 4.7 Illustration of size-dependence energy diagram of CdSe quantum dots (QDs) and various reduction reactions at pH 0. The inset table shows the free energy between the CB positions of CdSe QDs and the redox potentials [87].

level, which is beneficial for the occurrence of a reduction reaction because of the production of stronger reductive photoelectrons. In Cho's work, it was observed that the smaller the size of CdSe NCs, the larger the reaction rate constant k_{CH4} was. It was concluded that the higher CB resulted in the preferential occurrence of the CO_2 to CH_4 conversion reaction. Meanwhile, it was found that the rate constants of k_{H2} and k_{CO} were promoted with the increasing of k_{CH4}, but the amplitude was lower than that of k_{CH4}. This result mainly stemmed from the size-dependent free energy of the photoinduced ET from the CB of the CdSe NCs to the adsorbed CO_2 molecule based on the theoretical calculation results.

In addition to CdSe NCs, CdS NCs were also reported to have the ability to reduce CO_2 under visible light irradiation by Wang et al. [88]. They established a CdS/Co-ZIF-9 system for the capture, storage, and conversion of CO_2. Under visible light irradiation, the hybridized system converted CO_2 to carbon monoxide with the cooperation of bipyridine and triethanolamine, and it showed a high catalytic deoxygenation ability. In addition, a $^{13}CO_2$ isotopic experiment demonstrated that the produced carbon monoxide derived from the photoreduction of CO_2 rather than from organic compounds in the hybridized system. Jeon et al. [89] utilized the strong absorption ability of CdTe to visible light, even to near-infrared and combined a CdTe/fluorine-doped tin oxide (FTO) electrode with pyridine for the enhanced photoelectrocatalytic reduction of CO_2. Owing to the formation of a pyridine/CO_2 intermediate, the activation barrier of the CO_2 molecule was apparently decreased, and thus the selective conversion efficiency of CO_2 to formic acid was efficiently enhanced upon the incident of light irradiation. Meanwhile, compared to the CdTe/FTO in the absence of pyridine, the overall faradaic efficiency of pyridine–CdTe/FTO increased to 30%.

Although Cd-based NCs have been widely studied and discussed, Zn- and Pb-based NCs and III–VI NCs have also been explored to obtain a higher CO_2 conversion efficiency and selectivity by combining them with other metal oxides or noble metals such as ZnTe/ZnO [90], PbS/TiO_2 [91], InP/TiO_2 [92], and others. To further improve the selectivity and conversion efficiency for specific reduction reactions, many different functional components are assembled together, for example, Ag_2Se/graphene/TiO_2 [93] and CdSe/Pt/TiO_2 [94], in which the introduction of graphene was to endow the system with a more efficient separation of photoinduced electrons and holes. Recently, a three-component Au/ZnTe/ZnO system was designed to photoelectrocatalytically reduce CO_2, in which the Au functioned as an electrocatalytic "engine" owing to its high electrocatalytic activity and it increased over 10-fold and 7.5-fold the conversion efficiency and selectivity for CO, respectively, in comparison with the ZnTe/ZnO [95]. Besides, binary NCs are used to cooperate with metal complexes for increasing the charge transfer efficiency, which can effectively decrease the energy barrier of CO_2 reduction. For example, Kuehnel et al. reported a series of molecular Ni complex/QD photoelectrocatalysts for the selective photoreduction of CO_2 to CO in water and the competitive H_2 generation reaction was obviously suppressed under visible light irradiation [96,97]. Furthermore, it has been demonstrated that the immobilization efficiency of the Ni complex on the surface of the QDs was closely correlated with the production selectivity, which plays a vital role in picking CO_2 reduction rather than H_2 evolution in aqueous solutions.

Except binary NCs, ternary NCs are used in independent or combined ways to reduce CO_2 and produce hydrocarbons. For example, Xu et al. [98] coupled $CuInS_2$ NCs with TiO_2 nanofibers to form a direct Z-scheme hybrid system for improving the photoreduction conversion of CO_2 to CH_4 and CH_3OH under simulated solar light irradiation. Usually, *p*-type and *n*-type semiconductors can form type-II heterojunctions owing to the mismatching band alignment, while the excited electrons at the CB of *p*-type semiconductors with more negative positions move to the CB of *n*-type semiconductors with less negative positions, which results in the fading of the reductive ability of electrons and hardly triggers a CO_2 reduction reaction. Instead of a type-II heterojunction, the direct Z-scheme configuration not only forms a built-in electronic field for improving the separation of electrons and holes, but also keeps the reductive and oxidative ability of photoinduced electrons and holes [99]. Owing to this unique spatial configuration, the $CuInS_2$ NCs/TiO_2 nanofibers can efficiently convert CO_2 to CH_4 and CH_3OH. Therefore by constructing a Z-scheme heterojunction it is possible to achieve the dream of efficient CO_2 reduction. On the other hand, although the Z-scheme heterojunction has been intensively explored and discussed, finding direct and convincing evidences that verify the existence of a Z-scheme is still a challenge.

4.3.3 Metal Oxide Nanocrystals for CO_2 Reduction

Compared with large-size semiconductor photocatalysts, NCs have many obvious advantages such as their large surface area, thus many groups have made efforts to decrease the sizes of metal oxide semiconductors such as TiO_2, Co_3O_4 [100], ZnO, and Cu_2O, to obtain these unique properties and apply them to the photoreductive conversion of CO_2 to other high-value chemicals. Xu et al. [101] reported cube-like TiO_2 NCs with the exposure of the (100) and (001) facets through a facile hydrothermal method. Owing to the quantum confinement effect, the CB of TiO_2 moved to a more-negative position than that of bulk TiO_2, which was beneficial for enhancing the photocatalytic activity toward the reduction of CO_2 under UV irradiation. In addition to enlarging the bandgap, the exposed (100) facet has been demonstrated to be superior to other facets in electronic structure, which is easier to adsorb and activate CO_2 molecules. Based on these synergistic effects, the cube-like TiO_2 NCs with exposed (100) facet present enhanced photoreduction activity toward CO_2 to produce methane and methanol under UV light irradiation. However, this strategy improves the CO_2 photoreduction efficiency as well as the H_2 generation due to the competitive photoreduction of H^+/H_2O and various CO_2 reduction paths. Hence, the selective conversion of CO_2 into specific chemicals is difficult to achieve with a high yield. It is reported that MgO [102] and Cu_2O [103] were able to enhance the chemisorption of CO_2 and noble Pt increased the surface electron density, which synergistically improved the selection and yield of CO_2 photoreduction in comparison with H_2 evolution. Based on these considerations, Xiong and Zhao et al. [104–106] reported a series of TiO_2 NC-based photocatalysts for enhancing CO_2 photoreduction efficiency and selectivity such as GO/TiO_2 NCs, Pt/TiO_2 NCs, and Pt–Cu_2O/TiO_2 NCs. As compared to the bare TiO_2 NCs, Xiong's samples exhibited higher conversion efficiencies from CO_2 to CH_4 and CH_3OH. To further increase the efficient adsorption and activation of CO_2 on the surface of NCs, some groups have attempted to design superlattices or superparticles constituted by NC units for forming mesoporous internal cavities. Gondal et al. [43] successfully fabricated ordered mesoporous In_2O_3 NCs using SBA-15 as a hard template. The ordered mesoporous In_2O_3 NCs exhibited a great methanol conversion efficiency, achieving 481 $\mu mol\ g^{-1}\ h^{-1}$ after 150 min of illumination, which is higher than Ag–TiO_2 (130 $\mu mol\ g^{-1}\ h^{-1}$). The enhancement was attributed to the increasing of surface area, active sites, and defects deriving from the ordered mesoporous. In addition, Wang et al. [107] took advantage of MOFs as a template to fabricate a Co_3O_4-embedded porous ZnO dodecahedron system for selectively and efficiently converting CO_2 to CH_4 and CO under UV light irradiation, which was

66 times and 367 times higher than that of commercial ZnO and P25, respectively. Meanwhile, the encapsulated Co_3O_4 shell could significantly reduce the photocorrosion of ZnO.

Apart from these semiconductor-based NCs, another class of NC, metallic NCs, have attracted more and more attention in the catalytic reduction of CO_2 to hydrocarbons. Such metallic NCs have a great advantage, that is, they are rich in free electrons on the surface, which makes it easier to allow for NCs to adsorb and activate CO_2 molecules. Therefore the design and fabrication of metal NCs-containing catalysts will be paid more attention in future. Until now, Pt, Ag, Cu, Ni, Au, and Zn NCs have been developed [108], while the majority of metal NC catalysts mainly reduce CO_2 to CO rather than higher-value species, for example, methanol, ethylene, and others. Attempting to selectively and efficiently synthesize longer carbon-chain species is the main direction of future study.

4.3.4 Nanocrystals for Catalytic Pollutant Removal

It is well known that catalytic techniques are often used to mineralize organic pollutants including organic dyes, antibiotics, hormone, versatile organic compounds, and persistent organic pollutants. Following the above discussion, NC catalysts have a high surficial energy and are rich in electrons, which easily react with substances adsorbed on the surface of NCs. Once the reaction condition is satisfied, a redox reaction occurs which is associated with the transferring of excited electrons. Until now, external energy forms usually include photo, electronic, and thermal ways. Considering the physiochemical properties of semiconductor NCs, optical and electronic excitation are preferentially chosen to drive the redox occurrence on the surface of NCs.

4.3.5 Binary Nanocrystals

In general, binary NCs mainly denote compounds consisting of I and II transition metal and chalcogen elements such as CdX, ZnX, PbX, Ag_2X, and Cu_2X (where X = S, Se, or Te). For example, Samadi-Maybodi et al. extensively investigated water-soluble and stable CdS QDs and their applications in the photocatalytic degradation of organic pollutants through the facile aqueous phase synthetic route [109,110]. Superoxide and hydroxyl radicals have been demonstrated as the primary active species produced from excited electrons and holes, respectively. In addition, ZnS QDs have also attracted increasing attention for advanced oxidation processes in wastewater purification due to the uncertain toxicity of Cd-based QDs to environmental media [111]. For catalytic reactions, the surface of the catalyst is a vital site where the redox reaction and energy transfer between the catalyst and substances occur because the surface properties decide the reaction rate and activity. Wen et al., reported vesicle and granular CdSe photocatalysts for the photocatalytic removal of tetracycline hydrochloride using two synthesis methods. The vesicle CdSe NCs presented higher photocatalytic activity than the granular CdSe NCs because the vesicle CdSe NCs had a larger surface area thus providing more surficial active sites in comparison with the granular CdSe NCs [112].

To further enlarge a given reactive surface area, except when preparing porous NCs, combining NCs with porous or even mesoporous supporting materials including MOFs, hydrotalcites, silica, and carbon materials, provides a new path for improving the photocatalytic activity of NCs. For instance, Ke et al. [113] reported a facile preparation of CdS-MIL-100(Fe) composites for an improved photocatalytic performance of benzyl alcohol oxidation. Chikate et al., took advantage of porous montmorillonite by coupling it with CdSe NCs for enhancing the photocatalytic activity toward the removal of indigo carmine through a route similar to "ship in the bottle" [114]. It was found that a spontaneous and exothermic chemisorption process occurred due to the large surface area and rich porous nanostructure, followed by the photocatalytic reaction. In addition, the mesoporous-activated carbon was used to support the CdS for an enhanced photocatalytic performance, while the introduced CdS was not nanosized but instead in the form of microparticles, ascribed to the uncontrollable synthetic process, which resulted in the severe aggregation of CdS [115]. To obtain a discrete dispersion of NCs inside mesoporous carbon materials, Yang et al. [116] utilized MOFs as a precursor to carbon materials and deposited CdS inside the carbonized framework through amino groups functionalization. This unique nanostructure could achieve a reduction in the aggregation growth of CdS NCs, which facilitated the separation and transfer of photoinduced charge carriers. The prepared samples displayed good adsorption ability and degradation toward Cephalexin under visible light illumination. Besides, Hu et al. [70] prepared CdS NCs that were dispersed into mesoporous silica nanospheres with a Brunauer–Emmett–Teller (BET) surface area of 640 $m^2 g^{-1}$, which possessed a strong adsorption ability, removing 60% of RhB in the dark. After CdS NCs with a diameter of 5 nm were introduced, the hybridized system exhibited a rapid photodegradation ability toward the remaining RhB under visible light irradiation. It can be concluded that the introduction of mesoporous supporting materials not only facilitates the dispersion of NCs, by suppressing aggregation, but also improves the adsorbing of target pollutants for enhancing the removal efficiency by coupling adsorption and catalysis processes.

On the other hand, single semiconductors usually have one severe issue, that is the rapid recombination rate of photoinduced electrons and holes. As a nanosized semiconductor, NCs confront the problem due to the number of delocalized excited charge carriers inside NCs. Hence, constructing a hybridized system is an effective approach to mediating charge transfer for mitigating or even overcoming this problem. Tsai et al., combined graphene QDs with CdSe NCs using chitosan as a linker molecule [117]. From the PL results, it was observed that compared with bare CdSe NCs, the PL emission of CdSe NCs was totally quenched, which indicated that the irradiative recombination process inside CdSe NCs was blocked after the introduction of the graphene QDs. In graphene-containing photocatalytic systems, this blocking process has been observed frequently because graphene has a high conductive property that is beneficial for collecting photoinduced electrons and interrupting the irradiative recombination of electrons and holes. Hence, the quenching of PL intensity is often considered as an evidence to verify the enhancement of charge separation efficiency. Compared with other nanostructured graphene modified CdSe NCs such as nanosized and microsized sheets, the graphene QD–modified CdSe NCs exhibited strong photocurrent response under visible light irradiation, which indicated that the graphene QDs facilitated the production of more photogenerated charge carriers.

Generally, semiconductor NCs have narrow bandgap energy and are utilized as a sunlight sensitizer to modify wide-bandgap host nanomaterials for increasing solar energy harvesting in hybrid systems such as CdX/TiO_2 [118], CdX/ZnO [119], and others. After being modified by NCs, the photoresponse ranges are obviously extended to visible light and even to the near-infrared region so that the photoactivity can be apparently enhanced. In 2004 Lo et al. [120] coupled nanosized TiO_2 with CdSe QDs for enhancing the mineralization of 4-chlorophenol (4-CP) in comparison with bare TiO_2. However, the photodegradation and mineralization efficiencies of 4-CP barely reached 40% and 22%, respectively, which do not satisfying the requirements for its practical application. Fakhri et al. [121] reported the synthesis of CdTe QD decorated SnO_2 nanospheres through a facile hydrothermal method and its application for the degradation of two carcinogenic substances, ethylmethanesulfonate (EMS) and *N*-nitrosonornicotine (NNN). The CdTe QDs with a mean diameter of 3.75 nm were tightly anchored and dispersed at the surface of the SnO_2 nanospheres, which is beneficial for harvesting the visible light and charge transfer between CdTe and SnO_2. Compared with bare SnO_2 nanospheres, the photodegradation efficiencies of the CdTe QDs/SnO_2 nanospheres for both EMS and NNN were significantly enhanced, reaching over 90% in 90 min under UV light irradiation. It is clearly observed that QDs can be introduced into synthesized host materials through a hydrothermal route. QDs directly grow on the surface of host materials without the use of organic ligands, while the dispersion and quantum-size of QDs is difficult to control. Therefore synthesized QDs kept in solution phase are often used to modify other semiconductors.

Frankly speaking, high-quality QDs are prepared under elevated temperatures and are passivated by long-chain organic ligands where the surface ligands hamper charge transfer between the QDs and the host semiconductors; however, it is difficult to effectively separate photoinduced electrons and holes in QD/semiconductor systems. Therefore the removal or exchange of long-chain ligands is a practical barrier in improving photocatalytic performance using QDs to sensitize host semiconductors. To realize the application of QDs in modifying other photocatalysts, two strategies are used, namely (1) ligand exchange, which replaces the original long-chain ligands with short-chain and hydrophilic ligands and (2) directly preparing water-soluble QDs in an aqueous solution. For example, Bajorowicz et al. [122] first prepared $KTaO_3$ and thioglycolic acid (TGA) capped CdTe QDs by hydrothermal and aqueous synthetic routes, respectively. And then the two substances were mixed together, stirred, and the water was evaporated for obtaining the composites. The TGA contained thiol and carboxyl groups; the thiol group bonded with the CdTe QDs and the carboxyl group tightly adhere to the surface of the $KTaO_3$, which acted as linker molecule. When the CdTe QDs were excited by the incident of light, the photoinduced charges could transfer along with this linker molecule and achieve an efficient separation. This unique structure significantly enhanced the photocatalytic performance in the photodegradation of gaseous toluene using an LED light source ($\lambda_{max} = 415$ nm). The removal efficiency of CdTe QDs/$KTaO_3$ was 1.9 times higher than that of bare $KTiO_3$. For the ligand exchange methodology, the related operation procedure is facile, whereas the selection principle of desirable ligands is complicated due to the existence of many affecting factors such as the carbon-chain length, functional groups, the surface properties of NCs, the replacement efficiency, and feasibility. More importantly, ligand exchange has significant influence on the surface chemical properties of NCs, which might damage the stability, optical, and catalytic properties of NCs. Therefore a unified principle for the selection of ligands has been not reached up to now. In spite of these uncertainties during ligand exchange, some research groups have also made great efforts to utilize NCs in the catalytic field via the ligand exchange route. For example, rod-like TiO_2 NCs were prepared through a modified Schlenk technique under nitrogen and these were passivated by hydrophobic OA, where the ligand was not beneficial for applications in aqueous solutions [123]. Hence, the hydrophobic OA was substituted with an amphiphilic ligand, which could achieve a phase transfer of TiO_2 NCs from hexane to water.

It was reported that the 1D NCs have directional charge transfer properties in comparison with the isotropic properties of quasi sphere-like NCs. For example, NR NCs have a higher optical absorption cross section and exhibit an easier photoinduced charge carrier separation, which has been demonstrated in an Ni-decorated CdS system for photocatalytic H_2 production [124]. Therefore rod- and belt-like NCs have been widely used to construct NC-based photocatalysts. Ultrathin ZnSe nanobelts have been synthesized through the hydrothermal route in a mixed solvent of water and ethanol amine [125]. The obtained ultrathin ZnSe nanobelts displayed a higher photocatalytic removing efficiency of fuchsine acid than that of commercial ZnSe and TiO_2 under UV light irradiation. In addition, Laatar et al., combined CdSe NRs with TiO_2 NPs to enhance the photocatalytic activity toward RhB through a facile calcination route that ensures close contact between the CdSe NRs and the TiO_2 NPs [126]. Recently, more complicated shapes of NCs have been developed to regulate the photoexcited charge behaviors in NCs. Xiang et al. [127] fabricated CdSe-seeded CdS NRs by combining NCs with Au and PdS NPs for investigating the charge transfer behaviors in this hybridized system and the effects on the photocatalytic degradation of RhB. It was found that the introduction of Au NPs improved the collecting of photoexcited electrons and thus increased the separation efficiency of excited electrons and holes, while the addition of PdS resulted in a reduction of the separation efficiency, thus weakening the photocatalytic activity.

Apart from the traditional semiconductor NCs, various novel metal oxide NCs have also been fabricated to obtain better photocatalytic performance in water and air purification. For example, although the bandgap of bulk TiO_2 is large, some groups have tried to prepare nanosized TiO_2 NCs to narrow the bandgap and enhance the utilization of sunlight. NCs provide a large surface area that can be used to modify physiochemical properties and electronic structure. Wang et al., took advantage of modified hydrothermal and calcination routes to prepare TiO_2 NCs of different morphologies, which exhibited apparently different absorption ranges. It was observed that the novel nanostructured TiO_2 NCs could only absorb UV light and a small part of visible light at about 400 nm. Moreover, the obtained TiO_2 NCs were treated in a reductive H_2 atmosphere, resulting in the formation of defected black TiO_2 NCs, which significantly increased the visible light absorption efficiency of TiO_2 NCs. The black TiO_2 NCs exhibited enhanced photodegradation activity in comparison with other TiO_2 NCs under visible light illumination. The black TiO_2 NCs had rich surface defects due to the H_2 treatment, which could introduce a new energy level into the bandgap and act as a recombination center for enhancing the separation of photoinduced electrons and holes [128]. To further improve the photoactivity of TiO_2, narrow semiconductors have been introduced to construct a heterojunction, which not only increases the harvesting efficiency of sunlight, but also improves the separation efficiency of photoinduced electrons and holes. For instance, iron (II) sulfide (FeS_2) was used to modify TiO_2 to make the absorption range redshift and simultaneously improve the separation of photogenerated charge carriers [129]. Compared with bare TiO_2, the bandgap of FeS_2 NCs/TiO_2 decreased to 2.68 eV, which indicated that the heterojunction can absorb more visible light and exhibited enhanced photoactivity for organic dyes irradiated by sunlight.

Other transition metal oxide NCs including Co_3O_4, Fe_2O_3, NiO, Bi_2O_3, ZnO, In_2O_3, and Cu_2O have been intensively studied through similar Schlenk techniques in the past few decades. For obtaining a high photocatalytic degradation of water and air pollutants, well-developed strategies on enhancing the photocatalytic activity of TiO_2 have also been applied in these NCs such as combining with RGO or C_3N_4, coupling NCs with other semiconductors for forming a heterojunction, or tightly contacting with noble NPs (Ag, Au, Pt) relying on surface plasmon resonance. According to those strategies, many architectures have been studied such as ZnO NCs/RGO [130], Co_3O_4 QDs/g-C_3N_4

[131], $MnCO_3/Fe_2O_3$ NCs [132], and Fe_3O_4 NCs/Au [133] for wastewater purification.

4.3.6 Ternary and Quaternary Nanocrystals

Honestly, although a dozen binary NCs have been developed, the most widely used are Cd-based NCs because of their proper bandgap, high quantum efficiency, and well-developed synthesis routes. However, the uncertain toxicity of the Cd-containing NCs overshadows its practical applications. Except for their unique and tunable optical properties, this is one reason that low toxic and earth-abundant ternary NCs have been intensively investigated as a substitute for Cd-based NCs [134]. Among ternary NCs, $CuInS_2$, $CuInSe_2$, and $AgInS_2$ are the most studied material compositions, which are used to modify other wide-bandgap semiconductors for enhanced photocatalytic activities in environment treatment, energy conversion, and photovoltaics. Luo et al. [135] constructed $CuInS_2$ QDs/Bi_2WO_6 heterojunction by a facile adsorption-deposition route, which allows the prepared heterojunction to harvest a major part of visible light, thus resulting in apparent enhancement of photoresponse range. In addition, the *p*-type $CuInS_2$ QDs were tightly anchored at the surface of *n*-type hierarchical Bi_2WO_6 microflowers to efficiently form a *p-n* heterojunction at the porous surface, which endowed the photocatalyst with more active sites. The *p-n* heterojunction promoted the separation of the photogenerated electrons and holes and displayed an obvious enhancement of the photocatalytic degradation of toluene under visible light illumination. In addition, $AgInS_2$ NCs with a narrow direct bandgap of 1.8–2.2 eV were utilized to modify TiO_2 to enlarge the sunlight harvesting range and improve the charge separation [136]. The $AgInS_2/TiO_2$ heterojunction exhibited enhanced photooxidation activity toward gaseous 1,2-dichlorobenzene under visible light irradiation in comparison with bare TiO_2. Baek et al., took advantage of the large surface area of NRs and coupled $CuInS_2$ QDs with ZnO NRs through an in situ chemisorption and nucleation route for attempting to enhance the visible light absorption and increase the charge separation in the hybridized system [137]. In addition, $CuInSe_2$ has also been explored to modify other semiconductors for enhancing the photocatalytic activity toward organic pollutants [138] and water splitting [139] in the past, owing to its narrow bandgap of 1.04 eV, high absorption coefficient, and good photostability. For instance, Wu et al. [140] reported the synthesis of $CuInSe_2$ NC-coated TiO_2 nanotubes through a facile dropping routc, where the prepared heterojunction presented high photoelectrocatalytic efficiencies for 2,4-dichlorophenoxyacetic acid (2,4-D) and anthracene-9-carboxylic acid under visible light irradiation. Meanwhile, the complete elemental replacement of In with Ga is another interesting route that can be undertaken to form $CuGaS_2$ and $CuGaSe_2$ NCs [141].

Apart from ternary NCs, more complicated quaternary NCs also have been fabricated in photovoltaics, thermal electronics, and energy conversion at the initial stage, which mainly consist of Cu and chalcogens with In/Ga or Zn/Sn pairs. After the synthetic route and growth mechanism have been well obtained, their applications are widely spread to photo/photoelectrocatalytic abatement of environmental pollutants. Miyauchi et al. [142] took advantage of the high absorption coefficient of *p*-type Cu_2ZnSnS_4 to modify an *n*-type TiO_2 electrode and WO_3 powder through a solvothermal method, which showed a strong photoresponse transient current and significantly enhanced the photocatalytic decomposition of gaseous acetaldehyde under specific visible light in the range of 400–530 nm illumination.

4.4 The Future of Nanocrystals for Catalysis

The sections summarized in this chapter are intended to provide an introductory overview for new researchers in the design, preparation, and application of QD-based photocatalysts in energy and environment catalysis. Exciting examples have been provided to describe the basic fundamental principles, which can serve as a resource for researchers to contribute their efforts to the development of new structures or functions. Specifically, the attractive charge transfer feature of semiconductor QDs from the quantum confinement effect provides significant advantages as charge donors for catalytic reactions and sensitizers for photovoltaics. Unfortunately, the details regarding the procedures of charge transfer still remain undiscovered at present, which depend on applicable transient tools in catalytic systems including femto- or pico-second spectroscopy. Therefore the wide use of these tools facilitates the unveiling of the accurate reaction path, thereby optimizing QD-based catalysts and their performances in CO_2 reduction, water splitting, and environmental remediation. On the other hand, new system coupling, for example, photo-electro coupling and photothermal coupling, should be considered widely to provide more energy formats for improving catalytic performance.

Acknowledgment

This work was financially supported by the National Natural Science Foundation of China (21501138), the Natural Science Foundation of Hubei Province (2015CFB177), and the Science Research Foundation of Wuhan Institute of Technology (K201513).

References

[1] D. Astruc, F. Lu, J.R. Aranzaes, Nanoparticles as recyclable catalysts: the frontier between homogeneous and heterogeneous catalysis, Angew. Chem. Int. Ed. 44 (2005) 7852–7872.

[2] A.T. Bell, The impact of nanoscience on heterogeneous catalysis, Science 299 (2003) 1688–1691.

[3] H. Kotani, R. Hanazaki, K. Ohkubo, Y. Yamada, S. Fukuzumi, Size- and shape-dependent activity of metal nanoparticles as hydrogen-evolution catalysts: mechanistic insights into photocatalytic hydrogen evolution, Chemistry 17 (2011) 2777–2785.

[4] K.L. Sowers, B. Swartz, T.D. Krauss, Chemical mechanisms of semiconductor nanocrystal synthesis, Chem. Mater. 25 (2013) 1351–1362.

[5] J. Zhao, M.A. Holmes, F.E. Osterloh, Quantum confinement controls photocatalysis: a free energy analysis for photocatalytic proton reduction at CdSe nanocrystals, ACS Nano 7 (2013) 4316–4325.

[6] F. Scholz, L. Dworak, V.V. Matylitsky, J. Wachtveitl, Ultrafast electron transfer from photoexcited CdSe quantum dots to methylviologen, ChemPhysChem 12 (2011) 2255–2259.

[7] B.-R. Hyun, Z.Y. Wu, A.C. Bartnik, L.F. Sun, H.D. Abruna, F.W. Wise, et al., Electron injection from colloidal PbS quantum dots into titanium dioxide nanoparticles, ACS Nano 2 (2008) 2206–2212.

[8] J. Wang, T. Ding, K. Wu, Charge transfer from n-doped nanocrystals: mimicking intermediate events in multielectron photocatalysis, J. Am. Chem. Soc. 140 (2018) 7791–7794.

[9] C.B. Murray, D.J. Norris, M.G. Bawendi, Synthesis and characterization of nearly monodisperse CdE (E = sulfur, selenium, tellurium) semiconductor nanocrystallites, J. Am. Chem. Soc. 115 (1993) 8706–8715.

[10] A.A. Peng, X.G. Peng, Formation of high-quality CdTe, CdSe, and CdS nanocrystals using CdO as precursor, J. Am. Chem. Soc. 123 (2001) 183–184.

[11] L.H. Qu, A.A. Peng, X.G. Peng, Alternative routes toward high quality CdSe nanocrystals, Nano. Lett. 1 (2001) 333–337.

[12] M.A. Hines, P. Guyot-Sionnest, Bright UV-Blue luminescent colloidal ZnSe nanocrystals, J. Phys. Chem. B 102 (1998) 3655–3657.

[13] R.G. Xie, D. Battaglia, X.G. Peng, Colloidal InP nanocrystals as efficient emitters covering blue to near-infrared, J. Am. Chem. Soc. 129 (2007) 15432–15433.

[14] I. Moreels, K. Lambert, D.D. Muynck, F. Vanhaecke, D. Poelman, J.C. Martins, et al., Composition and size-dependent extinction coefficient of colloidal PbSe quantum dots, Chem. Mater. 19 (2007) 6101–6106.

[15] D. Battaglia, X.G. Peng, Formation of high quality InP and InAs nanocrystals in a noncoordinating solvent, Nano. Lett. 2 (2002) 1027–1030.

[16] G. Zhang, Y. Zhou, X. Fan, J. Zou, W. Dong, X. Xu, Efficient and robust visible light photocatalytic H_2 production based on CdSe quantum dots sensitized titania, Int. J. Hydrogen Energy 42 (2017) 19877–19884.
[17] S. Sardar, P. Kar, S. Sarkar, P. Lemmens, S.K. Pal, Interfacial carrier dynamics in PbS-ZnO light harvesting assemblies and their potential implication in photovoltaic/photocatalysis application, Sol. Energy Mater. Sol. C 134 (2015) 400–406.
[18] Z.T. Deng, L. Cao, F.Q. Tang, B.S. Zou, A new route to zinc-blende CdSe nanocrystals: mechanism and synthesis, J. Phys. Chem. B 109 (2005) 16671–16675.
[19] H. Shen, H. Wang, X. Chen, J.Z. Niu, W. Xu, X.M. Li, et al., Size- and shape-controlled synthesis of CdTe and PbTe nanocrystals using tellurium dioxide as the tellurium precursor, Chem. Mater. 22 (2010) 4756–4761.
[20] V. Lesnyak, N. Gaponik, A. Eychmuller, Colloidal semiconductor nanocrystals: the aqueous approach, Chem. Soc. Rev. 42 (2013) 2905–2929.
[21] C. Ma, D. Wu, X. Yao, X. Liu, M. Wei, Y. Liu, et al., Enhanced visible-light photocatalytic decomposition of organic dye over CdSe/Al_2TiO_5 heterojunction photocatalysts, J. Alloys Compd. 712 (2017) 486–493.
[22] L. Liu, Y. Wang, W. An, J. Hu, W. Cui, Y. Liang, Photocatalytic activity of PbS quantum dots sensitized flower-like Bi_2WO_6 for degradation of Rhodamine B under visible light irradiation, J. Mol. Catal. A: Chem. 394 (2014) 309–315.
[23] Y.R. Lv, R. Huo, S.Y. Yang, Y.Q. Liu, X.J. Li, Y.-H. Xu, Self-assembled synthesis of PbS quantum dots supported on polydopamine encapsulated $BiVO_4$ for enhanced visible-light-driven photocatalysis, Sep. Purif. Technol. 197 (2018) 281–288.
[24] J.J. Shi, S. Wang, T.T. He, E.S. Abdel-Halim, J.J. Zhu, Sonoelectrochemical synthesis of water-soluble CdTe quantum dots, Ultrason. Sonochem. 21 (2014) 493–498.
[25] A.E. Saunders, B. Koo, X. Wang, C.K. Shih, B.A. Korgel, Structural characterization and temperature-dependent photoluminescence of linear CdTe/CdSe/CdTe heterostructure nanorods, Chemphyschem 9 (2008) 1158–1163.
[26] D.V. Talapin, J.H. Nelson, E.V. Shevchenko, S. Aloni, B. Sadtler, A.P. Alivisatos, Seeded growth of highly luminescent CdSe/CdS nanoheterostructures with rod and tetrapod morphologies, Nano. Lett. 7 (2007) 2951–2959.
[27] S. Li, X. Tang, Z. Zang, Y. Yao, Z. Yao, H. Zhong, et al., I-III-VI chalcogenide semiconductor nanocrystals: synthesis, properties, and applications, Chin. J. Catal. 39 (2018) 590–605.
[28] D. Wang, W. Zheng, C. Hao, Q. Peng, Y. Li, General synthesis of I-III-VI_2 ternary semiconductor nanocrystals, Chem. Commun. 0 (2008) 2556–2558.
[29] H. Yang, L.A. Jauregui, G. Zhang, Y.P. Chen, Y. Wu, Nontoxic and abundant copper zinc tin sulfide nanocrystals for potential high-temperature thermoelectric energy harvesting, Nano Lett. 12 (2012) 540–545.
[30] M. Tabata, K. Maeda, T. Ishihara, T. Minegishi, T. Takata, K. Domen, Photocatalytic hydrogen evolution from water using copper gallium sulfide under visible-light irradiation, J. Phys. Chem. C 114 (2010) 11215–11220.
[31] I. Tsuji, H. Kato, H. Kobayashi, A. Kudo, Photocatalytic H_2 evolution under visible-light irradiation over band-structure-controlled $(CuIn)_xZn_{2(1-x)}S_2$ solid solutions, J. Phys. Chem. B 109 (2005) 7323–7329.
[32] J. Park, J. Joo, S.G. Kwon, Y. Jang, T. Hyeon, Synthesis of monodisperse spherical nanocrystals, Angew. Chem. Int. Ed. 46 (2007) 4630–4660.

[33] J. Rockenberger, E.C. Scher, A.P. Alivisatos, A new nonhydrolytic single-precursor approach to surfactant-capped nanocrystals of transition metal oxides, J. Am. Chem. Soc. 121 (1999) 11595–11596.
[34] M. Yin, C.K. Wu, Y.B. Lou, C. Burda, J. Koberstein, Y.M. Zhu, et al., Copper oxide nanocrystals, J. Am. Chem. Soc. 127 (2005) 9506–9511.
[35] X. Zhou, Q. Xu, W. Lei, T. Zhang, X. Qi, G. Liu, et al., Origin of tunable photocatalytic selectivity of well-defined alpha-Fe_2O_3 nanocrystals, Small 10 (2014) 674–679.
[36] X. Zhou, J. Lan, G. Liu, K. Deng, Y. Yang, G. Nie, et al., Facet-mediated photodegradation of organic dye over hematite architectures by visible light, Angew. Chem. Int. Ed. 51 (2012) 178–182.
[37] R. Xu, H.C. Zeng, Mechanistic investigation on salt-mediated formation of free-standing Co_3O_4 nanocubes at 95 °C, J. Phys. Chem. B 107 (2003) 926–930.
[38] T. He, D.R. Chen, X.L. Jiao, Controlled synthesis of Co_3O_4 nanoparticles through oriented aggregation, Chem. Mater. 16 (2004) 737–743.
[39] N. Shi, W. Cheng, H. Zhou, T. Fan, M. Niederberger, Facile synthesis of monodisperse Co_3O_4 quantum dots with efficient oxygen evolution activity, Chem. Commun. 51 (2015) 1338–1340.
[40] O. Yehezkeli, D.R. de Oliveira, J.N. Cha, Electrostatically assembled CdS-Co_3O_4 nanostructures for photo-assisted water oxidation and photocatalytic reduction of dye molecules, Small 11 (2015) 668–674.
[41] W.S. Seo, H.H. Jo, K. Lee, J.T. Park, Preparation and optical properties of highly crystalline, colloidal, and size-controlled Indium oxide nanoparticles, Adv. Mater. 15 (2003) 795–797.
[42] L.Y. Chen, Z.X. Wang, Z.D. Zhang, Corundum-type tubular and rod-like In_2O_3 nanocrystals: synthesis from designed InOOH and application in photocatalysis, New J. Chem. 33 (2009) 1109.
[43] M.A. Gondal, M.A. Dastageer, L.E. Oloore, U. Baig, S.G. Rashid, Enhanced photo-catalytic activity of ordered mesoporous indium oxide nanocrystals in the conversion of CO_2 into methanol, J. Environ. Sci. Health A 52 (2017) 785–793.
[44] J. Yin, H. Cao, Synthesis and photocatalytic activity of single-crystalline hollow rh-In_2O_3 nanocrystals, Inorg. Chem. 51 (2012) 6529–6536.
[45] S. Jana, B.B. Srivastava, N. Pradhan, A controlled growth process to design relatively larger size semiconductor nanocrystals, J. Phys. Chem. C 117 (2013) 1183–1188.
[46] M.A. Hines, P. Guyot-Sionnest, Synthesis and characterization of strongly luminescing ZnS-capped CdSe nanocrystals, J. Phys. Chem. B 100 (1996) 468–471.
[47] H.M. Zhu, N.H. Song, T.Q. Lian, Controlling charge separation and recombination rates in CdSe/ZnS type I core-shell quantum dots by shell thicknesses, J. Am. Chem. Soc. 132 (2010) 15038–15045.
[48] L. Huang, X. Wang, J. Yang, G. Liu, J. Han, C. Li, Dual cocatalysts loaded type I CdS/ZnS core/shell nanocrystals as effective and stable photocatalysts for H_2 evolution, J. Phys. Chem. C 117 (2013) 11584–11591.
[49] P.S. Nair, T. Radhakrishnan, N. Revaprasadu, G.A. Kolawole, P. O'Brien, A single-source route to CdS nanorods, Chem. Commun. (2002) 564–565.
[50] Q. Peng, Y.J. Dong, Z.X. Deng, Y.D. Li, Selective synthesis and characterization of CdSe nanorods and fractal nanocrystals, Inorg. Chem. 41 (2002) 5249–5254.

[51] J. Li, X. Hong, D. Li, K. Zhao, L. Wang, H. Wang, et al., Mixed ligand system of cysteine and thioglycolic acid assisting in the synthesis of highly luminescent water-soluble CdTe nanorods, Chem. Commun. (2004) 1740–1741.

[52] J.M. Luther, H.M. Zheng, B. Sadtler, A.P. Alivisatos, Synthesis of PbS nanorods and other ionic nanocrystals of complex morphology by sequential cation exchange reactions, J. Am. Chem. Soc. 131 (2009) 16851–16857.

[53] B. Sadtler, D.O. Demchenko, H.M. Zheng, S.M. Hughes, M.G. Merkle, U. Dahmen, et al., Selective facet reactivity during cation exchange in cadmium sulfide nanorods, J. Am. Chem. Soc. 131 (2009) 5285–5293.

[54] F. Chen, D. Jia, Y. Cao, X. Jin, A. Liu, Facile synthesis of CdS nanorods with enhanced photocatalytic activity, Ceram. Int. 41 (2015) 14604–14609.

[55] J. Ran, J. Yu, M. Jaroniec, $Ni(OH)_2$ modified CdS nanorods for highly efficient visible-light-driven photocatalytic H_2 generation, Green. Chem. 13 (2011) 2708.

[56] K. Wu, H. Zhu, T. Lian, Ultrafast exciton dynamics and light-driven H_2 evolution in colloidal semiconductor nanorods and Pt-tipped nanorods, Acc. Chem. Res. 48 (2015) 851–859.

[57] F. Qiu, Z. Han, J.J. Peterson, M.Y. Odoi, K.L. Sowers, T.D. Krauss, Photocatalytic hydrogen generation by CdSe/CdS nanoparticles, Nano Lett. 16 (2016) 5347–5352.

[58] M.T. Ng, C.B. Boothroyd, J.J. Vittal, One-pot synthesis of new-phase $AgInSe_2$ nanorods, J. Am. Chem. Soc. 128 (2006) 7118–7119.

[59] Q.H. Wang, K. Kalantar-Zadeh, A. Kis, J.N. Coleman, M.S. Strano, Electronics and optoelectronics of two-dimensional transition metal dichalcogenides, Nat. Nanotechnol. 7 (2012) 699–712.

[60] S. Ithurria, M.D. Tessier, B. Mahler, R.P. Lobo, B. Dubertret, A.L. Efros, Colloidal nanoplatelets with two-dimensional electronic structure, Nat. Mater. 10 (2011) 936–941.

[61] S. Ithurria, B. Dubertret, Quasi 2D colloidal CdSe platelets with thicknesses controlled at the atomic level, J. Am. Chem. Soc. 130 (2008) 16504–16505.

[62] S. Ithurria, G. Bousquet, B. Dubertret, Continuous transition from 3D to 1D confinement observed during the formation of CdSe nanoplatelets, J. Am. Chem. Soc. 133 (2011) 3070–3077.

[63] Y. Xu, W. Zhao, R. Xu, Y. Shi, B. Zhang, Synthesis of ultrathin CdS nanosheets as efficient visible-light-driven water splitting photocatalysts for hydrogen evolution, Chem. Commun. 49 (2013) 9803–9805.

[64] Y. Xu, X. Yin, Y. Huang, P. Du, B. Zhang, Hydrogen production on a hybrid photocatalytic system composed of ultrathin CdS nanosheets and a molecular nickel complex, Chem. 21 (2015) 4571–4575.

[65] K. Wu, Q. Li, Y. Du, Z. Chen, T. Lian, Ultrafast exciton quenching by energy and electron transfer in colloidal CdSe nanosheet–Pt heterostructures, Chem. Sci. 6 (2015) 1049–1054.

[66] M. Zhukovskyi, P. Tongying, H. Yashan, Y. Wang, M. Kuno, Efficient photocatalytic hydrogen generation from Ni nanoparticle decorated CdS nanosheets, ACS Catal. 5 (2015) 6615–6623.

[67] J. He, B. Li, J. Mao, Y. Liang, X. Yang, Z. Cui, et al., Four-electron oxygen reduction from mesoporous carbon modified with Fe_2O_3 nanocrystals, J. Mater. Sci. 52 (2017) 10938–10947.

[68] J. Hu, X. Li, L. Li, M. Qi, X. Chen, Q. Zuo, et al., In situ preparation of uniform and ultrafine SnO_2 nanocrystals anchored within a mesoporous carbon network as advanced anode materials, Inorg. Chem. Front. 5 (2018) 378–385.

[69] T. Qin, S. Zhang, P. Chen, F. Zhao, H. Gao, X. Chen, et al., Facile preparation of TiO_2 nanocrystals inserted in monodispersed mesoporous SiO_2 nanospheres for enhanced photocatalytic activity, J. Mater. Sci: Mater. Electron. 27 (2016) 13161–13170.
[70] J.L. Hu, Q.H. Yang, H. Lin, Y.P. Ye, Q. He, J.N. Zhang, et al., Mesoporous silica nanospheres decorated with CdS nanocrystals for enhanced photocatalytic and excellent antibacterial activities, Nanoscale 5 (2013) 6327–6332.
[71] R. Lin, L. Shen, Z. Ren, W. Wu, Y. Tan, H. Fu, et al., Enhanced photocatalytic hydrogen production activity via dual modification of MOF and reduced graphene oxide on CdS, Chem. Commun. 50 (2014) 8533–8535.
[72] J. He, Z. Yan, J. Wang, J. Xie, L. Jiang, Y. Shi, et al., Significantly enhanced photocatalytic hydrogen evolution under visible light over CdS embedded on metal-organic frameworks, Chem. Commun. 49 (2013) 6761–6763.
[73] L. Shen, S. Liang, W. Wu, R. Liang, L. Wu, CdS-decorated UiO–66(NH_2) nanocomposites fabricated by a facile photodeposition process: an efficient and stable visible-light-driven photocatalyst for selective oxidation of alcohols, J. Mater. Chem. A 1 (2013) 11473.
[74] H. Gan, Z. Wang, H. Li, Y. Wang, L. Sun, Y. Li, CdSe QDs@UIO-66 composite with enhanced photocatalytic activity towards RhB degradation under visible-light irradiation, RSC Adv. 6 (2016) 5192–5197.
[75] J. Aguilera-Sigalat, D. Bradshaw, Synthesis and applications of metal-organic framework–quantum dot (QD@MOF) composites, Coord. Chem. Rev. 307 (2016) 267–291.
[76] G. Lu, S. Li, Z. Guo, O.K. Farha, B.G. Hauser, X. Qi, et al., Imparting functionality to a metal-organic framework material by controlled nanoparticle encapsulation, Nat. Chem. 4 (2012) 310–316.
[77] X. Lin, G. Gao, L. Zheng, Y. Chi, G. Chen, Encapsulation of strongly fluorescent carbon quantum dots in metal-organic frameworks for enhancing chemical sensing, Anal. Chem. 86 (2014) 1223–1228.
[78] I.E. Rauda, R. Buonsanti, L.C. Saldarriaga-Lopez, K. Benjauthrit, L.T. Schelhas, M. Stefik, et al., General method for the synthesis of hierarchical nanocrystal-based mesoporous materials, ACS Nano 6 (2012) 6386–6399.
[79] I. Vamvasakis, K.S. Subrahmanyam, M.G. Kanatzidis, G.S. Armatas, Template-directed assembly of metal chalcogenide nanocrystals into ordered mesoporous networks, ACS Nano 9 (2015) 4419–4426.
[80] J. Feng, J. Liu, X. Cheng, J. Liu, M. Xu, J. Zhang, Hydrothermal cation exchange enabled gradual evolution of Au@ZnS-AgAuS yolk-shell nanocrystals and their visible light photocatalytic applications, Adv. Sci. 5 (2018) 1700376.
[81] D. Raciti, C. Wang, Recent advances in CO_2 reduction electrocatalysis on copper, ACS Energy Lett. 3 (2018) 1545–1556.
[82] J. Ran, M. Jaroniec, S.Z. Qiao, Cocatalysts in semiconductor-based photocatalytic CO_2 reduction: achievements, challenges, and opportunities, Adv. Mater. 30 (2018) 1704649.
[83] S.A. Razzak, S.A.M. Ali, M.M. Hossain, H. deLasa, Biological CO_2 fixation with production of microalgae in wastewater – a review, Renew. Sustain. Energy Rev. 76 (2017) 379–390.
[84] F.A. Costa Oliveira, M.A. Barreiros, S. Abanades, A.P.F. Caetano, R.M. Novais, R.C. Pullar, Solar thermochemical CO_2 splitting using cork-templated ceria ecoceramics, J. CO_2 Util. 26 (2018) 552–563.

[85] R.K. de Richter, T. Ming, S. Caillol, Fighting global warming by photocatalytic reduction of CO_2 using giant photocatalytic reactors, Renew. Sustain. Energy Rev. 19 (2013) 82–106.
[86] L.G. Wang, S.J. Pennycook, S.T. Pantelides, The role of the nanoscale in surface reactions: CO_2 on CdSe, Phys. Rev. Lett. 89 (2002) 075506.
[87] H. Cho, W.D. Kim, K. Lee, S. Lee, G.-S. Kang, H.I. Joh, et al., Selectivity of photoelectrochemical CO_2 reduction modulated with electron transfer from size-tunable quantized energy states of CdSe nanocrystals, Appl. Surf. Sci. 429 (2018) 2–8.
[88] S. Wang, X. Wang, Photocatalytic CO_2 reduction by CdS promoted with a zeolitic imidazolate framework, Appl. Catal. B: Environ. 162 (2015) 494–500.
[89] J.H. Jeon, P.M. Mareeswaran, C.H. Choi, S.I. Woo, Synergism between CdTe semiconductor and pyridine–photoenhanced electrocatalysis for CO_2 reduction to formic acid, RSC Adv. 4 (2014) 3016–3019.
[90] M.F. Ehsan, T. He, In situ synthesis of ZnO/ZnTe common cation heterostructure and its visible-light photocatalytic reduction of CO_2 into CH_4, Appl. Catal. B: Environ. 166-167 (2015) 345–352.
[91] C. Wang, R.L. Thompson, P. Ohodnicki, J. Baltrus, C. Matranga, Size-dependent photocatalytic reduction of CO_2 with PbS quantum dot sensitized TiO_2 heterostructured photocatalysts, J. Mater. Chem. 21 (2011) 13452.
[92] P. Li, X. Sui, J. Xu, H. Jing, C. Wu, H. Peng, et al., Worm-like InP/TiO_2 NTs heterojunction with unmatched energy band photo-enhanced electrocatalytic reduction of CO_2 to methanol, Chem. Eng. J. 247 (2014) 25–32.
[93] A. Ali, W.C. Oh, Synthesis of Ag_2Se–graphene–TiO_2 nanocomposite and analysis of photocatalytic activity of CO_2 reduction to CH_3OH, Bull. Mater. Sci. 40 (2017) 1319–1328.
[94] C. Wang, R.L. Thompson, J. Baltrus, C. Matranga, Visible light photoreduction of CO_2 using CdSe/Pt/TiO_2 heterostructured catalysts, J. Phys. Chem. Lett. 1 (2009) 48–53.
[95] Y.J. Jang, J.W. Jang, J. Lee, J.H. Kim, H. Kumagai, J. Lee, et al., Selective CO production by Au coupled ZnTe/ZnO in the photoelectrochemical CO_2 reduction system, Energy Environ. Sci. 8 (2015) 3597–3604.
[96] M.F. Kuehnel, C.D. Sahm, G. Neri, J.R. Lee, K.L. Orchard, A.J. Cowan, et al., ZnSe quantum dots modified with a Ni(cyclam) catalyst for efficient visible-light driven CO_2 reduction in water, Chem. Sci. 9 (2018) 2501–2509.
[97] M.F. Kuehnel, K.L. Orchard, K.E. Dalle, E. Reisner, Selective photocatalytic CO_2 reduction in water through anchoring of a molecular Ni catalyst on CdS nanocrystals, J. Am. Chem. Soc. 139 (2017) 7217–7223.
[98] F. Xu, J. Zhang, B. Zhu, J. Yu, J. Xu, $CuInS_2$ sensitized TiO_2 hybrid nanofibers for improved photocatalytic CO_2 reduction, Appl. Catal. B: Environ. 230 (2018) 194–202.
[99] Z. Jiang, W. Wan, H. Li, S. Yuan, H. Zhao, P.K. Wong, A hierarchical z-scheme alpha-Fe_2O_3/g-C_3 N_4 hybrid for enhanced photocatalytic CO_2 reduction, Adv. Mater. 30 (2018) 1706108.
[100] Y. Huang, C.F. Yan, C.Q. Guo, S.L. Huang, Enhanced photoreduction activity of carbon dioxide over Co_3O_4/CeO_2 catalysts under visible light irradiation, Int. J. Photoenergy 2015 (2015) 1–11.
[101] Q. Xu, J. Yu, J. Zhang, J. Zhang, G. Liu, Cubic anatase TiO_2 nanocrystals with enhanced photocatalytic CO_2 reduction activity, Chem. Commun. 51 (2015) 7950–7953.

[102] S. Xie, Y. Wang, Q. Zhang, W. Deng, Y. Wang, MgO- and Pt-promoted TiO_2 as an efficient photocatalyst for the preferential reduction of carbon dioxide in the presence of water, ACS Catal. 4 (2014) 3644–3653.
[103] E. Kecsenovity, B. Endrődi, Z. Pápa, K. Hernádi, K. Rajeshwar, C. Janáky, Decoration of ultra-long carbon nanotubes with Cu_2O nanocrystals: a hybrid platform for enhanced photoelectrochemical CO_2 reduction, J. Mater. Chem. A 4 (2016) 3139–3147.
[104] Z. Xiong, Z. Lei, C.-C. Kuang, X. Chen, B. Gong, Y. Zhao, et al., Selective photocatalytic reduction of CO_2 into CH_4 over Pt-Cu_2O/TiO_2 nanocrystals: the interaction between Pt and Cu_2O cocatalysts, Appl. Catal. B: Environ. 202 (2017) 695–703.
[105] Z. Xiong, Z. Lei, X. Chen, B. Gong, Y. Zhao, J. Zhang, et al., CO_2 photocatalytic reduction over Pt deposited TiO_2 nanocrystals with coexposed {101} and {001} facets: effect of deposition method and Pt precursors, Catal. Commun. 96 (2017) 1–5.
[106] Z. Xiong, Y. Luo, Y. Zhao, J. Zhang, C. Zheng, J.C. Wu, Synthesis, characterization and enhanced photocatalytic CO_2 reduction activity of graphene supported TiO_2 nanocrystals with coexposed {001} and {101} facets, Phys. Chem. Chem. Phys. 18 (2016) 13186–13195.
[107] T. Wang, L. Shi, J. Tang, V. Malgras, S. Asahina, G. Liu, et al., A Co_3O_4-embedded porous ZnO rhombic dodecahedron prepared using zeolitic imidazolate frameworks as precursors for CO_2 photoreduction, Nanoscale 8 (2016) 6712–6720.
[108] Yiliguma, Y. Tang, G. Zheng, Colloidal nanocrystals for electrochemical reduction reactions, J. Colloid Interface. Sci. 485 (2017) 308–327.
[109] A. Samadi-maybodi, F. Abbasi, R. Akhoondi, Aqueous synthesis and characterization of CdS quantum dots capped with some amino acids and investigations of their photocatalytic activities, Colloid Surf. A 447 (2014) 111–119.
[110] A. Samadi-Maybodi, M.R. Sadeghi-Maleki, In-situ synthesis of high stable CdS quantum dots and their application for photocatalytic degradation of dyes, Spectrochim. Acta A Mol. Biomol. Spectrosc. 152 (2016) 156–164.
[111] A.A.P. Mansur, H.S. Mansur, F.P. Ramanery, L.C. Oliveira, P.P. Souza, "Green" colloidal ZnS quantum dots/chitosan nano-photocatalysts for advanced oxidation processes: study of the photodegradation of organic dye pollutants, Appl. Catal. B: Environ. 158-159 (2014) 269–279.
[112] J. Wen, C. Ma, P. Huo, X. Liu, M. Wei, Y. Liu, et al., Construction of vesicle CdSe nano-semiconductors photocatalysts with improved photocatalytic activity: Enhanced photo induced carriers separation efficiency and mechanism insight, J. Environ. Sci. (China) 60 (2017) 98–107.
[113] F. Ke, L. Wang, J. Zhu, Facile fabrication of CdS-metal-organic framework nanocomposites with enhanced visible-light photocatalytic activity for organic transformation, Nano Res. 8 (2015) 1834–1846.
[114] R.C. Chikate, B.S. Kadu, Improved photocatalytic activity of CdSe-nanocomposites: effect of Montmorillonite support towards efficient removal of Indigo Carmine, Spectrochim. Acta A Mol. Biomol. Spectrosc. 124 (2014) 138–147.
[115] H.B. Huang, Y. Wang, W.B. Jiao, F.Y. Cai, M. Shen, S.G. Zhou, et al., Lotus–leaf–derived activated carbon supported nano–CdS as energy–efficient photocatalysts under visible irradiation, ACS Sustain. Chem. Eng. 6 (2018).

[116] C. Yang, J. Cheng, Y. Chen, Y. Hu, CdS nanoparticles immobilized on porous carbon polyhedrons derived from a metal-organic framework with enhanced visible light photocatalytic activity for antibiotic degradation, Appl. Surf. Sci. 420 (2017) 252–259.

[117] K.-A. Tsai, Y.-J. Hsu, Graphene quantum dots mediated charge transfer of CdSe nanocrystals for enhancing photoelectrochemical hydrogen production, Appl. Catal. B: Environ. 164 (2015) 271–278.

[118] P. Wang, D. Li, J. Chen, X. Zhang, J. Xian, X. Yang, et al., A novel and green method to synthesize CdSe quantum dots-modified TiO_2 and its enhanced visible light photocatalytic activity, Appl. Catal. B: Environ. 160-161 (2014) 217–226.

[119] C. Eley, T. Li, F. Liao, S.M. Fairclough, J.M. Smith, G. Smith, et al., Nanojunction-mediated photocatalytic enhancement in heterostructured CdS/ZnO, CdSe/ZnO, and CdTe/ZnO nanocrystals, Angew. Chem. Int. Ed. 53 (2014) 7838–7842.

[120] S.C. Lo, C.F. Lin, C.H. Wu, P.H. Hsieh, Capability of coupled CdSe/TiO_2 for photocatalytic degradation of 4-chlorophenol, J. Hazard. Mater. 114 (2004) 183–190.

[121] A. Fakhri, M. Naji, L. Fatolahi, P.A. Nejad, Synthesis and characterization of Fe_3O_4 and CdTe quantum dots anchored SnO_2 nanofibers and SnO_2 nanospheres for degradation and removal of two carcinogen substance, J. Mater. Sci: Mater. Electron. 28 (2017) 16484–16492.

[122] B. Bajorowicz, J. Nadolna, W. Lisowski, T. Klimczuk, A. Zaleska-Medynska, The effects of bifunctional linker and reflux time on the surface properties and photocatalytic activity of CdTe quantum dots decorated $KTaO_3$ composite photocatalysts, Appl. Catal. B: Environ. 203 (2017) 452–464.

[123] R.J. Wilkerson, T. Elder, O. Sowinksi, J.I. Fostvedt, J.D. Hoefelmeyer, Phase transfer of oleic acid stabilized rod-shaped anatase TiO_2 nanocrystals, Surf. Sci. 648 (2016) 333–338.

[124] T. Simon, N. Bouchonville, M.J. Berr, A. Vaneski, A. Adrovic, D. Volbers, et al., Redox shuttle mechanism enhances photocatalytic H_2 generation on Ni-decorated CdS nanorods, Nat. Mater. 13 (2014) 1013–1018.

[125] S. Xiong, B. Xi, C. Wang, G. Xi, X. Liu, Y. Qian, Solution-phase synthesis and high photocatalytic activity of wurtzite ZnSe ultrathin nanobelts: a general route to 1D semiconductor nanostructured materials, Chemistry 13 (2007) 7926–7932.

[126] F. Laatar, H. Moussa, H. Alem, L. Balan, E. Girot, G. Medjahdi, et al., CdSe nanorod/TiO_2 nanoparticle heterojunctions with enhanced solar- and visible-light photocatalytic activity, Beilstein J. Nanotechnol. 8 (2017) 2741–2752.

[127] X. Xiang, L. Chou, X. Li, Remarkable charge transfer between CdSe seeded CdS nanorods and its metallic and semiconducting tips and its effect on Rhodamine B photodegradation, Phys. Chem. Chem. Phys. 15 (2013) 19545.

[128] D. Wang, J. Yang, X. Li, J. Wang, J. Lang, J. Wang, et al., Preparation of morphology-controlled TiO_2 nanocrystals for the excellent photocatalytic activity under simulated solar irradiation, Mater. Res. Bull. 94 (2017) 38–44.

[129] J. Rashid, S. Saleem, S.U. Awan, A. Iqbal, R. Kumar, M.A. Barakat, et al., Stabilized fabrication of anatase-TiO_2/FeS_2 (pyrite) semiconductor composite nanocrystals for enhanced solar light-mediated photocatalytic degradation of methylene blue, RSC Adv. 8 (2018) 11935–11945.

[130] M. Darwish, A. Mohammadi, N. Assi, Q.S. Manuchehri, Y. Alahmad, S. Abuzerr, Shape-controlled ZnO nanocrystals synthesized via auto

combustion method and enhancement of the visible light catalytic activity by decoration on graphene, J. Alloys Compd. 703 (2017) 396–406.
[131] H. Zhang, W. Tian, L. Zhou, H. Sun, M. Tade, S. Wang, Monodisperse Co_3O_4 quantum dots on porous carbon nitride nanosheets for enhanced visible-light-driven water oxidation, Appl. Catal. B: Environ. 223 (2018) 2–9.
[132] L. Cui, M. Niu, G. Chen, Y. Wang, Fabrication and structure characterization of $MnCO_3/\alpha\text{-}Fe_2O_3$ nanocrystal heterostructures, Mater. Lett. 63 (2009) 2499–2502.
[133] J. Zhang, Q. Fang, J. Duan, H. Xu, H. Xu, S. Xuan, Magnetically separable nanocatalyst with the Fe_3O_4 core and polydopamine-sandwiched Au nanocrystal shell, Langmuir 34 (2018) 4298–4306.
[134] C. Coughlan, M. Ibanez, O. Dobrozhan, A. Singh, A. Cabot, K.M. Ryan, Compound copper chalcogenide nanocrystals, Chem. Rev. 117 (2017) 5865–6109.
[135] S. Luo, J. Ke, M. Yuan, Q. Zhang, P. Xie, L. Deng, et al., $CuInS_2$ quantum dots embedded in Bi_2WO_6 nanoflowers for enhanced visible light photocatalytic removal of contaminants, Appl. Catal. B: Environ. 221 (2018) 215–222.
[136] B. Liu, X. Li, Q. Zhao, J. Ke, M. Tadé, S. Liu, Preparation of $AgInS_2/TiO_2$ composites for enhanced photocatalytic degradation of gaseous o-dichlorobenzene under visible light, Appl. Catal. B: Environ. 185 (2016) 1–10.
[137] M. Baek, E.J. Kim, S.W. Hong, W. Kim, K. Yong, Environmentally benign synthesis of $CuInS_2/ZnO$ heteronanorods: visible light activated photocatalysis of organic pollutant/bacteria and study of its mechanism, Photochem. Photobiol. Sci. 16 (2017) 1792–1800.
[138] A.S. Kshirsagar, A. Gautam, P.K. Khanna, Efficient photo-catalytic oxidative degradation of organic dyes using $CuInSe_2/TiO_2$ hybrid hetero-nanostructures, J. Photochem. Photobiol. A: Chem. 349 (2017) 73–90.
[139] P. Sheng, W. Li, X. Tong, X. Wang, Q. Cai, Development of a high performance hollow $CuInSe_2$ nanospheres-based photoelectrochemical cell for hydrogen evolution, J. Mater. Chem. A 2 (2014) 18974–18987.
[140] Z. Wu, X. Tong, P. Sheng, W. Li, X. Yin, J. Zou, et al., Fabrication of high-performance $CuInSe_2$ nanocrystals-modified TiO_2 NTs for photocatalytic degradation applications, Appl. Surf. Sci. 351 (2015) 309–315.
[141] B. Han, Z. Yu, Y. Yang, L. Yang, F. Hu, Z. Li, et al., Synthesis of high-efficiency $CuGaS_2$ quantum dots for photocatalysis by fine modulation of the reaction temperature, Sci. Adv. Mater. 10 (2018) 808–817.
[142] M. Miyauchi, T. Hanayama, D. Atarashi, E. Sakai, Photoenergy conversion in p-type Cu_2ZnSnS_4 nanorods and n-type metal oxide composites, J. Phys. Chem. C 116 (2012) 23945–23950.

5

NANO-GOLD BOOSTED ENVIRONMENTAL CATALYSIS

Wei Xiong[1,2], Pancras Ndokoye[3] and Michael K.H. Leung[2]

[1]Key Laboratory of Industrial Ecology and Environmental Engineering (Ministry of Education), School of Environmental Science and Technology, Dalian University of Technology, Dalian, P.R. China [2]Ability R&D Energy Research Centre, School of Energy and Environment, City University of Hong Kong, Hong Kong, P.R. China [3]Science, Technology and Innovation Unit, Directorate of Education Policy and Planning, Rwanda Ministry of Education, Kigali, Rwanda

5.1 Introduction to the Catalytic Properties of Gold

Gold is a metal element that is usually used in our daily lives as the currency and jewelry. It has also been widely used in a lot of fields such as the electronics, instrumentation, and biomedical industries, attributing to its excellent ductility, heat resistance, corrosion resistance, electrical and thermal conductivity, good chemical stability, and biocompatibility. Due to its inherent chemical inertia, coupled with the high price and difficulty of dispersion, Au did not received sufficient attention in the field of catalysis until the beginning of this century. In the 1980s, Haruta et al. [1,2] unexpectedly discovered that highly dispersed gold nanoparticles (AuNPs) supported by metal oxide prepared by the coprecipitation method showed good low-temperature activity for CO catalytic oxidation with good stability and resistance at the low temperature of 77K. This research enabled the realization of the excellent catalytic activity of Au and led to great interest in CO catalytic oxidation by gold nanomaterials in the following decades. Valden et al. [3] studied the catalytic activity of AuNPs with different sizes deposited on TiO_2 and found that AuNPs with a diameter of 3 nm have the highest catalytic activity for CO oxidation, which can be attributed to the quantum size effect of the nanosized gold particles and the charge transfer effect with the support. The activity of supported gold nanocatalysts is also influenced by the support

Advanced Nanomaterials for Pollutant Sensing and Environmental Catalysis. DOI: https://doi.org/10.1016/B978-0-12-814796-2.00005-8

[4]. Different supports can affect the size of the deposited AuNPs and thus exhibit different catalytic properties. In recent years, studies have shown that the catalytic properties of gold nanomaterials may come from the gold particles themselves. Unsupported AuNPs have been proven to be an efficient catalyst for the aerobic oxidation of glucose, showing a similar activity to enzymatic systems [5]. An unsupported nanoporous gold film has also been fabricated via dealloying the Ag from an Au/Ag alloy film by a modified electrochemical method [6]. The gold film had a pore size of sub-10 nm and displayed excellent CO oxidation efficiency with a conversion of nearly 100% in 20 h at room temperature without active pretreatment. This outstanding catalytic activity is attributed to the nanostructured gold itself, rather than to the effective charge transfer and quantum size effect. The catalytic mechanism of this nanoporous gold film is much different from that of the supported ones. Fujita et al. [7] prepared porous gold nanofilms with a particle size of about 30 nm, which also showed excellent catalytic performance. The spherical-aberration-corrected transmission electron microscopy image indicates that the low-coordination crystal face [111] enriched on the surface of porous gold nanofilm. The crystal face [111] has a strong binding ability to the CO molecule and can reduce the activation energy barrier. In addition, a large number of surface lattice defects are also considered as the main reason for its high catalytic activity.

The catalytic performance of gold nanomaterials has become one of the fastest growing scientific research fields. Extensive explorations have been made for the preparation of these gold-based nanomaterials and their catalytic reaction mechanism. With the exploration of the relationship between its catalytic activity and structures, Au has been employed in many important reactions such as CO oxidation, water gas shift, and the reduction of nitrogen oxides, and shows huge potential applications in clean energy and pollutant control [8].

Gold materials can also be employed in photocatalysis. Photocatalysis technology plays an increasingly important role in solving the current severe energy shortage and environmental pollution, which uses solar energy by converting it into high-density chemical energy [9,10]. However, traditional photocatalysts cannot effectively utilize solar energy due to their wide bandgap, the recombination of photogenerated supports, and their low quantum efficiency. Therefore it is of great significance to develop new photocatalysts to achieve efficient use of solar energy. In 2008 Awazu et al. [11] designed Ag/TiO_2 photocatalysts to broaden the light absorption to the visible light region through the surface plasmon resonance (SPR) effect of AgNPs. Since then, Ag and Au have been widely employed to modify semiconductors toward their use as the plasmonic photocatalysts. Gold/semiconductor composite materials can effectively promote the light absorption of the semiconductor due to the excellent SPR performance of AuNPs. In addition, the deposition of Au nanomaterials can also promote the separation of electron holes to improve the quantum efficiency. Liu et al. [12] prepared TiO_2 nanotube arrays by anodic oxidation, and then deposited an Au film with a thickness of 5 nm onto the surface of the TiO_2 by electron beam evaporation. The photocatalytic activity of water splitting is 66 times higher than that of TiO_2 nanotube arrays alone under visible light irradiation. Electromagnetic simulation results show that the improved visible light photocatalytic activity is attributed to the local electric field enhancement near the TiO_2 surface. Benefitting from the SPR effect, Au can not only enhance the absorption range of incident light, but it can also effectively inhibit the recombination of photogenerated supports due to the Schottky junction at the metal–semiconductor interface, therefore increasing the energy conversion efficiency of the photocatalytic reaction obviously. Research on gold-based plasmonic photocatalysis involves the shape and size of NPs, the compound of gold and semiconductor, and the interface between them. Gold/semiconductor composites have been widely used in the application of environmental catalysis such as hydrogen generation, pollution control, and solar cells.

5.2 Synthesis and Characterization of Nano-Gold

5.2.1 Preparation of Supported Gold Nanoparticles

As early as 1973, Bond et al. [13] pointed out that the interaction between gold and the support is a key point for the gold to be deposited on the support. The promotion of the interaction between the gold and the support is an effective approach to obtain highly dispersed ultrafine AuNPs. Since the melting point of gold is 1336K, which is much lower than those of platinum (2042K) and palladium (1823K), it is much more difficult to obtain highly dispersed supported gold nanocatalysts. Therefore the selection and design of the preparation methods and processing conditions are critical to obtaining highly active supported AuNPs. Impregnation, coprecipitation, deposition-precipitation, chemical vapor deposition, and lots of other methods are used in the preparation of supported Au particles with different sizes and Au-supported interactions.

Impregnation. During the process of impregnation, the support is immersed in a salt solution containing gold precursor, followed by drying, calcining, and reduction, to obtain supported AuNPs on TiO_2, Al_2O_3, SiO_2, and other supports

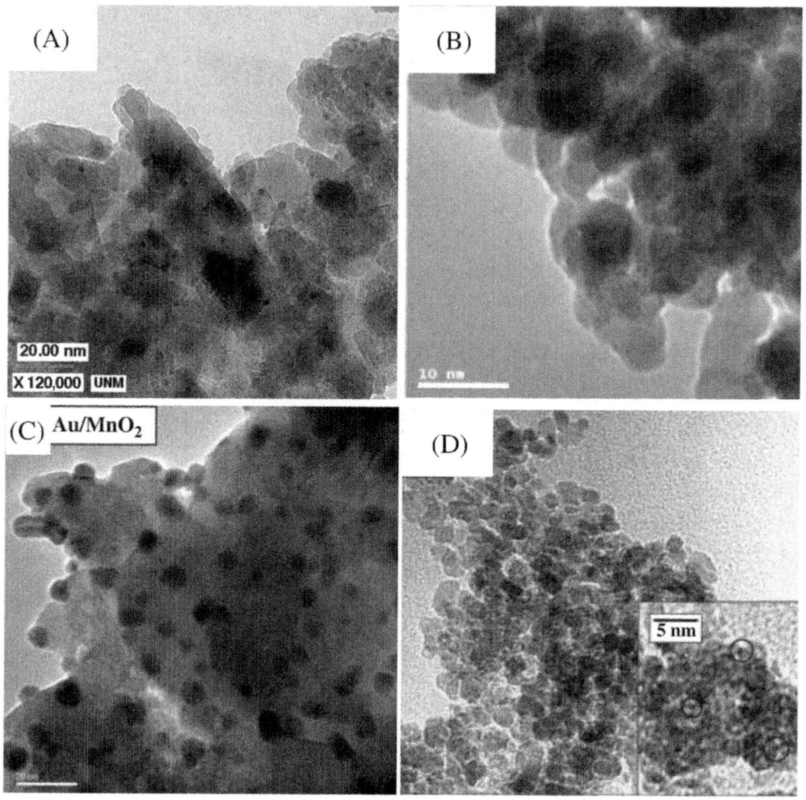

Figure 5.1 (A) TEM image of 1 wt.% Au/Al_2O_3 prepared by aqueous impregnation; (B) Au/SnO_2 prepared by coprecipitation; (C) Au/MnO_2 prepared by homogeneous deposition-precipitation; and (D) Au/TiO_2 prepared by chemical vapor deposition. (A) Reprinted by permission from Springer: Q. Xu, K.C.C. Kharas, A.K. Datye, The preparation of highly dispersed Au/Al_2O_3 by aqueous impregnation, Catal. Lett. 85 (2003) 229–235, Copyright 2003 [14]. (B) Reprinted from S. Wang, Y. Wang, J. Jiang, R. Liu, M. Li, Y. Wang, et al., A DRIFTS study of low-temperature Co oxidation over Au/SnO_2 catalyst prepared by co-precipitation method, Catal. Commun. 10 (2009) 640–644, Copyright (2009), with permission from Elsevier [15]. (C) Reprinted from N.S. Patil, B.S. Uphade, D.G. Mcculloh, S.K. Bhargava, V.R. Choudhary, Styrene epoxidation over gold supported on different transition metal oxides prepared by homogeneous deposition–precipitation, Catal. Commun. 5 (2004) 681–685, Copyright (2004), with permission from Elsevier [16]. (D) Reprinted from S. Schimpf, M. Lucas, C. Mohr, U. Rodemerck, A. Brückner, J. Radnik, et al., Supported gold nanoparticles: in-depth catalyst characterization and application in hydrogenation and oxidation reactions, Catal. Today 72 (2002) 63–78, Copyright (2002), with permission from Elsevier [17].

(Fig. 5.1A) [14,18,19]. It has been proven that Au catalysts prepared by impregnation under acidic conditions have poor dispersion, low loading, and large particle size, resulting in bad catalytic performance as compared with other methods. However, a series of treatments have been reported to improve its catalytic activity by changing the pH of the solution, aging time, pretreatment temperature, or using organic ligands.

Coprecipitation. Coprecipitation is performed by adding a chloroauric acid solution and a metal nitrate precursor together into a sodium carbonate solution, followed by washing, collecting, drying, and calcining (573K or above) the obtained precipitate to obtain active catalysts. The supports employed in coprecipitation are usually Co_3O_4, SnO_2, CeO_2, Fe_2O_3, NiO, ZnO, and others (Fig. 5.1B) [15,20]. Supported Au catalysts prepared by this method show good catalytic performance at low temperatures, even at 200K in the catalytic oxidation of CO. However, supported AuNPs prepared by this method usually suffer from the covering of the supports, which can lead to a decrease in catalytic activity.

Deposition-precipitation. As one of the most widely used and effective methods to prepare supported AuNPs, deposition-precipitation is operated in an alkaline solution of chloroauric acid precursor. The concentration of chloroauric acid in the solution is controlled to be slightly lower than that required for uniform precipitation, but sufficient to deposit on the surface of the supports. The supports deposited with Au hydroxide are washed, filtered, and calcined to obtain the desired catalysts. Controlling suitable pH can be conducive to obtain Au catalysts with low loading, high dispersion, and high activity. Since Au hydroxide cannot be deposited at a low pH, an acidic oxide cannot be used as the support in the deposition-precipitation method and MnO_2, Al_2O_3, TiO_2 are the most widely used (Fig. 5.1C) [16,21]. In addition, increasing the surface area of supports will facilitate the deposition of Au hydroxide.

Chemical vapor deposition. This method is used for the preparation of Au catalysts using a volatile precursor such as Au $(CH_3)_2$(acac) on various oxide supports [17,22]. Haruta et al. [22] synthesized Au catalysts on supports of Al_2O_3, SiO_2, and TiO_2 with an average particle size of smaller than 2 nm (Fig. 5.1D). The chemical vapor deposition method consists of two steps, namely (1) contacting Au$(CH_3)_2$(acac) volatile precursor with the supports at 306K overnight and (2) calcination in air at a temperature of between 473K and 773K to decompose the precursor into metallic Au particles. The prepared supported Au catalysts usually have relatively high dispersion and effective catalytic activity compared with the liquid-phase method using $HAuCl_4$ as the starting reagent [23]. Another major advantage of this method is that both basic and acidic metal oxides can be employed as supports.

In addition, there are a lot of other approaches to prepare supported Au nanocatalysts such as photochemical deposition, organic gold-complex grafting, cosputtering, liquid grafting, amorphous metal alloys, cation exchange, solvated metal atom impregnation, and others [24–26]. Different methods can obtain different particle sizes, thus affecting the interactions between AuNPs and supports, resulting in unexpected catalytic activity during the reaction. Therefore it is necessary to select an appropriate preparation method according to the physical and chemical properties of the Au precursor and the support, the purpose, and the operating conditions of the catalytic reaction.

5.2.2 Preparation of Surface Plasmon Resonance–Enhanced Photocatalysts

SPR-enhanced photocatalysts have become a hot research area recently due to the improvement of light absorption and the enhancement of electron–hole separation efficiency. A lot of recipes have been proposed to design Au-based plasmonic photocatalysts [27]. The most common type of SPR-enhanced photocatalysis system is the Au/semiconductor photocatalyst, which mainly has four types as can be seen in Fig. 5.2.

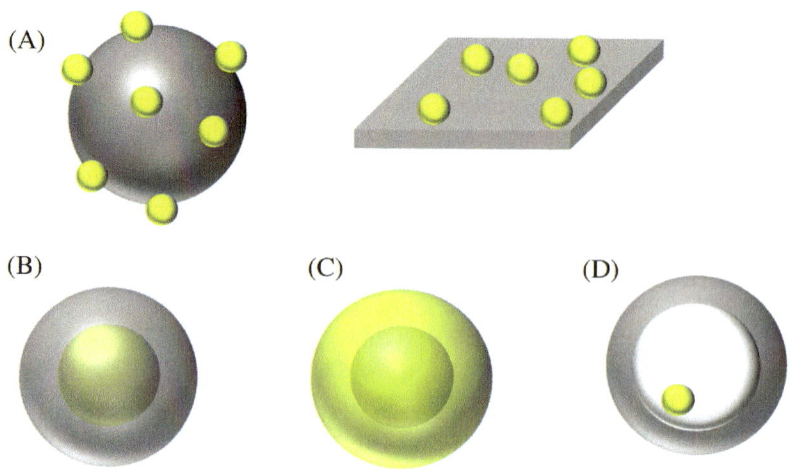

Figure 5.2 Scheme of different types of surface plasmon resonance–enhanced photocatalysts. (A) Supported Au/semiconductors, (B) Au core–semiconductor shell, (C) semiconductor core–Au shell, and (D) Au yolk–semiconductor shell composites.

Type A is the earliest and most reported type of SPR-enhanced photocatalyst (Fig. 5.2A). There are many classical synthetic methods to prepare these supported catalysts. By loading AuNPs on the surface of semiconductors such as titanium dioxide, cerium oxide, and cadmium sulfide, the photocatalytic activity has been obviously improved due to the SPR effect of AuNPs, which can broaden the range of light absorption from ultraviolet (UV) to visible light [28]. The morphology of semiconductors can change from sphere, rod, or wire to film and porous structure.

Type B has been developed on the basis of type A in recent years, for the purpose of clarifying the photocatalytic mechanism of SPR-enhanced photocatalysis (Fig. 5.2B). These photocatalysts generally appear in a core–shell structure. The semiconductor coatings used for the outer shell including titanium dioxide or zinc oxide can be expanded to other metal (e.g., Ag and Cu)/semiconductor systems [29]. As is well known, titanium dioxide does not absorb visible light. However, after coating gold nanorods (AuNRs) with a titanium dioxide film, Au/TiO_2 photocatalysts exhibit excellent photocatalytic activity under the irradiation of visible light, thus confirming the key role of AuNRs during the photocatalytic reaction [30]. In addition, this core–shell structure can effectively prevent the aggregation and oxidation of the metal NPs, and therefore improve the stability of the photocatalysts.

Type c is also a core–shell structure with the semiconductor as the core and Au as the shell (Fig. 5.2C). In general, semiconductor core–Au shell nanostructures are prepared via the chemical deposition of Au on premade semiconductor nanostructures. Fe_2O_3/Au, Fe_3O_4/Au, TiO_2/Au, Fe_3O_4/Ag/SiO_2/Au have been prepared by this strategy [28,31,32]. However, these semiconductor core–Au shell nanostructures are usually employed in applications of surface enhanced Raman scattering (SERS)-based sensing, photothermal therapy, and diagnostics. Few of them show applications in photocatalysis [32].

Type D is a kind of specially designed photocatalyst with a yolk–shell structure (Fig. 5.2D). The design of the yolk–shell structure is helpful to understand the mechanism of photocatalysts and to improve the stability of photocatalysts. Since there is no direct contact between the metal NPs and the semiconductor, the electron transfer between them is inhibited by the intermediate isolation layer. The Schottky barriers and the sensitization of AuNPs, can not affect the photocatalytic activity anymore. Therefore the photocatalytic mechanism is relatively simple to study. Similar to the core–shell structure, the yolk–shell structure can also effectively prevent the aggregation and oxidation of metal NPs with the protection of the semiconductor shell. In addition, the space between can be tuned in thickness, thus providing a powerful approach to investigate the effect of SPR during photocatalysis.

The most widely used kind among these is type a, Au supported on the semiconductor, which can be synthesized by the approaches mentioned such as impregnation or photochemical deposition, and others. The difference in preparing Au/semiconductor photocatalysts is that a reduction step is usually necessary. Reducing approaches include chemical reduction, photochemical reduction, and electronic reduction. Wark et al. [33] modified mesoporous titania films with Au nanostructures through impregnation and reduction by $NaBH_4$ or pulsed cathodic electrodeposition. With the incorporation of Au, a significant improvement in the photonic efficiency, which contributes to the generation of Schottky junction and the SPR effect, results in an increase in the concentration of photogenerated holes at the film surface, therefore providing enhanced photocatalytic activities in NO oxidation. Kowalska et al. [34] modified 15 commercial titania powders with Au by photochemical deposition. Benefitting from the SPR effect of Au, all tested samples showed obvious visible light absorption, with a maximum located between 530 and 600 nm. In addition, the absorption ranges are dependent on the size and shape of AuNPs. All Au/TiO_2 powders showed enhanced photocatalytic activity of acetic acid and 2-propanol photooxidation under UV and visible light irradiation (mainly >450 nm).

The chemical reduction and photoreduction methods are the most widely used in the preparation of Au/semiconductor composites. Pearson et al. [35] used 12-phosphotungstic acid as a reducing agent to prepare an Au-modified TiO_2 nanotube array through the in situ chemical reduction of $KAuBr_4$. During the preparation process, the TiO_2 nanotube array was first functionalized with 12-phosphoric acid, and therefore, AuNPs could grow in situ on the surface with the assistance of UV light. The photocatalytic degradation efficiency of Congo red was increased to 89% from 42% under the irradiation of simulated solar light after modifying with AuNPs. This in situ chemical reduction method can also be employed in the preparation of TiO_2 nanotube arrays modified with other metallic NPs using $CuCl_2$, Ag_2SO_4, or H_2PtCl_6 as precursor solutions. Xiong et al. [36] also developed an in situ chemical reduction method using polydopamine as the reducing agent in the deposition of AuNPs, benefitting from a range of catechol groups (Fig. 5.3A1).

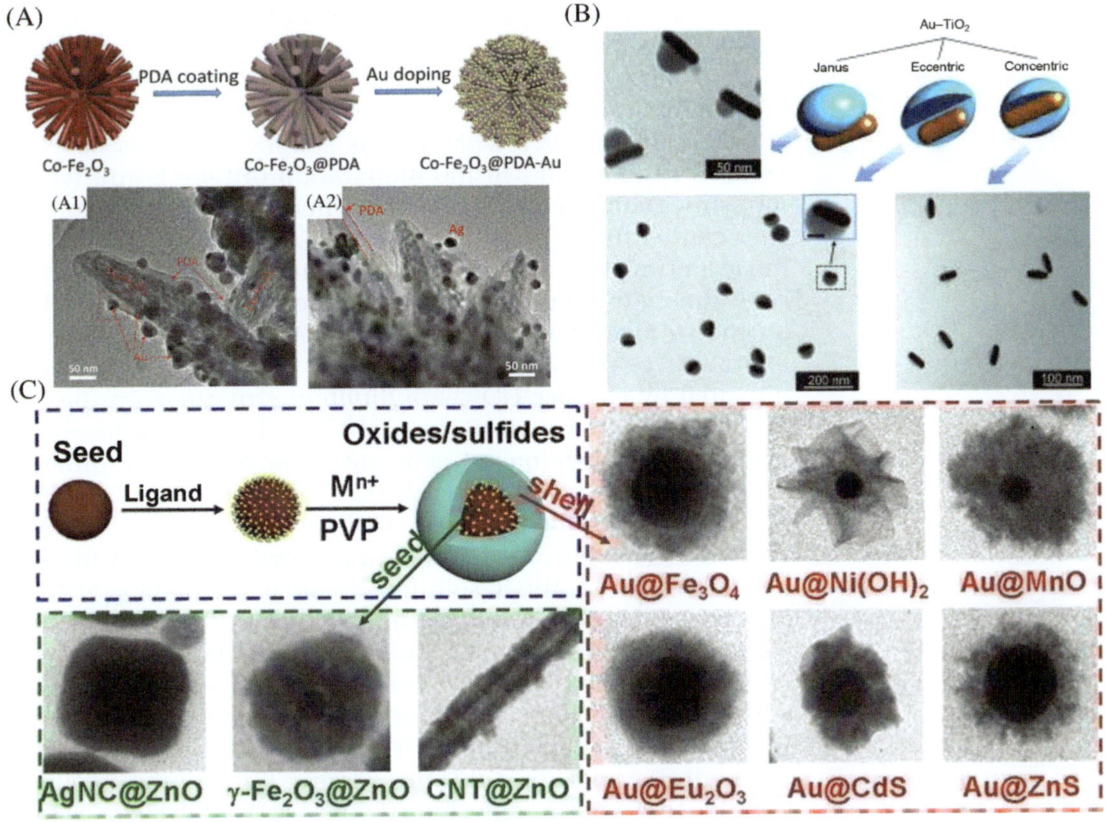

Figure 5.3 (A) In situ chemical reduction method for supported metal photocatalysts, (A1) synthesis scheme and TEM image of Co–Fe_2O_3/PDA–Au and (A2) TEM image of Fe_2O_3/PDA–Ag; (B) scheme and TEM images of TiO_2-coated short AuNRs with Janus, eccentric, and concentric geometries; and (C) scheme and TEM images of various core–shell nanostructures. (A1) Reprinted from W. Xiong, Q. Zhao, X. Li, L. Wang, Multifunctional plasmonic co-doped Fe2O3@polydopamine-Au for adsorption, photocatalysis, and SERS-based sensing, Part. Part. Syst. Char. 33 (2016) 602–609, Copyright (2016), with permission from Wiley-VCH Verlag & Co. KGaA. (A2) Reprinted from W. Xiong, X. Li, Q. Zhao, Y. Shi, C. Hao, Insight into the photocatalytic mineralization of short chain chlorinated paraffins boosted by polydopamine and Ag nanoparticles, J. Hazard. Mater. 359 (2018) 186–193, Copyright (2018), with permission from Elsevier. (B) Reprinted from W. Seh Zhi, S. Liu, S.-Y. Zhang, M.S. Bharathi, H. Ramanarayan, M. Low, et al., Anisotropic growth of titania onto various gold nanostructures: synthesis, theoretical understanding, and optimization for catalysis, Angew. Chem. Int. Ed. 50 (2011) 10140–10143, Copyright (2011), with permission from Wiley-VCH Verlag & Co. KGaA. (C) Reprinted with permission from H. Sun, J. He, J. Wang, S.-Y. Zhang, C. Liu, T. Sritharan, et al., Investigating the multiple roles of polyvinylpyrrolidone for a general methodology of oxide encapsulation, J. Am. Chem. Soc. 135 (2013) 9099–9110. Copyright (2013) American Chemical Society.

In addition, polydopamine has been widely used in the fabrication of photocatalytic composites due to its universal adhesive ability, broad spectrum light absorption from the visible to the near-infrared region, and good electrical conductivity. After coating a shell of polydopamine through the self-polymerization of dopamine, cobalt-doped Fe_2O_3 could be modified in situ with AuNPs. The average diameter of AuNPs is about 15 nm. This method can also be used for the deposition of AgNPs using a silver ammonia solution as the precursor (Fig. 5.3A2) [37].

Au nanostructures with various shapes and sizes can also be used as templates in the preparation of type B and type D Au/semiconductor nanomaterials. The morphology of composites can be tuned by controlling the amount of precursor and the experimental conditions. For example, Au/TiO_2 composites can be prepared by coating different structures of TiO_2 on the surface of AuNPs or AuNRs [38,40]. By controlling the hydrolysis of titanium precursors, Au/TiO_2 composites can be prepared by coating TiO_2 with different structures on the surface of Au nanomaterials. Han et al., used titanium diisopropoxide bis(acetylacetonate) as a precursor to prepare different morphologies by controlling the rate of hydrolysis of titanium precursor with the presence of dispersed AuNPs or AuNRs [38,40]. When the precursor solution was added into the dispersed AuNP or AuNR solution at once, Janus-like Au/TiO_2 NPs were obtained, and uniform, spherical Au/TiO_2 core–shell NPs can be obtained by adding the precursors three times (Fig. 5.3B). Changing the amount of the precursor solution added can also control the morphology of the semiconductor. When a small amount of TiF_4 precursor is added into the hydrothermal reaction, the coating outside the AuNPs is constructed with wedge-shaped TiO_2 [41]. However, wedge-shaped TiO_2 nanoshells gradually become uniform by increasing the amount of TiF_4, and the thickness can be tuned within the range of 200–400 nm by controlling the volume of TiF_4 precursors [42].

Although Au–TiO_2 can be successfully synthesized by controlling the hydrolysis of titanium precursors, there are still great challenges in the fabrication of Au/semiconductor NPs with a core–shell structure. Surfactants such as polyvinylpyrrolidone can be employed in the synthesis of these core–shell composites (Fig. 5.3C). During the synthesis of an Au–ZnO core–shell structure, the Au core and the ZnO particles can be connected by polyvinylpyrrolidone, which can promote the nucleation and growth of ZnO particles on the surface of the Au to finally form a ZnO shell. This method can be regarded as a common method to extend to other cores (polymer, metal oxide, carbon nanotubes, graphene oxide, etc.) and shells (Fe_3O_4, MnO, TiO_2, ZnS, CdS, etc.) used in core–shell nanostructures (Fig. 5.3C) [39].

5.2.3 Factors for Catalytic Activity

The effect of nanoparticle size. A large number of studies have shown that the size of Au has an important effect on its catalytic activity. It is generally believed that AuNPs with particle sizes smaller than 10 nm show high catalytic activity. Valden et al. [3] employed scanning tunneling microscopy/spectroscopy (STM/STS) to investigate the effect of NP size on catalytic activity using an Au/TiO_2 model and they found that AuNPs with a size of about 3.5 nm exhibit the highest catalytic activity. The AuNPs in this system showed a nonmetallic property and the special catalytic activity of AuNPs can be attributed to the quantum size effect. However, Haruta et al. [8] think that the STM/STS technique measures the changes of the electronic state for only one single size of AuNP, while the actual catalytic activity is developed on the entire catalytic system. Therefore they thought that the conclusion based on the results of STM/STS were not rigorous enough. Boyen et al. [43] studied the oxidation behavior of AuNPs with diameters ranging between 1 and 8 nm on a SiO_2/Si substrate and they found that AuNPs consisting of 55 atoms with a diameter of 1.4 nm show strong antioxidant performance. X-ray photoelectron spectroscopy (XPS) characterization results indicate that the AuNPs show a metallic property, which means that the electronic structure will not change due to the small particle size. Therefore it is believed that the stability of this AuNP is related to its special closed-shell geometry. It is also speculated that active Au species should be the active center of CO catalytic oxidation. In addition, active Au species consisting of 10 atoms are more favorable for CO activation from the simulation by the density functional theory [44].

The chemical state of gold. There are a few different opinions on the chemical state of active Au species (Au^0, Au^+, or Au^{3+}) during the reaction. Haruta et al. [45] believe that Au atoms with a positive charge at the metal–support interface play a crucial role in the catalytic oxidation of CO. Based on the in situ XPS characterization of Au/Fe_2O_3 catalysts, Makkee and Hutchings et al., suggested that Au^{3+} shows an important effect on the catalytic activity of CO oxidation [46,47]. Scurrell et al. [48] found that the active center of a catalyst should be Au^+ rather than Au^{3+} after investigating Au/TiO_2, Au/Fe_2O_3, and Au/HY by Mössbauer spectrum. Flytzani-Stephanopoulos et al. [49] washed approximately 90% of the elemental Au in an $Au/La_2O_3{-}CeO_2$ catalyst with an NaCN solution, and applied the catalyst with only the oxidized Au ions on the surface in a water gas shift reaction. It was found that the catalyst activity was unexpectedly improved significantly. It can be concluded that ionic Au should be the key active species of a catalyst for the water gas shift reaction. Using in situ X-ray fine absorption spectroscopy and infrared spectroscopy characterization techniques, Gates et al., found that Au^{3+} in Au/MgO and Au/zeolite catalysts is gradually reduced to Au^{+} during the reaction of CO catalytic oxidation, while no metal Au was detected [50–52]. It can be confirmed that Au^{3+} was the active center of the CO catalytic oxidation reaction. Okumura et al. [36] studied the effect of heat treatment on the stability of Au^{3+} on zeolite-supported AuNPs and found that oxidized Au species are highly sensitive to the processing conditions, especially the oxidation state of Au species at higher temperatures. This indicates that the valence state of active Au species of AuNPs is closely related to the reaction conditions [53].

The effect of supports. In the absence of supports, AuNPs themselves can still exhibit catalytic activity [5,6,54]. However, at present, most of the high-activity Au nanocatalysts reported in the literature are supported. As one of the most important compositions of most Au nanocatalysts, the presence of the support can facilitate the stabilization of active AuNPs. In addition, the interface between the support and the AuNPs can provide extra active sites for catalytic reactions, therefore enhancing the catalytic activity [55]. The most commonly used types of supports for supported Au nanocatalysts are metal hydroxides ($Al(OH)_3$, $Ti(OH)_4$, $Fe(OH)_3$, $Mn(OH)_2$, $Ce(OH)_2$, $Cr(OH)_2$, $Cu(OH)_2$, $Ni(OH)_2$, $Zn(OH)_2$, etc.), metal oxides (Al_2O_3, TiO_2, Fe_2O_3, CeO_2, Co_3O_4, ZnO, Cr_2O_3, CuO, In_2O_3, La_2O_3, ZrO_2, MnO_x, etc.), and metal salts ($BaCO_3$, $CaCO_3$, $SrCO_3$, aluminosilicate, phosphate, etc.). In addition, silicon-based materials such as SiO_2, zeolite, and molecular sieves and carbon materials such as activated carbon, graphene, carbon nanotubes are also used as supports because of their high specific surface area.

There has been considerable controversy about the contribution of the support to the activity of the catalyst. Davis and Flytzani-Stephanopoulos et al. [49,56] pointed out that the role of the support should be taken into account in explaining the essence for the high activity of Au nanocatalysts, rather than the size effect of Au alone. The presence of a support not only contributes to the stability of active Au species, but also plays an important role in promoting the activity of the entire catalyst due to the interaction between the support and the AuNPs.

The elemental composition on the surface of supports can also affect the catalytic activity of Au nanocatalysts. Okazaki et al. [57] found that the composition on the surface of TiO_2 supports has an important influence on the electronic structure of AuNPs using electron holography, STM, and the first-principles theory. TiO_2 with lattice defects on the surface shows a much stronger interface with AuNPs. It is further speculated that the activity of the Au/TiO_2 catalyst originates from electron transfer between the AuNPs and the TiO_2 support and the orbital hybridization at the contact interface. It needs to be pointed out that the interaction between the Au and the support has a particularly significant effect on the catalytic performance when the particle size of the Au is less than 2 nm.

In addition, the physical and chemical properties of the support also play a key role during the catalytic reaction. Rare-earth oxides (such as CeO_2, Y_2O_3, and La_2O_3) as supports can stabilize Au^{3+} effectively, and the concentration of Au^{3+} has a good linear correlation with the catalytic activity in the CO catalytic oxidation reaction [58,59]. In addition, the hydroxyl groups on the surface of oxide supports also has an important effect on the catalytic performance of Au nanocatalysts, manifesting as an activity that is sensitive to water or enhanced under humidity [60]. Daniells et al. [61] found that the deactivation of the Au/Fe_2O_3 catalyst was accompanied by the elimination of the surface hydroxyl groups and the reduction of Au^{3+} to Au^0, therefore it can be believed that hydroxyl groups play a key role in promoting the decomposition of carbonate intermediate on the surface of Au nanocatalysts.

5.3 Environmental Catalysis of Supported Gold Catalysts

5.3.1 Catalytic Oxidation of Carbon Monoxide

The CO concentration in the atmosphere has increased year by year due to exhaust emissions from automobiles and the incomplete combustion of fossil fuels and others, thus leading to serious air pollution. Dizziness and vomiting can occur within 2 h when the concentration of CO in the air is only 2.0×10^{-5} mol L^{-1}, and death within 1–3 min when the content reaches 1.2%. Air pollution caused by CO has received widespread attention. A series of technologies has been developed to control CO pollution in the atmosphere. CO oxidation under normal conditions usually requires high temperatures, large energy consumption, and may cause explosions. However, taking into account the advantages of simple operation, low cost, and high efficiency, the catalytic oxidation of CO has shown more practical significance in eliminating large amounts of low-concentration CO. Actually, the catalytic oxidation of CO by AuNPs is the most studied reaction so far. In this reaction, Au catalysts show much better catalytic performance at low temperatures and high humidities than other metal catalysts. It can completely catalyze the oxidation of CO at room temperature, with the lowest temperature being 203K.

An unsupported nanoporous gold film has been fabricated for the catalytic oxidation of CO to prove that the outstanding catalytic activity is ascribed to the nanostructured Au itself, rather than effective charge transfer and the quantum size effect [6]. However, supported AuNPs are another large kind of Au-based catalysts with a wide range of applications. Many researchers have shown that the catalytic activity of supported Au catalysts is much greater than that of unsupported ones. Supports are selected for improving the activity of Au catalysts, by activating the reactants or interacting with the Au to provide active sites. In addition, supports can also enhance the stability of AuNPs.

The catalytic activity of supported AuNPs significantly depends on the properties of the supports. 3D transition metal oxides, alkaline-earth metal oxides, and hydroxides present the most effective improvements in catalytic activity. Au catalysts supported on $Fe(OH)_3$, $Mn(OH)_2$, and $Co(OH)_2$ have been prepared using $Au(PPh_3)NO_3$ as a precursor and these show catalytic oxidation activity for CO at 203K, while those supported on $Ti(OH)_4$ and $Al(OH)_3$ are at 253K. Supports are generally classified into active supports and inert supports, according to their ability in catalyzing the oxidation of CO. TiO_2, Fe_2O_3, CeO_2, and Co_3O_4-supported AuNPs are usually smaller in size and higher in activity than Au loaded on SiO_2 and Al_2O_3, which can be classified as active supports and inert supports, respectively. Active supports usually show CO catalytic performance, while inert ones show poor activity, even bad oxygen adsorption capacities. The catalytic performance can be enhanced by employing mixed metal oxides as the support. For example, the catalytic activity of $Au/Fe_2O_3-TiO_2$ is greater than that of Au/TiO_2 and Au/Fe_2O_3, especially when the temperature is lower than 373K [62]. In the catalytic system of $Au/Zr_xCe_{1-x}O_2$, the combination of Ce in a ZrO_2 support significantly enhanced the catalytic activity, compared to that of Au/ZrO_2 [63].

Another important approach to improve the CO oxidation activity of Au catalysts is to introduce a small amount of cocatalyst into the catalytic systems. The most commonly used cocatalysts are usually transition metal oxides. The addition of cocatalysts can improve the Au loading amount, stability, and antisintering ability. Cocatalysts can also act with the supports to generate new species which can provide extra effective active sites. By adding CaO, ZnO, or doping with Fe, Ni, Cu, and other metal oxides, the CO oxidation activity of the Au/TiO_2 catalyst can be improved obviously [64,65]. However, using In_2O_3 as the cocatalyst would decrease the catalytic activity of Au/TiO_2 in the oxidation of CO. But it can significantly improve its antisintering ability and maintain good catalytic activity even at 773K because In_2O_3 is highly dispersible in the lattice matrix of TiO_2, which is beneficial to improve the stabilization of AuNPs [66]. In addition, due to the different physical and chemical properties of various supports, the addition of the same cocatalyst may result in different, even opposite effects on the catalytic activity. For example, La_2O_3, as a cocatalyst can improve the low-temperature catalytic activity of Au/TiO_2, while reducing that of Au/CeO_2 [67].

It can be seen that the catalytic oxidation of CO can be significantly increased by loading with an active or an inert support by introducing suitable active metal oxides as the cocatalyst. However, the mechanism of the interaction between the AuNPs, the supports, and the cocatalysts on the catalytic activity of CO remains to be further studied. In addition, due to the differences in the nature of supports and cocatalysts, it is currently expected that the desired CO catalytic oxidation activity will be obtained, and it is necessary to study a series of supports and cocatalysts to achieve this purpose.

Understanding the mechanism of CO catalytic oxidation is significant to the development of high activate catalysts. The catalytic oxidation reaction of CO is as follows:

$$CO + \frac{1}{2}O_2 \rightarrow CO_2, \ \Delta H = -284.09 \text{ kJ mol}^{-1} \tag{5.1}$$

The mechanism of the catalytic oxidation of CO by different catalysts is dependent on the catalytic system and the reaction conditions.

1. *Langmuir–Hinshelwood mechanism.* The oxidation of CO on noble-metal catalysts is generally considered to proceed according to the Langmuir–Hinshelwood mechanism under high vacuum conditions. According to this mechanism, the adsorbed CO reacts with the adsorbed O_2 to form CO_2 without the participation of lattice oxygen. Under the condition in the high-pressure chemical reactors, Pt, Pd, Rh, and Ir catalysts still meet the Langmuir–Hinshelwood mechanism under the ideal pressure and temperature [68–70]. However, when using Ru as the catalyst, the oxidation reaction is

accomplished by the diffusion of CO and the adsorbed O atoms on the catalysts.

2. *Redox mechanism*. CO oxidation on the surface of metal oxide catalysts usually proceeds according to the redox mechanism. In the redox procedure, the lattice oxygen of the catalyst directly participates in the reaction and reacts with activated CO molecules preferentially adsorbed on the surface, leading to the formation of lattice oxygen deficiency. Following this, the oxygen in the gas phase is adsorbed on the catalyst surface and becomes lattice oxygen. The cyclic process promotes the oxidation of CO on the metal oxide surface. However, the generated oxygen deficiency can be inactivated by the adsorption of CO, which makes the decreasing of the catalytic activity.

The mechanism of CO catalytic oxidation by supported Au is very complex and cannot be simply summed up by the Langmuir–Hinshelwood mechanism or the redox mechanism. During CO oxidation, the CO is adsorbed on the surface of the Au catalyst first. However, there is no consensus on the location where the oxidation reaction takes place. In particular, the mechanism of oxygen adsorption and activation is still not clear. There are two explanations for this, namely oxygen adsorption directly on the Au particles and oxygen adsorption on the supports or the interface between the Au and the supports.

Direct adsorption of oxygen on the gold particles. Based on this type of adsorption, there are usually three explanations, namely (1) CO and O_2 adsorb on the Au particles and produce a four-center surface complex, which is unstable and can decompose under mild conditions; (2) the adsorption of CO and O_2 takes place at the surface defects of the AuNPs; and (3) CO and O_2 are adsorbed on arc-shaped Au particles with special electronic structures. Because oxygen is adsorbed on the Au particles, the role of the support is to stabilize Au catalysts or increase the highly reactive state of Au catalysts.

Based on researches of many years, Haruta et al. [71] proposed the mechanism of catalytic oxidation of CO at 273K on supported AuNPs based on the direct adsorption of oxygen on the surface (Fig. 5.4). The adsorption of CO and O_2 on supported Au catalysts is noncompetitive, and there are carbonyl-like intermediates formed during the reaction. The mechanism for catalyzing CO oxidation is:

1. Reversible adsorption of CO at the surface or interface of Au catalysts:

$$Au_{s,p} + CO \leftrightarrow Au - C - O_{s,p} \tag{5.2}$$

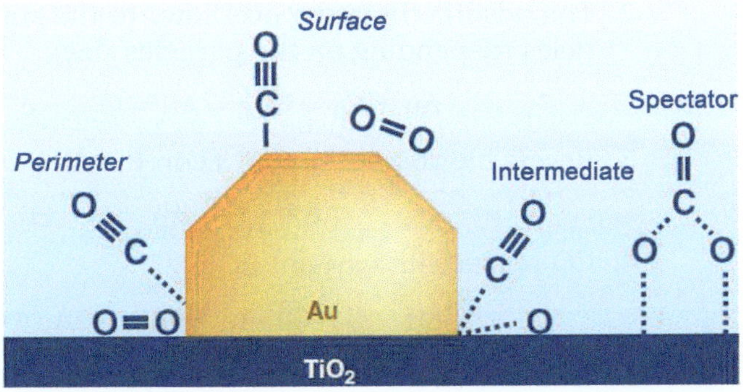

Figure 5.4 Scheme of the CO and O_2 adsorbed on the surface of Au/TiO$_2$ during the redox reaction. Reprinted by permission from Springer: M. Haruta, Catalysis of gold nanoparticles deposited on metal oxides, CATTECH 6 (2002) 102–115, Copyright 2002.

2. O_2 irreversible adsorption at Au/support interface (rate-determining step):

$$Au_{(p)}/TiO_2 + O_2 \rightarrow O_{ads} \quad (5.3)$$

3. The production of the intermediate:

$$Au - C - O_{s,p} + O_{ads} \rightarrow O - Au - C - O_{s,p} \quad (5.4)$$

4. The decomposition of the intermediate:

$$O - Au - C - O_{s,p} \rightarrow CO_2\uparrow \quad (5.5)$$

5. The decomposition of carbonyl intermediates:

$$Au - C - O_p - OTiO_2 \rightarrow O - COOTi + Au \rightarrow TiO + Au + CO_2\uparrow \quad (5.6)$$

Oxygen adsorption at the support or gold–support interface. If using a semiconductor material as the support such as TiO_2, Fe_2O_3, or ZnO, oxygen vacancies are generated at the Au–support interface due to the Schottky junction. The oxygen vacancies promote the adsorption of a large number of oxygens in the form of molecules peroxo ions (O_2^-). This adsorption mainly occurs at the interface between the Au clusters and supports, followed by the dissociation of molecules peroxo ions into atomic oxygen. The atomic oxygen would react with the adsorbed CO to produce at the same site. Since this adsorbed oxygen has a good fluidity, the diffusion of oxygen onto the Au particles is no longer a rate-determining step in this type of reaction.

Taking Au/TiO_2 as an example, the mechanism of catalytic CO oxidation with oxygen adsorbed on the support or at the Au–support interface is:

1. The irreversible adsorption of oxygen at the oxide vacancy:

$$2V_O + O_2 + 2e^- \rightarrow 2V_O - O_{ads}^- \quad (5.7)$$

where V_O is the oxygen vacancies.

2. The adsorbed oxygen overflows to the surface of the Au particles (depending on the particles size):

$$Au + V_O - O^-_{ads} \rightarrow Au - O_{ads} + e^- + V_O \quad (5.8)$$

3. Reversible adsorption of CO on the Au particles:

$$Au + CO \rightarrow Au - C - O \quad (5.9)$$

4. CO reacts with oxygen:

$$Au - C - O + Au - O_{ads} \rightarrow 2Au + CO_2\uparrow \quad (5.10)$$

The information on the reaction kinetics, in particular, the determination of the reaction order, is an essential element in the discussion of the reaction mechanism. At present, there are relatively few reports on the kinetics of CO oxidation. From the results of these studies, the reaction orders of the CO and O_2 reactants are both between 0.05 and 0.46, which means that they may exist from medium to strong adsorption. The reaction order is not negative, indicating that the adsorption of the reactants is not strong enough to hinder the reaction. Vannic et al. [72] studied the kinetics of CO catalytic oxidation systematically. They prepared AuNPs supported on TiO_2 by the impregnation method and measured the reaction kinetics data. The experimental results were compared with simulated reaction rates obtained from the chemical adsorption entropy and enthalpy calculations, and six possible mechanism models that had been set up and verified. The Eley–Rideal mechanism and the model involving rate control were first excluded because they were all first-order reaction kinetic models. The calculated enthalpy of dissociation oxygen is much larger than the entropy of gas molecules, so that the Langmuir–Hinshelwood mechanism involving competitive adsorption (one active site) and noncompetitive adsorption (two active sites) is also not suitable. This is because the values of entropy and enthalpy for competitive adsorption and noncompetitive adsorption should be consistent in the system with the reaction of CO and O_2 acting as the rate control step. Using diffuse reflection Fourier transform infrared spectroscopy (DRIFTS) observations, it was confirmed that the adsorption was noncompetitive, which means that the adsorption of oxygen did not affect the adsorption of CO. This calculation assumes that the adsorption occurs on the surface of the support near the supported metal. Their kinetic studies give the reaction order of CO (ranging from 0.05 to 0.85) and O_2 (ranging from 0.07 to 0.46) in the catalytic oxidation of CO over AuNPs with various sizes and supports in a temperature range of 273K–333K. In this way, other kinetic results are also understandable except for some systems with completely different reaction orders.

At present, Au catalysts with good catalytic activity of CO oxidation have been used in or shown potential applications in gas purification, CO_2 lasers, gas sensors, and others. Gas purification includes air purification and industrial gas purification. Air purification processes mainly utilize supported Au catalysts in the oxidation of CO at room temperature, thereby achieving the purposes of eliminating CO and air purification. Industrial gas purification is performed to eliminate trace amounts of CO ($2\ mg\ m^{-3}$) in mixed gas using a supported Au catalyst, for the purpose of producing high-purity nitrogen–oxygen mixed gas. Au/Fe_2O_3 and Au/Al_2O_3 are the most common commercial catalysts and have been widely used in this area. Supported Au catalysts can also be used in gas sensors for CO. Gas sensors usually use *n*-type semiconductors as supports. This kind of supported Au catalyst is highly sensitive in the detection of CO. For example, $Au/\alpha\text{-}Fe_2O_3$ doped with TiO_2 can detect trace CO in hydrogen at room temperature, which has been widely used in industrial applications [73]. In CO_2 lasers, CO and O_2 from CO_2 decomposition are detrimental to CO_2 lasers. Therefore it is necessary to reform CO and O_2 into CO_2 at room temperature. As early as 1983, Pt/SnO_x had already been proven to be an effective catalyst in this application. In the 1990s, Au/MnO_x was found to show better catalytic performance than that of Pt/SnO_x under the condition where CO_2 laser worked [74]. Au/Fe_2O_3 and $Au/NiFe_2O_4$ can also convert CO and O_2 into CO_2 at room temperature, even at 263K, while the conversion rate for Pt/SnO_x is only 20% at 323K.

5.3.2 Catalytic Decomposition of Volatile Organic Compounds

Most volatile organic compounds (VOCs) are toxic and even carcinogenic and can cause great harm to human health. VOCs can also react with nitrogen oxides in the atmosphere under sunlight to produce highly toxic photochemical smog. VOCs have become one of the major atmospheric pollutants and their purification technology has become a research hotspot in environmental catalysis.

The combustion method is usually employed for the control of VOCs. The combustion method refers to the complete combustion and oxidation of VOCs to CO_2 and H_2O under certain conditions, which can treat VOCs of various concentrations. The removal rate can reach as high as 95% without secondary pollution. In addition, the heat generated by combustion can also be used as waste heat. Combustion is widely used in the control of industrial VOCs. The direct combustion temperature is usually between 973K and 1173K, so that the carbon-containing organics can be completely burned and oxidized into CO_2. The use of a catalyst can reduce the activation energy of the reaction obviously. The reaction temperature of the catalytic decomposition of VOCs can be reduced to 473K–673K, which can avoid secondary pollution from nitrogen oxides and reduce the addition of auxiliary heat. However, it may lead to catalyst poisoning when the VOCs contain S, Cl, and other elements.

The catalytic combustion reaction of VOCs on the surface of catalysts is:

$$C_xH_yO_z + \left(x + \frac{y}{4} - \frac{z}{2}\right)O_2 \rightarrow x\,CO_2 + \frac{y}{2}H_2O \tag{5.11}$$

In the catalytic decomposition of alcohols and their derivatives, Au/Fe_2O_3 catalysts show high activity in complete oxidation [75]. These supported Au catalysts show comparable catalytic activity with Pd/Al_2O_3 and Pt/Al_2O_3, but better moisture resistance. The activity of Au/Fe_2O_3 can be improved by increasing the loading amount of Au [76]. With an 8.2 wt.% Au loading, the oxidation temperature of acetone and toluene began from 353K and 593K, respectively, while the temperature was only 313K for alcohol. It needs to be pointed out that the reaction temperature of 8.2 wt.% Au/Fe_2O_3 is much lower than that of other catalysts. In the catalytic oxidation of ethanol and 2-propanol, acetaldehyde and acetone would be formed as intermediates. However, further increasing the reaction temperature helps to obtain a complete oxidized production (CO_2), while propionic acid and propylene would be generated with the traditional Mn_3O_4, TiO_2, and V_2O_5 as catalysts.

In the catalytic decomposition of chlorinated hydrocarbon, supported Au catalysts also show catalytic activity. Au/Co_3O_4 and Au/Al_2O_3 have been proven to show good performance in the catalytic oxidation of chloromethane [77]. They show comparable activity with Pt-based catalytic performance and better stability and water resistance. Au/Co_3O_4 prepared by coprecipitation can be used in the catalytic decomposition of CH_2Cl_2 to CO_2 and HCl at a temperature of 623K without any byproducts ($CHCl_3$, CCl_4) and incompletely oxidized products (CO, Cl, $COCl_2$, HCHO). In this catalytic system, the activity is independent of the loading amount of Au, and the stability is not influenced by the CH_2Cl_2, HCl, or H_2O.

5.3.3 Removal of Nitrogen Oxides

Nitrogen oxides are primary atmospheric pollutants, which can dissolve in water to form nitrous and nitric acid or combine with unburnt hydrocarbons to produce photochemical smog. Studies with supported Au catalysts on the removal of nitrogen oxides have focused on the catalytic oxidation of NO and the reduction of NO_x with alkanes, carbon monoxide, or hydrogen.

In the process of flue gas denitrification, the selective oxidation of NO is of great importance to improve the removal efficiency. Yuan et al. [78] investigated catalytic NO oxidation by ultrafine AuNPs supported on Co_3O_4. The AuNPs supported on the Co_3O_4 were prepared by modified coprecipitation with urea as the precipitant. After calcination at 653K for 4 h, the $Au(OH)_3$ and $Co(OH)_3$ transfer to Au/Co_3O_4 with a loading amount of 5 wt.% and average particle size of 4 nm. Au/Co_3O_4 catalyst is applied in the catalytic oxidation of NO with a selectivity of over 50% and satisfactory sulfur resistance and water resistance at 393K, while the selectivity of Co_3O_4 is below 10%. Yang et al. [79] used a titanium silicate molecular sieve, TS-1 as the support to support Au and check the effect of the NO and O_2 feed concentration, the space velocity, and the Au loading on the NO conversion. It was found that the conversion of NO is independent of its initial concentration and the selective catalytic oxidation of NO is a first-order reaction. However, increasing the initial O_2 concentration can first promote the conversion efficiency because it can improve the content of adsorbed oxygen, and then the NO conversion tends to be stable when the oxygen adsorption on the surface of the catalyst reaches saturation. The efficiency of a gas–solid heterocatalytic reaction is related with the transfer process, which means that the reactant must reach the active site efficiently through external and internal diffusions. In the system of Au/TS-1, NO conversion can reach as high as 78% with a space velocity under 5000 h^{-1} and the conversion decreases to 54% when the space velocity increases to 7000 h^{-1}. The relatively low space velocity can provide a longer contact time between the reactant and the catalysts, thus benefiting the diffusion, adsorption, and reaction of reactants inside the catalyst pores. As the active site in NO catalytic oxidation, the loading amount of Au can also affect the NO conversion. The conversion reaches a maximum with an Au loading of 1.0%. A higher loading of Au ($>$1.0%) may cause the aggregation of Au and a lower loading amount ($<$1.0%) may cause a decrease in active sites for the oxidative reaction, leading to a decreasing in the NO oxidation.

1. *Reduction of NO_x with alkanes.* A lot of supported Au catalysts have shown the ability to catalytically reduce NO by propylene. Ueda et al. [80] checked the catalytic reduction of NO by AuNPs on various supports of Fe_2O_3, ZnO, MgO, TiO_2, and Al_2O_3. With a loading amount of 0.1–0.2 wt.%, Au/Al_2O_3 shows excellent catalytic activity in the reduction of NO, with a conversion of nearly 70% at 700K. They found that the reaction temperature varied with the metal oxide supports. There is rare N_2O generated. Au/ZnO showed the best catalytic activity. At 523K, the maximum conversion of NO on Au/ZnO reached 25%. The programmed temperature desorption results indicated that during the catalytic reduction of NO by propylene on Au/Al_2O_3, NO is first oxidized by O_2 to NO_2, which is a slow reaction, and then reduced by C_3H_6 to N_2.

$$NO + O_2 \rightarrow NO_2 \tag{5.12}$$

$$NO_2 + C_3H_6 \rightarrow N_2 \tag{5.13}$$

However, the formation rate of NO_2 can be accelerated by doping Au/Al_2O_3 with Mn_2O_3, therefore significantly improving the conversion of NO to NO_2 over a wide range of temperatures [81]. The maximum catalytic conversion of NO to N_2 reaches 88% at 698K and the reduction of NO on the surface of Au/Al_2O_3 (0.17 wt.%) takes place even in the presence of water and oxygen. However, using Au/Mn_2O_3/Al_2O_3 (with an Mn_2O_3:Au/Al_2O_3 ratio of 1:19) prepared by mechanical mixing as the catalysts can increase the conversion from 88% to 98% at 698K. In addition, the addition of Mn_2O_3 can also enhance the catalytic reduction efficiency of NO even at a temperature as low as 523K.

2. *Reduction of NO_x with carbon monoxide.* The reduction of NO_x with carbon monoxide usually applies in the treatment of engine exhaust gases by catalytic converters of NO_x by CO produced from the incomplete combustion of fuel.

$$2NO + 2CO \rightarrow N_2 + 2CO_2 \tag{5.14}$$

However, the reduction of NO by CO in the absence of oxygen will lead to the production of nitrous oxide.

$$2NO + CO \rightarrow N_2O + CO_2 \tag{5.15}$$

The catalytic activity of supported AuNPs depends strongly on the supports used. Among them, iron-based supports provide the highest activity for supported Au catalysts in the reduction of NO by CO. For example, Au/Fe_2O_3, Au/$NiFe_2O_4$, and Au/$MnFe_2O_4$ can catalyze the reduction of NO at temperatures below 423K and even at room temperature [82], which indicates that supported Au catalysts have superior low-temperature catalytic activity than that of rhodium, platinum, and palladium [83,84]. The main reduced product is N_2, while the main product of the corresponding unsupported catalyst is N_2O. The high catalytic reduction activity of supported AuNPs is maintained in the presence of water, while the presence of O_2 can affect the activity. The proper amount of O_2 can promote the selectivity of N_2, while an excess will hinder the catalytic reduction of NO by CO due to the catalytic oxidation of CO taking place in preference.

Because the temperature of engine exhaust gases is usually lower than 423K, the low-temperature catalytic activity of supported AuNPs shows an attractive potential commercial application. However, with the increasing commercial interest in the lean-burn engine, it becomes more difficult for supported AuNPs to catalyze the reduction of NO. Under this condition, Au catalysts show poorer low-temperature catalytic activity than the platinum group metals. In order to improve the catalytic activity, the mechanical mixing of Mn_2O_3 can also achieve this goal. Mn_2O_3 can catalyze the oxidation of NO to NO_2, and then the NO_2 reacts with the carbon monoxide adsorbed to form N_2.

$$NO + O_2 \rightarrow NO_2 \quad (5.16)$$

$$NO_2 + CO \rightarrow N_2 + CO_2 \quad (5.17)$$

3. *Reduction of NO_x with hydrogen.* The nitrogen oxides can also be reduced by H_2 with the catalysis of supported AuNPs.

$$2NO + H_2 \rightarrow N_2O + H_2O \quad (5.18)$$

$$2NO + 5H_2 \rightarrow 2NH_3 + 2H_2O \quad (5.19)$$

$$N_2O + H_2 \rightarrow N_2 + H_2O \quad (5.20)$$

AuNPs supported by NaY and ZSM-5 are active for the reduction of NO by H_2 and a significant enhancement can be produced in the presence of oxygen by forming NO_2 and N_2O_4 [85]. The oxidation of NO is very fast, and then NO_2 and N_2O_4 intermediates are eventually reduced to N_2. The promotion is proposed as:

$$2NO + O_2 \rightarrow 2NO_2 \leftrightarrow N_2O_4 \quad (5.21)$$

$$N_2O_4(NO_2) + 2H_2 \rightarrow N_2 + 2H_2O \quad (5.22)$$

AuNPs supported on Al_2O_3, TiO_2, SiO_2, and other metal oxides also active for this reaction of NO reduction by H_2 [82,86,87]. The presence of an additional metal oxide (such as CoO_x, CeO_x, LaO_x) is usually beneficial to the selectivity of N_2 and improves the catalytic activity of NO reduction.

5.3.4 Ozone Decomposition

The ozone in the upper atmosphere is beneficial to humans due to the absorption of UV radiation. However, ozone in the living environment is harmful as it causes respiratory diseases and enhances photochemical pollution. The concentration of ozone in the working environment is limited to below 0.1 ppm. Therefore ozone from the recycled air of aircraft cabins and from electrostatic copiers and lasers in office environments is necessary to be removed. It is tempting to develop catalysts for the decomposition of ozone.

Hao et al. [88] employed Fe_2O_3 as the support to synthesize supported Au catalysts with a high specific surface area of 158.31 $m^2\,g^{-1}$. Au/Fe_2O_3 (1 at.%) catalysts are highly effective at 273K, giving a 98% conversion in the decomposition of ozone, which is superior to those of Ag- and Ni-based catalysts with the same conditions. Au/Fe_2O_3 catalysts also show excellent performance in the resistance to moisture and ozone corrosion. In addition, Au/Fe_2O_3 catalysts are also employed in the simultaneous removal of ozone and CO. In this process, Au catalysts can catalyze the decomposition of O_3 and the oxidation of CO efficiently, which is not limited by the stoichiometric ratio.

5.4 Surface Plasmon Resonance–Enhanced Photocatalysis

The rapid development of SPR has offered a new approach to improve the efficiency of photocatalysis. SPR-enhanced photocatalysis is usually based on metal–semiconductor composite materials. Benefitting from the SPR effect of metal NPs (such as Au, Ag, and Cu), the free electrons of the metal are collectively oscillated by the irradiation of incident light with the same vibration frequency of metal NPs, causing the migration of electron clouds on the surface of the metal NPs. The induced transfer of the electrons can lead to the redox reaction. Noble-metal NPs can increase the energy conversion efficiency of photocatalytic materials by extending the absorption range of incident light, increasing the light scattering, and effectively exciting photogenerated electron–hole pairs in the semiconductor by transferring the plasmonic energy to the semiconductor.

5.4.1 Design of Surface Plasmon Resonance–Enhanced Photocatalysts

Because the SPR effect can produce photogenerated electron–hole pairs under the irradiation of light, nonsemiconductor supported AuNPs can also be employed for photocatalytic applications. Silica and zirconia have been widely used in these systems due to the bandgaps of 9.0 and 5.0 eV. These wide bandgaps mean that silica and zirconia show poor activity under the irradiation of visible light. However, silica- and zirconia-supported AuNPs showed effective photocatalytic activity for the degradation of polar organic molecules such as formaldehyde and methanol, while being less effective in the degradation of nonpolar molecular cyclohexane. In addition, zirconia as a support shows more efficient activity than that of silica, as zirconia shows effective adsorption for organic molecules. The catalytic effect in this system can be attributed to the thermal effect caused by the SPR effect.

Usually, the crystal structure, band structure, and surface atomic configuration of SPR-enhanced photocatalysts have a significant effect on photocatalytic performance. To claim the effect of the band structure on the photocatalysis, anatase and rutile TiO_2 have been employed to support AuNPs, respectively [89]. It was found that the size of AuNPs shows a negligible effect on the photocatalytic activity above the 3–12 nm range. However, after loading AuNPs with a similar size, anatase TiO_2 delivered a photocatalytic rate two orders of magnitude higher than that recorded for rutile. In this SPR-enhanced photocatalysis system of Au/TiO_2, anatase and rutile TiO_2 have the same elemental composition but different crystalline forms and band structures, which results in differences in light absorption and electron–hole recombination rate. The electron–hole recombination rate of rutile TiO_2 is 1–3 orders of magnitude faster than that of anatase, which leads to a poorer photocatalytic performance. Therefore selecting and designing the semiconductor are key concerns for the construction of SPR-enhanced photocatalytic materials.

5.4.2 Mechanism of Surface Plasmon Resonance–Enhanced Photocatalysis

A great deal of research has been done on the synthesis and activity of SPR-enhanced photocatalysts. As mentioned previously, AuNPs can increase the photocatalytic performance by: (1) extending the absorption range of incident light; (2) increasing the light scattering; and (3) effectively exciting photogenerated electron–hole pairs by transferring plasmonic energy to the semiconductor. Process (1) extends solar light absorption from the visible to near-infrared light range by concentrating the incident photon energy in plasmon oscillations, while process (2) can originate from the large scattering cross section associated with SPR due to the scattering of incident light and amplification of the electromagnetic field by the metallic NPs. In process (3), concentrated plasmonic energy is transferred to the semiconductor, inducing the generation and separation of the electron–hole pairs in the semiconductor. However, the mechanism of SPR-induced charge generation and separation in the photocatalysts is still not clear, which limits its application in environmental catalysis.

Different semiconductors would construct totally different interfaces with noble metals, thus affecting the photocatalytic reaction process. As a comparison, the charge separation mechanisms for various photocatalysts are shown in Fig. 5.5 [90], where (A) traditional semiconductor photocatalysts absorb photons with energy greater than the bandgap *Eg* and produce electron–hole pairs. The electrons (holes) in the conduction (valence) band [CB (VB)], could contribute to chemical redox reactions ($X + e^- = X^-$) and ($Y + h^+ = Y^+$) after transferring to their surface. (B) Early research suggests that the reason for the enhanced photocatalytic activity of noble-metal NPs on the semiconductor surface is that the metal NPs act as cocatalysts that can form a Schottky junction with the semiconductor interface. It can provide additional surface sites via the trapping of electrons. (C) Direct electron transfer (DET) from the plasmonic metal to the conduction band of the semiconductor possibly occurs when they are in direct contact and the band levels of the semiconductor and Fermi level of the metal matches (types A, B, and C). Most reported SPR-enhanced photocatalysis processes in the literature follow this mechanism. This DET process can be experimentally proven. Furube et al. [91] directly observed SPR excited electrons transfer from AuNPs with a size of 10 nm to a TiO_2 nanocrystalline film using femtosecond transient absorption spectroscopy. The reaction time in this system is within 240 fs and the yield is about 40%. (D) If the noble metal and the semiconductor are not in direct contact, the local electromagnetic field (LEMF) generated by the SPR effect can promote the generation of electron–hole pairs in the semiconductor (type D). Torimoto et al. [92] constructed an Au/SiO_2/

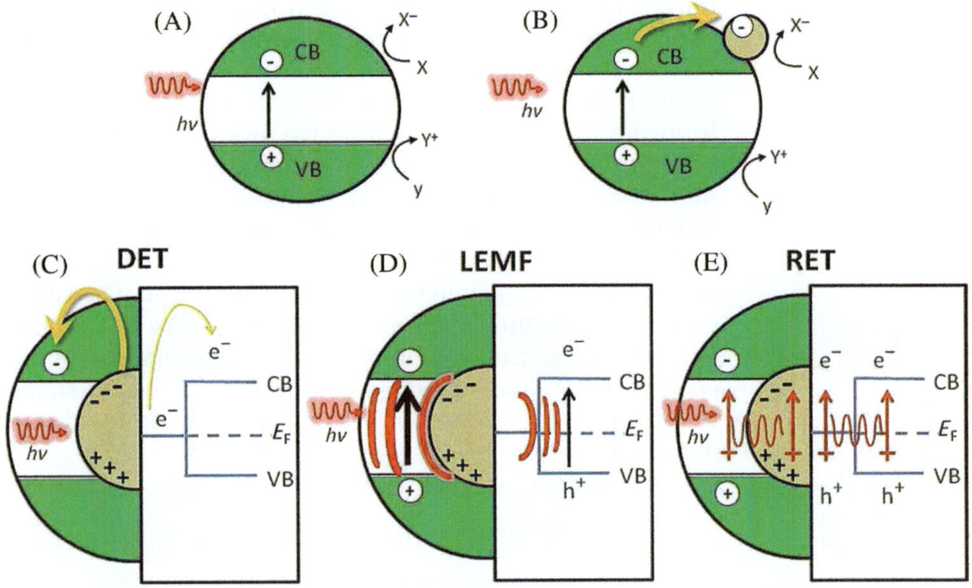

Figure 5.5 Charge separation mechanisms in various photocatalytic nanostructures. (A) Photoexcited semiconductors produce electrons (holes) in the conduction (valence) band [CB (VB)], each contributing to chemical reactions at their surface. (B) Metal NPs can act as cocatalysts to provide additional surface sites via the trapping of electrons. Au/semiconductor structures can increase charge separation by (C) direct electron transfer (DET) of hot electrons contained in surface plasmon resonance (SPR) to the semiconductor, by (D) local electromagnetic field (LEMF) of the semiconductor charge separation process, and (E) resonant energy transfer (RET) from the SPR dipole to the electron–hole pair in the semiconductor shell. Reprinted with permission from S.K. Cushing, J. Li, F. Meng, T.R. Senty, S. Suri, M. Zhi, et al., Photocatalytic activity enhanced by plasmonic resonant energy transfer from metal to semiconductor, J. Am. Chem. Soc. 134 (2012) 15033–15041. Copyright (2012) American Chemical Society.

CdS photocatalytic system and isolated the noble metal from the semiconductor with a SiO_2 layer. It was demonstrated that the enhanced photocatalytic activity of the system was attracted to the effective excitation of CdS by the LEMF induced by the SPR effect of Au. (E) Resonant energy transfer based on dipole–dipole energy transfer can directly excite semiconductors to generate electron–hole pairs. The $Au/SiO_2/Cu_2O$ system reported by Cushing et al., is based on this mechanism [90].

Although these three mechanisms for SPR-enhanced photocatalysis are quite different, the semiconductor plays a key role in each of them, by acting as an active center of photocatalytic reaction or receiving the electrons from the metal. However, Chen et al. [93] found AuNPs supported by oxides with wide bandgaps such as ZrO_2, SiO_2 that also have high visible light photocatalytic activity. There is no charge transfer between the

AuNPs and the oxides, and the LEMF generated by the SPR effect is not sufficient to generate electron–hole pairs due to the large bandgap of the semiconductors. Therefore the photocatalytic mechanism in this system is significantly different from the mechanism shown above. Because AuNPs can rapidly generate heat once absorbing visible light, which is called the photothermal effect [94] it is believed that the thermal effect induced by the SPR can activate and further degrade organic molecules. The AuNPs in this system act as the active center during the photocatalytic reaction. Subsequently, Christopher et al. [95] further demonstrated that there is a correlation between photocatalysis and thermal catalysis for noble-metal NPs/wide-bandgap oxide systems, which can significantly reduce the temperature for the catalytic reaction with the aid of light fields.

5.4.3 Surface Plasmon Resonance–Enhanced Photocatalysis for Air Purification

SPR-enhanced photocatalysts show excellent activity in air purification applications such as the decomposition of ozone, the mineralization of VOCs, the conversion of inorganic gases (NO_x, SO_x, CO, H_2S, etc.).

Au/TiO_2 was prepared by photochemical deposition for the photocatalytic decomposition of ozone [96]. Metallic AuNPs deposited on the TiO_2 by UV light irradiation can be regarded as an electronegative center. The photocatalytic decomposition of ozone can be improved by increasing the photodeposition time of Au/TiO_2. When the Au loading was 1.0% with a photodeposition of 120 min, the elimination rate of ozone generated by the germicidal lamp was over 96%. The interface between the Au cluster and the TiO_2 is the main active site. The photogenerated electrons from the excitation of TiO_2 by the irradiation of the UV light transfer to the Au cluster causes a negative shift of the Fermi level of the Au cluster to be a strong negative center. Therefore metallic Au deposited on the TiO_2 not only serves as a trapping center for electrons, but also promotes the effective separation of electron–hole pairs, and increases the adsorption active center of O_3, thereby promoting the decomposition of O_3. During photocatalytic decomposition, electrons and hydroxyl radicals work as the reactive species and the productions include O_2, OH^-, and others.

$$O_3 + e^-(\bullet OH) \rightarrow O_2 \quad (5.23)$$

AuNPs can also be employed for improving the photocatalytic activity of semiconductors in the oxidation of organic pollutants such as CO and VOCs in the air under visible light irradiation. Once supported by four different structures of TiO_2, Au-based photocatalysts with a loading amount of 3% can catalyze the oxidation of HCHO in an air atmosphere under blue light (400–500 nm) [97]. The effect of hydroxyl groups on the surface of the catalyst during the photooxidation process was investigated. It was found that hydroxyl groups and adsorbed water can promote the adsorption and enhance the activity of molecular oxygen on the surface of TiO_2. The SPR effect of AuNPs can induce electron transitions on Au 6sp to high energy levels under the irradiation of visible light. These electrons with high energy can migrate into the conduction band of TiO_2; therefore leading to the generation of superoxide radicals and hydroxyl radicals. In addition, supported Au can improve the adsorption of HCHO. With the reaction of reactive oxygen species, HCHO and other VOCs can be completely decomposed into water, carbon dioxide, and other small molecules.

$$\mathrm{VOCs} + (\bullet\mathrm{OH}^-, \ \bullet\mathrm{O}_2^-) \rightarrow \mathrm{CO}_2 + \mathrm{H}_2\mathrm{O} + \cdots \qquad (5.24)$$

With the presence of O_2, SPR-enhanced photocatalysts can also be used for the removal of NO_x, SO_x, and CO. For example, after loading AuNPs onto the surface of TiO_2 by the sol–gel method, the photocatalytic conversion of NO can reach 85% in 60 min under the irradiation of UV light at atmospheric pressure and room temperature [98]. The final productions were usually N_2 and O_2. The effect of the Au particles in the improvement of the photocatalytic activity can be attributed to a diminution of the e^-/h^+ pair recombination rate.

$$\mathrm{NO} + \mathrm{e}^- \rightarrow \mathrm{N}_2 + \mathrm{O}_2 \qquad (5.25)$$

In the photocatalytic oxidation of CO, the SPR effect can also induce the formation of superoxide by accepting hot electrons generated from the absorption of visible light by the AuNPs, therefore enhancing the activity of CO conversion [99]. The mechanism of photocatalytic oxidation of CO can be summarized as:

$$\mathrm{O}_2 + \mathrm{e}^- \rightarrow \bullet\mathrm{O}_2^- \qquad (5.26)$$

$$\mathrm{CO} + \bullet\mathrm{O}_2^- \rightarrow \mathrm{CO}_2 \qquad (5.27)$$

5.4.4 Photocatalytic Treatment of Wastewater

SPR-enhanced photocatalysis has also been employed in the oxidation of a range of organic pollutants. By loading AuNPs on the surface, P25 (TiO_2 with a ratio of anatase to rutile of 83:17) demonstrated an improved photocatalytic performance of alcohol oxidation under visible light irradiation (Fig. 5.6) [100]. The photocatalytic process of alcohol on the surface of Au/TiO_2 can be generalized as: (1) due to the SPR effect, AuNPs absorb visible light and transfer excited electrons to the conduction band of the rutile TiO_2; (2) the electrons are further transferred from the rutile to the anatase TiO_2; and (3) oxygen is reduced by electrons. From the thermodynamic theory, since the conduction band bottom of anatase TiO_2 (−0.25 V vs normal Hydrogen electrode, NHE) is more negative than that of rutile TiO_2

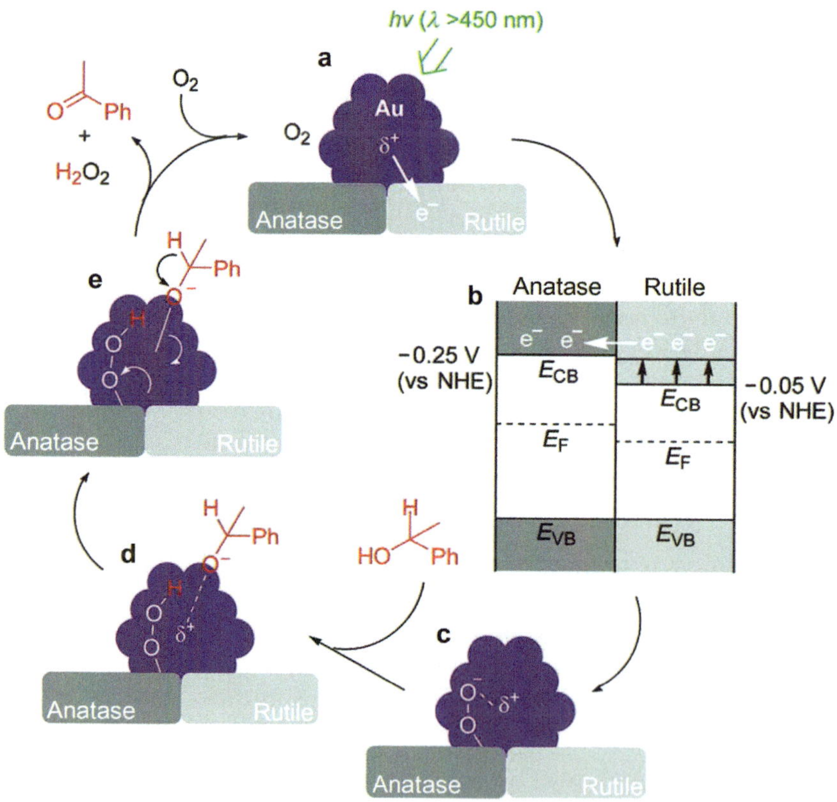

Figure 5.6 Proposed mechanism for aerobic oxidation of alcohol on Au/P25 under visible light irradiation. Reprinted with permission from D. Tsukamoto, Y. Shiraishi, Y. Sugano, S. Ichikawa, S. Tanaka, T. Hirai, Gold Nanoparticles located at the interface of anatase/rutile TiO2 particles as active plasmonic photocatalysts for aerobic oxidation, J. Am. Chem. Soc. 134 (2012) 6309–6315. Copyright (2012) American Chemical Society.

(−0.05 V vs NHE), the electrons are difficult to move from rutile to anatase. However, the electrons in the conduction band of rutile TiO_2 that migrate from the AuNPs can negatively shift to the conduction band of the anatase TiO_2. Therefore the active sites of Au/rutile TiO_2/anatase TiO_2 could enhance the photocatalytic performance of alcohol oxidation.

The photogenerated holes induced by the SPR show good reduction performance, which can be used in the removal of heavy metals and nitrate. Semiconductors, graphene, and metal–organic frameworks are employed as supports [101–103]. The enhanced removal of heavy metals and nitrate can also be attracted to the enhanced generation of reactive radicals and more efficient separation of the charge supports when excited by visible light due to the SPR effect of AuNPs. In addition, with the addition of organics, the photocatalytic reduction efficiency can be improved due to the synergistic effect of photocatalytic oxidation and reduction.

5.5 Conclusion and Future Perspectives

In summary, two kinds of nanosized Au-based catalysts have been introduced, namely (1) catalysts with activity arising from effective charge transfer, the quantum size effect, and the nanostructured Au itself and (2) SPR-enhanced photocatalysts. The preparation methods, environmental application in catalytic oxidation of CO, catalytic decomposition of VOCs, removal of nitrogen oxides, and O_2 decomposition have also been summarized.

Researches on Au nanocatalysts have indicated their potential application in the commercially viable environment such as for the elimination of CO, formaldehyde, and O_2 in air purification systems, and the elimination of ethylene in fruit storage warehouses. The research efforts in this area are so for at an impressive level, but further studies on Au catalysts regarding issues such as cost effectiveness, longevity, stability in storage, and high activity are still essential for the commercial application of Au catalysts as an environment-friendly technology with great advantages. The development of Au nanocatalysts includes (1) large-scale synthesis methods for uniform size and shape; (2) advanced characterization for the reaction mechanism; and (3) the design of reaction devices. Obtaining large-scale Au nanocatalysts with uniform size and shape is essential for exploring the mechanisms and reducing the cost for the intended applications. While investigating the role of Au, their polydispersity makes it difficult to clarify the structure–activity relationship between structure and catalytic performance. Among the "top-down" or "bottom-up" approaches for producing NPs, "bottom-up" approaches tend to produce AuNPs on a large scale with uniform size and shape. The initial quantities of Au seeds and the strength of reducing reagents must be considered carefully to achieve the large-scale preparation of AuNPs. In addition, the growth conditions such as the pH, temperature, or the stirring rate must also be controlled and maintained when aiming for a high monodispersity.

The development of advanced characterization, especially the "in situ/operando" technology for the reaction mechanism also greatly expands the potential application of Au nanocatalysts. The "in situ/operando" technology indicates real-time monitoring for the entire process. The employment of the "in situ/operando" technology makes it possible to investigate changes in the morphology, crystal structure, or electronic valence of catalysts during the reaction. It is highly significant for the improvement of the activity of nanomaterials. For supported Au catalysts, the focus is on stable Au cluster or single Au atom catalysts and the strong metal–support interactions. To explore the catalytic mechanism, the active sites of supported Au materials can be investigated by advanced characterization techniques such as spherical-aberration-corrected transmission electron microscopy, cryo-electron microscopy, X-ray absorption spectroscopy, etc. The use of these advanced characterization techniques plays an important role in exploring the catalytic mechanism of Au catalysts. In addition, as a kind of supported catalyst, nanoreactors have opened up a new field of research on supported catalysts and have recently received increasing attention. In a nanoreactor, all of the metal NPs are supported on and isolated by the metal oxide shell, and thus have a relatively uniform environment. In addition, the shell formed by metal oxides also prevents metal NPs from aggregating. Moreover, the interaction between the metal NPs and the shell can increase the catalytic activity of catalysts. Considering these advantages, nanoreactors show great potential compared to traditionally supported catalysts. It is believed that with the development of nanoscience and nanotechnology, catalytic technologies based on Au nanomaterials will have significant applications in environmental remediation.

Acknowledgments

The authors are grateful for support from the Hong Kong Scholars Program (No. XJ2017051).

References

[1] M. Haruta, T. Kobayashi, H. Sano, N. Yamada, Novel gold catalysts for the oxidation of carbon monoxide at a temperature far below 0°C, Chem. Lett. 16 (1987) 405–408.

[2] M. Haruta, N. Yamada, T. Kobayashi, S. Iijima, Gold catalysts prepared by coprecipitation for low-temperature oxidation of hydrogen and of carbon monoxide, J. Catal. 115 (1989) 301–309.

[3] M. Valden, X. Lai, D.W. Goodman, Onset of catalytic activity of gold clusters on titania with the appearance of nonmetallic properties, Science 281 (1998) 1647–1650.

[4] M. Comotti, W.-C. Li, B. Spliethoff, F. Schüth, Support effect in high activity gold catalysts for CO oxidation, J. Am. Chem. Soc. 128 (2006) 917–924.

[5] M. Comotti, C. Della Pina, R. Matarrese, M. Rossi, The catalytic activity of "naked" gold particles, Angew. Chem. Int. Ed. 43 (2004) 5812–5815.

[6] C. Xu, J. Su, X. Xu, P. Liu, H. Zhao, F. Tian, et al., Low temperature CO oxidation over unsupported nanoporous gold, J. Am. Chem. Soc. 129 (2007) 42–43.

[7] T. Fujita, P. Guan, K. Mckenna, X. Lang, A. Hirata, L. Zhang, et al., Atomic origins of the high catalytic activity of nanoporous gold, Nat. Mater. 11 (2012) 775.

[8] M. Haruta, Gold as a novel catalyst in the 21st century: preparation, working mechanism and applications, Gold Bull. 37 (2004) 27–36.

[9] S. Linic, P. Christopher, D.B. Ingram, Plasmonic-metal nanostructures for efficient conversion of solar to chemical energy, Nat. Mater. 10 (2011) 911.

[10] W. Hou, S.B. Cronin, A review of surface plasmon resonance-enhanced photocatalysis, Adv. Funct. Mater. 23 (2013) 1612–1619.

[11] K. Awazu, M. Fujimaki, C. Rockstuhl, J. Tominaga, H. Murakami, Y. Ohki, et al., A plasmonic photocatalyst consisting of silver nanoparticles embedded in titanium dioxide, J. Am. Chem. Soc. 130 (2008) 1676–1680.

[12] Z. Liu, W. Hou, P. Pavaskar, M. Aykol, S.B. Cronin, Plasmon resonant enhancement of photocatalytic water splitting under visible illumination, Nano. Lett. 11 (2011) 1111–1116.

[13] G.C. Bond, P.A. Sermon, Gold catalysts for olefin hydrogenation, Gold Bull. 6 (1973) 102–105.

[14] Q. Xu, K.C.C. Kharas, A.K. Datye, *The preparation of highly dispersed Au/ Al_2O_3 by aqueous impregnation*, Catal. Lett. 85 (2003) 229–235.

[15] S. Wang, Y. Wang, J. Jiang, R. Liu, M. Li, Y. Wang, et al., *A DRIFTS study of low-temperature CO oxidation over Au/SnO_2 catalyst prepared by co-precipitation method*, Catal. Commun. 10 (2009) 640–644.

[16] N.S. Patil, B.S. Uphade, D.G. Mcculloh, S.K. Bhargava, V.R. Choudhary, Styrene epoxidation over gold supported on different transition metal oxides prepared by homogeneous deposition–precipitation, Catal. Commun. 5 (2004) 681–685.

[17] S. Schimpf, M. Lucas, C. Mohr, U. Rodemerck, A. Brückner, J. Radnik, et al., Supported gold nanoparticles: in-depth catalyst characterization and application in hydrogenation and oxidation reactions, Catal. Today 72 (2002) 63–78.

[18] S.-J. Lee, A. Gavriilidis, Supported Au catalysts for low-temperature CO oxidation prepared by impregnation, J. Catal. 206 (2002) 305–313.

[19] S.H. Overbury, L. Ortiz-Soto, H. Zhu, B. Lee, M.D. Amiridis, S. Dai, Comparison of Au catalysts supported on mesoporous titania and silica:

investigation of Au particle size effects and metal-support interactions, Catal. Lett. 95 (2004) 99–106.
[20] A. Luengnaruemitchai, S. Osuwan, E. Gulari, *Comparative studies of low-temperature water–gas shift reaction over Pt/*CeO_2*, Au/*CeO_2*, and Au/*Fe_2O_3 *catalysts*, Catal. Commun. 4 (2003) 215–221.
[21] R. Zanella, S. Giorgio, C.-H. Shin, C.R. Henry, C. Louis, *Characterization and Reactivity in CO oxidation of gold nanoparticles supported on* TiO_2 *prepared by deposition-precipitation with NaOH and urea*, J. Catal. 222 (2004) 357–367.
[22] M. Okumura, S. Nakamura, S. Tsubota, T. Nakamura, M. Azuma, M. Haruta, *Chemical vapor deposition of gold on* Al_2O_3*,* SiO_2*, and* TiO_2 *for the oxidation of CO and of* H_2, Catal. Lett. 51 (1998) 53–58.
[23] P. Serp, P. Kalck, R. Feurer, Chemical vapor deposition methods for the controlled preparation of supported catalytic materials, Chem. Rev. 102 (2002) 3085–3128.
[24] Z. Ma, S. Dai, Development of novel supported gold catalysts: a materials perspective, Nano Res. 4 (2011) 3–32.
[25] Z. Ma, S. Dai, Design of novel structured gold nanocatalysts, ACS Catal. 1 (2011) 805–818.
[26] X.Y. Liu, A. Wang, T. Zhang, C.-Y. Mou, Catalysis by gold: new insights into the support effect, Nano Today 8 (2013) 403–416.
[27] C. Wang, D. Astruc, Nanogold plasmonic photocatalysis for organic synthesis and clean energy conversion, Chem. Soc. Rev. 43 (2014) 7188–7216.
[28] R. Jiang, B. Li, C. Fang, J. Wang, Metal/semiconductor hybrid nanostructures for plasmon-enhanced applications, Adv. Mater. 26 (2014) 5274–5309.
[29] T. Hirakawa, P.V. Kamat, *Charge separation and catalytic activity of Ag@* TiO_2 *core – shell composite clusters under UV – irradiation*, J. Am. Chem. Soc. 127 (2005) 3928–3934.
[30] B. Wu, D. Liu, S. Mubeen, T.T. Chuong, M. Moskovits, G.D. Stucky, *Anisotropic growth of* TiO_2 *onto gold nanorods for plasmon-enhanced hydrogen production from water reduction*, J. Am. Chem. Soc. 138 (2016) 1114–1117.
[31] W. Li, Y. Guo, P. Zhang, *General strategy to prepare* TiO_2*-core gold-shell nanoparticles as SERS-tags*, J. Phys. Chem. C 114 (2010) 7263–7268.
[32] A. Dawson, P.V. Kamat, *Semiconductor – metal nanocomposites. Photoinduced fusion and photocatalysis of gold-capped* TiO_2 *(*TiO_2*/gold) nanoparticles*, J. Phys. Chem. B 105 (2001) 960–966.
[33] I. Bannat, K. Wessels, T. Oekermann, J. Rathousky, D. Bahnemann, M. Wark, Improving the photocatalytic performance of mesoporous titania films by modification with gold nanostructures, Chem. Mater. 21 (2009) 1645–1653.
[34] E. Kowalska, O.O.P. Mahaney, R. Abe, B. Ohtani, Visible-light-induced photocatalysis through surface plasmon excitation of gold on titania surfaces, Phys. Chem. Chem. Phys. 12 (2010) 2344–2355.
[35] A. Pearson, H. Zheng, K. Kalantar-Zadeh, S.K. Bhargava, V. Bansal, *Decoration of* TiO_2 *nanotubes with metal nanoparticles using polyoxometalate as a UV-switchable reducing agent for enhanced visible and solar light photocatalysis*, Langmuir 28 (2012) 14470–14475.
[36] W. Xiong, Q. Zhao, X. Li, L. Wang, Multifunctional plasmonic Co-doped Fe_2O_3@polydopamine-Au for adsorption, photocatalysis, and SERS-based sensing, Part. Part. Syst. Char. 33 (2016) 602–609.

[37] W. Xiong, X. Li, Q. Zhao, Y. Shi, C. Hao, Insight into the photocatalytic mineralization of short chain chlorinated paraffins boosted by polydopamine and Ag nanoparticles, J. Hazard. Mater. 359 (2018) 186–193.

[38] W. Seh Zhi, S. Liu, S.-Y. Zhang, M.S. Bharathi, H. Ramanarayan, M. Low, et al., Anisotropic growth of titania onto various gold nanostructures: synthesis, theoretical understanding, and optimization for catalysis, Angew. Chem. Int. Ed. 50 (2011) 10140–10143.

[39] H. Sun, J. He, J. Wang, S.-Y. Zhang, C. Liu, T. Sritharan, et al., Investigating the multiple roles of polyvinylpyrrolidone for a general methodology of oxide encapsulation, J. Am. Chem. Soc. 135 (2013) 9099–9110.

[40] Z.W. Seh, S. Liu, M. Low, S.Y. Zhang, Z. Liu, A. Mlayah, et al., *Janus Au-TiO_2 photocatalysts with strong localization of plasmonic near-fields for efficient visible-light hydrogen generation*, Adv. Mater. 24 (2012) 2310–2314.

[41] X.-F. Wu, H.-Y. Song, J.-M. Yoon, Y.-T. Yu, Y.-F. Chen, *Synthesis of core – shell Au@TiO_2 nanoparticles with truncated wedge-shaped morphology and their photocatalytic properties*, Langmuir 25 (2009) 6438–6447.

[42] N. Zhang, S. Liu, X. Fu, Y.-J. Xu, *Synthesis of M@TiO_2 (M = Au, Pd, Pt) core–shell nanocomposites with tunable photoreactivity*, J. Phys. Chem. C 115 (2011) 9136–9145.

[43] H.-G. Boyen, G. Kästle, F. Weigl, B. Koslowski, C. Dietrich, P. Ziemann, et al., Oxidation-resistant gold-55 clusters, Science 297 (2002) 1533–1536.

[44] N. Lopez, J.K. Nørskov, Catalytic Co oxidation by a gold nanoparticle: a density functional study, J. Am. Chem. Soc. 124 (2002) 11262–11263.

[45] A.K. Sinha, S. Seelan, S. Tsubota, M. Haruta, A three-dimensional mesoporous titanosilicate support for gold nanoparticles: vapor-phase epoxidation of propene with high conversion, Angew. Chem. Int. Ed. 43 (2004) 1546–1548.

[46] S. Daniells, A. Overweg, M. Makkee, J. Moulijn, *The mechanism of low-temperature CO oxidation with Au/Fe_2O_3 catalysts: a combined Mössbauer, FT-IR, and tap reactor study*, J. Catal. 230 (2005) 52–65.

[47] G.J. Hutchings, M.S. Hall, A.F. Carley, P. Landon, B.E. Solsona, C.J. Kiely, et al., Role of gold cations in the oxidation of carbon monoxide catalyzed by iron oxide-supported gold, J. Catal. 242 (2006) 71–81.

[48] D. Boyd, S. Golunski, G. Hearne, T. Magadzu, K. Mallick, M. Raphulu, et al., Reductive routes to stabilized nanogold and relation to catalysis by supported gold, Appl. Catal., A 292 (2005) 76–81.

[49] Q. Fu, H. Saltsburg, M. Flytzani-Stephanopoulos, Active nonmetallic Au and Pt species on ceria-based water-gas shift catalysts, Science 301 (2003) 935–938.

[50] J. Guzman, B.C. Gates, Structure and reactivity of a mononuclear gold-complex catalyst supported on magnesium oxide, Angew. Chem. 115 (2003) 714–717.

[51] J. Guzman, B.C. Gates, A mononuclear gold complex catalyst supported on MgO: spectroscopic characterization during ethylene hydrogenation catalysis, J. Catal. 226 (2004) 111–119.

[52] J.C. Fierro-Gonzalez, B.C. Gates, *Mononuclear Au^{III} and Au^{I} complexes bonded to zeolite NaY: catalysts for CO oxidation at 298 K*, J. Phys. Chem. B 108 (2004) 16999–17002.

[53] K. Okumura, K. Yoshino, K. Kato, M. Niwa, *Quick XAFS studies on the Y-type zeolite supported Au catalysts for CO-O_2 reaction*, J. Phys. Chem. B 109 (2005) 12380–12386.

[54] V. Zielasek, B. Jürgens, C. Schulz, J. Biener, M.M. Biener, A.V. Hamza, et al., Gold catalysts: nanoporous gold foams, Angew. Chem. Int. Ed. 45 (2006) 8241–8244.

[55] M. Haruta, Size- and support-dependency in the catalysis of gold, Catal. Today 36 (1997) 153–166.
[56] R.J. Davis, All that glitters is not AuO, Science 301 (2003) 926–927.
[57] K. Okazaki, S. Ichikawa, Y. Maeda, M. Haruta, M. Kohyama, *Electronic structures of Au supported on* TiO_2, Appl. Catal., A 291 (2005) 45–54.
[58] P. Concepción, S. Carrettin, A. Corma, *Stabilization of cationic gold species on Au/*CeO_2 *catalysts under working conditions*, Appl. Catal., A 307 (2006) 42–45.
[59] J.C. Fierro-Gonzalez, V.A. Bhirud, B.C. Gates, *A highly active catalyst for CO oxidation at 298 K: mononuclear* Au^{III} *complexes anchored to* La_2O_3 *nanoparticles*, Chem. Commun. (2005) 5275–5277.
[60] M. Daté, M. Okumura, S. Tsubota, M. Haruta, Vital role of moisture in the catalytic activity of supported gold nanoparticles, Angew. Chem. Int. Ed. 43 (2004) 2129–2132.
[61] S.T. Daniells, M. Makkee, J.A. Moulijn, *The effect of high-temperature pre-treatment and water on the low temperature CO oxidation with Au/*Fe_2O_3 *catalysts*, Catal. Lett. 100 (2005) 39–47.
[62] A. Luengnaruemitchai, K. Srihamat, C. Pojanavaraphan, R. Wanchanthuek, *Activity of Au/*Fe_2O_3*–*TiO_2 *catalyst for preferential CO oxidation*, Int. J. Hydrogen Energy 40 (2015) 13443–13455.
[63] F. Morfin, A. Ait-Chaou, M. Lomello, J.-L. Rousset, *Influence of the partner oxide on the catalytic properties of Au/*Ce_xZr_{1-x} *highly loaded gold catalysts*, J. Catal. 331 (2015) 210–216.
[64] Z. Ma, S.H. Overbury, S. Dai, *Au/*M_xO_y*/*TiO_2 *catalysts for CO oxidation: promotional effect of main-group, transition, and rare-earth metal oxide additives*, J. Mol. Catal. A: Chem. 273 (2007) 186–197.
[65] J. Mizera, N. Spiridis, R.P. Socha, M. Zimowska, R. Grabowski, K. Samson, et al., *The influence of base metal (M) oxidation state in Au-MO/*TiO_2 *systems on their catalytic activity in carbon monoxide oxidation*, Catalysts 2 (2011) 38–55.
[66] M. Debeila, R.P.K. Wells, J.A. Anderson, *Influence of water and pretreatment conditions on CO oxidation over Au/*TiO_2*–*In_2O_3 *catalysts*, J. Catal. 239 (2006) 162–172.
[67] F. Moreau, G.C. Bond, B. Van Der Linden, B.A. Silberova, M. Makkee, Gold supported on mixed oxides for the oxidation of carbon monoxide, Appl. Catal., A 347 (2008) 208–215.
[68] R.J. Baxter, P. Hu, Insight into why the Langmuir–Hinshelwood mechanism is generally preferred, J. Chem. Phys. 116 (2002) 4379–4381.
[69] K. Ramesh, L. Chen, F. Chen, Y. Liu, Z. Wang, Y.-F. Han, *Re-investigating the CO oxidation mechanism over unsupported MnO,* Mn_2O_3 *and* MnO_2 *catalysts*, Catal. Today 131 (2008) 477–482.
[70] A. Bongiorno, U. Landman, Water-enhanced catalysis of co oxidation on free and supported gold nanoclusters, Phys. Rev. Lett. 95 (2005) 106102.
[71] M. Haruta, Catalysis of gold nanoparticles deposited on metal oxides, CATTECH 6 (2002) 102–115.
[72] M.A. Bollinger, M.A. Vannice, *A kinetic and drifts study of low-temperature carbon monoxide oxidation over Au-*TiO_2 *catalysts*, Appl. Catal., B 8 (1996) 417–443.
[73] N. Funazaki, A. Hemmi, S. Ito, Y. Asano, S. Yamashita, T. Kobayashi, et al., *Development of carbon monoxide detector using Au fine particles-doped A-*Fe_2O_3, Sens. Actuators, B 14 (1993) 536–538.
[74] S.D. Gardner, G.B. Hoflund, B.T. Upchurch, D.R. Schryer, E.J. Kielin, J. Schryer, *Comparison of the performance characteristics of Pt/*SnO_x *and*

Au/MnO$_x$ catalysts for low-temperature CO oxidation, J. Catal. 129 (1991) 114–120.

[75] M. Haruta, A. Ueda, S. Tsubota, R.T. Sanchez, Low-temperature catalytic combustion of methanol and its decomposed derivatives over supported gold catalysts, Catal. Today 29 (1996) 443–447.

[76] S. Minicò, S. Scirè, C. Crisafulli, R. Maggiore, S. Galvagno, Catalytic combustion of volatile organic compounds on gold/iron oxide catalysts, Appl. Catal., B 28 (2000) 245–251.

[77] B. Chen, C. Bai, R. Cook, J. Wright, C. Wang, Gold/cobalt oxide catalysts for oxidative destruction of dichloromethane, Catal. Today 30 (1996) 15–20.

[78] H. Wang, J. Wang, W.-D. Xiao, W.-K. Yuan, *Preparation and catalytic activity of nanometer gold ultrafines supported on Co_3O_4*, Powder Technol. 111 (2000) 175–178.

[79] P.-F. Yang, J.-C. Zhou, D.-H. Li, W.-M. Ren, Catalytic performance of Au/TS-1 in selective oxidation of nitrogen monoxide, J. Fuel Chem. Technol. 38 (2010) 80–84.

[80] A. Ueda, T. Oshima, M. Haruta, Reduction of nitrogen monoxide with propene in the presence of oxygen and moisture over gold supported on metal oxides, Appl. Catal., B 12 (1997) 81–93.

[81] A. Ueda, M. Haruta, *Reduction of nitrogen monoxide with propene over Au/Al_2O_3 mixed mechanically with Mn_2O_3*, Appl. Catal., B 18 (1998) 115–121.

[82] A. Ueda, M. Haruta, Nitric oxide reduction with hydrogen, carbon monoxide, and hydrocarbons over gold catalysts, Gold Bull. 32 (1999) 3–11.

[83] J.M. Schwartz, L.D. Schmidt, Reactivity and microstructure of Rh and Rh-Ce in NO + CO, J. Catal. 148 (1994) 22–29.

[84] A. Elhamdaoui, G. Bergeret, J. Massardier, M. Primet, A. Renouprez, CO and NO interaction with Pd-Ag and Pd-Cr bimetallic catalysts: 1. X-ray-diffraction, infrared-spectroscopy, and thermoreaction, J. Catal. 148 (1994) 47–55.

[85] T.M. Salama, R. Ohnishi, M. Ichikawa, Remarkable oxygen promotion of the selective reduction of nitric oxide by hydrogen over Au/NaY and Au/ZSM-5 zeolite catalysts, Chem. Commun. (1997) 105–106.

[86] Ma.P. Dekkers, M.J. Lippits, B.E. Nieuwenhuys, *Supported gold/Mo$_x$ catalysts for NO/H_2 and CO/O_2 reactions*, Catal. Today 54 (1999) 381–390.

[87] A.C. Gluhoi, Ma.P. Dekkers, B.E. Nieuwenhuys, *Comparative studies of the N_2O/H_2, N_2O/CO, H_2/O_2 and CO/O_2 reactions on supported gold catalysts: effect of the addition of various oxides*, J. Catal. 219 (2003) 197–205.

[88] Z. Hao, D. Cheng, Y. Guo, Y. Liang, Supported gold catalysts used for ozone decomposition and simultaneous elimination of ozone and carbon monoxide at ambient temperature, Appl. Catal., B 33 (2001) 217–222.

[89] M. Murdoch, G.I.N. Waterhouse, M.A. Nadeem, J.B. Metson, M.A. Keane, R. F. Howe, et al., *The effect of gold loading and particle size on photocatalytic hydrogen production from ethanol over Au/TiO_2 nanoparticles*, Nat. Chem. 3 (2011) 489.

[90] S.K. Cushing, J. Li, F. Meng, T.R. Senty, S. Suri, M. Zhi, et al., Photocatalytic activity enhanced by plasmonic resonant energy transfer from metal to semiconductor, J. Am. Chem. Soc. 134 (2012) 15033–15041.

[91] A. Furube, L. Du, K. Hara, R. Katoh, M. Tachiya, *Ultrafast plasmon-induced electron transfer from gold nanodots into TiO_2 nanoparticles*, J. Am. Chem. Soc. 129 (2007) 14852–14853.

[92] T. Torimoto, H. Horibe, T. Kameyama, K.-I. Okazaki, S. Ikeda, M. Matsumura, et al., Plasmon-enhanced photocatalytic activity of cadmium sulfide nanoparticle immobilized on silica-coated gold particles, J. Phys. Chem. Lett. 2 (2011) 2057–2062.
[93] X. Chen, H.-Y. Zhu, J.-C. Zhao, Z.-F. Zheng, X.-P. Gao, Visible-light-driven oxidation of organic contaminants in air with gold nanoparticle catalysts on oxide supports, Angew. Chem. Int. Ed. 47 (2008) 5353–5356.
[94] J.R. Adleman, D.A. Boyd, D.G. Goodwin, D. Psaltis, Heterogenous catalysis mediated by plasmon heating, Nano. Lett. 9 (2009) 4417–4423.
[95] P. Christopher, H. Xin, S. Linic, Visible-light-enhanced catalytic oxidation reactions on plasmonic silver nanostructures, Nat. Chem. 3 (2011) 467.
[96] P. He, M. Zhang, D. Yang, J. Yang, *Preparation of Au-loaded* TiO_2 *by photochemical deposition and ozone photocatalytic decomposition*, Surf. Rev. Lett. 13 (2006) 51–55.
[97] Z. Zheng, J. Teo, X. Chen, H. Liu, Y. Yuan, E.R. Waclawik, et al., *Correlation of the catalytic activity for oxidation taking place on various* TiO_2 *surfaces with surface OH groups and surface oxygen vacancies*, Chem. Eur. J. 16 (2010) 1202–1211.
[98] J. Hernández-Fernández, A. Aguilar-Elguezabal, S. Castillo, B. Ceron-Ceron, R.D. Arizabalo, M. Moran-Pineda, *Oxidation of NO in gas phase by Au–*TiO_2 *photocatalysts prepared by the sol–gel method*, Catal. Today 148 (2009) 115–118.
[99] X.-Q. Deng, B. Zhu, X.-S. Li, J.-L. Liu, X. Zhu, A.-M. Zhu, *Visible-light photocatalytic oxidation of CO over plasmonic Au/*TiO_2*: unusual features of oxygen plasma activation*, Appl. Catal., B 188 (2016) 48–55.
[100] D. Tsukamoto, Y. Shiraishi, Y. Sugano, S. Ichikawa, S. Tanaka, T. Hirai, *Gold* nanoparticles located at the interface of anatase/rutile TiO_2 particles as active plasmonic photocatalysts for aerobic oxidation, J. Am. Chem. Soc. 134 (2012) 6309–6315.
[101] R. Liang, F. Jing, L. Shen, N. Qin, L. Wu, M@Mil-100(Fe) (M = Au, Pd, Pt) nanocomposites fabricated by a facile photodeposition process: efficient visible-light photocatalysts for redox reactions in water, Nano Res. 8 (2015) 3237–3249.
[102] J. Yang, X. Wang, Y. Chen, J. Dai, S. Sun, *Enhanced photocatalytic activities of visible-light driven green synthesis in water and environmental remediation on Au/*Bi_2WO_6 *hybrid nanostructures*, RSC Adv. 5 (2015) 9771–9782.
[103] M.V. Dozzi, A. Saccomanni, E. Selli, *Cr(Vi) photocatalytic reduction: effects of simultaneous organics oxidation and of gold nanoparticles photodeposition on* TiO_2, J. Hazard. Mater. 211-212 (2012) 188–195.

6

NANOMATERIALS DEVELOPED FOR REMOVING AIR POLLUTANTS

Qidong Zhao, Xinyong Li, Qiang Zhou, Dan Wang and Huixin Xu
Dalian University of Technology, Dalian, P.R. China

6.1 Introduction to the General Principles of Air Pollutants Removal by Nanomaterials

Rapid population growth coupled with an increase in industrial outputs and the depletion of natural resources are causing the deterioration of the environment. Air pollutants have always been paid special attention because of their harm and great threat to life. Poor air quality always poses a threat to human health by possibly causing various types of diseases. Thus it is necessary to acquire enough knowledge on the sources of air pollutants and develop advanced technologies for air purification.

Pollutants are chemicals that cause environmental harm [1]. It is a fact that any chemical can be a pollutant because organisms are essentially constructed with chemicals. Air pollution refers to alterations in the natural composition of the atmosphere caused by the introduction of chemical, physical, or biological substances that are being emitted from anthropogenic, geogenic, or biogenic sources. Indoor and outdoor air pollutants may exist in gaseous or particulate form. The former form includes various chemical molecules such as carbon monoxide (CO), sulfur dioxide (SO_2) and ozone (O_3), whereas the latter refers to tiny-sized aggregates of complex chemical components with sizes varying from nanometers to micrometers including aerosols of biological origin such as viruses, bacteria, and fungi. Most indoor air pollution is caused by household items that could emit harmful chemicals. Usually, the indoor air quality could be improved to some extent by storing chemical products

Advanced Nanomaterials for Pollutant Sensing and Environmental Catalysis. DOI: https://doi.org/10.1016/B978-0-12-814796-2.00006-X

safely, trying to use low-emitting products, and improving the indoor ventilation. However, the requirement for removing some air pollutants by artificial air cleaners is growing all the time under many circumstances when an indoor environment is preferentially isolated from the outdoors for the sake of maintaining temperature, safety, comfort, and privacy.

As a major source of outdoor air pollution, unsustainable fossil fuels currently meet most of the world energy demands. Both industry production and daily life activities rely on the utilization of fuels. The adverse environmental consequences are the emission of many pollutants including but not limited to CO_2, CO, NO_x, SO_2, Hg^0, and volatile organic compounds (VOCs) [2], among others. Industrial gaseous wastes contribute to a major part of harmful gases such as CO and VOCs. In addition, on many occasions, ash can also be generated. Outdoor air pollution leads to a most serious consequence, that is, global warming which could induce many life-threatening global changes in the atmosphere, on land and in water sources. As direct contributors to global warming, well-recognized greenhouse gases include carbon dioxide, methane, nitrous oxide, and fluorinated gases. Fortunately, all these pollutant species can actually be controlled technically before being discharged into the environment as there are plenty of possible methods for their conversion into other less harmful forms.

Successful control of air pollution can be generally achieved by two strategies. One is to control the pollutant sources by reducing the production and discharging of waste gas; the other is an end-of-pipe approach by environmental remedy [1]. The former way focuses on avoiding the production or release of certain chemicals from becoming potential environmental pollutants, whereas the latter deals with capturing, removing, or converting pollutant chemicals that have already entered into the environment.

Traditional air cleaning technologies are limited to physical methods such as ventilation, adsorption, and filtration, which transfer pollutants away from the original space without destruction. Developing effective strategies for further decomposing these harmful contaminants is more favorable. Appropriate treatment technologies for wastes depend on the nature of the wastes including the state of the matter, their solubility in solvents, density, volatility, boiling point, and melting point [3,4]. The treatment of industrially discharged gaseous chemicals usually utilizes methods of physical separation processes involving absorption, adsorption, membrane processes, and phase transformations, as well as various chemical reaction routes. Commonly, the most practical treatment measures have both physical and chemical aspects.

For instance, VOCs abatement is achieved through technologies of either destruction or recovery [3]. Recovery technologies mainly separate contaminants from the exhaust gas stream for recovery or further treatment. Destruction technologies involve chemical oxidation of the VOCs to their most oxidized form, namely carbon dioxide and water (for hydrocarbons containing chlorine or sulfur, the exhaust will also include HCl and SO_2).

As early as 1987, Glaze et al. [5] introduced the term "advanced oxidation processes" (AOPs) for water treatment processes performed at room temperature based on the in situ generation of a series of powerful oxidizing agents such as hydroxyl radicals at a sufficient concentration to effectively decontaminate waters. Nowadays AOPs are considered as viable technologies for environmental remediation of wastewaters containing recalcitrant compounds that cannot be easily destroyed by conventional treatments, through the generation of reactive oxygen species (ROS) [6]. Many gaseous pollutants can be absorbed by media of condensed phase and transferred into liquid for further treatment by AOPs, which is an effective strategy of removing gaseous pollutants. Among the various forms of AOPs with different mechanisms, the catalysis-based mechanism has proven to be a viable and sustainable technique to remove vast categories of undesirable chemical contaminants including low-concentration air pollutants. Both thermal and nonthermal catalysis technologies have been adopted in converting pollutants into environmentally benign substances in practical [7].

As a branch of catalysis-derived AOPs, gas phase photocatalysis presents additional advantages over its conventional counterparts such as adsorption or filtration [8]. Through photocatalytic processes, organic pollutants could be completely oxidized to CO_2 and H_2O, instead of being merely transferred from one place to another, which therefore avoids the disposal issue. The process could be operated at ambient conditions, making it suitable for integration into existing heating, ventilation, and air conditioning equipment. Furthermore, photocatalysis works best at low concentration levels (ppb or ppm), which are typical loadings for polluted air in offices and buildings.

In principle, a catalyst is a special substance that can modulate the speed of chemical reactions without being consumed itself in the reaction process [7]. The power of a catalyst lies in its capability of accelerating chemical reactions by reducing the energy barrier (i.e., activation energy) for the transition state and in controlling reaction pathways for the desired product. With respect to materials possessing catalytic power, it is an interesting fact that almost all types of substances (e.g., acids, bases, metals, semiconductors, clays, carbon, organometallic complexes, nucleic acids, proteins, etc.) can serve as catalysts for certain chemical processes. Industrial catalysis has been practiced for more than a century. The importance of catalysis is also reflected in environmental protection and public health; a well-known example is the catalytic converters for removing toxic emissions from automobiles that were first developed by General Motors Corporation and Ford Motor Company as early as in 1974.

As an important and massively discharged gaseous pollutant, CO_2 represents about 75% of the greenhouse gases in the environment [9]. Some strategies have been proposed to control its emission by either separation or capturing such as filtration, absorption in liquids, adsorption on solids, or a combination of these processes. Post sequestration or chemical conversion has been regarded as a major alternative for reducing CO_2 emissions in the atmosphere. Electrocatalytic CO_2 reduction to useful chemical fuels represents an attractive route for the capture and utilization of atmospheric CO_2 [10]. When coupled with renewable energy sources such as solar energy, this process could potentially enable a sustainable energy economy and chemical industry. In this aspect, studies of various catalysts for photocatalytic or electrocatalytic CO_2 reduction have made great progress in the past few decades [11].

Nowadays, growing concerns regarding environmental hazards and the treatment of toxic chemicals have resulted in promoted research activities to question efficient and cost-effective decontamination and remediation technologies. It is essential to understand why nanomaterials could play a more efficient role in pollution control. Theoretically, nanostructured materials have been used for environmental remediation and green chemistry mainly due to the following reasons [12–14]. (1) They possess high specific surface areas and have a large surface to bulk ratio compared to bulk materials; (2) they have a flexible textile property and a high number of reactive edge, corner, and defect sites which lead to intrinsically higher surface reactivity; and (3) chemical properties such as Lewis acid and Lewis base properties, oxidation, and reduction potential can be tailored or "tuned" for a specific reaction. Regarding nanostructured catalyst materials such as nanoaerogels, nanotubes/rods, nanoplates/sheets, nanospheres, and so on, their unique properties such as high specific surface area can provide easy transport/diffusion pathways for substrates to access, leading to faster kinetics, more efficient contact for reactants, and more active sites for the catalytic process. Furthermore, these nanostructured materials can be used as catalyst supports to provide a synergistic effect between the catalyst and the support particles, resulting in highly active and stable catalysts. In comparison to bulk materials, the surface area of nanomaterials of a given mass grows exponentially as the diameter shrinks [15]. The high surface area-to-mass ratio of nanomaterials can greatly improve the adsorption capacities of sorbent materials. Because of their reduced size and large radii of curvature, nanomaterials have a surface that is especially reactive (mainly due to the high density of low-coordinated atoms at the surface, edges, and vortices). These unique properties can be applied to degrade and scavenge pollutants in water or air.

Heterogeneous catalysis occurs on the surface of the nanoparticles of a catalyst, more accurately at the interface between solid catalyst nanoparticles and gaseous or/and liquid reactants. The catalytic reactivity of nanoparticles is highly dependent on their size and composition, as well as on some other parameters. The high surface-to-volume ratio of nanocatalysts is one of the primary factors in catalysis, which largely enhances atom efficiency and reduces the cost of precious-metal catalysts. It is useful that the reduction potential of metal particles becomes progressively negative as the size goes down. A significant fraction of all the atoms in nanostructured materials are coordinatively undersaturated with respect to the equilibrium bulk structure. Those coordinatively undersaturated sites exhibit local electronic structures that are decoupled from the band structure of the interior, rendering these sites more reactive [16,17]. Furthermore, in numerous chemical processes involving the use of nanocatalysts, additives such as ligands are frequently used to improve the selectivity toward desired products by modifying the catalyst surface.

The catalytic reactivity of nanoparticles is also related to the shapes of the particles. The shapes of nanocatalysts are related to their crystalline structures, including crystallinity, terminating facets, and anisotropy [16]. These parameters strongly affect the properties of nanocatalysts. Faceted nanomaterials are of two types, namely low-index-faceted nanomaterials and high-index-faceted nanomaterials. During the synthesis of faceted metal or metal oxide nanoparticles, if the growth is under thermodynamic control, the product will be bound by low-index facets with lower surface energy. In this case, the stabilizing or capping agent used plays an imperative role in determining the product shape because different facets have selective adsorption properties that are also reflected in their growth rates. In contrast, when the growth is governed by kinetically controlled conditions, then the product can drastically diverge from the thermodynamically favored structure. Metal and metal oxide nanostructures with high-index facets possess more active catalytic sites than usual due to the presence of a high density of low-coordinated atoms, steps, edges, and kinks [17].

In the subsequent sections of the chapter, mainly some research progresses are described relating to the separation and conversion of various air pollutants over specific nanomaterials.

6.2 Reactive Nanomaterials With Well-Defined Physical and Chemical Structures

6.2.1 Nanostructured Adsorbents

Calcium (Ca)-based nanoadsorbents have been used to capture CO_2 at high temperatures based on the reversible carbonation reaction of calcium oxides (CaO). The serious disadvantage of using high-temperature adsorbents lies in their ability to aggregate easily leading to a sintering problem during the carbonation/calcination cycles [18]. As a result, the surface coating of Ca-based nanoadsorbents is used to prevent the aggregation of these adsorbents and consequently avoid the sintering problem. Wang et al. reported that coating nanoscale calcium carbonate with titanium dioxide (TiO_2) can prevent the sintering of nanoscale calcium carbonate and the yielded composite could more effectively capture carbon dioxide using the adsorption phase technique [19]. Another example of CO_2 adsorbents is carbon-based materials at low temperature, which are widely used due to their high surface and high amenability to pore structure modification and surface functionalization [20–23].

Graphene has a large theoretical specific surface area and graphene oxide has functional groups, indicating their potential for adsorption processes. In the past few years, many investigations have been focused on the applications of graphene or graphene composites in the removal of pollutants from air and water [20–22]. Graphene oxide possesses several functional groups and strong acidity, exhibiting high adsorption for basic compounds and cations while graphene shows a hydrophobic surface and presents high adsorption to chemicals due to strong $\pi-\pi$ interaction. The modification of graphene oxide or graphene with metal oxides or organics can produce various nanocomposites, thus enhancing the adsorption capacity and separation efficiency [23]. Assembling graphene oxide or graphene into a porous carbonaceous material with controlled oxygen species will be a promising way to further enhance the adsorption capacity.

Mercury emission from combustion sources such as coal-fired boilers, municipal waste combustors, and medical waste incinerators, has become a great public concern due to its high toxicity, environmental persistency, bioaccumulation, and detrimental effects on human health and ecosystems [24]. Depending on combustion conditions and flue gas chemistry, mercury exists in three forms in typical flue gas, namely elemental mercury (Hg^0), oxidized mercury, and particulate-bound mercury. Both the latter two forms of mercury species are easy to remove from flue gas using conventional air pollution control devices.

TiO_2 from different synthesis methods often shows different Hg^0 removal performances [25]. Suriyawong et al. [26] tested the performance of Hg^0 capture by nanostructured TiO_2 with different synthesis methods under UV irradiation, and they found that presynthesized nanostructured TiO_2 demonstrated the highest Hg^0 capture efficiency because of its larger surface area and higher proportion of anatase to rutile, followed by in situ–generated and commercial TiO_2 (Degussa, P25). Wang et al. [27] prepared a novel titania nanotube (TNT) with a vast surface area and high porosity by the hydrothermal method to remove Hg^0 in flue gas, and their results showed that the TNT exhibited an excellent Hg^0 removal efficiency.

In order to avoid the loss and agglomeration of TiO_2 powders and to provide a stronger adsorption capacity, TiO_2 powders typically need to be coated on a variety of support materials with a larger specific surface area and a stronger adsorption capacity, also referred to as carriers, to be more adaptable for future industrial applications. Common TiO_2 support materials or carriers include reactor walls, glass beads, metal oxides, carbon-based materials, zeolites, silicone, natural mineral materials, and even some organic materials [28].

6.2.2 Metallic Nanostructured Catalyst

Advanced techniques for preparing well-defined nanoparticles, especially solution phase synthesis of precious-metal nanoparticles with excellent control over particle size, shape, morphology, and others have been well-developed in recent years [29]. Ligand-protected nanoparticles of extremely narrow size distribution (e.g., 5% standard deviation) can now be readily obtained in solution phase. For example, uniform Au nanoparticles with sizes ranging from ~1 to ~100 nm can now be routinely made [30].

The function of a nanometal cocatalyst in semiconductor photocatalysis is equivalent to that of the cathode of an electrochemical system; therefore discussions on electrocatalysts for CO_2 reduction are also applicable to cocatalysts used for photocatalytic CO_2 reduction. The catalytic reaction rate and the selectivity of different electrocatalysts for CO_2 reduction vary largely [31]. Group IB metals (Cu, Ag, and Au) and Zn are excellent electrocatalysts for CO_2 reduction with a high selectivity for CO (especially for Au, Ag, and Zn) and hydrocarbon products (especially for Cu). Group IIB and p-block metals (e.g., Cd, Hg, In, Sn, Tl, Pb, and Bi) mainly generate formate in aqueous conditions.

Bi is located close to traditional formate-producing metals on the periodic table. It is therefore suggested to also be active for CO_2 reduction to formate, yet it is significantly less toxic and more environmentally benign than many of its neighbors. Further improving its performance requires structural engineering at the nanoscale to enlarge its surface areas. Bi consists of stacked layers in a buckled honeycomb structure similar to that of black phosphorus. This structure permits Bi to be potentially exfoliated to its two-dimensional (2D) mono- or few-layers with enlarged surface area and enhanced electrochemical activity. Current attention is mostly focused on tin-based materials, which, unfortunately, often suffer from limited Faradaic efficiencies. Han et al. reported that ultrathin bismuth nanosheets prepared from the in situ topotactic transformation of bismuth oxyiodide nanosheets under cathodic electrochemical environments possess single crystallinity and enlarged surface areas [32]. A high selectivity (~100%) for CO_2 reduction to formate with an excellent durability for >10 h was observed with this well-defined nanocatalyst.

6.2.3 Nonmetallic Nanostructured Catalysts

The category of nonmetallic catalysts includes but is not limited to metal oxides, salts, solid acids, and other catalysts containing no zero-valent metals, most of which possess semiconductor characteristics. Among the various types of semiconductors, oxide-based ones are the most ideal because of their stability, cost effectiveness, high activity, and environmental compatibility [16,17]. Until now, the most commonly used semiconductors as photocatalysts for the photoinduced degradation of organic pollutants are TiO_2 and ZnO.

For instance, Evonik's P25 TiO_2 produced by Evonik Degussa GmbH, which is one of the most widely used commercial photocatalysts and the benchmark for most comparative studies, is a nanosized powder with an average primary particle size of approximately 21 nm. The major drawback for these photocatalysts is related to intrinsic relatively wide bandgap energy, that is, 3.2 eV for anatase TiO_2, 3.02 eV for rutile TiO_2 [33], and 3.2 eV for ZnO [34]. These semiconductors can only be excited by photons which are close to the UV region and utilize only 4%–6% of solar light, which limits their practical applications.

Semiconductor catalysts with various nanostructures have been produced through strategies such as the hydro/solvothermal process, the sol–gel process, coprecipitation, chemical polymerization, emulsion, the sonochemical method, electrospinning, and electrochemical deposition. Based on the well-established top-down and bottom-up strategies, the synthesis of TiO_2 micro- and nanostructures with controllable parameters such as size, morphology, composition, as well as assembly can be achieved. TiO_2 has been successfully synthesized as nanoparticles, core–shells, nanotubes (Fig. 6.1), nanorods, nanofibers, nanocubes, and porous spheres using relatively low temperatures and inexpensive methods [10,35–37].

Industrially, the fabrication of catalyst powders with small particle sizes is a well-known strategy. Maira et al. demonstrated that 7 nm is the optimal size of TiO_2 nanoparticles for the photocatalytic oxidation (PCO) of gaseous trichloroethylene [38]. Larger individual particles are deemed to be less efficient because less of the incident light is absorbed due to the larger scattered fraction, but this can be circumvented by the ultimate catalyst and/or reactor design. Multiple scattering can eventually lead to improved light utilization and the associated high efficiencies [39]. On the other hand, the activity also drops when decreasing the nanoparticle size to below 7 nm. This is due to the fact that the TiO_2 semiconductor displays discretization of its band structure for such small particles. Consequently, the bandgap experiences a blueshift, leading to a less efficient utilization of incoming photons [40–42].

With regard to the degradation of organic contaminants, one generally accepted reaction mechanism is a radical pathway. Radical chain reactions are typically initiated by the formation

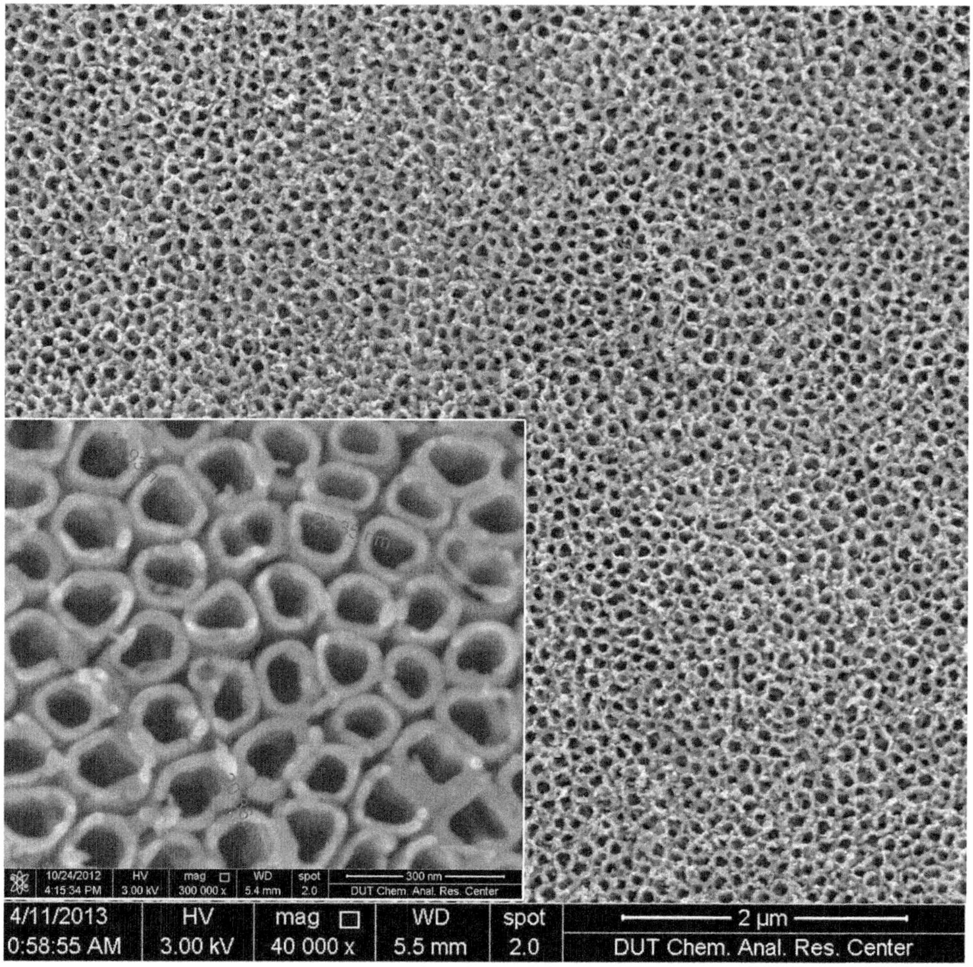

Figure 6.1 A typical SEM image of a TiO_2 nanotube array grown on a Ti sheet through anodized oxidation and annealing at 450°C with enlarged SEM image of local nanotubes (inset).

of hydroxyl radicals (•OH) from H_2O (+2.27 V vs standard hydrogen electrode, SHE) and superoxide anion radicals ($\bullet O_2^-$) from O_2 (−0.28 V vs SHE) with the aid of photogenerated h^+ and e^-, respectively (Eqs. 6.1 and 6.2) [43]. This is justified in view of the relative positions of the redox potentials. Subsequent reactions of $\bullet O_2^-$ with H^+ can yield other ROSs such as hydroperoxyl radicals (•OOH), H_2O_2, and finally •OH (Eqs. 6.3–6.5) that may participate in the further photocatalytic process [44,45].

$$H_2O + h^+_{VB} \rightarrow \bullet OH + H^+ \tag{6.1}$$

$$O_2 + e^-_{CB} \rightarrow \bullet O_2^- \tag{6.2}$$

$$\bullet O_2^- + H^+ \rightarrow \bullet OOH \tag{6.3}$$

$$\bullet OOH + \bullet OOH \rightarrow O_2 + H_2O_2 \tag{6.4}$$

$$H_2O_2 + e^-_{CB} \rightarrow \bullet OH + OH^- \tag{6.5}$$

Singlet oxygen ($^1O_2$) is another specific oxygen species that can be formed through the photooxidation of $\bullet O_2^-$on the TiO_2 surface, often in sensitized reactions [46]. The direct oxidation of adsorbed organic species by photogenerated holes has also been reported, especially at low water contents or high surface coverage of pollutants [44]. Apart from initiating a radical chain reaction, the possibility also exists for photogenerated charge carriers to neutralize one another. This process is known as recombination [47]. It is facilitated by lattice defects, crystal imperfections, and impurities. According to Hoffmann et al. [48,49], charge recombination is a semifast process occurring on a time scale of 10–100 ns, whereas the initial charge generation is very fast (fs time scale) and the interfacial charge transfer is rather slow (100 ns–ms time scale). Recombination should be avoided as much as possible because it results in a drastic efficiency decrease.

Photogenerated charge carriers can initiate the reduction/oxidation of species adsorbed on the catalyst surface, depending on the relative positions of their redox potentials. At neutral pH, the redox potential of the hole on TiO_2 is +2.53 V, whereas that of the excited electron is −0.52 V (both vs SHE) [43]. Hence, reductions can only occur when the redox potential of the e^- is negative enough to reduce the oxidant, while oxidations only take place when the redox potential of the h^+ is more positive than that of the reductant.

Many research efforts have attempted to reduce the TiO_2 bandgap by doping or bandgap engineering. These efforts have resulted in various TiO_2 modifications, which may be categorized by their colored appearance such as yellow, green, red, blue, black, and numerous shades of gray [35–37]. While cation substitution allows for the position of the CB of inorganic semiconductors to be controlled, the VB can be tuned by anion substitution.

Similar to TiO_2, ZnO nanostructures with different morphologies and properties have attracted much attention for photocatalytic applications [34]. ZnO is an oxide semiconductor with a direct wide bandgap of 3.37 eV. After excitation by proper photons, the highly reactive electrons and holes at the surface of ZnO photocatalysts tend to perform reduction and oxidation reactions to produce •OH and $•O_2^-$, respectively. In ZnO, the bottom level of the conduction band potential (−0.5 V vs normal hydrogen electrode, NHE) is more negative than the redox potential of $O_2/•O_2^-$ (−0.33 V vs NHE); therefore, $•O_2^-$ can be produced by electrons. In contrast, the top of the valence band potential (+2.7 V vs NHE) is more positive than the redox potential of $•OH/H_2O$ (+2.53 V vs NHE), so water molecules can be oxidized by holes to form •OH. ZnO-based nanostructures have been utilized as photocatalysts for the solar-driven degradation of various organic pollutants [50].

Solution-based approaches are favorable due to their ability to provide a good platform to control the growth of ZnO nanostructures, which has been demonstrated experimentally through well-controlled molar ratios of precursors. There are various methods similar to those for TiO_2 that enhance the photoresponse of ZnO nanostructures. In recent years, one-dimensional (1D) nanostructures have raised significant attention owing to their wide-spread applications in heterogeneous photocatalysis [36]. 1D ZnO nanostructures (Figs. 6.2 and 6.3) as photocatalysts also exhibit substantial advantages as compared to bulk materials [34].

The abundance of the iron element is most attractive to researchers in investigating pollutants removal. It has been reported that iron oxide could absorb and utilize about 40% of the incident solar spectra [51]. However, pure iron oxide exhibits a miserably short excited state lifetime, a short hole diffusion length, and a high recombination rate of photogenerated electron–hole pairs. Iron oxide can exist in several forms such as iron(II) oxide (FeO), amorphous hydrous ferric oxide (FeOOH), goethite (α-FeOOH), lepidocrocite (γ-FeOOH), Fe_3O_4(a mixture of Fe(II) and Fe(III)), and iron(III) oxide (Fe_2O_3) phases such as α-Fe_2O_3 and γ-Fe_2O_3. Although amorphous FeOOH, α-FeOOH, and γ-FeOOH have high surface areas, which is beneficial to the adsorption process, they are not stable and could easily decompose or form low surface-area crystalline iron oxides during synthesis and usage. Iron oxide can be synthesized in various morphologies such as zero-dimensional (0D) nanocrystals (particles, cubes), 1D

Figure 6.2 SEM image of sol–gel method derived ZnO nanorods.

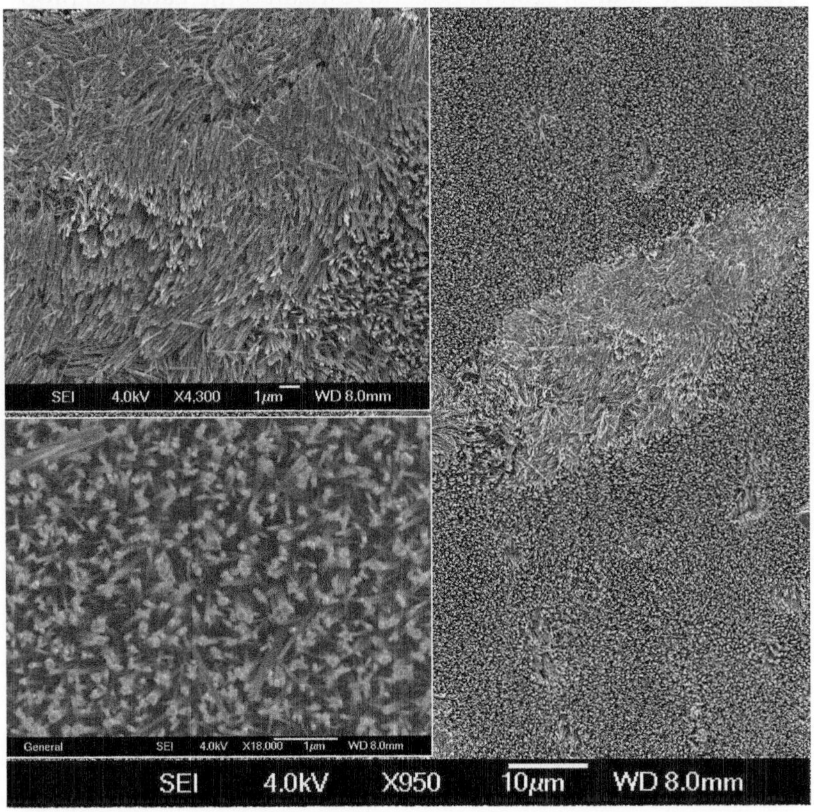

Figure 6.3 SEM image of ZnO nanorod array grown on ITO conductive glass.

nanocrystals (rods, wires, tubes, and belts), two-dimensional (2D) nanocrystals (disks, platelets, sheets, and films), and three-dimensional (3D) nanocrystals (dendrites, flowers, sea-urchin-like, and spheres). 1D nanostructures with an increasing ratio of length to diameter could restrict electrons flow in the radial direction and instead guide the movement of electrons through the axial direction. An α-Fe_2O_3 nanorod array could be obtained using a general solution preparation strategy, as shown in Fig. 6.4. Furthermore, it has been reported that aligned α-Fe_2O_3 nanotubes were able to achieve an enhancement of surface area without an increase of the geometric area and the aligned nanotubes could reduce the scattering of free electrons, thereby, enhancing the electrons mobility [52]. 3D structures of iron oxide photocatalysts assembled by lower dimensional units with large surface area and spatial channels could allow for a high mass transfer rate of reactants and products.

Sometimes the exposed crystal planes play a more important role than the surface area for determining the catalytic efficiency of a catalyst [16]. Li et al. investigated the catalytic properties of concave nanocube-like α-Fe_2O_3 with high-index facets

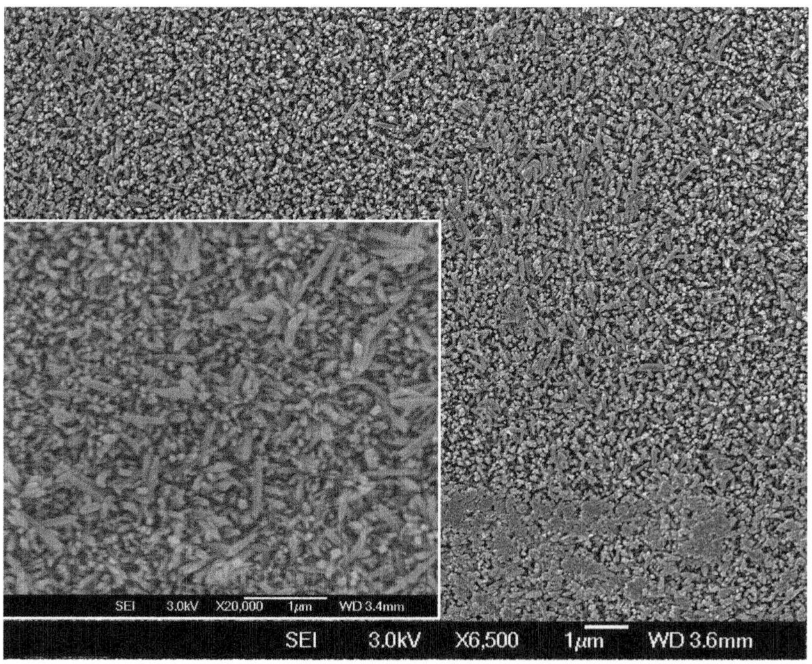

Figure 6.4 SEM image of α-Fe_2O_3 nanorod array grown on FTO conductive glass.

for low-temperature CO oxidation. 100% CO conversion was achieved at 160°C when high-index facets exposed α-Fe_2O_3 was used as a catalyst. However, for α-Fe_2O_3 nanorods, CO conversion was only 10.8% even though they had a higher surface area ($39.3\ m^2\ g^{-1}$) than the α-Fe_2O_3 nanocubes ($13.7\ m^2\ g^{-1}$). This enhanced efficiency can be attributed to the higher reactivity of the high-index facets.

Spinels, mostly with the composition AB_2O_4 (where A and B are metal ions), generally have a composition formed of $A-O$ tetrahedrons and $B-O$ octahedrons [53]. Spinels form a very large family, and they can contain one or more metal elements. Nearly all of the main group metals and transition metals have been observed in spinels. The traditional synthesis of spinels generally follows a high-temperature solid-state route. In the past years, many low-temperature synthesis methods have been developed to fabricate spinels with different sizes and morphologies. The benefits of spinel compounds such as their controllable composition, structure, valence, and morphology have made them suitable as catalysts in various reactions. Spinel catalysts have been used to facilitate NO_x reduction, CO oxidation, CO_2 reduction, NH_3 oxidation, formaldehyde oxidation, methane combustion, alcohols oxidation, and others.

Ozone is a ubiquitous pollutant and catalytic materials explored for eliminating O_3 include noble metals and metal oxides [54]. Among various supported transition metal oxides, manganese oxides, especially MnO_2, are the most frequently studied. The activity of three MnO_2 polymorphs for O_3 decomposition followed the order of α-> γ-> β-MnO_2. The α-MnO_2 owned the largest specific surface area and lowest average oxidation state of Mn. Furthermore, the adsorbed oxygen species on the surface of α-MnO_2 were more easily reduced. It was found that the catalytic activity of MnO_2 strongly depended on the density of oxygen vacancies.

Cerium oxide (CeO_2) is widely used in many areas of heterogeneous catalysis [55]. It has received a lot of attention due to its ability to switch between Ce^{4+} and Ce^{3+} oxidation states. Noble metal–free catalysts have been explored lately. In particular, copper and copper-based catalysts have been the focus of much attention because of their superior catalytic activity toward the oxidation of CO in regular and hydrogen-rich streams. The Cu(111) surface displays low activity for the oxidation of CO. The addition of ceria nanoparticles to Cu(111) produces a substantial enhancement in the catalytic activity of the system. The results of theoretical calculations indicate that the Ce^{3+} sites in a CeO_x/Cu(111) system are shown adsorb O_2, dissociate the molecule, and release atomic O for reaction with CO in an efficient way. The inverse CeO_x/Cu(111) catalysts display activities for the CO oxidation process that are comparable with or larger than those reported on the surfaces of expensive noble metals such as Rh(111), Pd(110), and Pt(100).

Tungsten oxide (WO_3)-based nanomaterials have been widely studied for various applications such as solar energy harvesting, sensors, heterogeneous catalysis, and others. The monoclinic I (γ-WO_3) phase is the most stable phase at room temperature and always shows photocatalytic activity among various crystal phases of WO_3. WO_3 endows some intriguing advantages such as low cost, harmlessness, and stability in acidic and oxidative conditions. The experimental bandgap energy (E_g) of WO_3 from optical, photocurrent, and photoemission measurements varies from 2.5 to 3.0 eV, enabling it to be a visible-light-driven photocatalyst. It has been reported that the bandgap of WO_3 ultrathin nanosheets could be altered due to size quantization effects. In the photocatalytic reduction of CO_2 to hydrocarbon fuels, WO_3 ultrathin nanosheets has exhibited outstanding performance in the presence of water with an increase in the generation of CH_4 under continuous visible light illumination while commercial WO_3 was unable to reduce CO_2.

C_3N_4 has been used in the photocatalytic reduction of CO_2, the photocatalytic degradation of pollutants, photocatalytic bacteria disinfection, and other important catalytic reactions [56]. Designing the molecular structure of C_3N_4 materials is an effective way to control their properties and improve their performance in various advanced applications. C_3N_4 materials are easily synthesized from abundant and inexpensive starting materials, which allows for the modification of any desired molecule, element, or functional group onto the final C_3N_4 framework. Specifically, the nitrogen-rich structure and abundant pores provide generally ideal sites and space for the inclusion of cations/single atoms through a coordination route into the C_3N_4 framework. Moreover, similar to graphene and its analogues, both the noncovalent and covalent functionalization of C_3N_4 provide powerful tools for controlling its structure and properties to meet the requirements of its applications thanks to the highly conjugated π-systems, unique 2D structure, as well as relatively easy chemical modification. The available strategies for designing molecular structures including mainly hydrogen bonding/coordination interaction engineering, polymerization degree modulation, doping and copolymerization, and covalent/noncovalent modification have been established.

6.2.4 Nanocomposite Catalysts

Composite catalyst materials should have different properties and catalytic performances from their individual components because the individual substances in the composites experience a synergistic effect; this comes about through optimizing particle size, specific surface area, porosity, and active sites, thus preventing particles from agglomerating, facilitating electron and proton conduction, and protecting active materials from chemical and mechanical degradation. As a result, the obtained composites may have high catalytic activity, high product selectivity, and high catalytic stability [57–60].

Bare semiconductors cannot satisfy the demand for high-performance photocatalysis in terms of charge kinetics. For this reason, different hybrid catalytic systems based on semiconductors have been developed. One typical strategy is to employ certain cocatalysts to work together with a host semiconductor. In this system, the photoexcited semiconductor is the only source of charge. The cocatalysts are usually not involved in light absorption, and their sizes generally should be considerably smaller than that of the light-harvesting semiconductor for reducing the shielding effect. The cocatalysts can make contact with the semiconductor directly, or in other cases, conductive materials such as graphene serve as the charge bridge between the cocatalyst and the semiconductor [57]. Cocatalysts mainly play two roles in the enhancement of photocatalytic performance. One is to promote the charge separation and transport through the formation of junctions/interfaces between the cocatalyst and the light-harvesting semiconductor, whereas the other is to serve as reaction sites to consume the separated charges for surface reactions.

To improve the photocatalytic activity of a semiconductor, a widely employed strategy is to decrease its particle size so as to increase the surface area, even to the nanoscale. However, for nanosized semiconductor particles, they are usually characterized by many surface defects and limited volume of aggregated atoms, which is usually unfavorable for photogenerated charge separation in space. This might greatly limit the photoactivity improvement of a nanosized semiconductor. Several surface tuning strategies for forming nanocomposite catalysts are successful via suitable functional molecules to achieve surface binding, surface deposition, or surface modification agents to alter the surface structures and surface properties of semiconductor crystallite units [61]. Coupling different components with matching electronic structures is also an attractive strategy.

Specifically, an improvement of photocatalysis kinetics can be obtained from the wide 2D surface area of inorganic nanosheets by the provision of many surface reaction sites [62]. The well-controlled immobilization of cocatalysts on the defect-free surface of 2D nanosheets is also useful in improving the reaction kinetics of photocatalysts. Also, the very thin nanometer-scale thickness of 2D nanosheets can enhance charge transport kinetics by minimizing the diffusion path of charge carriers. In addition, 3D nanosheets as a catalyst or catalyst support (Fig. 6.5) have already fulfilled some of the requirements necessary for being considered an advanced catalyst [63]. The confinement effect of catalytic components within 3D graphene could stabilize active sites in the catalytic process. Another unique advantage is the integrated appearance of these 3D graphene monoliths, which makes them convenient for manipulation and collection in use, as well as preventing environmental risk induced by the toxicity of the release of graphene nanosheets. Defect-related or heteroatom-doped 3D graphene displays good performance in catalytic reactions, especially in electrocatalysis. 3D graphene is not only regarded as a support for catalytic functionalities, but also as a cocatalyst to facilitate photocatalytic reactions after certain modifications.

A specific strategy for developing highly efficient and stable hybrid photocatalysts is constructing multicomponent microjunctions [59] including (1) the coupling of semiconductors with other semiconductors to satisfy the high absorption of solar energy and to create sufficient built-in potential for charge

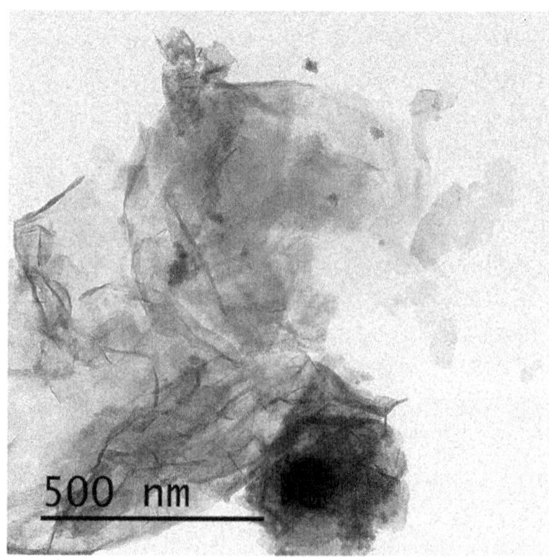

Figure 6.5 TEM image of graphene sheets prepared as catalyst support.

separation and redox reactions; (2) the formation of heterostructured junctions with carbon materials to effectively drive the separation and transportation of electron–hole pairs; (3) the deposition of metal to enhance the utilization of sunlight or improve the separation and transportation of electron–hole pairs; and (4) the formation of multicomponent heterojunctions for enhancing the utilization of sunlight and improving the separation/transportation of electron–hole pairs. Hybrid photocatalysts integrate the synergistic effects of the individual components for increased light harvesting, prolonged lifetimes, enhanced photocatalytic performance as well as higher chemical and environmental stability. For the optimization of the photocatalytic activity of semiconductor-based hybrids, it is highly critical to obtain the best match in the band structures of hybridized species for designed target photocatalytic reactions, which could be achieved by fine-control of the band positions of components with optimized electronic structures. TiO_2–polymer- and TiO_2–carbon-based hybrid catalysts have recently drawn considerable interest due to their exceptional photocatalytic activity. By coupling TiO_2 with nanocarbons or polymers having suitable bandgaps, composite nanocatalysts were formed with desirable bandgaps exhibiting greatly enhanced visible light photocatalytic activities.

Wang et al. [64] studied the effect of the graphene content on the photocatalytic degradation of acetone in air, resulting from an optimal concentration of 0.05 wt.% in hierarchical macro/mesoporous TiO_2 composites, at which the prepared materials improved the photocatalytic activity of bare TiO_2 and commercial P25 by a factor of 1.7 and 1.6, respectively. The stability and activity of TiO_2 (P25)–graphene nanocomposites were much higher than those of bare TiO_2. The increase of graphene or carbon nanotubes ratio in the composites resulted in higher adsorptivity of pollutants, but adversely, the exposed surface of the TiO_2 particles to light irradiation became reduced, which would provoke lower photocatalytic activity. Optimization of the graphene oxide/TiO_2 ratio is an important issue, and the reported optimal loadings were lower than 5% of graphene oxide [58].

Graphitic carbon nitride/titania (g-C_3N_4/TiO_2) composite photocatalysts with different C_3N_4/TiO_2 ratios could be synthesized by a simple preparation route through annealing mixtures of melamine and commercial P25 TiO_2 powder at 550°C for 3 h under Ar flow [65]. Under visible light, the g-C_3N_4/TiO_2 composite with an initial ratio of 1:4 exhibited superior photocatalytic activity in NO oxidation in comparison to the pure semiconductors g-C_3N_4 and TiO_2.

To achieve multiple functions of self-cleaning, antibacterial, antistatic, and UV resistance, efforts have been made to precipitate TiO_2 onto textile fabrics. Flexible organic fabrics coated with titania find wide applications in pollutant degradation and antibiosis. The sol–gel method is the most commonly adopted technique to coat nanoparticulate TiO_2 onto fabrics [59]. An alternative route is electrospinning of polymer melts mixed with TiO_2 nanoparticles.

Jin-Ming Wu et al. adopted polyester (PET, polyethylene glycol terephthalate) fabrics as a feasible template to fabricate TiO_2 microtubes consisting of radially aligned TiO_2 nanowires through multiple steps including surface roughening, sol–gel TiO_2 seeding, hydrogen titanate precipitation, and finally calcination [66]. The hydrogen titanate was subjected to a H_2SO_4 treatment for keeping the PET substrates. Interestingly, the achieved PET fabrics coated with TiO_2 nanowires exhibited excellent UV photocatalytic activity for the removal of organic pollutants and 100% sterilization rate of either *Escherichia coli* or *Staphylococcus epidermidis* within 15 min of visible light irradiation. The excellent photocatalytic and antibacterial performances can be attributed to the abundant surface hydroxyl groups, the phase junctions of anatase/rutile and rutile/brookite, the unique mixed 1D nanostructures, and the narrowed bandgap of 2.5 eV due to nitrogen doping.

6.3 Common Air Pollutants and Challenges in Air Purification

Although clean air is considered to be a basic requirement for human health and well-being, economic development and population growth have resulted in a considerable deterioration of air quality. Human activities like the intensification of agriculture, industrialization, increasing energy use, the burning of fossil fuels, and the increase in transportation have resulted to a rising cocktail of poisonous pollutants which impose many adverse effects on the environment as a whole, our human health and life expectancy, ecosystems services, biodiversity, agricultural crops, and building structures. Air pollutants are continuously released from numerous sources into the atmosphere. Air pollution can also arise from natural causes such as volcanic eruptions, whirlwinds, earthquakes, the decay of vegetation, pollen dispersal, as well as forest fires caused by lightning [67,68].

Air pollution is basically made up of three components and these are the source of the pollutants, the transporting medium, which is air, and the target or receptor which could be humans, animals, plants, and/or structural facilities [4]. Pollutants are often referred to as primary pollutants, if they exert harmful effects in the original form in which they enter the atmosphere, for example, CO, NO_x, HCs, SO_x, particulate matter, and so on. On the other hand, secondary pollutants are products of chemical reactions. The classification of pollutants can also be done according to their chemical compositions, that is, organic or inorganic pollutants or according to the state of matter, that is, gaseous or particulate pollutants [68].

Particulate matter is traditionally referred to and regulated using the operationally defined concepts of PM10, PM2.5 (fine particulate matter), and sometimes PM1 (ultrafine PM), which refer to the mass concentrations of aerosol particles with aerodynamic diameters of less than 10, 2.5, or 1 μm, respectively [69]. It is clear that fine PM in different global urban areas exhibit distinct characteristics in particle properties, dependent on the emission sources, the formation mechanisms, removal, and the meteorological conditions [70].

Publicly, US citizens depend on air pollution information on a more daily basis than satellite data by referring to their local air quality index (AQI) produced by the EPA. Other countries have similar indices. The EPA's system measures five air components to calculate the AQI, namely ozone, carbon monoxide, sulfur dioxide, PM2.5, and PM1.0 [67].

The greenhouse effect and climate change evoke a special interest since they are considered to be human hazards. The greenhouse effect is produced by infrared radiations, imprisoned between the earth and a thin layer of greenhouse gases, which get reflected and heat up the Earth's surface. Though the greenhouse effect helps life on Earth, too much warming has led to unhealthy conditions now recognized as global warming and increased extreme weather events. Greenhouse gases include carbon dioxide, methane, nitrous oxide, and chlorofluorocarbons. Carbon dioxide receives the most blame for global warming, but other substances contribute as much or even more to global warming [67].

People can use at least three tactics to improve their indoor air quality, namely (1) source control, (2) improving ventilation, and (3) air cleaners. Air cleaning devices collect pollutants from the air and trap them on a filter, which the owner simply changes when full. These precautions are especially important in newer buildings in which strong window and door seals create an airtight environment [4,71].

6.3.1 Typical Inorganic Air Pollutants

Nitrogen oxides (NO_x) are considered as a major factor contributing to acid rain, photochemical smog, and the greenhouse effect, which seriously threaten human health and ecosystems. More than 90% of NO_x in the atmosphere have been derived from the combustion of fuels from stationary and mobile sources [1]. Emissions from vehicles include a mixture of PM, NO_x, CO, and CO_2. The vast majority of NO_x ($NO_2 + NO$) emitted into the atmosphere come from fossil fuel combustion with the remaining emission sources, namely biomass burning, soils, and lightning, contributing to roughly one-third of the total amount of present day anthropogenic and natural emissions.

O_3 is a secondary air pollutant that is formed in the atmosphere from a combination of NO_x, VOCs, CO, and CH_4 in the presence of sunlight. In the presence of NO_x, VOC emissions contribute to the formation of tropospheric (surface) O_3 [67].

CO_2 makes up only 0.04% of the atmosphere but it is essential for photosynthesis performed by plants, algae, and some bacteria. Photosynthesis converts carbon dioxide into oxygen and it is the main source of oxygen in the atmosphere. Natural levels of carbon dioxide in the atmosphere contribute to the use and reuse of carbon on Earth, known as the carbon cycle. The rise in the Earth's average temperature has been correlated with this rise in carbon dioxide levels.

6.3.2 Organic Air Pollutants

As a major category of organic air pollutants, VOCs are organic volatile chemicals that have high vapor pressure and will easily form vapor at standard ambient temperature and pressure [72]. The term is generally applied to organic aromatic compounds such as benzene, toluene, ethylbenzene, *m*/*p*-xylene and *o*-xylene, organic solvents, aerosol spray can propellants, fuels (gasoline, kerosene), and petroleum distillates. VOCs are also naturally emitted by a number of plants and trees. Many VOCs are flammable. VOCs are an important health and environmental concern for several reasons. Some VOCs can be hazardous when inhaled. Benzene is a known human carcinogen and is toxic. Likewise, formaldehyde is an irritant and a sensitizer as well as being toxic. Some VOCs such as methyl-tertbutyl-ether (MTBE) are gasoline additives that are fairly soluble in water leading to the contamination of water. VOCs can form particulate matter if condensation of the gas occurs.

Synthetic materials release hundreds of VOCs into indoor air [73]. Compounds that may be found in the air in indoor environments such as houses, buildings, and offices may be formaldehyde, xylene, toluene, benzene, chloroform, alcohols, acetone, and others. Humans are also a source of indoor air pollutants especially in closed and poorly ventilated areas. In addition, mite and animal allergens, human- and animal-associated bacteria and fungi, and semivolatile organic compounds (SVOCs) can accumulate in settled dust on mattresses, pillows, and bed sheets. The raw materials utilized to manufacture mattresses and bedding products such as polyurethane foam and vinyl mattress covers, are possible sources of a myriad of chemical contaminants including VOCs, plasticizers, flame and retardants. Nowadays, there are increasing trends of avoiding the production of those pollutant-releasing materials and implementing actions of indoor air purification.

6.4 Nanomaterials for Eliminating Air Pollutants Through Adsorption and Separation

6.4.1 Air Pollutants Adsorption by Nanomaterials

The phenomenon by which molecules of a fluid adhere to the surface of a solid is known as adsorption. Through this process, solids or adsorbents can be selectively captured or removed from an airstream, gas, liquid, or solid, even at very small concentrations. The material being adsorbed is called the adsorbate and the adsorption system is called the adsorbent [3,4,71]. A fluid's composition will change when it comes into contact with an adsorbent and when one or more components in the fluid are adsorbed by the adsorbent. At all solid interfaces, adsorption can occur, but it is usually small unless the solid is highly porous and possesses fine capillaries. For a solid adsorbent to be effective, it should possess these characteristics: large surface-to-volume ratio and a preferential affinity for the individual component of concern. Three operations are commonly found in most processes involving adsorption, namely contact, separation, and regeneration. Adsorption techniques are usually simple and work effectively. However, the adsorption capacities of materials depend on their porous structure and surface properties [7].

The physical adsorption (physisorption) of CO_2 in porous materials is an attractive alternative because the process is clean and reversible and has small energy requirements due to the lower adsorption enthalpy in comparison to scrubbing [9]. Research on nanoporous materials with tailored properties for efficient CO_2 physisorption has accelerated over recent years, mainly including nanocarbons, metal–organic frameworks (MOFs), zeolites, zeolitic imidazole frameworks (ZIFs), porous silica, and combinations of these. By far too many porous materials have been analyzed for selective CO_2 capture to be able to provide a complete overview. Each of the selected materials have their own particular advantages and disadvantages, depending on the specifications of the gas mixture that CO_2 has to be removed from as well as the temperature and pressure during adsorption. The contact time and diffusion issues in dynamic processes are also relevant. One of the most widely applicable classes of materials for CO_2 capture is zeolites. Zeolites generally have very high CO_2 uptakes at low pressures due to their basicity and the polar fields in their cavities. It is a general trend that zeolites with a low Si/Al ratio are promising CO_2 adsorbents because they have a higher number of extra-framework cations that can promote adsorption by charge–quadrupole interactions.

MOFs are a class of porous materials with comparable or even better characteristics (i.e., high surface areas and uniform micropore size), but far lower stability against heat and water when compared to zeolites [74]. These organic–inorganic hybrid materials can reach ultrahigh specific surface areas and micropore volumes and they can exhibit remarkable structural flexibility which can lead to unique adsorption properties. Hence, MOFs are a class of materials that has been excessively studied for selective CO_2 capture and gas adsorption/separation in general. It is noteworthy that the amenable properties of MOFs offer a way toward the development of advanced catalysts suitable for the degradation of harmful gases and vapors into nontoxic substances. The integration of these catalysts into composite materials may lead to advanced filters, textiles, and surfaces with self-cleaning properties.

Up to now, numerous VOCs treatment technologies have emerged such as incineration, condensation, biological degradation, absorption, adsorption, catalysis oxidation, and others. Among these, adsorption technology has been recognized as an efficient and economical control strategy because it has the potential to recover and reuse both adsorbent and adsorbate [23]. Due to their large specific surface areas, rich porous structures, and high adsorption capacities, carbonaceous adsorbents are widely used in gas purification, especially with respect to VOCs treatment and recovery. A variety of engineered carbonaceous adsorbents have been developed, including activated carbon, biochar, activated carbon fiber, carbon nanotubes, graphene and its derivatives, carbon–silica composites, ordered mesoporous carbon, and others.

Alternative advanced adsorbents have been reported such as highly porous boron nitride (BN) composed of a flexible network of hexagonal BN nanosheets synthesized via thermal treatment of a boric acid/urea mixture [75]. Experimentally sponge-like BN adsorbents displayed fast adsorption rates and ultrahigh adsorption capacities for gaseous formaldehyde (HCHO), for example, 19 mg g^{-1} in equilibrium with approximately 20-ppm of HCHO in air, which is an order of magnitude higher than those of other tested materials including commercial hexagonal BN and various metal oxides. The superb HCHO adsorption performance of the porous BN is mainly due to its large specific surface area (627 m^2 g^{-1}), as well as the abundant surface hydroxyl and amine groups. Moreover, chemisorption can occur on the BN layers and contribute to the high HCHO uptake capacity via Cannizzaro-type disproportionation reactions during which HCHO is transformed into less toxic formic acid and methanol.

6.4.2 Air Pollutants Separation Through Nanostructured Membranes

Membrane gas separation has been widely applied industrially, which is based on the selective permeation of one component in a gas mixture across a membrane [76]. The pressurized feed gas is put into contact with the surface of the membrane inside a membrane module conceived for the given application. Alternatively, vacuum can be applied on the other side of the membrane to create the driving force necessary for mass transportation. Two outlet streams are recovered after treatment in a continuous process, namely a permeate stream, consisting of gas that has traversed the membrane and is thus enriched in more permeable components, and a retentate stream, consisting of the residual gas that did not traverse the membrane and is hence enriched in less permeable components. Membranes in the case of gas separation considered here consist mainly of dense layers without significant porosity, unlike ultrafiltration or nanofiltration membranes, where the separation capabilities are based on pore size. Organic constituents are concentrated by the membrane module because the membranes that are selected are more permeable to organic constituents than to air. The driving force that causes the separation of the organic constituents from the air emission stream is the pressure difference across the membrane.

Membrane materials typically consist of rubber, Buna-*n*-nitrile, polyvinyl chloride (PVC), neoprene, silicone polycarbonate, and other polymeric compounds. Some manufacturers produce a spiral-wound membrane module, in which the layers of the polymer are supported on a macroporous structure. Others produce hollow-tube membrane modules. Regardless of what type of membrane module is used, a compressor or vacuum pump is required to supply the pressure difference required for concentrating the organic contaminants.

A wide variety of membrane materials have been developed and are widely used in CO_2 separation on account of their high energy efficiency and low capital costs [77]. Among these various membranes, polymeric membranes dominate the current market owing to their good mechanical properties and easy processability. Although 2D material–based membranes exhibit promising gas separation performance as well as high thermal and chemical stability, their large-scale application is restricted by their high costs and demanding processability. As a kind of typical 2D material, nanoporous-derived membranes with defects of well-defined pores could be used for gas separation as theoretically predicted. However, there were limited successful experimental results because membranes of high quality with uniform pores could hardly be obtained with pure graphene and graphene oxide (Fig. 6.6) at low cost.

Another promising strategy to obtain good membranes for the separation of gases is the preparation of mixed-matrix membranes (MMMs), which consist of a blend of filler particles within continuous polymer matrices, aiming to increase the CO_2 separation performance of the resultant membranes while preserving their attractive features such as good mechanical properties, thermal/chemical resistance, and the excellent processability of polymeric matrices [78]. Numerous inorganic fillers such as silica and zeolites have been incorporated into polymeric backbones to increase either the permeability or selectivity. 2D materials such as graphene oxide [79], MOF-derived nanosheets, covalent organic frameworks (COFs), and transition metal dichalcogenides (TMDs), have attracted tremendous attention as new versatile fillers for the generation of MMMs [80]. Compared with conventional polymeric membranes, these newly developed, 2D material–based MMMs

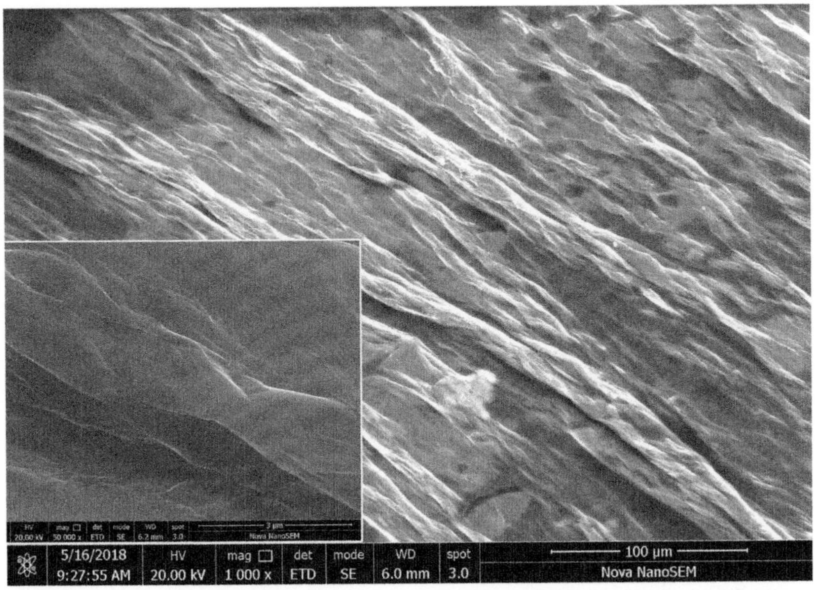

Figure 6.6 Local SEM image of a piece of graphene oxide membrane prepared by chemical oxidation of graphite, which does not possess regular pores.

exhibit extraordinary separation performance in postcombustion CO_2 capture processes.

6.5 Converting Air Pollutants Through Catalytic Pathways of Nanomaterials

Heterogeneous catalysis is a feasible alternative to reduce the air pollution impact [71], competing in many applications with conventional air treatment technologies such as adsorption, filtering, combustion, or thermal catalysis, particularly for low-flow and low-concentration emissions. Adsorbent–catalyst hybrids are promising bifunctional immobilized catalysts for environmental applications. The synergy between adsorption and catalytic activity may lead to composites with improved performance including superior conversion and selectivity to the desired reaction products. Here, the adsorbent acts as a support, immobilizing and dispersing the active phase, thereby increasing the surface area of the final solid and facilitating the shaping of the material. Additionally, it may also induce modifications that may promote physicochemical processes, for example, in the acid–base character or UV light absorption properties or the crystallinity of the semiconductor.

6.5.1 Reductive Catalysis Over Nanomaterials

Since CO_2 molecules are highly stable, only those electrons with sufficient reduction potential can be utilized to trigger CO_2 reduction reactions and a suitable catalyst is required to decrease the high reaction barrier [11]. It is generally accepted that the product selectivity of CO_2 reduction over the catalysts is determined by the chemisorption strength of the metallic catalyst surface to the $\bullet CO_2^-$ anion radical formed by one-electron transfer to the CO_2 molecule at the initial step of CO_2 reduction. CO is favorably produced on metals that can stabilize $\bullet CO_2^-$ effectively, while metals that can only weakly adsorb $\bullet CO_2^-$ tend to produce formate.

Since proton reduction for H_2 generation in aqueous solutions tends to compete with CO_2 reduction into the desired products, group VIII metals with a low overpotential of the hydrogen evolution reaction (HER) are generally not favorable for CO_2 conversion. Noble metals with a large work function and good activation ability are favorable for charge separation and to drive some photocatalytic reactions. However, taking photocatalytic CO_2 reduction as an example, although Pt has the largest work function and lowest overpotential for CO_2 reduction [81], the HER is very competitive over Pt and will decrease the selectivity of CO_2 reduction. The deactivation of Pt is another serious problem in CO_2 reduction due to the strong adsorption of CO in the active reaction sites, which has been observed in photocatalytic and electrochemical studies [31].

Besides the metallic catalysts previously mentioned for catalytic CO_2 reduction, TiO_2-based nanomaterials have attracted much attention due to their advantages of high reduction potential, low cost, and high stability [11]. The activity, selectivity, and durability of TiO_2 photocatalysts for CO_2 reduction are related to the efficiency of electron–hole separation and light utilization ability, which are highly sensitive to the surface structure, atomic configuration, and chemical composition of the photocatalysts [61]. Generally, bulk oxygen vacancies form a middle subband in the forbidden gap, which make TiO_2 respond to visible light, and those bulk oxygen vacancies also act as electron–hole recombination centers. The surface oxygen vacancies not only showed a strong response to visible light, but also acted as capture traps to inhibit electron–hole recombination. By adjusting the concentration ratio of the surface and bulk oxygen vacancies, it is possible to improve the photocatalytic efficiency of TiO_2 nanostructures. Li et al. [81] examined the effects of oxygen vacancies in TiO_2 nanocrystals on the photoreduction of CO_2. By analyzing the lifetime and intensity by positron annihilation, the efficiency of photocatalytic CO_2 reduction improved with an increase of the ratio of surface oxygen vacancies to bulk ones.

Various NO_x control methods cover the chemical processes on NO_x reduction, oxidation [82], photochemical reaction [83], and even biological processes [84]. The NO_x storage reduction (NSR) technique is more suited for passenger cars and works under cyclic conditions alternating between long periods during which NO_x is stored as nitrate on the catalyst and shorter rich periods during which nitrates are desorbed and reduced to N_2. Additionally, the number of methods proposed for NO_x emission reduction is increased by the development of hybrid technologies [84] like selective catalytic reduction (SCR)/nonthermal plasma, absorption–oxidation with different oxidizing agents, among others, or technologies in which NO_x can be removed simultaneously with other pollutants like SO_x, Hg, and VOCs.

The most promising technique to convert NO_x into N_2 from a stationary source (incinerators) is SCR with NH_3 as a reducing agent for its high efficiency and low cost. Various catalysts have been developed for the SCR purpose. As a widely investigated catalyst system, tungsten (W)-modified Ce-based catalysts have been developed from various aspects. Highly-dispersed W elements might bring about remarkable SCR activity for catalysts [85]. The addition of W as a stabilizer and promoter also significantly increases the surface area, the Ce^{3+}/Ce ratio, surface acidity, and the amount of active sites. In a recent work, a series of $Ce_xW_{1-x}O_y$ catalysts were synthesized utilizing cetyltrimethyl ammonium bromide (CTAB) as a soft template. The highest catalytic efficiency was observed with a Ce/W ratio of 3/1. With CTAB, the formation of mesoporous catalysts significantly increased the surface-active species including surface chemisorbed oxygen, and broadened the temperature window (175°C – 400°C), thus benefiting NO_x abatement. The good performance of $Ce_{0.75}W_{0.25}O_x$ was correlated with its lower crystallinity, smaller grain size, abundant oxygen vacancies, lattice defects, and the enrichment of the surface chemisorbed oxygen.

The TEM results shown in Fig. 6.7 indicate the morphology and microstructure of the $Ce_{0.75}W_{0.25}O_x$ catalyst. The $Ce_{0.75}W_{0.25}O_x$ catalyst dominated by the aggregation of nanoparticles possessed slit-like mesopores. Fig. 6.8A and B shows the conversion curves of NO_x as a function of the reaction temperature over different catalysts. Under the same experimental

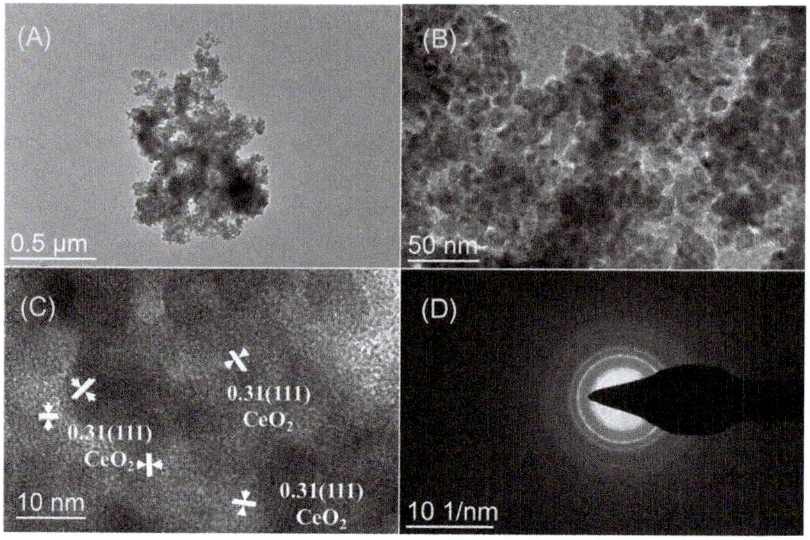

Figure 6.7 (A, B) TEM images, (C) HRTEM, and (D) the SAED pattern of the composite catalyst $Ce_{0.75}W_{0.25}O_x$.

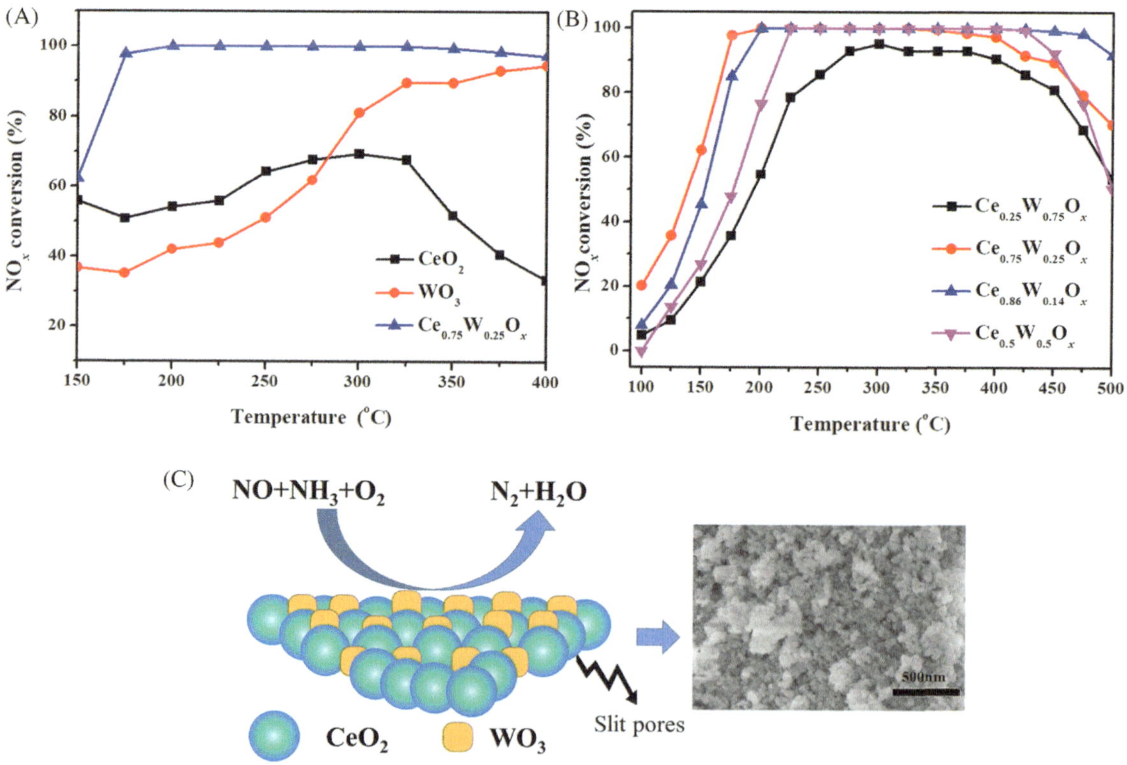

Figure 6.8 (A) NO_x conversion ratio of different kinds of SCR catalysts and (B) different Ce/W ratios of SCR catalysts. Reaction conditions: $[NO] = [NH_3] = 500$ ppm, $[O_2] = 5$ vol.%, balance He and GHSV = 24,000 h^{-1}. (C) Schematic illustration of NO_x SCR abatement of the nanocomposite of $Ce_{0.75}W_{0.25}O_x$.

conditions, pure CeO_2 showed lower NO reduction activity with a maximum conversion efficiency of NO of about 65% (Fig. 6.8). As for the WO_3 catalyst, at a low temperature range, the catalytic activity increased slowly and the NO_x conversion percentage approached 80%, which could only be obtained in a narrow temperature window of 320°C–400°C. A maximum NO conversion percentage of about 90% was obtained at 400°C. This result indicated that WO_3 had good high-temperature activity, which was in accordance with the previous report [86]. After the addition of W to CeO_2, the activities increased sharply. A higher W loading enhanced the NO conversion efficiency and widened the temperature window. Certain Ce species affected by W modification was likely a dominant component contributing to the low-temperature SCR, while W species played as a promoter at high temperature, thus making the temperature window even wider. In a wide temperature range of 175°C–425°C, 95.0%–100% NO_x conversion was obtained at a gas hourly space velocity (GHSV) of 24,000 h^{-1} with the $Ce_{0.75}W_{0.25}O_x$ catalyst. The varying dispersion degrees of the W species in these catalysts might account for the varying performances in NO_x conversion.

6.5.2 Oxidative Catalysis

The catalytic oxidation of CO is significant in the context of clean air technologies and automotive emission control. The low-temperature oxidation of CO is one of the most extensively investigated reactions with respect to heterogeneous catalysis owing to the strict regulations that require low CO emissions from the automobile industry. Literature reports have shown that by introducing graphene sheets, some metal catalysts can achieve highly efficient CO oxidation at low temperatures. Long et al. described a hydrothermal self-assembly protocol for obtaining a flexible Ru/graphene aerogel (Ru/GA) [87]. In the case of this composite material, the surface chemistry of the metallic Ru can be easily modulated by pretreatment in either oxidative or reductive atmospheres at moderate temperatures. These pretreatment conditions did not result in substantial changes to the bulk structure of the RuNPs. As a result, the catalytic activity of Ru/GA toward CO oxidation is very impressive, which exhibited 100% CO oxidation at room temperature and excellent long-term catalytic stability.

Catalytic incineration is the most suitable for the treatment of emission streams containing low concentrations of VOCs as it may allow for a more cost-effective operation compared to thermal incineration processes. However, it is not as broadly used as thermal incineration because of its greater sensitivity to pollutant characteristics and process conditions [3,4,71]. Catalytic oxidation is a well-established method for controlling VOCs emissions in waste gases. The control efficiency (also referred to as destruction efficiency or DE) for catalytic oxidation is typically 90%–95%. In some cases, the efficiency can be significantly lower, particularly when poisons/inhibitors exist that can significantly degrade the catalyst activity such as sulfur, chlorine, chloride salts, heavy metals (e.g., lead, arsenic), and particulate matter. Catalysts now exist that are relatively tolerant of compounds containing sulfur or chlorine. These new catalysts are often single or mixed-metal oxides and are supported by mechanically strong carriers [72].

Both CO and formaldehyde can be decomposed catalytically into CO_2 at room temperature, but the majority of pollutants consisting of aromatic ring(s) cannot. Thus the photocatalytic degradation of such pollutants (e.g., toluene) has received significant attention with the overall aim being total pollutant removal. For such applications, photocatalysts with strong visible light activity are required to effectively utilize solar light. Several kinds of nanomaterial-based catalysts have been developed and tested for their performance toward catalytically degrading air pollutants consisting of aromatic ring(s) [88–102]. A recent case introduced herein is about the photocatalytic degradation of the air pollutant toluene using nanostructured zinc cobaltate [102].

As a photocatalyst material, spinel zinc cobaltate possesses characteristics of visible light response, stable structural and chemical properties, low cost, availability, and nontoxicity. Experimentally, hollow porous $Zn_xCo_{3-x}O_4$ nanocubes and $ZnCo_2O_4$ nanoparticulate catalysts (Figs. 6.9 and 6.10) could be prepared by the self-sacrificial template method and the coprecipitation pyrolysis method, respectively. Under the irradiation of visible light for 6 h, the degradation rate of toluene by the hollow porous $Zn_xCo_{3-x}O_4$ was about 79%, which was 13% higher than that of the $ZnCo_2O_4$ nanoparticles under the same reaction conditions (Fig. 6.11). Compared with pure $Zn_xCo_{3-x}O_4$, the surface of rGO (reduced graphene oxide)/$Zn_xCo_{3-x}O_4$ is rougher with more sites available for substrate adsorption. Toward the photocatalytic degradation of gaseous toluene, the efficiency of rGO/$Zn_xCo_{3-x}O_4$ after 6 h of exposure to visible light reached 84%, which is 9% higher than that of pure $Zn_xCo_{3-x}O_4$ under the same experimental conditions.

Figure 6.9 SEM images of rGO/$Zn_xCo_{3-x}O_4$ (A, B) and $Zn_xCo_{3-x}O_4$ (C, D) samples.

6.6 Technical Aspects and Practical Applications

6.6.1 Device Performance and Economics

Typically, VOCs are treated in photocatalytic reactors with the catalyst being immobilized on the surface. Reactor designs are required that can offer the best conditions for photocatalytic reactions, namely compact size, large throughput, low pressure drop, optimal use of incident radiation, easy maintenance, and reduced catalyst loss. Examples of some reactor types are briefly introduced here [103]. Packed bed reactors are of simple construction and can have high conversion per unit mass of catalyst; however, high radial radiation gradients can occur and the maintenance of the unit can be difficult. Fluidized bed reactors allow for high throughputs and low pressure drops, but are difficult to control and tend to suffer from catalyst loss in entrained air, which means that either catalyst replacement is needed or additional equipment such as cyclones may be required to separate and return the entrained catalyst to the reactor. Monolith-type reactors are compact, have high throughputs, low pressure drops, and can easily be incorporated into a heating ventilation air conditioning system. However, the light intensity quickly declines through the monoliths. Such a problem could be solved by the

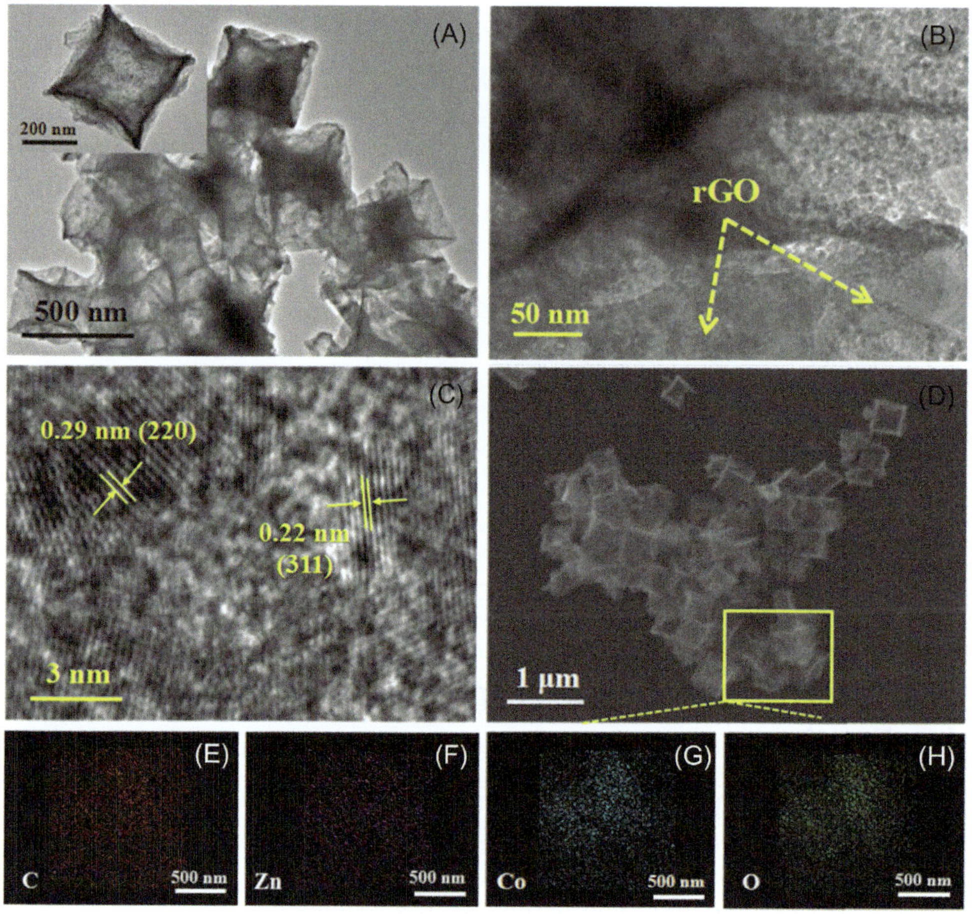

Figure 6.10 rGO/$Zn_xCo_{3-x}O_4$ catalyst. (A, B) TEM images (inset showing the TEM image of a single rGO/$Zn_xCo_{3-x}O_4$ nanocube), (C) HRTEM image and (D–H) elemental mapping images.

use of individual optical fibers passing through each monolith. Today many brands of commercial air cleaners powered by catalytic techniques for indoor use are available. The market is growing fast because there is increasing concern about air quality and health beyond economic income throughout the world.

Light sources other than conventional tubular lamps such as optical fibers and LEDs act as miniature lamps for photocatalytic air purification and therefore offer better illumination in confined spaces [103]. Under realistic environmental conditions, parameters such as light intensity levels, relative air humidity, pollutant concentration, and air flow rate affect the photocatalytic rate. The prevailing environmental conditions

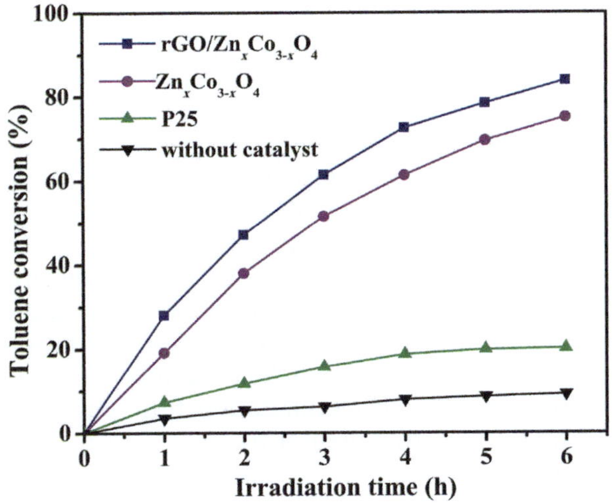

Figure 6.11 Photocatalytic degradation curves of toluene under different catalysts.

can be used advantageously for optimal photocatalytic degradation. For instance, NO_x photodegradation increased with increasing light intensity and was more efficient under UVA rather than visible light. Humid air was found to be a good means for self-cleaning (catalyst regeneration) by assisting with the oxidation of organic pollutants and carbon deposits formed during the decomposition of pollutants.

For practical solar photoreactors, as the reactor walls must be able to transmit solar radiation, materials must be transparent, consequently leading to size limitations, sealing problems, and the risk of breakage [104]. Low-iron borosilicate glass has good transmittance in the solar range to about 285 nm (Pyrex or Duran glass). The main factor affecting solar photocatalytic reactor technology costs is its scale-up. Scaling up solar photocatalytic reactors is considerably more complicated than scaling up conventional chemical reactors. In addition to conventional reactor complications, reagent and catalyst contact, flow patterns, mixing, mass transfer, and temperature control must be calculated to achieve efficient exposure of the catalyst to solar irradiation, therefore axial and radial scale-up are essential parameters for maximizing the surface areas exposed per unit of reactor volume, and ensuring that the distribution of sunlight inside the reactor is uniform. Higher illuminated surface-to-volume ratios reduce the reactor dimensions, and thereby, the capital and operating costs.

Over the past few decades, photocatalysts have been mainly limited to outdoor applications aimed at self-cleaning. However, the unique properties of abundant and safe elements like Cu(II) and Fe(III) nanocluster-grafted TiO_2 have the potential to be applied in various products including air filters, respiratory face masks, and antifungal fabrics for creating safe and secure indoor environments [105]. Nanoclusters can be facilely grafted onto TiO_2 particles or films by a simple wet chemical method, which is readily applicable for large-scale production processes. In a recent field test [105], nanocluster-grafted TiO_2 photocatalyst products installed in a washroom confirmed that films and tiles coated with nanocluster-grafted TiO_2 exhibited excellent antibacterial and deodorization functions, even in an indoor environment, with a greater than 90% decrease of bacteria and ammonia levels.

6.6.2 Mechanisms Limiting Performance in Practical Applications

Some important factors have an impact on device performance such as space velocity, which is defined as the volumetric flow rate of the combined gas stream entering the catalyst bed divided by the volume of the catalyst bed [3,7]. Space velocity depends on the type of catalyst and catalyst bed used. At a given space velocity, increasing the operating temperature at the inlet of the catalyst bed increases the conversion efficiency. At a given operating temperature, as the space velocity is decreased (i.e., as residence time in the catalyst bed increases), the conversion efficiency increases.

Regarding the factors hampering long-term performance, surface coking of catalysts is the primary cause of the deactivation of catalysts [3,71]. An appropriate design of catalyst structure may enable a better exposure to the active sites for the components in tar to suit specific conversion reactions. Because of the complex nature of tar, one active site alone in a catalyst may be insufficient for the conversion of all the compounds in tar. Therefore a better option is to design and synthesize composite catalysts with several active sites to facilitate the conversion of real tar. In addition, apart from coking, impurities such as N, S, P, Si, and metals in tar may also aggravate the deactivation of catalysts.

Membrane separations are often limited by the available driving force, so increases in membrane material selectivity result in little or no gain in product purity. In gas separation membranes, this concept is quantified by the pressure ratio (i.e., the ratio of feed to permeate pressure) divided by membrane selectivity. Pressure ratios may be set by economic considerations that are largely dependent on process conditions (i.e., independent of membrane properties). Most importantly, all synthetic membranes are subject to a trade-off between permeability and selectivity as well as practical challenges such as fouling, degradation, and material failure which limit their use. One popular approach to address difficulties in preparing large-surface-area, defect-free, ultrathin membranes of promising nanomaterials (e.g., carbon nanotubes, graphene, zeolites, and MOFs) is the use of MMMs [106].

6.7 Challenges and Perspective

Growing research efforts are focused on developing various advanced nanostructured adsorbents and catalysts with excellent performance toward removing various air pollutants from different sites. One typical challenge in air pollution control from the chemical industry is the extremely complex components, where more chemical species have to be removed than in other anthropogenic sources of NO_x. The composition of flue gases from the chemical industry is different for each type of chemical plant and additionally it varies according to process parameters [4].

Among the multiple approaches to lower atmospheric CO_2 levels, the reutilization of CO_2 in an electrochemical process, preferably using renewable and sustainable energy sources such as solar energy, is deemed as one of the key challenges to steer toward a more sustainable future. The highly selective and efficient photoelectroreduction of CO_2 into value-added products is much needed. Due to the complex reaction pathways, high-yield selective product generation remains challenging, and has been achieved only for CO and formic acid in aqueous solutions so far. CO is of particular interest, as together with hydrogen, it can be valorized to synthesis gas (syngas) [10]. Solar-powered fuel production that consumes CO_2 as a feedstock is an ideal concept that shows great potential for simultaneous CO_2 mitigation and renewable fuel production.

Catalytic oxidation at ambient temperature has been identified as a new promising area for indoor hazard abatement, converting harmful compounds into nonhazardous ones (e.g., CO_2 and water) without providing additional energy. Producing catalyst-loaded functional materials that are active and effective in reducing the concentrations of hazardous materials at ambient temperature on a large scale is posing a challenge to research scientists and air cleaning engineers. Regarding PCO techniques that work at room temperature, there are several challenges in photocatalyst design for PCO air purification. To maximize the number of photons converted into electron–hole pairs, developing catalysts that absorb a broader spectrum of solar energy is the main pursuit for researchers to optimize the charge generation step. In indoor air purification, the visible light response of materials also promises the prevention of biohazardous UV exposure and the possibility of the utilization of room illumination. For the charge-transfer step, it is imperative to suppress detrimental electron–hole recombination so as to allow more electrons and holes to arrive at the catalyst surface. Certainly, having a sufficient number of charge carriers on the surface does not necessarily ensure a high efficiency of ongoing redox reactions. The emphasis of the charge-consumption step is to enhance the surface adsorption and activation on the catalyst surface, thereby efficiently coupling more surface charges into a specific reduction or oxidation reaction. Proper band structures for the formation of reactive species, and stability of the photocatalysts are also important aspects. Moreover, unveiling the mechanistic aspect of various complex catalyst systems in depth on a molecular level is necessary to understand the exact processes happening during catalysis [107]. For example, some novel phenomena could be induced by the surface chemistry and catalysis confined under 2D nanomaterials [62].

The use of various AOPs has shown vigorous growth in recent years driven and enforced by advanced nanomaterials and today it represents an important field of research concerning pollutants treatment. However, the cost of nanomaterials in practical air purification applications is a great challenge that is limiting the scaling up. AOPs economic assessment methodologies could be based on different parameters [108]. In general, cost calculation is similar to any engineering project. Some costs can change from one country to other according to the prices of equipment, materials, products, electricity, and others. Consequently, their percentage in the total cost can also change. A balance between the cost and the purification performance of developed air cleaning devices is becoming more and more acceptable with the persistent efforts of investigators. With delicately tailored nanomaterials, the performance of developed air cleaners would be promoted further to meet the critical demands of air purification.

Acknowledgment

The authors appreciate the support provided by the Key Project of the National Ministry of Science and Technology (No. 2016YFC0204204), the Fundamental Research Funds for the Central Universities (DUT17LK55), National University Student Innovation Program, and National Natural Science Foundation of China (Nos. 21577012 and 21677022).

References

[1] R.M. Harrison (Ed.), Pollution: Causes, Effects and Control, fourth ed., Royal Society of Chemistry, Cambridge, 2001.

[2] C.E. Baukal Jr., Industrial Combustion Pollution and Control, Marcel Dekker, Inc, New York, 2004.

[3] L.K. Wang, N.C. Pereira, Y.-T. Hung, K.H. Li (Eds.), Air Pollution Control Engineering, Humana Press Inc., New Jersey, 2004.

[4] N.P. Cheremisinoff (Ed.), Handbook of Air Pollution Prevention and Control, Elsevier Science, Butterworth-Heinemann, 2002.

[5] W.H. Glaze, J.W. Kang, D.H. Hapin, The chemistry of water-treatment processes involving ozone, hydrogen-peroxide and ultraviolet radiation, Ozone Sci. Eng. 9 (1987) 335–352.

[6] G. Boczkaj, A. Fernandes, Wastewater treatment by means of advanced oxidation processes at basic pH conditions: a review, Chem. Eng. J. 320 (2017) 608–633.

[7] second ed., G. Ertl, H. Knözinger, F. Schüth, J. Weitkamp (Eds.), Handbook of Heterogeneous Catalysis, vols. 1–8, Wiley-VCH, Weinheim, 2008.

[8] A.H. Mamaghani, F. Haghighat, C.-S. Lee, Photocatalytic oxidation technology for indoor environment air purification: the state-of-the-art, Appl. Catal. B: Environ. 203 (2017) 247–269.

[9] M. Oschatz, M. Antonietti, A search for selectivity to enable CO_2 capture with porous adsorbents, Energy Environ. Sci. 11 (2018) 57–70.

[10] F. Marques Mota, D.L.T. Nguyen, J.-E. Lee, H. Piao, J.-H. Choy, Y.J. Hwang, et al., Toward an effective control of the H_2 to CO ratio of syngas through CO_2 electroreduction over immobilized gold nanoparticles on layered titanate nanosheets, ACS Catal. 8 (2018) 4364–4374.

[11] S.R. Lingampalli, M.M. Ayyub, C.N.R. Rao, Recent progress in the photocatalytic reduction of carbon dioxide, ACS Omega 2 (2017) 2740–2748.

[12] M.M. Khin, A.S. Nair, V.J. Babu, R. Murugan, S. Ramakrishna, A review on nanomaterials for environmental remediation, Energy Environ. Sci. 5 (2012) 8075–8109.

[13] J.C. Colmenares, R. Luque, Heterogeneous photocatalytic nanomaterials: prospects and challenges in selective transformations of biomass-derived compounds, Chem. Soc. Rev. 43 (2014) 765–778.

[14] R.K. Ibrahim, M. Hayyan, M.A. AlSaadi, A. Hayyan, S. Ibrahim, Environmental application of nanotechnology: air, soil, and water, Environ. Sci. Pollut. Res. Int. 23 (2016) 13754–13788.

[15] K.W. Kolasinski, Surface Science: Foundations of Catalysis and Nanoscience, second ed., John Wiley & Sons Ltd, England, 2008.

[16] J. Pal, T. Pal, Faceted metal and metal oxide nanoparticles: design, fabrication and catalysis, Nanoscale 7 (2015) 14159–14190.
[17] Y.K. Peng, S.C.E. Tsang, Facet-dependent photocatalysis of nanosize semiconductive metal oxides and progress of their characterization, Nano Today 18 (2018) 15–34.
[18] J.C. Abanades, D. Alvarez, Conversion limits in the reaction of CO_2 with lime, Energy Fuel 17 (2003) 308–315.
[19] Y. Wang, Y. Zhu, S. Wu, A new nano CaO-based CO_2 adsorbent prepared using an adsorption phase technique, Chem. Eng. J. 218 (2013) 39–45.
[20] C.H.A. Tsang, H.Y.H. Kwok, Z. Cheng, D.Y.C. Leung, The applications of graphene-based materials in pollutant control and disinfection, Prog. Solid State Chem. 45–46 (2017) 1–8.
[21] F. Perreault, A. Fonseca de Faria, M. Elimelech, Environmental applications of graphene-based nanomaterials, Chem. Soc. Rev. 44 (2015) 5861–5896.
[22] V. Kumar, K.-H. Kim, J.-W. Park, J. Hong, S. Kumar, Graphene and its nanocomposites as a platform for environmental applications, Chem. Eng. J. 315 (2017) 210–232.
[23] X. Zhang, B. Gao, A.E. Creamer, C. Cao, Y. Li, Adsorption of VOCs onto engineered carbon materials: a review, J. Hazard. Mater. 338 (2017) 102–123.
[24] Y. Liu, Y.G. Adewuyi, A review on removal of elemental mercury from flue gas using advanced oxidation process: chemistry and process, Chem. Eng. Res. Des. 112 (2016) 199–250.
[25] A.L. Linsebigler, G.Q. Lu, J.T. Yates, Photocatalysis on TiO_2 surfaces, principles, mechanisms, and selected results, Chem. Rev. 95 (1995) 735–758.
[26] A. Suriyawong, M. Smallwood, Y. Li, Y. Zhuang, P. Biswas, Mercury capture by nano-structured titanium dioxide sorbent during coal combustion. Lab-scale to pilot-scale studies, Aerosol Air Qual. Res. 9 (2009) 394–403.
[27] H.Q. Wang, S.Y. Zhou, L. Xiao, Y.J. Wang, Y. Liu, Z.B. Wu, Titania nanotubes – a unique photocatalyst and adsorbent for elemental mercury removal, Catal. Today 175 (2011) 202–208.
[28] S. MiarAlipour, D. Friedmann, J. Scott, R. Amal, TiO_2/porous adsorbents: recent advances and novel applications, J. Hazard. Mater. 341 (2018) 404–423.
[29] L. Liu, A. Corma, Metal catalysts for heterogeneous catalysis: from single atoms to nanoclusters and nanoparticles, Chem. Rev. 118 (2018) 4981–5079.
[30] P. Liu, R. Qin, G. Fu, N. Zheng, Surface coordination chemistry of metal nanomaterials, J. Am. Chem. Soc. 139 (2017) 2122–2131.
[31] G.O. Larrazábal, A.J. Martín, J. Pérez-Ramírez, Building blocks for high performance in electrocatalytic CO_2 reduction: materials, optimization strategies, and device engineering, J. Phys. Chem. Lett. 8 (2017) 3933–3944.
[32] N. Han, Y. Wang, H. Yang, J. Deng, J. Wu, Y. Li, et al., Ultrathin bismuth nanosheets from in situ topotactic transformation for selective electrocatalytic CO_2 reduction to formate, Nat. Commun. 9 (2018) 1320.
[33] Y. Liu, Z. Li, M. Green, M. Just, Y.Y. Li, X. Chen, Titanium dioxide nanomaterials for photocatalysis, J. Phys. D: Appl. Phys. 50 (2017) 193003.
[34] C.B. Ong, L.Y. Ng, A.W. Mohammad, A review of ZnO nanoparticles as solar photocatalysts: synthesis, mechanisms and applications, Renew. Sustain. Energy Rev. 81 (2018) 536–551.
[35] K. Lee, A. Mazare, P. Schmuki, One-dimensional titanium dioxide nanomaterials: nanotubes, Chem. Rev. 114 (2014) 9385–9454.

[36] X. Wang, Z. Li, J. Shi, Y. Yu, One-dimensional titanium dioxide nanomaterials: nanowires, nanorods, and nanobelts, Chem. Rev. 114 (2014) 9346–9384.
[37] X. Zhou, N. Liu, P. Schmuki, Photocatalysis with TiO_2 nanotubes: "colorful" reactivity and designing site-specific photocatalytic centers into TiO_2 nanotubes, ACS Catal. 7 (2017) 3210–3235.
[38] A.J. Maira, K.L. Yeung, C.Y. Lee, P.L. Yue, C.K. Chan, Size effects in gas-phase photo-oxidation of trichloroethylene using nanometer-sized TiO_2 catalysts, J. Catal. 192 (2000) 185–196.
[39] S.W. Verbruggen, S. Ribbens, T. Tytgat, B. Hauchecorne, M. Smits, V. Meynen, et al., The benefit of glass bead supports for efficient gas phase photocatalysis: case study of a commercial and a synthesised photocatalyst, Chem. Eng. J. 174 (2011) 318–325.
[40] K. Nakata, A. Fujishima, TiO_2 photocatalysis: design and applications, J. Photochem. Photobiol. C 13 (2012) 169–189.
[41] L. Brus, Electronic wave-functions in semiconductor clusters – experiment and theory, J. Phys. Chem. 90 (1986) 2555–2560.
[42] S.W. Verbruggen, TiO_2 photocatalysis for the degradation of pollutants in gas phase: from morphological design to plasmonic enhancement, J. Photochem. Photobiol. C 24 (2015) 64–82.
[43] A. Fujishima, T.N. Rao, D.A. Tryk, Titanium dioxide photocatalysis, J. Photochem. Photobiol. C (2000) 1–21.
[44] O. Carp, C.L. Huisman, A. Reller, Photoinduced reactivity of titanium dioxide, Prog. Solid State Chem. 32 (2004) 33–177.
[45] M. Pelaez, N.T. Nolan, S.C. Pillai, M.K. Seery, P. Falaras, A.G. Kontos, et al., A review on the visible light active titanium dioxide photocatalysts for environmental applications, Appl. Catal. B 125 (2012) 331–349.
[46] Y. Nosaka, T. Daimon, A.Y. Nosaka, Y. Murakami, Singlet oxygen formation in photocatalytic TiO_2 aqueous suspension, Phys. Chem. Chem. Phys. 6 (2004) 2917–2918.
[47] N. Serpone, Relative photonic efficiencies and quantum yields in heterogeneous photocatalysis, J. Photochem. Photobiol. A 104 (1997) 1–12.
[48] M.R. Hoffmann, S.T. Martin, W.Y. Choi, D.W. Bahnemann, Environmental applications of semiconductor photocatalysis, Chem. Rev. 95 (1995) 69–96.
[49] J. Schneider, M. Matsuoka, M. Takeuchi, J. Zhang, Y. Horiuchi, M. Anpo, et al., Understanding TiO_2 photocatalysis: mechanisms and materials, Chem. Rev. 114 (2014) 9919–9986.
[50] L.Y. Wan, X.Y. Li, Z.P. Qu, Y. Shi, H. Li, Q.D. Zhao, et al., Facile synthesis of ZnO/Zn_2TiO_4 core/shell nanowires for photocatalytic oxidation of acetone, J. Hazard. Mater. 184 (2010) 864–868.
[51] Y.L. Pang, S. Lim, H.C. Ong, W.T. Chong, Research progress on iron oxide-based magnetic materials: synthesis techniques and photocatalytic applications, Ceram. Int. 42 (2016) 9–34.
[52] L. Vayssieres, An aqueous solution approach to advanced metal oxide arrays on substrates, Appl. Phys. A 89 (2007) 1–8.
[53] Q. Zhao, Z. Yan, C. Chen, J. Chen, Spinels: controlled preparation, oxygen reduction/evolution reaction application, and beyond, Chem. Rev. 117 (2017) 10121–10211.
[54] J. Jia, P. Zhang, L. Chen, Catalytic decomposition of gaseous ozone over manganese dioxides with different crystal structures, Appl. Catal. B: Environ. 189 (2016) 210–218.
[55] J.A. Rodriguez, D.C. Grinter, Z. Liu, R.M. Palomino, S.D. Senanayake, Ceria-based model catalysts: fundamental studies on the importance of the

metal-ceria interface in CO oxidation, the water-gas shift, CO_2 hydrogenation, and methane and alcohol reforming, Chem. Soc. Rev. 46 (2017) 1824–1841.

[56] W.-J. Ong, L.-L. Tan, Y.H. Ng, S.-T. Yong, S.-P. Chai, Graphitic carbon nitride (g-C_3N_4)-based photocatalysts for artificial photosynthesis and environmental remediation: are we a step closer to achieving sustainability? Chem. Rev. 116 (2016) 7159–7329.

[57] S. Bai, J. Jiang, Q. Zhang, Y. Xiong, Steering charge kinetics in photocatalysis: intersection of materials syntheses, characterization techniques and theoretical simulations, Chem. Soc. Rev. 44 (2015) 2893–2939.

[58] M. Faraldos, A. Bahamonde, Environmental applications of titania-graphene photocatalysts, Catal. Today 285 (2017) 13–28.

[59] K.R. Reddy, M. Hassan, V.G. Gomes, Hybrid nanostructures based on titanium dioxide for enhanced photocatalysis, Appl. Catal. A: General 489 (2015) 1–16.

[60] V.C. Padmanaban, M.S. Giri Nandagopal, G. Madhangi Priyadharshini, N. Maheswari, G. Janani Sree, N. Selvaraju, Advanced approach for degradation of recalcitrant by nanophotocatalysis using nanocomposites and their future perspectives, Int. J. Environ. Sci. Technol. 13 (2016) 1591–1606.

[61] L. Jing, W. Zhou, G. Tian, H. Fu, Surface tuning for oxide-based nanomaterials as efficient photocatalysts, Chem. Soc. Rev. 42 (2013) 9509–9549.

[62] Q. Fu, X. Bao, Surface chemistry and catalysis confined under two-dimensional materials, Chem. Soc. Rev. 46 (2017) 1842–1874.

[63] B. Qiu, M. Xing, J. Zhang, Recent advances in three-dimensional graphene based materials for catalysis applications, Chem. Soc. Rev. 47 (2018) 2165–2216.

[64] W. Wang, J. Yu, Q. Xiang, B. Cheng, Enhanced photocatalytic activity of hierarchical macro/mesoporous TiO_2–graphene composites for photodegradation of acetone in air, Appl. Catal. B: Environ. 119–120 (2012) 109–116.

[65] T. Giannakopoulou, I. Papailias, N. Todorova, N. Boukos, Y. Liu, J. Yu, et al., Tailoring the energy band gap and edges' potentials of g-C_3N_4 /TiO_2 composite photocatalysts for NO_x removal, Chem. Eng. J. 310 (2017) 571–580.

[66] Y. Xu, W. Wen, J.-M. Wu, Titania nanowires functionalized polyester fabrics with enhanced photocatalytic and antibacterial performances, J. Hazard. Mater. 343 (2018) 285–297.

[67] A. Maczulak, Pollution: Treating Environmental Toxins, Facts on File, New York, 2009.

[68] D.A. Vallero, Fundamentals of Air Pollution, fourth ed., Elsevier, Academic Press, 2008.

[69] C.S. Liang, F.K. Duan, K.B. He, Y.L. Ma, Review on recent progress in observations, source identifications and countermeasures of PM2.5, Environ. Int. 86 (2016) 150–170.

[70] R. Zhang, G. Wang, S. Guo, M.L. Zamora, Q. Ying, Y. Lin, et al., Formation of urban fine particulate matter, Chem. Rev. 115 (2015) 3803–3855.

[71] R.M. Heck, R.J. Farrauto, S.T. Gulati, Catalytic Air Pollution Control: Commercial Technology, third ed., John Wiley & Sons, Hoboken, NJ, 2009.

[72] M.S. Kamal, S.A. Razzak, M.M. Hossain, Catalytic oxidation of volatile organic compounds (VOCs) – a review, Atmos. Environ. 140 (2016) 117–134.

[73] B.E. Boor, M.P. Spilak, J. Laverge, A. Novoselac, Y. Xu, Human exposure to indoor air pollutants in sleep microenvironments: a literature review, Build. Environ. 125 (2017) 528–555.
[74] J. Yu, L.-H. Xie, J.-R. Li, Y. Ma, J.M. Seminario, P.B. Balbuena, CO_2 capture and separations using MOFs: computational and experimental studies, Chem. Rev. 117 (2017) 9674–9754.
[75] J. Ye, X. Zhu, B. Cheng, J. Yu, C. Jiang, Few-layered graphene-like boron nitride: a highly efficient adsorbent for indoor formaldehyde removal, Environ. Sci. Technol. Lett. 4 (2016) 20–25.
[76] R.W. Baker, B.T. Low, Gas separation membrane materials: a perspective, Macromolecules 47 (2014) 6999–7013.
[77] K. Goh, H.E. Karahan, L. Wei, T.-H. Bae, A.G. Fane, R. Wang, et al., Carbon nanomaterials for advancing separation membranes: a strategic perspective, Carbon 109 (2016) 694–710.
[78] M. Wang, Z. Wang, S. Zhao, J. Wang, S. Wang, Recent advances on mixed matrix membranes for CO_2 separation, Chin. J. Chem. Eng. 25 (2017) 1581–1597.
[79] X. Zhu, K. Yang, B. Chen, Membranes prepared from graphene-based nanomaterials for sustainable applications: a review, Environ. Sci. Nano 4 (2017) 2267–2285.
[80] S.S. Swain, L. Unnikrishnan, S. Mohanty, S.K. Nayak, Effect of nanofillers on selectivity of high performance mixed matrix membranes for separating gas mixtures, Korean J. Chem. Eng. 34 (2017) 2119–2134.
[81] J. Li, M. Zhang, Z. Guan, Q. Li, C. He, J. Yang, Synergistic effect of surface and bulk single-electron-trapped oxygen vacancy of TiO_2 in the photocatalytic reduction of CO_2, Appl. Catal. B: Environ. 206 (2017) 300–307.
[82] P. Granger, V.I. Parvulescu, Catalytic NO_x abatement systems for mobile sources: from three-way to lean burn after-treatment technologies, Chem. Rev. 111 (2011) 3155–3207.
[83] J. Ma, H. Wu, Y. Liu, H. He, Photocatalytic removal of NO_x over visible light responsive oxygen-deficient TiO_2, J. Phys. Chem. C 118 (2014) 7434–7441.
[84] K. Skalska, J.S. Miller, S. Ledakowicz, Trends in NO_x abatement: a review, Sci. Total Environ. 408 (2010) 3976–3989.
[85] L. Zhang, J. Sun, Y. Xiong, X. Zeng, C. Tang, L. Dong, Catalytic performance of highly dispersed WO_3 loaded on CeO_2 in the selective catalytic reduction of NO by NH_3, Chinese J. Catal. 38 (2017) 1749–1758.
[86] M. Kobayashi, K. Miyoshi, WO_3–TiO_2 monolithic catalysts for high temperature SCR of NO by NH_3: influence of preparation method on structural and physico-chemical properties, activity and durability, Appl. Catal. B: Environ. 72 (2007) 253–261.
[87] W. Li, H. Zhang, J. Wang, W. Qiao, L. Ling, D. Long, Flexible Ru/graphene aerogel with switchable surface chemistry: highly efficient catalyst for room-temperature CO oxidation, Adv. Mater. Interf. 3 (2016) 1500711.
[88] Z.R. Zhu, X.Y. Li, Q.D. Zhao, H. Li, Y. Shen, G.H. Chen, Porous "brick-like" $NiFe_2O_4$ nanocrystals loaded with Ag species towards effective degradation of toluene, Chem. Eng. J. 165 (2010) 64–70.
[89] Z.R. Zhu, X.Y. Li, Q.D. Zhao, Z.P. Qu, Y. Hou, L. Zhao, et al., FTIR study of the photocatalytic degradation of gaseous benzene over UV-irradiated TiO_2 nanoballs synthesized by hydrothermal treatment in alkaline solution, Mater. Res. Bull. 45 (2010) 1889–1893.
[90] X.Y. Li, X.J. Zou, Z.P. Qu, Q.D. Zhao, L.Z. Wang, Photocatalytic degradation of gaseous toluene over Ag-doping TiO_2 nanotube powder prepared by

anodization coupled with impregnation method, Chemosphere 83 (2011) 674–679.

[91] W.H. Zhang, X.Y. Li, Q.D. Zhao, Y. Hou, Y. Shen, G.H. Chen, Uniform alpha-Fe_2O_3 nanotubes fabricated for adsorption and photocatalytic oxidation of naphthalene, Mater. Chem. Phys. 129 (2011) 683–687.

[92] H. Li, Q.D. Zhao, X.Y. Li, Y. Shi, Z.R. Zhu, M. Tade, et al., Photocatalytic degradation of gaseous toluene over hollow "spindle-like" alpha-Fe_2O_3 loaded with Ag, Mater. Res. Bull. 47 (2012) 1459–1466.

[93] Y. Shen, Q.D. Zhao, X.Y. Li, D.L. Yuan, Y. Hou, S.M. Liu, Enhanced visible-light induced degradation of benzene on Mg-ferrite/hematite/PANI nanospheres: in situ FTIR investigation, J. Hazard. Mater. 241 (2012) 472–477.

[94] Q.Y. Yan, X.Y. Li, Q.D. Zhao, G.H. Chen, Shape-controlled fabrication of the porous Co_3O_4 nanoflower clusters for efficient catalytic oxidation of gaseous toluene, J. Hazard. Mater. 209 (2012) 385–391.

[95] Z.R. Zhu, Q.D. Zhao, X.Y. Li, Y.H. Li, C. Sun, G. Zhang, et al., Photocatalytic performances and activities of $ZnAl_2O_4$ nanorods loaded with Ag towards toluene, Chem. Eng. J. 203 (2012) 43–51.

[96] J.J. Sun, X.Y. Li, Q.D. Zhao, M.O. Tade, S.M. Liu, Quantum-sized $BiVO_4$ modified TiO_2 microflower composite heterostructures: efficient production of hydroxyl radicals towards visible light-driven degradation of gaseous toluene, J. Mater. Chem. A 3 (2015) 21655–21663.

[97] X.J. Zou, Y.Y. Dong, X.Y. Li, Q.D. Zhao, Y.B. Cui, G. Lu, Inorganic-organic photocatalyst $BiPO_4$/g-C_3N_4 for efficient removal of gaseous toluene under visible light irradiation, Catal. Commun. 69 (2015) 109–113.

[98] F. Zhang, X.Y. Li, Q.D. Zhao, A.C. Chen, Facile and controllable modification of 3D In_2O_3 microflowers with In_2S_3 nanoflakes for efficient photocatalytic degradation of gaseous ortho-dichlorobenzene, J. Phys. Chem. C 120 (2016) 19113–19123.

[99] F. Zhang, X.Y. Li, Q.D. Zhao, D.K. Zhang, Rational design of $ZnFe_2O_4/In_2O_3$ nanoheterostructures: efficient photocatalyst for gaseous 1,2-dichlorobenzene degradation and mechanistic insight, ACS Sustain. Chem. Eng. 4 (2016) 4554–4562.

[100] Z.G. Zhang, X.Y. Li, B.J. Liu, Q.D. Zhao, G.H. Chen, Hexagonal microspindle of NH_2-MIL-101(Fe) metal-organic frameworks with visible-light-induced photocatalytic activity for the degradation of toluene, RSC Adv. 6 (2016) 4289–4295.

[101] J.J. Sun, X.Y. Li, Q.D. Zhao, M.O. Tade, S.M. Liu, Construction of p-n heterojunction beta-$Bi_2O_3/BiVO_4$ nanocomposite with improved photoinduced charge transfer property and enhanced activity in degradation of ortho-dichlorobenzene, Appl. Catal. B-Environ. 219 (2017) 259–268.

[102] Q. Zhou, Q.D. Zhao, W. Xiong, X.Y. Li, J.N. Li, L.B. Zeng, Hollow porous zinc cobaltate nanocubes photocatalyst derived from bimetallic zeolitic imidazolate frameworks towards enhanced gaseous toluene degradation, J. Colloid Interface Sci. 516 (2018) 76–85.

[103] Y. Boyjoo, H. Sun, J. Liu, V.K. Pareek, S. Wang, A review on photocatalysis for air treatment: from catalyst development to reactor design, Chem. Eng. J. 310 (2017) 537–559.

[104] D. Spasiano, R. Marotta, S. Malato, P. Fernandez-Ibañez, I. Di Somma, Solar photocatalysis: materials, reactors, some commercial, and pre-industrialized applications. A comprehensive approach, Appl. Catal. B: Environ. 170–171 (2015) 90–123.

[105] M. Miyauchi, H. Irie, M. Liu, X. Qiu, H. Yu, K. Sunada, et al., Visible-light-sensitive photocatalysts: nanocluster-grafted titanium dioxide for indoor environmental remediation, J. Phys. Chem. Lett. 7 (2016) 75–84.
[106] H.B. Park, J. Kamcev, L.M. Robeson, M. Elimelech, B.D. Freeman, Maximizing the right stuff: the trade-off between membrane permeability and selectivity, Science 356 (2017) eaab0530.
[107] W. Gao, Z.D. Hood, M. Chi, Interfaces in heterogeneous catalysts: advancing mechanistic understanding through atomic-scale measurements, Acc. Chem. Res. 50 (2017) 787–795.
[108] J. Giménez, B. Bayarri, Ó. González, S. Malato, J. Peral, S. Esplugas, Advanced oxidation processes at laboratory scale: environmental and economic impacts, ACS Sustain. Chem. Eng. 3 (2015) 3188–3196.

Further Reading

P. Dong, G. Hou, X. Xi, R. Shao, F. Dong, WO_3-based photocatalysts: morphology control, activity enhancement and multifunctional applications, Environ. Sci. Nano 4 (2017) 539–557.

7

ADVANCED NANOMATERIALS FOR DEGRADING PERSISTENT ORGANIC POLLUTANTS

Qidong Zhao, Xiuming Zhao and Jingjing Cao
Dalian University of Technology, Dalian, P.R. China

With the rapidly growing production of artificially manufactured chemicals and their release due to human activity, various micropollutants emerging in the environment have attracted special attention because of their persistent accumulation in organic tissues and long-term toxicity to life. Persistent organic pollutants (POPs) as typical micropollutants are chemicals that are resistant to environmental degradation, which can accumulate through food chains and induce toxic effects in humans and wildlife, thus threatening the survival of all living things on Earth [1–3]. Most POPs are toxic substances which are produced intentionally for various applications or created as byproducts during combustion or industrial manufacturing processes. For example, some POPs such as aldrin, chlordane, dichlorodiphenyltrichloroethane (DDT), mirex, and toxaphene are used as pesticides. Polychlorinated biphenyls (PCBs) have been used as electric insulating ingredients. On addressing these POPs globally, to date, 29 nonpolar organic molecules have been classified as POPs by the Stockholm Convention, to be banned or restricted globally [4,5]. Meanwhile, the Stockholm Convention treaty includes measures to provide aid to countries to eliminate the use and production of POPs.

POPs break down very slowly in soil, air, and water and therefore remain in the environment for a long time. Because they persist in the environment, they can be transported over long distances through wind or water media before they are deposited. This is true even if there are no local sources of pollutants. For example, varying levels of PCBs and DDT have been found in the Arctic even though they have never been used there [6]. Nowadays, the control and degradation of POPs has

Advanced Nanomaterials for Pollutant Sensing and Environmental Catalysis. DOI: https://doi.org/10.1016/B978-0-12-814796-2.00007-1

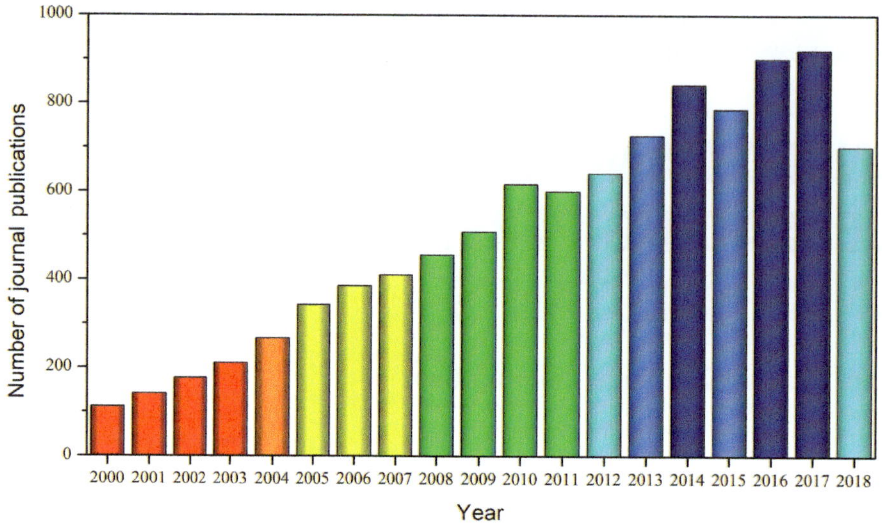

Figure 7.1 Number of publications investigating POPs associated with their control and degradation. *Source*: Data derived from the Web of Science (2018/10).

aroused worldwide and growing research interest, as indicated by recent journal publications on the topic of POPs (Fig. 7.1).

Nanomaterials for environmental protection have been successfully developed over the past few decades. Some extraordinary properties of nanomaterials such as their large surface area, electrochemical and magnetic properties, as well as other size-dependent physical and chemical properties, endow them with significant advantages in pollution control. To date, nanomaterials have been extensively applied or investigated in adsorption, chemical reduction, catalytic oxidation, photocatalysis, electrochemical, and filtration processes with promising application prospects. This chapter mainly describes recent progress in the removal of POPs by utilizing advanced nanomaterials with sufficient efficacy in these processes, wherein the catalytic strategies are highlighted.

7.1 Introduction to the Advantages of Nanomaterials Toward Persistent Organic Pollutants Removal

The application of nanotechnology in environmental purification and remediation has considerable potential. Traditional treatment technologies do not always offer the most cost-effective solution for the removal of several common pollutants, and in particular are they are not cost-effective for the removal of recalcitrant pollutants present in low concentrations. In addition, many of these techniques are already stretched to their limits and may be unable to meet the increasingly stringent environmental quality standards. Conventional treatment methods are often energy-intensive while generating a considerable amount of hazardous wastes like sludge. Nanobased techniques may thus become important, especially for the removal of emerging pollutants and low levels of contaminants [7–10]. Many nanobased treatment technologies have been shown to perform well as complements to or substitutes for traditional treatment technologies. Regarding POPs treatment, some nanomaterial-based treatment technologies have emerged as more efficient and promising strategies.

7.1.1 Adsorption Capability of Specific Nanomaterials Toward Persistent Organic Pollutants

Most adsorption methods use a carbonaceous material to trap pollutant molecules inside its pore structure. While the raw materials for activated carbon can be inexpensive, the energy required to produce high-quality activated carbon has been shown to have significant environmental impacts if nonrenewable energy sources are used. In activated carbon treatment, the carbon material eventually has to be regenerated to remove the adsorbed organic compounds. Other adsorption media are often expensive to produce with a limited regeneration lifespan for which the regeneration is also quite expensive. Though lower cost alternatives have been found, many are not as efficient and clogging of the pores is a common issue. Adsorption also only removes the pollutant; it does not transform it, which generates a hazardous waste stream that has to be dealt with. If the adsorbent is not regenerated on site, it must be handled as a hazardous waste, requiring special disposal and added costs [7].

The following are some basic innate factors which influence the function of nanoparticles (NPs) as adsorbents in solutions or substrates: the location of the most atoms on the surface, high surface area, high chemical activity, high adsorptive capacity, lack of internal diffusional resistance, and high surface binding energy [11]. Each of these properties successively lead to a significant fraction of atoms or molecules being associated with surfaces and interfaces and increase the potential impact of surface accessibility and affinity, surface enrichment, number of active sorption sites, and accordingly, surface energy toward specific analytes. Essentially, the physical, material, and chemical properties of NPs are directly related to their intrinsic compositions, apparent sizes, and extrinsic surface structures. Because of capillary and sorption effects, the high surface area present in NPs may retain solvents in circumstances that surprise researchers.

Coupling the wide variety of nanomaterials with different external functionalization methods will result in more excellent adsorption properties. Further functionalization of the surface may prevent NPs from aggregating and enable their selectivity. Intended coatings may have a significant impact on a variety of their properties. Functionalized groups induce important characteristics to adsorbents such as high absorption capacity (often measured as the breakthrough volume for a flowing system) and rapid desorption. More detailed knowledge on the absorptive properties of nanomaterials could be learned from literatures such as Ref. [11].

Among various nanomaterials that have been developed for removing organic pollutants, nanofibers, carbon nanomaterials, molecular sieves (such as nanoscale zeolites), metal-organic frameworks (MOFs), and magnetic nanomaterials are the most popular [11–17]. Besides, those nanomaterials used in pollution trace detection have been continuously and actively explored [10,18].

Nanofiber-based materials are important candidates for POPs in water remediation by adsorption, as they can be woven, adapted, and integrated with various devices and instruments more easily [12]. Nanofibrous media have a low basis weight, high permeability, and small pore size that make them appropriate for a wide range of filtration applications.

Carbon nanomaterials are composed entirely (or mainly) of carbon atoms. They include carbon nanotubes (single-walled or multiwalled, MWCNTs), carbon nanofibers, fullerene, graphene and its derivatives, and amorphous carbonaceous NPs or their composites [7,13,14]. Carbon nanomaterials characteristically have exceptionally high surface areas, which make them ideal candidates for the adsorption of pollutants. In addition, the surfaces of these inherently hydrophobic materials can be functionalized to target specific pollutants via chemical or electrical interactions. Some carbon nanomaterials may be aligned to form efficient filters or incorporated into conventional membranes for the removal of pollutants [11–14]. Carbonaceous nanomaterials and their composites have been shown in lab-scale studies to be good adsorbents for halogenated organics [7] including PCBs [13].

Magnetic nanoparticles (MNPs) such as iron oxide–nanostructured materials have been considered as potential sorbents in applications of environmental pollution cleanup. The advantages of MNPs are attributed to their high specific surface area, easy separation under external magnetic fields, and ease of surface functionalization. Magnetic-core NPs/microparticles (MPs) have cores made with magnetic elements such as iron, nickel, cobalt, or their oxides and alloys with ferromagnetic or superparamagnetic properties, and shells [7,16–18]. Shells can be made of inorganic components (e.g., silica or alumina) or organic molecules such as polymers or surfactants. These coatings can improve particle chemical stability, prevent oxidation, and also provide specific functionalities like selectivity for ion uptake or enhancing the water solubility of hydrophobic contaminants. MPs can be produced in sizes ranging from the nanoscale to several microns. One of the major benefits of MPs is that their superparamagnetism facilitates the rapid separation of pollutant-laden particles from treated water via an external magnetic field, requiring less energy to achieve a given level of separation than nonmagnetic particles.

Magnetic cyclodextrin nanocomposites are of particular interest in environmental applications. β-Cyclodextrin (β-CD) is a cyclic oligosaccharide composed of α-(1,4)-linked D-glucopyranoses, which have a hydrophobic inner cavity with a hydrophilic external surface. The introduction of cyclodextrin, which is known to form host – guest inclusion complexes with various nonpolar species onto the surfaces of MNPs, may render them water dispersible and capable of host – guest chemistry, thereby increasing the surface areas and improving their sorption capacities. Therefore cyclodextrin complexation has become a promising choice of procedure for decontaminating techniques [19]. Badruddoza et al. developed an environment-friendly β-cyclodextrin – ionic liquid (β-CD-IL) polyurethane-modified magnetic sorbent (Fe_3O_4-CDI-IL MNPs) for the removal of two typical POPs, that is, perfluorooctanoic acid (PFOA) and perfluorooctane sulfonate (PFOS), and Cr(VI) from water [19]. The solution pH had more obvious effects on the sorption of PFOA and Cr(VI) than that of PFOS. The coupling of ionic liquid with the β-CD polymer backbone could significantly enhance the removal efficiencies of both PFOS and PFOA. Sorption isotherms indicated that the heterogeneous sorption capacities of the Fe_3O_4-CDI-IL MNPs were 13,200 and 2500 $\mu g\ g^{-1}$ for PFOS and PFOA, respectively, and the monolayer sorption capacity was 2600 $\mu g\,g^{-1}$ for Cr(VI) ions. Cr(VI)-perfluorinated compound (PFC) binary sorption experiments exhibited a decrease in the sorption capacities for PFCs, but the removal of Cr(VI) was unaffected by the introduction of PFCs as cocontaminants. Hydrophobic interactions and electrostatic attraction were mainly involved in the PFC sorption process, whereas ion exchange and reduction was responsible for Cr(VI) sorption. In addition, the Fe_3O_4-CDI-IL MNPs could be readily recovered with a permanent magnet, regenerated, and reused at least 10 times without any significant efficiency loss, which shows potential in the removal of coexisting toxic contaminants from water or wastewater.

7.1.2 Activity of Specific Nanomaterials in Degrading Persistent Organic Pollutants

Among metal and metal oxide engineered nanomaterials, the most commonly applied for water treatment/environmental remediation is nanoscale zero-valent iron (nZVI) [7,20–30]. Partial mechanisms for water treatment or remediation with nZVI include adsorption, chemical degradation, and photoinduced degradation. Some pilot- and full-scale field studies applying nZVI-based remediation methods have been reported [21,29,30]. Effective degradation by nZVI and its derivatives has been shown for a variety of compounds, including chlorinated organic solvents, PCBs, organochlorine pesticides, and organic dyes [20–32].

Over the past decade, growing attention has been paid to clay-supported nanoscale nZVI composite materials as efficient and promising remediation materials in wastewater treatment and groundwater remediation technologies. This chapter gives an overview of the clay minerals, zero-valent iron materials, clay-supported nZVI composites, and the progress obtained during the remediation of contaminated aqueous solutions utilizing clay-supported nZVI composites for the removal of heavy metals, nitrate, selenate, dyes, phenolic compounds, chlorinated organic compounds (COCs), nitroaromatic compounds, and polybrominated diphenyl ethers (PBDEs) [20]. The removal of contaminants by clay–nZVI composites is mainly attributed to the synergetic effect between adsorption by the clay mineral and removal by the nZVI particles. For this reason, these composite materials have higher removal efficiencies compared to nZVI particles.

Freshly prepared nZVI is highly reactive due to its high specific surface area and strong reducing power. Over the years, the classic borohydride reduction method for preparing nZVI has been modified by the use of various stabilizers or surface modifiers to acquire more stable and soil deliverable nZVI for the treatment of different organic and inorganic contaminants in water and soil. While most studies have been focused on testing nZVI for water treatment, the greater potential or advantage of nZVI appears to be for the in situ remediation of contaminated soil and groundwater by directly delivering stabilized-nZVI into the contaminated subsurface as was proposed from the beginning. Compared to conventional remediation practices, the in situ remediation technique using stabilized-nZVI offers some unique advantages [21].

nZVI has been explored for the reductive dehalogenation of short-chain chlorinated paraffins (SCCPs) at a laboratory scale [22]. The results showed that the dechlorination rate of chlorinated n-decane (CP10) by nZVI increased with decreases in the solution pH. Increasing the loading of nZVI enhanced the dechlorination rate of CP10. With an increase in temperature, the degradation rate increased. The reduction of CP10 by nZVI was accelerated with increases in the concentration of humic acid up to 15 $mg\ L^{-1}$ after which it was inhibited. The dechlorination of CP10 within the initial 18 h followed a pseudo–first order rate model. The formation of intermediate products indicates a stepwise dechlorination pathway of SCCPs by nZVI. The carbon chain length and chlorination degree of SCCPs have a polynomial impact on dechlorination reactions. However, all dechlorination reactions were incomplete. One of the reasons for this incompleteness of the reactions may be the existence of a nonreactive fraction in the SCCPs due to their low solubility. In a previous study [22], acetone was added to enhance the solubility of SCCPs in an acetone–water mixture. However, it is likely that a fraction of the targeted chlorinated paraffins (CPs) was not dissolved in the aqueous phase. That insoluble fraction may be stuck on the surfaces of the NPs and the walls of the reactors. The chemicals stuck on the walls may not participate in the reaction, but they were extracted from the reactor (three times with hexane).

PCBs are carcinogenic and can accumulate in soils and sediments. Currently, there is no cost-effective and sustainable remediation technology for these contaminants. Gomes et al. developed a combination of electrodialytic remediation and zero-valent iron NPs in a two-compartment cell and compared it with a more conventional combination of electrokinetic (EK) remediation and nZVI in a three-compartment cell [23]. In the new two-compartment cell, the soil is suspended and stirred simultaneously with the addition of nZVI. Their results showed that soil characteristics like pH and buffer capacity are important and affect the reaction between nZVI and the target contaminant of PCBs.

PBDEs and some of their degradation products are persistent toxic contaminants, but Kim et al. [24] showed that a sequential nano–biotreatment using nZVI and diphenyl ether–degrading bacteria successfully degraded PBDEs to bromophenols and other less toxic metabolites.

Recently, nZVI, bimetallic NPs (nZVI/Pd), and nZVI/Pd-impregnated activated carbon (nZVI/Pd–AC) composite particles were synthesized and investigated by Zhuang et al. for their effectiveness in the removal of PBDEs and/or PCBs [25]. The palladization of nZVI promoted the dehalogenation kinetics for mono- to tri-BDEs and 2,3,4-trichlorobiphenyl (PCB 21). Compared to nZVI, the iron-normalized rate constants for nZVI/Pd were about 2, 3, and 4 orders of magnitude greater for tri-, di-, and mono-BDEs, respectively, with diphenyl ether as a main reaction product. The reaction kinetics and pathways suggest an H-atom transfer mechanism. The reaction pathways with nZVI/Pd favor the preferential removal of para-halogens on PBDEs and PCBs. Related factors that likely hindered the reaction with nZVI/Pd–AC are the heterogeneous distribution of nZVI and Pd on the activated carbon and/or the immobilization of hydrophobic organic contaminants at the adsorption sites thereby inhibiting contact with the nZVI.

Bimetallic Ni/Fe NPs were also found to be efficient in the degradation of DDT and PBDEs in aqueous solutions at weakly acidic or alkaline conditions [25,26]. Acidic media promote the degradation of DDT more effectively because the production of protons helps in generating hydrogen [15].

PFCs are extremely persistent micropollutants that are detected worldwide. The removal of PFCs from water by different types of nZVI has been attempted [27]. Batch experiments showed that an iron dose of $1\,g\,L^{-1}$ in the form of Mg-aminoclay (MgAC)-coated nZVI, at an initial pH of 3.0 effectively removed 38%–96% of individual PFCs. A maximum removal was observed for all PFCs with a high nZVI concentration, freshly synthesized nZVI, low pH, and low temperature. A mass balance experiment with PFOS in a higher concentration of nZVI revealed that the removal was due to both sorption and degradation. A maximum removal was observed for all PFCs with a high nZVI concentration, freshly synthesized nZVI, low pH, and low temperature.

Major limitations of ZVI include low reactivity due to its intrinsic passive layer, narrow working pH, reactivity loss with time due to the precipitation of metal hydroxides and metal carbonates, low selectivity for target contaminants, especially under oxic conditions, limited efficacy for the treatment of some refractory contaminants, and passivity of ZVI arising from certain contaminants [28]. Countermeasures to these limitations can be divided into seven categories, namely the pretreatment of pristine ZVI to remove the passive layer, the fabrication of nanosized ZVI to increase the surface area, the synthesis of ZVI-based bimetals taking advantage of the catalytic ability of the noble metal, employing physical methods to enhance the performance of ZVI, coupling ZVI with other adsorptive materials and chemically enhanced ZVI technology, as well as methods to recover the reactivity of aged ZVI. The key to improving the rate of contaminants removal by ZVI and broadening the applicable pH range is to enhance ZVI corrosion and to enhance the mass transfer of reactants including oxygen and H^+ to the ZVI surface. The safe disposal or potential utilization of iron-containing sludge generated from the process of contaminants sequestration by ZVI is necessary.

Some investigation has indicated that modifications of the nZVI surface can help in its stabilization, the reduction of aggregation, and the reduction of toxicity [29]. However, the improvement of those properties may simultaneously lead to the creation of other threats related with the application of nZVI. Research demonstrates that nZVI can be an effective and versatile tool for the purification of different matrices and especially groundwater. The application of NPs in the remediation of the environment not only reduces the concentration of potential noxious substances, but also reduces the costs of large-scale remediation and of the duration of the process. Nevertheless, it is necessary to conduct investigations that will take into account not only the toxicity of nZVI prior to the process of remediation, but also to address the consequences resulting from the application of nZVI [30].

Besides nZVI, copper NPs (CuNPs) can be also utilized in POPs treatment. Lee et al. studied the mechanism of the reductive defluorination of PFOA by titanium(III) citrate in the presence of catalysts with an initial pH of 9.0 at 70°C [31]. Vitamin B_{12} was used to catalyze reduction reactions by shuttling electrons from a reducing agent (electron donor) to PFOA to produce Co–carbon bond intermediates. In the presence of CuNPs, a precursor complex, $B_{12}-C_7F_{14}COOH$, was adsorbed on the metal surface, followed by a hydrogenolytic reaction to form less-fluorinated products. The synergistic effect between vitamin B_{12} and CuNPs enhances the reductive activities by electron transfer reactions and hydrogenolysis. In the anoxic aqueous solution, the biomimetic reductive system effectively removed 65% of PFOA, while no short-chain intermediates were detected.

PBDEs, as widely used additive flame retardants, have attracted attention owing to their ubiquitous distribution in the environment and toxicity to organisms. Potential methods for PBDE removal in contaminated environmental systems have been reported, with a strong focus on reductive debromination. However, the complete structure decomposition of PBDEs cannot be achieved by reductive debromination alone. Alternatively, oxidation processes are more suitable as higher debromination and mineralization degrees can be achieved [32]. In this regard, flower-like LiαTiO_x micro/nanostructures were successfully synthesized by Li et al. [32] to degrade tetrabromodiphenyl ether (BDE-47) at 250°C – 350°C. The excellent performance attained by LiαTiO_x was attributed to the Li dopant having an electron-donating effect, which enhanced the oxygen species' mobility. The oxidative reaction was believed to be the dominant degradation pathway. In the oxidative reaction, a series of oxidative products such as OH-tri-BDEs and OH-tetra-BDEs were first formed via nucleophilic O^{2-} attack and subsequently transformed into dibromophenol, tribromophenol, benzenedicarboxylic, and benzoic acids, among others. They could be further attacked by electrophilic O_2^- and O^- and completely cracked to small molecules such as formic, acetic, propionic, and butyric acids.

7.2 Current Status and Challenges in Degrading Anthropogenic Persistent Organic Pollutants

7.2.1 Current Status on Identifying and Eliminating Anthropogenic Persistent Organic Pollutants

Halogenated organic contaminants including perfluorinated or partially fluorinated chemicals, COCs, and brominated organic compounds, show the characteristics of POPs after they are discharged into the environment. The chemicals targeted by the Stockholm Convention are listed in the annexes of the convention text. There have been 29 types of organic chemicals listed in the Stockholm Convention up until 2018 since the initial 12 POPs in 2001 were targeted. In addition, other chemicals are under review for adding to the list including dicofol, PFOA, its salts, and PFOA-related compounds, and perfluorohexane sulfonic acid (PFHxS), its salts, and PFHxS-related compounds [33].

Selective descriptions of some of those chemicals are introduced below with reference referring to the Stockholm Convention. Parties must take measures to eliminate the production and use of the 24 groups of chemicals listed under Annex A of the Stockholm Convention. For Annex B (Restriction), parties must take measures to restrict the production and use of the chemicals listed under Annex B, the first group is perfluorooctane sulfonic acid, its salts, and perfluorooctane sulfonyl fluoride (POSF), while the other is DDT. As for Annex C (Unintentional production), it has been suggested that parties must take measures to reduce the unintentional release of chemicals listed under Annex C with the goal of continuing minimization and, where feasible, ultimate elimination. Polychlorinated dibenzo-*p*-dioxins (PCDDs) and polychlorinated dibenzofurans (PCDFs) are on the list.

CPs are complex mixtures of certain organic compounds containing chloride such as polychlorinated n-alkanes. The chlorination degree of CPs can vary between 30 and 70 wt.%. SCCPs are between C10 and C13, which were added to the list in 2017. SCCPs can be used as a plasticizer in rubber, paints, adhesives, and flame retardants for plastics as well as an extreme pressure lubricant in metal working fluids. SCCPs are sufficiently persistent in air for long-range transport to occur and they appear to be hydrolytically stable. Many SCCPs can accumulate in biota. It is concluded that SCCPs are likely, as a result of their long-range environmental transport, to lead to significant adverse environmental and human health effects.

Due to their broad industrial production and application, SCCPs have been found in high levels in different environmental matrices. The toxicity of SCCPs and potent techniques for their elimination are attracting growing research interest. SCCPs, as a class of emerging organic pollutants, have even been listed among the candidates of potential POPs under the Stockholm Convention in 2006, and subsequently the POPs review committee came to an agreement that SCCPs fulfill the criteria of Annex A in 2017. The production of SCCPs has decreased globally as jurisdictions have established control measures.

Decabromodiphenyl ether (decaBDE; commercial mixture, c-decaBDE) is used as an additive flame retardant, and has a variety of applications including in plastics/polymers/composites, textiles, adhesives, sealants, coatings, and inks. DecaBDE containing plastics are used in housings of computers and TVs, wires and cables, pipes, and carpets. Commercially available decaBDE consumption peaked in the early 2000s, but c-decaBDE is still extensively used worldwide. DecaBDE is highly persistent and has a high potential for bioaccumulation and food-web biomagnification as well as for long-range transport. Adverse effects are reported for soil organisms, birds, fish, frogs, rats, mice, and humans.

PFOA is identified as a substance of high concern with a persistent, bioaccumulative, and toxic structure to the environment and living organisms. PFOA-related compounds are released into air, water, soil, and solid waste, and these degrade into PFOA in the environment and in organisms. Major health issues such as kidney cancer, testicular cancer, thyroid disease, pregnancy-induced hypertension, and high cholesterol have been linked to PFOA. PFOA, its salts, and PFOA-related compounds are used widely in the production of fluoroelastomers and fluoropolymers, for the production of nonstick kitchenware and food processing equipment. PFOA-related compounds including side-chain fluorinated polymers are used as surfactants and surface treatment agents in textiles, paper and paints, and firefighting foams. PFOA has been detected in industrial waste, stain resistant carpets, carpet cleaning liquids, house dust, microwave popcorn bags, water, food, and Teflon. The unintentional formation of PFOA is created from the inadequate incineration of fluoropolymers from municipal solid waste incineration with inappropriate incineration or open burning facilities at moderate temperatures [34].

Endeavors toward identifying various POPs in the environment have been made globally in recent years. For example, Cao et al. demonstrated the body burden of SCCPs in human breastmilk in Japan, Korea, and China by analyzing human breastmilk and pooled breastmilk samples comparatively [35]. They also estimated the potential lactational exposure to SCCPs in infants, suggesting that the average risk is low based on the available data. Electronic waste (e-waste) recycling sites have been demonstrated as hot spots for SCCP contamination. In a study by Sun et al. [36], wild aquatic organisms (fish and invertebrates), water, and sediment collected from an enclosed freshwater pond contaminated by e-waste were analyzed to investigate the bioaccumulation, distribution, and trophic transfer of SCCPs in the aquatic ecosystem. SCCPs were detected in all of the investigated aquatic species at concentrations of

1700 – 95,000 ng g^{-1} lipid weight. The calculated bioaccumulation factors varied from 2.46 to 3.49. SCCP congeners with higher log K_{OW} were preferentially distributed in the liver of the fish species, and the pathway (i.e., water or sediment) of bioaccumulation could affect the tissue distribution of SCCP congeners.

CPs generally function as flame retardants and plasticizers in various materials. They are most likely to be processed by thermal processes during their entire life cycle. However, data on the formation and emission of CPs during thermal processes are still not fully understood. Xin et al. studied the synergistic emission of SCCPs and chlorinated polycyclic aromatic hydrocarbons (Cl-PAHs) during some common thermal processes of CPs such as incineration, open burning, or accidental fire in a department [37]. These investigators obtained the conclusion that higher temperatures can minimize the formation of SCCPs, but will generate Cl-PAHs.

Gallistl et al. obtained fat from wipe tests of the inner surface of 21 baking ovens from Stuttgart (Germany) and analyzed the samples for halogenated flame retardants (HFRs) as well as PCBs [38]. The exceptionally high concentrations and exclusive presence of CPs in half of the samples produced strong evidence that these compounds were released from the baking ovens themselves. This hypothesis was supported by the detection of medium-chain chlorinated paraffins (MCCPs) at even higher concentrations in the inner components of one dismantled baking oven. The release of substantial amounts of HFRs from oven casings during their use may contribute to human exposure to these compounds, especially MCCPs and SCCPs.

The widespread environmental contamination and well documented toxic effects of PBDEs have led to bans and voluntary withdrawals in many jurisdictions [39]. Replacement novel brominated flame retardants (NBFRs) have, however, exhibited many of the same toxic characteristics as PBDEs and appear to share a similar environmental fate, posing a similar risk to land contamination as PBDEs. Evidence suggests that PBDEs and NBFRs are also released from flame-retarded products during disposal via landfill, dumping, incineration, and recycling. E-waste appears to be one of the greatest contributors to such contamination in regions where the practice is widespread. High levels of contamination have been identified in China and other parts of Asia and Africa where informal methods such as burning or acid-stripping of electrical components may enhance the release of PBDEs and NBFRs. PBDEs have also been determined in almost all background soils assessed including in remote areas of Antarctica and northern polar regions. Sources to these regions are likely to be a combination of long-range atmospheric transport and emissions from flame-retarded goods onsite.

Beyond the category of POPs listed in the Stockholm Convention in 2009, chlorinated polyfluoroalkyl ether sulfonic acids (Cl-PFESAs), with the trade name F-53B, have been applied as a mist suppressant to prevent the emissions of the toxic substance, chromium, for decades in the Chinese chrome plating industry [40]. Cl-PFESAs showed similar distribution profiles, enrichment mechanisms, biological toxicity, and elimination kinetics to the well-known per- and polyfluoroalkyl substances (PFASs) like PFOS.

Aromatic hydrocarbons including benzene, toluene, ethyl benzene, xylene, and PAHs are considered to be POPs, released into the environment mainly by the exploration activities of the petroleum industry [41]. These pollutants are mutagenic, carcinogenic, immunotoxic, and teratogenic to lower and higher forms of life, that is, from microorganisms to humans. The treatment and remediation of petroleum refinery waste have been shown to either reduce or eliminate the genotoxicity of these pollutants. Bioremediation using microorganisms to treat this waste is a promising technology as it is a safe and cost-effective option among various technologies tested. Many microorganisms produce biosurfactants that increase the bioavailability of these hydrophobic substrates for biodegradation. There is scarce knowledge on genes, enzymes as well as the molecular mechanism of PAHs degradation in anaerobic environments.

A new global emission inventory of $C_4 - C_{10}$ perfluoroalkane sulfonic acids from the life cycle of POSF-based products from 1958 to 2030 is presented by Wang et al. [42]. Between 1958 and 2015, the total direct and indirect emissions of PFOS are estimated to be 1228–4930 tonnes, and the emissions of PFOS precursors are estimated to be 1230–8738 tonnes and approximately 670 tonnes for x-perfluorooctanesulfonamides/sulfonamido ethanols (xFOSA/Es) and POSF, respectively. The degradation of side-chain fluorinated polymers in the environment and landfills can be a long-term (potentially) substantial source of PFOS.

Various POPs, especially flame retardants from sediments [43], aquaculture [44], marine [45], aquatic, and terrestrial ecosystems [46] have been actively investigated recently. Depending on the matrix, a sample preparation step is necessary for the extraction/enrichment/cleanup of trace compounds in aqueous samples, for example, environmental waters or in solid samples like food. A powerful, smart generation of selective sorbent phases is represented by molecularly imprinted polymers (MIPs), a category of synthetic materials able to specifically rebind a target molecule in preference to other closely related compounds [47]. The ultimate goal in developing molecular imprinting is that the synthesized materials can provide binding properties comparable to those of natural receptors. Experimentally, MIPs are prepared by the copolymerization of functional and cross-linking monomers in the presence of a template molecule (i.e., the target analyte). The reaction between the monomers, chosen based on their affinity with the functional groups of the template, yields highly cross-linked three-dimensional (3D) network polymer binding sites with shape, size, and functionalities complementary to the target analyte. MIPs are not exclusively used for preconcentration purposes, but often their major merit is sample cleanup, especially in the case of highly complex matrices (e.g., meat tissues).

Compared to other recognition systems, MIPs, which possess three major unique features of structure predictability, recognition specificity, and application universality, have received widespread attention and have become attractive in many fields such as purification and separation, chemo/biosensing, artificial antibodies, drug delivery, and catalysis and degradation owing to their high physical stability, straightforward preparation, remarkable robustness, and low cost [48]. Various contaminants such as bisphenol A, dicofol, and organophosphorus or organochlorine pesticides including PCBs could be successfully extracted from food and environmental matrices.

Subsequently, multiple strategies toward eliminating POPs from environmental media will be introduced here based on literature survey, some of which utilize nanomaterials.

Many soils are impacted by pesticide pollution including POPs due to their massive use worldwide in history. Some effective technologies such as containment-immobilization, separation, or destruction are generally preferred [49]. The suitability depends on the extension of contamination (point-source or diffuse pollution), type, and concentration of the pesticide to be removed as well as the type of soil, climatic conditions, and the presence of other potential contaminants. The majority of scientific studies are still at the lab-scale stage. There is a lack of full-scale remediation technologies based on their complete degradation of contaminants as well as their real cost and efficacy.

The use of nZVI for the in situ remediation of soil contamination caused by heavy metals and organic pollutants has drawn great attention primarily owing to its potential for excellent activity, low cost, and low toxicity [50]. Towards a high-efficient and environment-friendly technique, biological method enhanced nZVI technology will lead the future research direction. Nanoremediation using nZVI could be applied for the transformation and detoxification of combined pollutants from soil. However, the complex and variable soil environment affects the performance of nZVI. A combination of physical, chemical, and biological methods with nZVI was developed to enhance the performance of nZVI. Unfortunately, the negative impacts of nZVI on organisms will inevitably induce adverse effects on soil bioremediation processes.

Zero-valent metals have proven effective in destroy POPs through chemical reduction. PBDEs undergo debromination when they are exposed to zero-valent metal or bimetallic systems. Yet the debromination pathways and mechanisms in these systems are not well understood. The debromination of BDE-21 in Pd/H_2 systems as well as the solvent kinetic isotope effect in single metal and bimetallic systems suggest that H-atom transfer is the dominant mechanism in Fe/Pd systems, while e-transfer is still the dominant mechanism in Fe/Ag systems [51]. The debromination of PBDEs by zero-valent zinc was achieved by Tang et al. [52]. The debromination pathways are the same as the debromination pathways of tetrabromodiphenyl ether in ZVI.

Fenton oxidation is a widely used technique to remediate PAHs in contaminated soils, but various limitations are associated with its environmental application [53]. Considering the main drawback that its optimal pH is around 3, the traditional Fenton treatment is costly and impractical due to the high buffering capacity of soils and the associated hazardous effects. The use of modified Fenton oxidation through the utilization of chelating agents can allow for its use at circumneutral pH, but chelating agents can also compete for •OH radicals with PAHs and thus significantly affect the oxidation efficiency. A recent approach in this regard is the use of inorganic chelating agents that offer advantages over organic chelating agents for oxidant consumption. Another possibility is the use of iron minerals, which are able to catalyze a Fenton-like oxidation over a wide range of pH. The use of iron minerals or NPs of ZVI could provide a cost-effective alternative to the use of more than one reagent.

Hydroxyl radical (•OH) is one of the most powerful oxidizing agents, able to react unselectively and instantaneously with the surrounding chemicals including organic pollutants and inhibitors [54]. •OH radicals are omnipresent in the environment (natural waters, the atmosphere, interstellar space, etc.) including in biological systems where •OH has an important role in immunity metabolism. Its reactivity with various water pollutants, which include bacteria, organic compounds, and inorganic compounds, continues to be a subject of scientific and governing agency pollution prevention interest in wastewater treatment processes. Various advanced treatment technologies utilizing the synergistic efficiency of combined advanced oxidation processes (AOPs) in the treatment of wastewater containing POPs have been reported [55].

Roth et al. reported a novel tubular electrochemical cell which is operated in a cyclic adsorption–electro-Fenton (EF) process and by this means overcomes the drawbacks of the traditional EF process [56]. A microtube made only of MWCNTs functions as a gas diffusion electrode (GDE) and a highly porous adsorbent. In the process, the pollutants were first removed electroless from the wastewater by adsorption on the MWCNT-GDE. Subsequently, the pollutants were electrochemically degraded in a defined volume of electrolyte solution using the EF process.

A procedure combining reductive and oxidative stages has been demonstrated to be promising to deal with pollutants that are difficult to oxidize by Santos-Juanes et al. [57]. It has the advantage of performing the reducing and oxidizing stages in different parts of a reactor, making it possible to favor each stage or even the overall performance of the combined process by tuning the experimental conditions. The application of this methodology to other families of compounds that cannot be easily oxidized (e.g., perfluorinated molecules such as flame retardants or COCs) requires future research.

The emissions of POPs from diesel engines are becoming more important due to their proximity to human beings compared to stationary sources whose emissions have been well controlled over the years [58]. Selective catalytic reduction (SCR) techniques have been adopted in the heavy-duty diesel engine to reduce nitrogen oxide (NO_x). SCR has also been reported to reduce PAHs, PCDD/PCDFs, and PCBs limitedly under proper working conditions. However, SCR can significantly increase several POPs emissions including PCDD/PCDFs, PCBs, and polybrominated dibenzo-*p*-dioxins and dibenzofurans unexpectedly from heavy-duty engines at mid- and high loads under unfavorable conditions, except polybrominated biphenyls.

PFOA is one of the most commonly found PFCs in the environment. The strong C–F bond in PFOA is extremely difficult to degrade; therefore AOPs at room temperature and pressure are not able to oxidize them, as was noticed by Santos et al. using a Fenton-like reagent or persulfate at 25°C [59]. On the contrary, using persulfate activated by heat (100 mM at 70°C) a complete defluorination of PFOA (0.1 mM) was noticed after 18 h, with a sequential degradation mechanism of losing one CF_2 unit from PFOA and its intermediates [perfluoroheptanoic acid, perfluorohexanoic acid (PFHxA), perfluoropentanoic acid, and perfluorobutanoic acid].

Recent decades have brought, in water management, an intense progress in AOPs based on the decomposition of pollutants by free radicals, which can be produced in different ways [60]. Particular methods have been developed based on the oxidation and reduction of target pollutants. Radical reactions can be effectively carried out through different methods both in oxidative and reductive modes for the decomposition of PFOA and PFOS where different molecular mechanisms of decomposition/degradation take place.

Multiple PFASs are often found together in the environment due to product manufacturing methods and abiotic and biotic transformations. Treatment methods are needed to effectively sequester or destroy a variety of PFASs from groundwater, drinking water, and wastewater. Hoffmann et al. presented a comprehensive summary of several categories of treatment approaches including [61] (1) sorption using activated carbon, ion exchange, or other sorbents, (2) AOPs including electrochemical oxidation, photolysis, and photocatalysis, (3) advanced reduction processes using aqueous iodide or dithionite and sulfite, (4) thermal and nonthermal destruction including incineration, sonochemical degradation, sub- or supercritical treatment, microwave hydrothermal treatment, and high-voltage electric discharge, (5) microbial treatment, and (6) other treatment processes including ozonation under alkaline conditions, permanganate oxidation, vitamin B12 and Ti(III) citrate reductive defluorination, and ball milling.

Sulfate radicals can oxidize a group of PFC, that is, perfluorinated carboxylic acids (PFCAs) [62]. It has been proven that PFCAs can be partly mineralized in chain reactions initiated by sulfate radicals ($SO_4^{\bullet -}$). Perfluorinated acetic acid, propionic acid, and butanoic acid are largely degraded in a primary reaction with sulfate radicals. In the case of PFCA with a chain length of >4 carbons, low yields of PFCA products were observed.

Graphene quantum dots (GQDs) were synthesized by Niu et al. via the oxygen-driven unzipping of graphene under ultrahigh frequency ultrasonication, and then these were attached to SiC NPs by the hydrothermal method to form SiC/GQD nanocomposites [63]. The SiC/GQDs exhibited superior photoactivity in the decomposition of PFOS ($C_8F_{17}SO_3H$), which was even harder to decompose than PFOA. This work presented the first instance of employing photoexcited semiconductor nanomaterials to realize the improvement from the activation of the $-F_2C-COOH$ bond in PFOA to the activation of $-F_2C-SO_3H$ in PFOS. The decomposition mechanisms of PFOS involved the cleavage of the ionic headgroup, hydrolysis, hydrodefluorination, and C−C bond scission.

PFCs have been found to be completely defluorinated in a UV/organoclay/3-indole acetic acid system [64]. The effect of different indole derivatives and organo-modified montmorillonite on the degradation of PFOA was comprehensively studied. Varying degradation performance could be ascribed to the independent contribution by the hydrated electron yields of indole derivatives, adsorption of PFOA and indole derivatives on organo-montmorillonite. Moreover, it has been found that the presence of humic substance would not significantly suppress the degradation process due to the strong adsorption of humic substance on the organo-montmorillonite surface.

MOFs could behave as sorbents for the removal of PFOA from aqueous solutions. Chen et al. investigated the effects of topology and surface functionality on PFOA sorption and the uptake kinetics of MOF materials [65]. Zeolitic imidazolate frameworks (ZIFs) were shown to outperform two commercial sorbents, zeolite 13X and activated carbon. The PFOA sorption capacity and kinetics of ZIF-L are comparable to the benchmark values achieved for PFOA. It appears that the interlayer spacing within ZIF-L plays a key role in sorption performance by reducing the structural restrictions found in most 3D porous materials and thereby allowing for a faster diffusion of PFOA.

The electrochemical oxidation of PFOA in an aqueous solution using highly hydrophobic modified PbO_2 electrodes has been reported by Zhuo et al. [66]. Approximately 92.1% of PFOA degradation was achieved in 180 min at initial pH 3. The efficient photocatalytic defluorination of PFOA by BiOCl nanosheets via a hole direct oxidation mechanism has been achieved by Song et al. [67]. It was demonstrated that the PFOA degradation was positively correlated with the amount of oxygen vacancies in the BiOCl. The oxygen vacancies not only acted as electron scavengers to suppress charge recombination, but also favored the binding of PFOA tightly onto the BiOCl surface through a special monodentate coordination, which is beneficial for the hole oxidation of PFOA.

Environmental microorganisms can sometimes aid the degradation of POPs [68]. Nutrition and pollution stress stimulate genetic adaptation in microorganisms and assist in the evolution of diverse metabolic pathways for their survival under several complex organic compounds. Diverse microorganisms, harboring numerous plasmids and catabolic genes, acclimatize to these environmentally unfavorable conditions by gene duplication, mutational drift, hypermutation, and recombination. The genetic aspects of the catabolic genes of some major POPs such as biphenyl dioxygenase (bph), DDT 2,3-dioxygenase, and angular dioxygenase assist in the degradation of biphenyl, organochlorine pesticides, and dioxins/furans, respectively. A major endeavor in today's scientific world is to characterize the exact genetic mechanisms of microbes for the bioremediation of these toxic compounds by excavating into the uncultured plethora.

Recently, mechanochemical destruction has been proposed as a promising, noncombustion technology for the disposal of toxic, halogenated, and organic pollutants [69–72].

In the past 20 years, mechanochemical destruction has shown potential to achieve pollutant degradation, both of pure substances and in contaminated soils. This capability has been tested for many halogenated pollutants with various reagents under different milling conditions [70]. For instance, hexachlorobenzene (HCB) was chosen as a model pollutant with aluminum and alumina ($Al + Al_2O_3$) powders as comilling regents [71]. Both the intermediate analysis and the quantum chemical calculations were adopted to elucidate the free radical dechlorination mechanism of HCB. It was found that the intermediates and radical-related reactions in the mechanochemical dechlorination of HCB undergo quite different intermediates from those that happen in a typical photocatalytic dechlorination process. Cagnetta et al. reported a new defect engineering strategy to improve the effectiveness of metal oxides as comilling reagents for halogenated organic pollutant destruction [72]. High-valent metal doping of a commonly employed comilling reagent such as CaO determines a 2.5 times faster pollutant degradation rate. This enhancement is due to electron-rich defects generated by the dopant; electrons are transferred to the organic pollutant thus causing its mineralization.

Among the remediation technologies cited in the literature, removal by activated carbon has been the most widely used, with several successful field tests being reported [73]. However, a number of limitations to the use of activated carbon exist such as it being ineffective at removing PFOA and other PFCs. Other adsorbents that have the potential to treat aqueous PFOS and PFOA include organoclays, clay minerals, and carbon nanotubes.

The use of metal oxide semiconductor nanostructures for the photocatalytic reductive debromination of PBDEs using visible light has been recently reported by Miller et al. [74]. They found that Cu_2O/Pd is a promising photocatalyst for the reductive dehalogenation of halogenated organic compounds.

The combined use of plants and bacteria is a promising approach for the remediation of soil and water contaminated with POPs [75]. Plants provide residency and nutrients to their associated rhizosphere and endophytic bacteria. In return, the bacteria support plant growth by the degradation and detoxification of POPs. Moreover, they improve plant growth and health due to their innate plant growth–promoting mechanisms. Many POPs, especially PCBs, have been remediated from a wide range of ecosystems through bacterial-assisted phytoremediation. A biochar–plant tandem approach could be an effective strategy for remediating soils contaminated with semi/volatile organic contaminants such as HCB [76].

POPs as hydrophobic substances can accumulate easily in oil media. Nowadays, with the increasing amount of oily wastewater being discharged, oil/water separation has become an urgent world challenge [77]. Porous materials with special wettability are most popular since this kind of material is easy to fabricate, cost effective, and time saving. Moreover, by combining the design of special wettability with the proper pore size, porous materials could achieve the separation of sundry oil/water mixtures. Two emerging categories of superwetting porous materials for oil/water separation and emulsion separation are very promising, namely water blocking porous materials with superhydrophobic/superoleophilic wettability including filtration membranes and adsorption sponges and oil blocking porous materials with superhydrophilic/underwater superoleophobic wettability.

Because they transfer over various media in the environment with time, POPs such as PCB can be found in contaminated soils and sediments. The most frequent remediation solutions adopted are "dig and dump" and "dig and incinerate," but there are currently new methods that could be more sustainable alternatives [78]. There is no single, portable technology that is applicable to both ex situ and in situ remediation of PCB in contaminated soils and sediments. Each case is unique and several factors must be considered. The successful treatment of a site depends on the proper selection, design, and adjustment of the remediation technology based on the congeners present, soil properties, and the performance of the system. More recently, the combined use of remediation technologies and the so-called treatment trains is a promising approach for persistent contaminants. Typically, the catalytic hydrodechlorination of PCB is usually performed with transition metals (e.g., Ni or Pd) and H_2 as heterogeneous catalysts and reducing agents, respectively, in aqueous or organic solvents. Photocatalytic dechlorination has been also reported.

7.2.2 Challenges in Degrading Anthropogenic Persistent Organic Pollutants in the Environment

Traditional wastewater management methods using biological microorganisms (biodegradation) and/or physicochemical processes (flocculation, chlorination, and ozonation), subsequently followed by filtration and adsorption-based separations are able to treat the majority of anthropogenic-polluted water sources. However, no method described above is efficient enough to produce water with legally and practically trace levels of refractory POPs chemicals.

Considering the main drawback that its optimal pH is around 3, the traditional Fenton treatment is costly and impractical due to the high buffering capacity of soils and the associated hazardous effects. The use of modified Fenton oxidation using chelating agents can allow for its use at neutral pH, but chelating agents can also compete for •OH radical with PAHs and thus significantly affect the oxidation efficiency [53].

A complete photodegradation of some POPs through photocatalysis is still difficult to achieve. More studies should be carried out with the aim of enhancing the reactivity of photocatalysts. In addition, it is necessary to explore the possibility of the combined use of titanium dioxide (TiO_2)-based technologies with other technologies (e.g., biological means and electrodynamics) to expand the scope of application. Besides, the final or intermediate products of photocatalytic degradation may not be innocuous substances. The degradation products can be more dangerous than the parent compounds. Harmful by-products may cause decreases in reaction rates and secondary pollution. There has been scant literature on the toxicity of photocatalysts or the overall photocatalytic process, especially for modified photocatalysts [7]. Fundamental characterization and quantification of the reactivity, toxicity, and fate of different modified photocatalysts are needed.

There is a continued need to create improved AOPs and other treatment technologies that can better remove these contaminants, where advanced materials are one of the key elements. In this regard, membranes hold a lot of promise, especially with new research in the development of antifouling and self-healing membranes. One critical aspect to consider with these treatment technologies (especially those utilizing AOPs) is to ensure that human or ecological health are not impacted through the creation of toxic by-products from these processes. It is thus critical to combine target and nontarget chemical analyses with toxicology studies with a variety of endpoints (e.g., endocrine disruption, antibiotic resistance, cytotoxicity, and genotoxicity) that are relevant to potential human health or ecological health effects [79].

Pariatamby proposed that guar gum and xanthan gum are highly recommended as options for treating POPs because they are biodegradable biopolymers, nontoxic, involve low treatment costs, are easily available, and can be produced in abundance [80]. Natural coagulants are better options as compared to chemical coagulants in treating POPs in landfill leachate and agricultural wastewater due to the minimal coagulant dosage requirement, efficiency at low temperature, and the production of only a small volume of sludge. Chemical coagulants are generally more expensive and toxic and the exhibit low biodegradability.

Many POPs such as PAHs, PCDD, and PCDF get trapped in the particulate phase deposited in flue transfer lines and air pollution control systems (equivalent to storage in the memory of a system) and are subsequently released. The so-called memory effect herein is the delayed emission of certain POPs off the particulate media [81]. Memory effect–driven emission is a combination of real time emission and the emission of stored compounds and as such is not a true measure of actual real time emission. Memory effect is now realized to have existed for a long time but was not identified and understood until recently.

PCDD/PCDFs are a family of unintentionally produced POPs that have received considerable public and scientific attention due to the toxicity of some of their congeners, more specifically those with chlorine substitution in the 2, 3, 7, and 8 positions [82]. The environmental management and control of PCDD/PCDFs has been addressed at a global level through the Stockholm Convention which establishes that POPs should be destroyed or irreversibly transformed in order to reduce or eliminate their release into the environment. Several technologies including AOPs such as photolysis, photocatalysis, and Fenton oxidation have been considered as effective methods for destroying PCDD/PCDFs in polluted waters. Nevertheless, during the remediation of wastewaters it is critical that the treatment technologies applied should not lead to the formation of by-products that are themselves POPs, especially if PCDD/PCDF precursors or chlorine are present in the reaction medium. The assessment of transformation by-products and their toxicity as well as the correct selection of operating conditions in extending the understanding of the overall effectiveness of AOPs for pollutant degradation is highly necessary; especially the effect of coexisting organic and/or inorganic species on the generation of transformation products should be taken into account.

Brominated and fluorinated compounds are commonly used in a wide range of consumer goods, and as consumer products reach the end of their useful lives, they ultimately enter waste recycling and disposal systems, particularly municipal landfills [83]. Because of their very slow or lack of degradability, POPs will persist in landfills for many decades and possibly centuries. Over these extended time periods engineered landfill systems and their liners are likely to degrade, thus posing a contemporary and future risk of releasing large contaminant loads into the environment. The challenges posed by POPs in transition/developing countries are from the risk of the increased leaching of POPs from landfills due to climate change, and the possible negative impact by natural attenuation processes.

Due to the climate characteristics, "cold trapping" by the Tibetan Plateau can happen following emission–transport–deposition events, leading to the enrichment of POPs in the environment [84]. The bioaccumulation of DDTs and high chlorinated PCBs have been found in Tibetan terrestrial and aquatic food chains, and newly emerging compounds such as PFASs and hexabromocyclododecanes have been widely detected in wild fish species. The corresponding ecological risks should be of great concern. Climate change such as increased temperatures and the changing coverage of snow and glaciers has the potential to affect the behavior and distribution of POPs.

Luek et al. recently studied POPs in the Atlantic and Southern Oceans and oceanic atmosphere [85]. They found that sample concentrations for most POPs in the air were higher in the northern hemisphere with the exception of HCB, which had high gas phase concentrations in the northern and southern latitudes and low concentrations near the equator. South Atlantic and Southern Ocean seawater had a high ratio of α-hexachlorocyclohexane (HCH) to γ-HCH, indicating persisting levels from technical grade sources. The Atlantic and Southern Oceans continue to be net sinks for atmospheric α-, γ-HCH, and endosulfan despite their declining usage.

While the current generation continues to make efforts to remediate and minimize traditional pollutants in the environment, other "emerging" environmental contaminants are now warranting attention [79]. These include PFCs, nanomaterials, pharmaceuticals, illicit drugs, antibacterials, hormones, flame retardants, disinfection by-products (DBPs), artificial sweeteners, benzotriazoles, 1,4-dioxane, and algal toxins as well as emerging contaminants on the horizon such as prions and ionic liquids. Wastewater effluents are a major source of many of these emerging contaminants due to their use in products that are used every day in households including pharmaceuticals, detergents, fabric coatings, foam cushions, lotions, sunscreens, cosmetics, hair products, foods and beverages, and food packaging. After use, these chemicals are released into wastewater, and because many are incompletely removed in wastewater treatment, they enter rivers and drinking water supplies. Surface run-off and agricultural run-off can also be important sources of their entry into the environment. Moreover, many of these contaminants can transform in the environment from such processes as microbial degradation, photolysis, and hydrolysis, and they can also react with disinfectants in drinking water or wastewater treatment to form DBPs. Issues surrounding these emerging contaminants include their widespread occurrence, bioaccumulation, persistence, and toxicity. Climate change can also serve to exasperate their effects by concentrating them in rivers during times of drought and by causing the resuspension of some (like nanomaterials) during floods. Due to their widespread distribution in various media, the elimination of POPs in the environment will be thus a tough and global challenge with many generations of human beings having to deal with it.

7.3 Degrading Persistent Organic Pollutants by Electrochemical and Photocatalytic Techniques Enhanced With Nanomaterials

7.3.1 Electrochemical Techniques

The electrochemical reduction of halogenated organic compounds is gaining increasing attention as a strategy for the remediation of environmental pollutants. At this moment, there are some rivals to electrochemistry as possible methodologies for the remediation of halogenated environmental pollutants

[86]. With a few references to recent literature, these include biological or microbial, chemical, photochemical, thermal, and mechanochemical strategies and all of these options including electrochemistry appear to have advantages and drawbacks. Microbial degradation has a tendency to be time-consuming and expensive and to afford incomplete dehalogenation. Chemical methods of dehalogenation can involve the use of large volumes of solvents that might be harsh or toxic. Thermal decomposition or incineration of halogenated compounds is wasteful of energy and can produce carbon dioxide and hydrogen halides as by-products.

Electrochemistry is not without its advantages and challenges. Optimally, a lower consumption of energy and time is possible, and desired products might be attainable because of the selectivity offered by the precise control of the potential of an electrode. On the other hand, the development of any successful and cost-effective electrochemical process requires monumental attention to the complexity of the process which ordinarily involves both electron transfer and accompanying chemical events. One must choose appropriate electrode materials, solvents, and supporting electrolytes, along with designing, engineering, and refining large and efficient electrochemical cells (reactors) and performing the often-laborious tasks of product isolation and purification, the recovery and recycling of solvents and electrolytes, and cleaning and preparing the reactors for reuse. Electrochemical oxidation also requires significant energy inputs as well as frequent replacement of the electrodes as a result of corrosion. For some persistent pollutants and/or at high concentrations in wastewater, single electrochemical oxidation can be an inefficient and ineffective removal technology.

In a real situation, the nature of a pollutant (and its matrix) dictates the choices of solvent, supporting electrolyte, electrode material, and procedure [86]. As an example, a nonaqueous solvent is needed for the electroreductive remediation of DDT because of solubility considerations. In addition, this process can be accomplished by direct reduction at a carbon or silver cathode in a divided cell or it can be performed with an electrogenerated complex such as nickel(I) salen. On the other hand, for pesticides in water, electrooxidative remediation can be carried out by means of an EF process. Alternatively, halogenated pollutants in aquatic environments can be extracted into an appropriate organic solvent and recovered as pure liquids or solids, which can then be subjected to electrochemical reduction.

PCBs can be strongly adsorbed by soils and sediments. There is a need to develop new and cost-effective solutions for the remediation of PCB-contaminated soils. Suspended electrodialytic remediation combined with nZVI could be a competitive alternative to the commonly adapted solutions of incineration or landfilling. Surfactants can enhance PCB desorption, dechlorination, and contaminated soil cleanup. The electrodechlorination of PCB with surfactants and nZVI showed encouraging tendencies, where the use of a direct current allowed the highest PCB removal rates [87]. Encouragingly, electrodialytic remediation of sediments was found to be effective for the simultaneous removal of heavy metals and organic pollutants in sediments from Arctic regions (Sisimiut in Greenland and Hammerfest in Norway) [88].

The mercury (Hg) and Hg/PAH removal effectiveness of several EK decontamination treatments including the combined use of ecofriendly enhancing agents, was investigated by Falciglia et al. as a potential remedial treatment of heavily contaminated marine sediments [89]. Similar to Hg, because of their high hydrophobicity, PAHs strongly interact with sediments and organic matter, being easily adsorbed onto grain surface, and due to their long-term effects, persistence, and potential mutagenic effects on human health, they have attracted particular concern. The EK technique involves the application of an electrical difference potential between couples of electrodes inserted across a confined contaminated area. In the presence of an electric field, the mobilization of contaminants through the media takes place.

In recent years, new AOPs based on electrochemical technology, so-called electrochemical advanced oxidation processes (EAOPs), have been developed for the prevention and remediation of environmental pollution, especially focusing on water streams [90]. These methods are based on the electrochemical generation of a very powerful oxidizing agent such as •OH in solution, which is then able to destroy organics up to their mineralization. EAOPs include heterogeneous processes like anodic oxidation and photoelectrocatalysis (PEC) methods in which •OH are generated at the anode surface either electrochemically or photochemically, and homogeneous processes like EF, photoelectro-Fenton (PEF), and sonoelectrolysis in which •OH are produced in a bulk solution. Two main challenges should be prioritized, namely (1) the drop in electrode prices, particularly BDD, and (2) the use of renewable energy sources to power these processes, thus enhancing the sustainability of all these EAOPs. The most popular technique among them is the EF process in which H_2O_2 is generated at the cathode with O_2 or air feeding while an iron catalyst (Fe^{2+}, Fe^{3+}, or iron oxides) is added to the effluent.

Ganiyu et al. summarized the fundamental principles and applications of heterogeneous electrochemical wastewater treatment based on Fenton's chemistry reaction [91]. The advantages of using insoluble solids as heterogeneous catalysts rather than soluble iron salts (heterogeneous EF process) were highlighted. The dissolution/excessive leaching of the catalyst at strong acidic pH are the most detrimental challenges that affect both the catalytic stability and reusability efficiency of heterogeneous catalysts.

Ren et al. developed a novel combined EF system for the treatment of wastewater containing biological recalcitrant using electric field–induced nanoscale CeO_2 as a synergistic catalyst for H_2O_2 generation [92]. Its current efficiency was greatly improved when compared with the traditional EF system. The remarkable efficiency of the CeO_2 EF system was caused by the generation of •OOH through the lattice oxygen and strong Brønsted acid sites on the sulfate-functionalized CeO_2 with an electric field–induction effect. The excellent performance of the CeO_2 EF system in the decomposition and mineralization of organic pollutants indicates its promising application.

Popescu et al. studied the EF process with different carbonaceous cathode configurations including taffeta carbon fiber, unidirectional carbon fiber (UCF), and graphite felt [93]. They concluded that UCF provided the best performance.

The main drawbacks of conventional regeneration methods could be avoided using the EF process [94]. In comparison to chemical oxidation, a much higher regeneration effectiveness of a microporous adsorbent can be achieved. This process can also mineralize organic pollutants, while thermal treatment under nonoxidizing conditions only leads to the desorption of organic compounds. Moreover, the adsorption capacity of AC tissue is not affected, contrary to chemical oxidation and thermal treatments under oxidizing conditions. The selection of AC tissue as an adsorbent is crucial for the effectiveness of the process because this material presents suitable features for both the adsorption and regeneration steps. One of the main drawbacks of EAOPs for the removal of low concentrations of organic compounds is the loss of efficiency caused by mass-transfer limitations. Therefore combining a preconcentration step using AC tissue is a promising way to improve the cost-efficiency of these processes.

Gozzi et al. treated single and mixed pesticide formulations by solar PEF using a flow plant [95]. Under comparable conditions, solar photoelectro-Fenton (SPEF) was much more powerful than AO-H_2O_2 for the remediation of wastewater polluted with single and mixed herbicides because it yielded a quicker substrate decay and greater mineralization, ending in a higher current efficiency.

Combined EF methods have also been reported for wastewater remediation [96]. These include peroxi-coagulation, PEF, SPEF, photoperoxi-coagulation, photoelectrochemical electro-Fenton, sonoelectro-Fenton, and cathodic generation of Fe^{2+} as well as electrochemical and combined Fenton processes with H_2O_2 added to the solution or produced indirectly including Fered-Fenton, electrochemical peroxidation (ECP), anodic Fenton treatment (AFT), photo-combined methods, bio-ECP remediation, and plasma-assisted treatments. EF and related EAOPs have been developed to provide alternative, clean, and effective technologies to be applied when conventional processes become ineffective, as is the case for wastewater containing refractory POPs. Four basic types of EAOPs based on Fenton's reaction chemistry, namely EF, Fered-Fenton, ECP, and AFT, have been by far the most applied processes. Recent work has shown that the alternative use of sunlight ($\lambda > 300$ nm) as a free and renewable energy source in the so-called solar photo-Fenton process is also useful for wastewater remediation. In fact, the determination of a parameter called the "environmental-economic index" has shown that solar driven photo-Fenton is preferable to lamp-driven photo-Fenton if both environmental and economic aspects must be satisfied in a well-balanced manner.

With the development of the degradation of POPs in wastewater, antibiotics have been considered as emerging pollutants [97]. Their lasting effects on entire ecosystems have directly resulted in irreversible damage all over the world. EAOPs including SPEF processes have been adopted toward the elimination of antibiotics for their considerable efficiency.

Bio-Fenton (BF) and bio-electro-Fenton (BEF) are new sustainable methods for the treatment of POPs from many sources of wastewaters [98]. They have been proven to be efficient in various pollutant degradation processes, along with reducing the costs and improving the safety conditions of the working environment. For the BF process, biomaterials are used to produce H_2O_2 in situ, which reduces the danger of stocking and transferring. Meanwhile for the BEF process, the production of electricity from microbial activity is the main advantage and it provides a proper solution for the high power consumption in EF-based systems without compromising the degradation efficiency of organic pollutants. However, the main challenge in these systems is the low power density generated in the anodic compartments to stimulate the Fenton's reaction in the cathodic compartments. Also, the stability and lifetime of these systems are still questionable.

Electrooxidation processes are promising options for the removal of organic pollutants from water [99]. The major appeal of these technologies is the possibility to avoid the addition of chemical reagents. However, a major limitation is associated with slow mass transfer which reduces the efficiency and hinders the potential for the large-scale application of these technologies. Therefore improving the reactor configuration is currently one of the most important areas for research and development. The recent development of a reactive electrochemical membrane (REM) as a flow-through electrode has proven to be a breakthrough innovation, leading to both a high electrochemically active surface area and convection-enhanced mass transport of pollutants. Recent advances in the development of substoichiometric titanium oxide REMs as anodes have been made. These electrodes possess high electrical conductivity, reactivity (generation of •OH), chemical/electrochemical stability, and suitable pore structure, which allows for efficient mass transport. Further development of REMs strongly relies on the development of materials with suitable physicochemical characteristics that produce electrodes with efficient mass transport properties, high electroactive surface area, high reactivity, and long-term stability. Nanostructured electrodes could meet these criteria with great flexibility.

Various studies on landfill leachate treatment by electrochemical oxidation have indicated that this process can effectively reduce two major pollutants present in landfill leachate: organic matter and ammonium nitrogen [100]. In addition, the process is able to enhance the biodegradability index of landfill leachate, which matures or stabilizes landfill leachate sufficiently for biological treatment. The elevated concentrations of ammonium nitrogen, especially observed in bioreactor landfill leachate, can also be reduced by electrochemical oxidation. The pollutant removal efficiency of the system depends upon the mechanism of oxidation (direct or indirect oxidation), which depends upon the property of the anode material. Applied current density, pH, type and concentration of electrolyte, interelectrode gap, mass-transfer mode, ratio of total anode area to volume of effluent to be treated, temperature, flow rate or flow velocity, reactor geometry, cathode material, and lamp power during photoelectrochemical oxidation may also influence the performance of the system. Electrochemical oxidation can be employed as a complementary treatment system to biological processes for conventional landfill leachate treatment as well as a standalone system for ammonium nitrogen removal from bioreactor landfill leachate.

Mixed metal oxides (MMOs) have been extensively employed for heterogeneous catalysis. In recent years, MMOs have received intensive interest as anode materials for the electrochemical treatment of wastewaters that contain recalcitrant organics [101]. Intensive research efforts have been devoted to the surface modification of MMO anodes, with the aim of increasing the surface area or introducing expected properties into the MMO layer. Such approaches include nano- and microstructure deposits, element doping, and polymer composites. Compared with single-component metal oxide anodes, they exhibit better electrocatalytic activity resulting from an increase of active sites and a change in the oxidation states of the metal ions.

The effect of electrode materials on the abatement of lindane by electrooxidation processes has been studied [102]. Comparative performances of different anodic [platinum (Pt), dimensionally stable anode, and boron-doped diamond (BDD)] and cathodic [carbon sponge, carbon felt (CF), and stainless steel] materials on lindane electrooxidation and mineralization were investigated. Special attention was paid to determine the role of chlorine active species during the electrooxidation process. The results showed that better performances were obtained when using a BDD anode and CF cathode cell. The formation of chlorinated and hydroxylated intermediates and carboxylic acids during the treatment were identified and a plausible mineralization pathway of lindane by hydroxyl radicals.

Dominguez et al. performed an effective removal of recalcitrant organochlorine pesticides including HCH present in real groundwater coming from a landfill of an old lindane (g-HCH) factory by electrochemical oxidation using a BDD anode and a CF cathode [103]. The oxidation rate of chloride (and its oxidized intermediates) depends on the applied current value. Although some of the species generated from them are active oxidants, the presence of inorganic salts is detrimental to the efficiency of the electrochemical process.

PFASs have recently been listed as emerging contaminants and POPs due to their human and environmental health concerns [104]. In the past 10 years, their detection and remediation have progressed significantly. Due to the persistence of PFASs, other remediation approaches may experience difficulty in degrading them, whilst EAOPs have demonstrated success. Traditional AOPs require chemicals or other components (such as UV or catalysis) to generate radicals, which usually means harsh conditions and a low environmental compatibility. EAOP overcomes some of these limitations because radicals can be generated in situ via electrochemistry, which means the oxidation process can be driven by electricity rather than by chemicals to produce radicals. The low consumption of chemicals in EAOP promises a more environment-friendly approach.

Over the past decades, research efforts have been made toward developing more effective technologies for the remediation of waters containing POPs. Among the various technologies, so-called EAOPs have gained increasing interest. These technologies are based on the electrochemical generation of strong oxidants such as •OH. There are five key EAOPs, that is, anodic oxidation (AO), anodic oxidation with electrogenerated H_2O_2 (AO-H_2O_2), EF, PEF, and SPEF, alone and in combination with other methods like biological treatment, electrocoagulation, coagulation, and membrane filtration processes. Meanwhile, the fundamentals of each EAOP and their performance in treating wastewaters have been also summarized by Moreira et al. [100].

Gomez-Ruiz et al. reported the electrochemical treatment of PFASs in effluent from an industrial wastewater treatment plant with a removal rate of 99.7% for practical PFASs (1652 $\mu g\ L^{-1}$) [101]. While most previous research has focused on the electrochemical degradation of PFOA and PFOS in model solutions, this work dealt with the simultaneous removal of 8 PFASs at environmentally relevant concentrations in real industrial emissions, which also contained organic matter and inorganic anions.

Soriano et al. developed a strategy for the removal and degradation of PFHxA from industrial process waters at concentrations in the range of 60–200 $mg\ L^{-1}$ [102]. The treatment train consisted of nanofiltration (NF) separation followed by the electrochemical degradation of the NF concentrate. They highlighted the potential of combining membrane separation and electrochemical oxidation for the efficient treatment of PFAS-impacted waters.

Zr-doped nanocrystalline PbO_2 (Zr-PbO_2) film electrodes were prepared at different bath temperatures and introduced to sequentially treat PFOA wastewater [103]. The removal efficiency and defluorination ratio of PFOA reached 97.0% and 88.1% after 90 min of electrolysis, respectively, demonstrating the promising application of the sequential treatment system for the treatment of PFOA wastewater.

The removal of persistent organic contaminants by electrochemically activated sulfate could be achieved due to the formation of strong sulfate-derived oxidant species at BDD anodes when polarized at high potentials [104]. The sulfate yielded electrooxidation rates 10 – 15 times higher for all target contaminants than those of nitrate anolyte. This may have positive implications in the electrooxidation of wastewaters containing sulfate.

The application of heterogeneous catalysis to EAOPs is especially useful due to its efficiency and environmental safety [105]. Among those EAOPs, EF stands out as one in which heterogeneous catalysis has been broadly applied. Recent advances highlight the use of different catalysts such as iron minerals (pyrite, magnetite, or goethite), prepared catalysts by the loading of metals in inorganic and organic materials, NPs (carbon nanotubes and derivatives, aerogel, reduced graphene oxide (RGO), and hierarchical CoFe-layered double hydroxide), and the inclusion of catalysts on the cathode.

7.3.2 Photocatalytic Techniques

Conventional Fenton and photo-Fenton reactions suffer many of the same limitations as hydrogen peroxide oxidation and UV photolysis. There are significant expenses associated with the energy required to manufacture hydrogen peroxide and the energy needed to power UV lights. Especially, the use of Fenton reagents requires the handling of highly reactive materials. The use of sunlight to produce •OH radicals for the photocatalytic treatment of various organic pollutants has been attempted by many countries [106].

Lindane is a highly persistent chlorinated pesticide and a potent endocrine disruptor. The strong electrons with the drawing property of chlorine atoms result in a relatively low reactivity of lindane with •OH in conventional AOPs. Khan et al. reported the kinetics and mechanism of sulfate radical- and hydroxyl radical-induced degradation of the highly chlorinated pesticide lindane in a UV/peroxymonosulfate system [107], which can generate both •OH and $SO_4^{\bullet -}$. Compared to common hydroxyl radical reactions, $SO_4^{\bullet -}$ exhibited a slightly higher reactivity toward lindane. Various degradation by-products as identified by gas chromatograph and mass spectra revealed dechlorination, chlorination, dehydrogenation, and hydroxylation to be potential transformation steps. Ring opening and cleavage could also be achieved as demonstrated indirectly by the significant decrease in total organic carbon.

Photocatalysis is often not effective in the long term and is limited by water chemistry (e.g., hardness) and the presence of cocontaminants. Although some nanomaterials can easily degrade POPs photocatalytically, more efforts are required in order to achieve efficient technology that can be employed at an industrial scale. Canle et al. suggested that more research is needed to design efficient nanostructured photocatalysts able to be visible light and sunlight-responsive materials [108]. The photodegradation pathways are strongly dependent on the photocatalyst and, usually, complete mineralization is not achieved. Much research is needed on visible light–harvesting photocatalysts with enhanced electron–hole pair lifetimes, which are thermally stable, reusable, and able to be supported on a surface. On the other hand, extensive research should be conducted on the degradation pathways involved in the photocatalyzed degradation of POPs, on the identification and estrogenecity/toxicity of photogenerated intermediates, and on the optimal operational parameters to achieve the greatest mineralization.

Chen et al. first attempted a method using a RGO/$CoFe_2O_4$/Ag nanocomposite-based photocatalyst toward SCCPs degradation [109]. Xiong et al. further developed the technique using well-designed polydopamine and Ag NPs toward SCCPs degradation and revealed the mechanism for the degradation pathways through density functional theory calculation [110].

Zhao et al. reported that decabromodiphenyl ether (BDE-209) undergoes efficient reductive debromination reactions under visible light irradiation ($\geq$420 nm) in the presence of various carboxylate anions that are common in environmental media [111]. The debromination reactions occur in a stepwise manner, producing a series of lower brominated PBDE congeners. The formation of a halogen-binding-based complex between PBDE and carboxylate enables the visible light absorption and debromination of PBDEs; although neither PBDEs nor carboxylates have visible light absorption.

Wang et al. investigated the photodebromination behaviors of PBDEs in methanol or methanol/water systems [112]. The kinetics of three sets of BDE isomers showed that the PBDE isomers with lower energy of lowest unoccupied molecular orbital and higher energy of highest occupied molecular orbital will be degraded faster by ultraviolet light than other BDE isomers. The bromine substituents with higher mulliken charges were preferentially removed, as pointed out by those authors.

Highly brominated PBDEs are easily photocatalytically reduced to lower brominated congeners, and lower brominated congeners are relatively susceptible to photocatalytic oxidation but not to further reduction. Lei et al. developed a one-pot photocatalytic consecutive reduction and oxidation method

(O-CRO) using RGO/TiO_2 containing a small amount of methanol for the complete debromination of 2,2′,4,4′,-tetrabromodiphenyl ether (BDE-47) [113]. It was observed that BDE-47 was resistant to photocatalytic reduction on TiO_2, but underwent a slow oxidation in aerobic aqueous solutions. This system yielded the complete debromination and mineralization of BDE-47 within 14 h. In this O-CRO system, the photogenerated electrons on TiO_2 could not directly reduce BDE-47, but the accumulated electrons on the RGO layer initiated a rapid reduction of BDE-47 into tri-BDEs, which were immediately oxidized into CO_2 and Br^- by trapping photogenerated holes on TiO_2. The significantly improved debromination in the O-CRO system was attributed to the fact that BDE-47 was easily reduced to tri-BDEs, which are susceptible to oxidation but not further reduction.

Chen et al. reported a series of graphene oxide (GO)/Ag_3PO_4 composites synthesized using ion exchange–chemical precipitation method with different GO contents, and the photocatalytic reduction degradation effect on BDE-209 of the nanocomposites was studied [114]. When the content of GO was 7%, the degradation efficiency of BDE-209 could reach 97.33% in anoxic water with methanol as an electron donor after 8 h of visible light irradiation, being three times that achieved by pure Ag_3PO_4, and the BDE-209 reduction generated 3Br-8Br PBDEs congeners. The photocatalytic degradation of BDE-209 through a reduction mechanism was ascribed to the multielectron reduction process. GO not only trapped electrons to improve the charge separation on Ag_3PO_4, but also fleetingly transferred the accumulated electrons to BDE-209.

Shao et al. reported the enhanced photoreduction degradation of polybromodiphenyl ethers with Fe_3O_4-g-C_3N_4 under visible light irradiation (>420 nm) [115]. A series of high-activity Fe_3O_4-g-C_3N_4 (named FeOCN-x) photocatalysts have been synthesized by an in situ growth method. The FeOCN-x nanocomposites not only exhibited good photostability, but could also be easily recovered by magnetism. Amongst hybrids, FeOCN-4 with a 4 wt.% Fe_3O_4 content shows the highest reaction rate, with a rate up to 6.7 times higher than that of pure g-C_3N_4.

Compared with photooxidation, photoreduction appears to be more suitable for PFOX (X = A or S) removal since it is more favorable for the defluorination of PFOX and further complete mineralization [116]. The key parameters in the photocatalytic degradation and defluorination process of PFOX are light wavelength, photocatalyst concentration, initial PFOX concentration, pH, reaction atmosphere, temperature, and coexisting organic or inorganic matter. The development of new integrated or coupling systems for the enhanced photomineralization of PFOX such as in situ photodegradation systems and photoelectrochemical techniques, among others, is encouraged.

Nanostructured modified TiO_2, In_2O_3, and Ga_2O_3 are potent catalysts for the photoinduced degradation of PFOX [117]. The decomposition performance of PFAS through heterogeneous photocatalysis with these catalysts in water follows the order: $In_2O_3 > Ga_2O_3 > TiO_2$. In_2O_3 porous nanoplates were found to have the best performance with 100% PFAS decomposition comparatively under UV light under the same conditions.

Overall, redox strategies including photocatalysis, sonochemical, electrochemical, radiochemical, thermochemical, subcritical, and plasma treatment processes, are becoming promising technologies for PFOA removal due to their ability for complete mineralization [118]. Photocatalytic deposition has proven to be an inexpensive, effective, and sustainable technology for the removal of PFOA in the aqueous phase. Since the C – F bond of fluorocarbon is one of the strongest bonds in nature (approximately 110 kcal mol^{-1}), the most conventional technologies are ineffective for PFOA degradations. GO/TiO_2 nanotube arrays have been developed by Park et al. for the photocatalytic degradation of PFOA. In the photoelectrocatalytic process, at a 2.0 V external bias, 97% of PFOA (concentration scale of $\mu g\ L^{-1}$) was degraded in 4 h.

As a class of recently developed versatile porous materials, MOFs have shown huge potential and a bright perspective in the adsorptive removal and photocatalytic degradation of POPs for water remediation [119]. A different strategy is that rather than applying adsorption and photocatalysis separately, adsorption and photocatalysis can be integrated together for the effective treatment of POPs with less energy consumption using MOFs. There are three distinct advantages for the integration of adsorption and photocatalysis using MOFs. First, the porous structure and the volume of micro/mesoporosity of MOFs can adsorb target pollutants fast and increase the local concentration of substrates on the surface, resulting in the promotion of the subsequent photocatalytic degradation. Second, the photocatalytic active sites in MOFs can aid the generation of reactive oxygen species (ROS), which chemically attack and break bulky pollutant molecules down into smaller ones, which can then access the pores of MOFs readily, and thus enhance the accessibility of MOFs to pollutants. Third, owing to their potential as both adsorbents and photocatalysts, uniform bifunctional MOFs can behave as an adsorbent for the accumulation of trace amounts of POPs, and consequently as a photocatalyst in the degradation of POPs under irradiation, meanwhile the MOF itself is regenerated for the cyclic adsorption/photodegradation of POPs, if complete mineralization was achieved. With the effective photodegradation of organic pollutants, MOFs are also regenerated simultaneously and thus recycled.

Sonophotocatalysis (SPC) is considered to be one of the most important wastewater treatment techniques and hence has attracted the attention of researchers to eliminate recalcitrant hazardous organic pollutants from aqueous phase [120]. In general, SPC refers to the integrated use of ultrasonic sound waves, ultraviolet radiation, and the addition of a semiconductor material which functions as a photocatalyst. Current research has led to numerous improvements to the SPC-based treatment by adding visible light irradiation, nanocomposite catalysts, and numerous catalyst supports for better stability and performance. A higher synergistic pollutant removal rate has been reported during SPC treatment as compared to individual methods. The application of SPC-based treatment methods enhance the efficiency of the removal of pollutants in aqueous media and the doping of catalysts along with the simultaneous use of ultrasound seems to resolve the limitations related to rapid photogenerated electron–hole recombination, clogging of the photocatalyst surface, among others. The evaluation of a successful SPC-based treatment method should be decided based on the mineralization efficiency rather than the extent of decolorization or degradation of a recalcitrant pollutant. SPC-based treatments should be explored for the mineralization of scarce water volatile POPs.

7.3.3 Synergistic Methods

Beyond conventional EAOPs, novel combined methods involving photocatalysis, adsorption, NF, microwave, and ultrasound, among others and the use of microbial fuel cells have been explored to deal with nonreadily biodegradable and toxic pollutants [121]. PEC has emerged as a promising and powerful EAOP with a combination of photocatalytic and electrolytic processes [122]. It involves the promotion of electrons from the valence band to the conduction band of a semiconductor photocatalyst upon light irradiation, with the production of positive holes. The fast recombination of the electron–hole pairs formed is avoided in PEC by applying an external bias potential to the photocatalyst that extracts the photogenerated electrons up to the cathode of the electrolytic cell [123]. By increasing the applied potential, different synergetic mechanisms occur for the degradation of organic contaminants. The development of UV-responsive and visible light–responsive photoanode materials and the application of PEC in the removal of organics such as dyes, pharmaceuticals and personal care products, and endocrine disrupting chemicals are attracting growing research interest. Organics can be oxidized directly by the holes, •OH formed from water oxidation with the holes, and other ROS produced between the electrons and the dissolved O_2.

PEC involves a photocatalytic system to which an external positive bias is applied, which can significantly increase the rates of photocatalytic reactions by driving the photogenerated electron–hole pairs in opposite directions, reducing their recombination rates [124]. The design of a cost-efficient photoelectrocatalytic reactor plays a critical role in the ultimate acceptance of this promising technology in industry for environmental remediation as well as other applications. Due to the limitations of the slurry system, an immobilized photoelectrode may still be the major option for reactor design. In our opinion, the ideal reactor configuration should make use of a highly active and visible light–active photocatalyst, allow for the efficient absorption of photons by the photocatalyst, have a photoelectrode that effectively suppresses the recombination of photogenerated electron–hole pairs, boast a high mass-transfer capacity, and be easy to scale up, fabricate, and operate.

PEC takes advantage of the heterogeneous photocatalytic process by applying a biased potential on a photoelectrode in which the catalyst is supported [125]. This configuration allows for a greater effectiveness in the separation of photogenerated charges due to light irradiation with energy being higher compared to that of the band gap energy of the semiconductor, which thereby leads to an increase in the lifetime of the electron–hole pairs. Nanostructured photoanodes have mainly two advantages over thin films, namely (1) a high surface area (this improves reaction/interaction between the material and interacting media) and (2) efficient charge transfer, thereby reducing electron–hole recombination The main disadvantage of photoelectrocatalytic treatment lies in efficiency losses in highly concentrated matrices (mainly highly colored media) due to light absorption by the medium resulting in the reduction of photoanode activation. Sometimes it is necessary to align it with other techniques such as physical treatments (to remove matter particles), biological treatment, or other AOPs. Finally, it is worth pointing out the relevant nature of the development of efficient reactors capable of working on a scale-up level in industrial plant treatment.

PEC technologies have received particular attention due to their potential and effectiveness in the photodegradation of refractory organic and microbial pollutants present in water and wastewater [126]. TiO_2 plays the most important role compared to other semiconductor photocatalysts (CdS, ZnO, SnO_2, etc.) due to its excellent chemical and physical properties [127].

The recent advancements in synthetic processes and photoelectrocatalytic applications continuously increase the understanding of the interaction mechanisms among the constituents of composites [128]. Detailed study and optimization of the features related to the interface are still very much sought after to efficiently design photocatalysts with the aim of their eventual commercialization. It is imperative to fully understand the interaction mechanisms between the important two-dimensional (2D) nanomaterial, graphene (GR) and inorganic semiconductors. The photocatalytic activity of such composites depends not only on high GR electron mobility and charge transfer, but also on the properties of the interface (such as interface morphology, size, crystal phases and facets, and dimensionality of composites). The dimensionality of composites is also important to their performance; the higher the dimensionality of the semiconductor, the higher the interfacial contact area and the better the electron charge transfer, resulting in an enhanced photocatalytic activity. Three-dimensional structures can also help to increase specific surface area, thus improving photocatalytic performance by generating multidimensional electron transfer pathways and elongating the lifespan of electron–hole pairs.

The architecture of nanomaterials can directly control both photon absorption and electrochemical catalysis [129]. The nanoarchitecture has a larger impact on the electrocatalytic properties of the photoelectrode than its light-harvesting capability, but this feature is usually neglected. In fact, designing an appropriate nanoarchitecture is a tuning process to balance a series of different processes, which are in competition in a complicated system, for optimizing the PEC performance. The most important tasks are maximizing the number of electrocatalytic sites and forming a more effective electrode/electrolyte interface while reducing the number of charge recombination centers. Nanostructuring is normally in favor of the former while unfavorably supporting the latter.

Nanostructuring is now widely used to overcome the trade-off between optical absorption and minority-carrier diffusion lengths in photoelectrochemical electrodes, and to increase catalytic activity through an increased surface area [130]. For Fe_2O_3 photoanodes, nanostructuring has afforded an increase in performance by a factor of 10 compared to the single-crystal film-based electrode. The best-performing metal-oxide photoanodes reported so far are nanostructured WO_3–$BiVO_4$ heterojunctions. In general, however, a higher surface area leads to lower intrinsic photovoltage and increased surface recombination, which are both important loss mechanisms. To minimize surface recombination, it may be sufficient to nanostructure only the cocatalyst, especially if the system is limited by surface reaction kinetics. Moreover, *nanostructuring is not imperative for all semiconductors* utilized in PEC because it only benefits materials for which the carrier diffusion length falls short of the optical absorption depth. These properties should thus be well established before any attempts at nanostructuring a particular semiconductor are made.

The effect of various affective factors such as photoanode type, light source and its intensity, pH solution value, type and concentration of supporting electrolyte, type of cathode electrode, the choice of photoanode or solution, thicknesses of semiconductor film on the electrode surface, and applied potential to the destruction of pollutants was discussed by Zarei et al. [131]. It should be mentioned that TiO_2 nanotube arrays are widely used in the photoelectrocatalytic degradation of pollutants due to their excellent physical and chemical properties. Quan et al. used an electrochemically assisted photocatalytic method for the degradation of pentachlorophenol in an aqueous solution using a TiO_2 nanotube film electrode as well as UV irradiation [132]. A degradation rate of 82% was obtained after 2 h of treatment.

Shen et al. reviewed the design and modification of TiO_2 nanostructures as photoelectrode materials for photoelectrochemistry applications [133]. Various kinds of TiO_2 nanostructures including nanoparticulate, one-dimensional (1D) (nanowires, nanorods, nanotubes), 2D (nanobelt, nanoribbon, nanosheet), 3D (meso/nanoporous, branched nanostructures, etc.), and crystal-facet tailored TiO_2 nanostructures were surveyed. Surface modification approaches (surface disordering, passivation, and decoration) to TiO_2 nanostructures are highlighted for developing efficient TiO_2-based PEC cells. Though significant advances have been achieved with plasmonic Au and Ag nanostructures on TiO_2 nanostructured photoelectrodes, the usage of noble metals in this field could be a serious issue; therefore further developing plasmonic nonnoble-metal nanostructures (such as in doped SnO_2, Cu, and Al) is in high demand. 1D and 2D TiO_2 nanostructures are well known for their direct charge-transfer pathways; whereas 3D TiO_2 exhibit the combined advantages of adequate specific surface areas and excellent charge-transfer ability. TiO_2 photoelectrodes with exposed highly reactive facets possess promoted charge separation, thus enhanced efficiencies for PEC applications. Different approaches for sensitizing various TiO_2 nanostructures to the visible light region were also elaborated and discussed. Doping foreign elements into TiO_2 nanostructures induced additional impurity levels within the bandgap of TiO_2 for extended optical absorption. Surface disorder and passivation engineered TiO_2 nanostructures showed promoted surface and/or bulk charge separation, leading to improved PEC activities. Nanocarbon, QD materials, and plasmonic metal decoration efficiently improved the optical absorption and/or charge separation of TiO_2, resulting in higher PEC performances in the UV and/or visible light regions. Although significant advances have been made in TiO_2 nanostructures for various PEC applications, high-efficiency TiO_2-based PEC systems are still limited and there still lacks sufficient understanding of the intrinsic relationship between TiO_2 nanostructures and their PEC performances.

Nair et al. recently explored the destructive removal of 2,4-dichlorophenol and 2,4-dichlorophenoxy acetic acid from water through the wet peroxide oxidation method at ambient conditions using cobalt substituted zinc ferrite nanocomposite catalysts [134]. The effect of the reaction variables like reaction time, pollutant concentration, catalyst composition and its dosage, temperature, and oxidant concentration on the removal efficiency was optimized. $Co_xZn_{1-x}Fe_2O_4$ was found to be an effective catalyst for the destructive removal of pollutants and the composite, $CoFe_2O_4$, showed the highest activity.

Two-dimensional graphitic materials have started being applied in PEC in the past decade; it will be still a popular research area in the near future [135]. To date, the main 2D graphitic materials applied in PEC are GR, layered transition–metal dichalcogenides, and g-C_3N_4. Electrodes applied in PEC are generally composed of photocatalyst(s) and cocatalyst(s) deposited on a conductive glass or metal mesh. In this case, various photocatalytic materials have been extended into the application of PEC. TiO_2, as one of the most popular photocatalysts, has been extensively studied for PEC as the photoanode. Other visible light–responsive materials such as bismuth-based semiconductors have also been implemented in PEC. However, one-component photoanode was generally with high recombination rate of photogenerated charge carriers, and proper modifications could help improve the PEC activities. One of the approaches is to combine with a cocatalyst. Graphitic materials can be promising candidates as cocatalysts to improve PEC activity, with superiorities compared to other materials. Graphitic materials with 2D structures exhibit excellent electrical conductivity and may serve as an electron transporter to improve the separation of photogenerated charge carriers and in turn improve PEC activity. These superiorities include the high electrical conductivity of 2D graphitic materials, which can make them an excellent electron transporter, so as to reduce the recombination of photogenerated charge carriers.

Interfacial charge transfer (IFCT) plays a significant role in electron–hole separation considering the energy barriers of the energy levels at semiconductor–semiconductor, semiconductor–metal, semiconductor–molecule, and semiconductor–electrolyte interfaces [136]. Both electron and hole transport across the interface via IFCT, with comparable rates, are important for maintaining the enhanced photocatalytic efficiency and stability of the catalysts. The catalytic performance of photocatalysts can be further controlled by tuning these IFCT processes. Hence an in-depth analysis of such processes is necessary for the advancement of photoelectrochemical and related PEC applications. The development of photocatalysts with precise morphology, size, and directional growth leading to faceted surfaces may yield efficient photoelectrochemical devices. Most PEC cells are designed for either electricity or hydrogen production as important means of solar energy conversion [137]. Water splitting by oxidation, reduction, or both as well as the oxidation of organic or inorganic wastes is the main target of PEC cells. The combination of a PEC cell with a commercial photovoltaic cell is preferable in terms of construction cost.

7.4 Perspective on Developing Efficient Nanomaterials for Removing Persistent Organic Pollutants

Some challenges such as high cost, poor separation performance, and environmental risks are still impeding the engineering application of nanomaterials in removing POPs. Based on the current knowledge of the working principles of catalytic materials for environmental purification (Fig. 7.2), the key sessions are directly correlated with the efficacy of nanomaterials

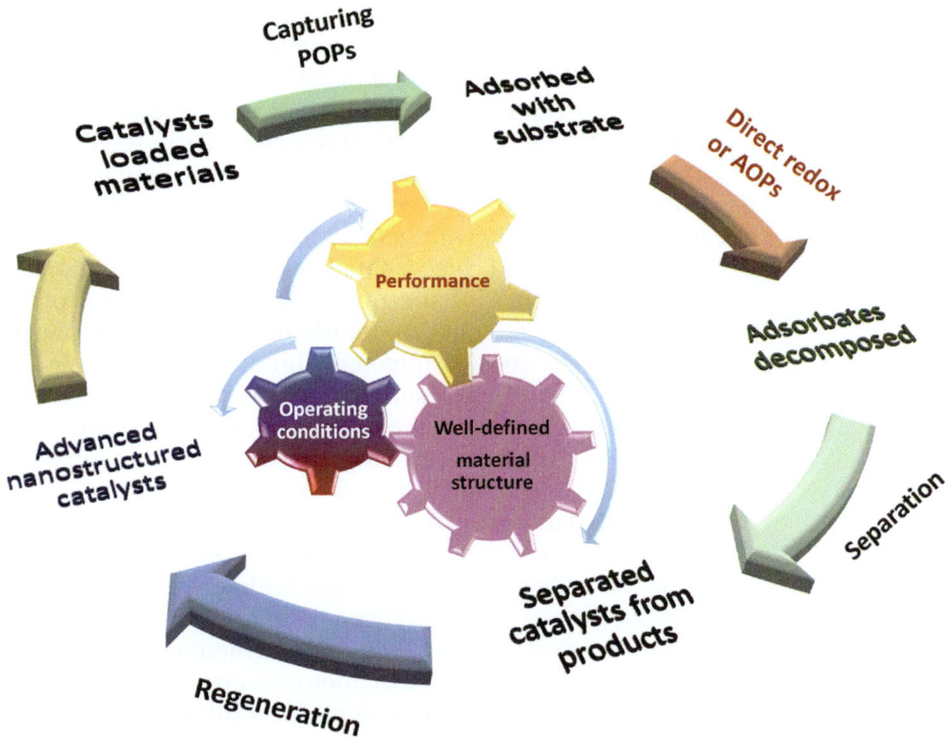

Figure 7.2 Schematic illustration of key operation sessions concerned with (nano)catalysts toward efficiently removing POPs from the environment.

under certain operating conditions. Some important designing aspects are intuitive such as an advanced structure facilitating the adsorption of targeted POPs, interfacial electron transfer, the formation of an optimum band structure for energy conversion, and the coordination of surface reactions with products detached from the catalyst.

For further performance improvement with the aim of developing efficient, robust, and economically viable photo(electro) catalytic systems, there is a ubiquitous need for the surface modification of semiconductor particles with other aid agents (such as plasmonic metals or cocatalysts) in order to increase light absorption efficiency, decrease electron – hole recombination, and enhance stability in humid or liquid environments. Notably, these modifications are usually performed at the nanoscale, thus leading to an increased structural complexity of each individual particle, which in turn exacerbates the heterogeneity issue. Hence it is even more difficult to decipher the charge carrier properties of these surface-modified semiconductor particles.

Understanding the fundamental properties of charge carriers on the surface of semiconductor photo(electro)catalysts is key to the rational design of efficient photo(electro)catalytic devices for sunlight-driven energy conversion. Establishing quantitative relations between the desired functions of photo(electro)catalysts and their surface charge carrier activities can provide guiding principles for rationally engineering photo(electro)catalytic systems, for example, with cocatalysts, for a broad range of applications. To this end, high spatial resolution information is always desirable because of the ubiquitous heterogeneity of semiconductor particles. Direct, operando functional assessments of their performances toward targeted photo(electro)catalytic processes through single- and subparticle photocurrent measurements have been recently highlighted by Hesari et al. [138].

Due to the inherent complexity of environmental problems, especially water and air pollution, single-function environmental nanomaterials used in conventional and unconventional environmental treatment technologies are gradually reaching their limits. Intelligent nanomaterials with environmentally responsive functionalities have shown potential to improve the performance of existing and new environmental technologies [139]. Through rational design of their structures and functionalities, intelligent nanomaterials can perform different tasks in response to varying application scenarios for the purpose of achieving the best performance. The latest research progress has envisioned intelligent environmental nanomaterials in filtration membranes with responsive gates, materials with switchable wettability for selective and on-demand oil/water separation, and environmental materials with self-healing capabilities.

Beyond the mentioned POPs, there has been a considerable increase in research on the ecological consequences of artificial microplastics released into the environment, but only a handful of works have focused on the nanosized particles of polymer-based materials [140]. Though their presence has been difficult to adequately ascertain due to the inherent technical difficulties of isolating and quantifying them, there is an overall consensus that they are not only present in the environment, but also pose a significant threat to the environment and human health. The reduced size of these particulates ($<1\ \mu m$) makes them susceptible to ingestion by organisms that are at the base of the food chain. Moreover, the characteristic high surface area-to-volume ratio of NPs enabling strong adsorptive capabilities may add to their potential hazardous effects, as other contaminants such as various POPs could be adsorbed and undergo bioaccumulation and bioamplification.

Nanotechnology is playing an increasingly important role in addressing innovative and effective solutions to a vast range of environmental challenges [141]. In recent years, nZVI, carbon nanotubes, and nanofibers have been applied for the remediation of a variety of contaminants including chlorinated compounds, hydrocarbons, organic compounds, and heavy metals. The use and development of nanomaterials are understandably heralded as an environmentally beneficial technology; however, the ecological risks associated with their use have only begun to be assessed. Although there is undoubtedly cause for concern among the scientific community over nanotechnology risk management, the efficacy of environmental nanotechnology is unquestionable. Hence, to advance this field in a rational manner without causing further environmental damage, an integrated preventive approach is suggested. This includes innovative and greener routes for NPs synthesis, advanced engineering methods for manufacturing smarter and more degradable nanomaterials, and governing local and international legislations to monitor NPs released into the environment.

PCBs and PCDDs/PCDFs have been unexpectedly found recently in commercially available (nano) particulate TiO_2 as a result of the TiO_2 fabrication process [142]. The toxicity of these new classes of nanoparticles carried POPs is related to their nanoscale-related properties such as surface-to-volume ratio, photocatalytic activity, polarity shifts, and stealth effect. Such new classes of POPs and their toxicologic effects are related to their size but are not a result of nanotechnology itself.

The integration of magnetic nanophotocatalysts (MNPCs) in water treatment technologies has recently received great attention due to the enhanced physical and chemical properties of these materials [143]. Thanks to the introduction of a magnetic material into nanocomposites, their separation can be easily performed through the application of external magnetic fields, thus obtaining more efficient, economic, and environment-friendly water purification processes. MNPCs could be potentially utilized in modified-AOPs to enhance the recyclability of catalysts [144].

The surface and interface are considered as crucial features that can be engineered to improve the performance of catalysts [145]. The great advancements in both controlled syntheses and catalytic mechanism studies have paved the way for the rational surface and interface design of catalysts. The synergistic effects of photocatalysis and electrocatalysis could be enhanced through the simultaneous control of the surface and the interface. Novel nanomaterial-based catalysis should address the evolving energy and chemical scenario for a cleaner and more sustainable future. In addressing the complexity of catalysis toward the advanced design of novel catalysts, understanding catalysts from the molecular level to the macroscopic material scale is essential [146]. Facet-engineered surface and interface design for photocatalytic materials has recently proved to be a versatile approach to enhance their photocatalytic performance [147].

As the source of highly active reducing species for reductive POPs degradation, solvated electron states at the oxide/aqueous interface represent the lowest energy charge-transfer pathways, thereby playing an important role in photocatalysis and electronic device applications. However, their energies are usually higher than the conduction band minimum, which makes solvated electrons difficult to utilize in charge-transfer processes. Thus it is essential to stabilize the energy of solvated electron states. It has been found that the modulation of solvated electron states can be achieved on polar – nonpolar oxide heterostructure surfaces as well as on ferroelectric oxides, which is important for charge and proton transfer at oxide/aqueous interfaces [148].

POPs released due to human activity will definitely exist in the environment in the long run, causing persistent harm to life. There is a lack of adequately potent natural processes to degrade POPs, which has already attracted great attention globally. Advanced photocatalysts with nanostructured heterojunctions are required in order to address the low efficiency of photochemical POPs degradation. Some interesting ideas could be developed based on the creatively combined principles of semiconductor physics and photochemistry. For instance, one could utilize the internal electric field induced by ferroelectric nanostructure to force the separation of the photoinduced charges to material surface locally. Then the locally raised photovoltage intensity on the nanometer scale will drive the redox reaction toward POPs degradation more smoothly with an enhanced energy conversion efficiency.

A recent report shows that some regular massive Janus heteronanostructure arrays for efficient plasmonic photocatalysis could be achieved by a template-guided programmable technique [149], which would help enforce the photochemical degradation of PFCs by the photoactivated surface charges. Certain nanoheterojunctions may also be constructed based on photoactive ferroelectric semiconductors coupled with traditional TiO_2. The regulation of photoinduced charge separation can promote degradation reactions over the ferroelectric nanojunctions toward the PFCs molecules adsorbed at the surface of the nanojunctions. Such design with the merit of facile material recycling could provide promising alternatives for combining photochemical techniques and ferroelectric materials to reduce refractory POPs in the environment more efficiently.

Finally, roadblocks to the advancement in the practical use of nanomaterials for environmental remediation toward POPs mainly include regulatory challenges, technical hurdles, uncertainties about the fate of certain nanomaterials in the environment, and the insufficiency in detailed cost–benefit analyses compared to existing technologies [7]. Nevertheless, it could be envisioned that environmentally benign nanomaterials with satisfactory recyclability will work in coordination and play critical roles for POPs capture and treatment in the near future.

Acknowledgments

The authors appreciate the support of the Fundamental Research Funds for the Central Universities (DUT17LK55) and National Natural Science Foundation of China (Nos. 21577012 and 21207015).

References

[1] M.A. Ashraf, Persistent organic pollutants (POPs): a global issue, a global challenge, Environ. Sci. Pollut. Res. Int. 24 (2017) 4223–4227.

[2] N. Oturan, E.D. van Hullebusch, H. Zhang, L. Mazeas, H. Budzinski, K. Le Menach, et al., Occurrence and removal of organic micropollutants in landfill leachates treated by electrochemical advanced oxidation processes, Environ. Sci. Technol. 49 (2015) 12187–12196.

[3] J. Wilkinson, P.S. Hooda, J. Barker, S. Barton, J. Swinden, Occurrence, fate and transformation of emerging contaminants in water: an overarching review of the field, Environ. Pollut. 231 (2017) 954–970.

[4] All POPs listed in the Stockholm Convention. <http://chm.pops.int/TheConvention/ThePOPs/ListingofPOPs/tabid/2509/Default.aspx>, 2017 (accessed 18.10.31).

[5] H. Hung, A.A. Katsoyiannis, R. Guardans, Ten years of global monitoring under the Stockholm Convention on Persistent Organic Pollutants (POPs): trends, sources and transport modelling, Environ. Pollut. 217 (2016) 1–3.

[6] H. Hung, A.A. Katsoyiannis, E. Brorstrom-Lunden, K. Olafsdottir, W. Aas, K. Breivik, et al., Temporal trends of Persistent Organic Pollutants (POPs) in arctic air: 20 years of monitoring under the Arctic Monitoring and Assessment Programme (AMAP), Environ. Pollut. 217 (2016) 52–61.

[7] A.S. Adeleye, J.R. Conway, K. Garner, Y. Huang, Y. Su, A.A. Keller, Engineered nanomaterials for water treatment and remediation: costs, benefits, and applicability, Chem. Eng. J. 286 (2016) 640–662.

[8] M.M. Khin, A.S. Nair, V.J. Babu, R. Murugan, S. Ramakrishna, A review on nanomaterials for environmental remediation, Energy Environ. Sci. 5 (2012) 8075–8109.
[9] Z. Cai, A.D. Dwivedi, W.-N. Lee, X. Zhao, W. Liu, M. Sillanpää, et al., Application of nanotechnologies for removing pharmaceutically active compounds from water: development and future trends, Environ. Sci. Nano 5 (2018) 27–47.
[10] R. Das, C.D. Vecitis, A. Schulze, B. Cao, A.F. Ismail, X. Lu, et al., Recent advances in nanomaterials for water protection and monitoring, Chem. Soc. Rev. 46 (2017) 6946–7020.
[11] M. Khajeh, S. Laurent, K. Dastafkan, Nanoadsorbents: classification, preparation, and applications (with emphasis on aqueous media), Chem. Rev. 113 (2013) 7728–7768.
[12] E. des Ligneris, L. Dumée, L. Kong, Nanofiber-based materials for persistent organic pollutants in water remediation by adsorption, Appl. Sci. 8 (2018) 166.
[13] A. Balasubramani, H.S. Rifai, Efficacy of carbon-based materials for remediating polychlorinated biphenyls (PCBs) in sediment, Sci. Total Environ. 644 (2018) 398–405.
[14] J. Wang, B. Chen, B. Xing, Wrinkles and folds of activated graphene nanosheets as fast and efficient adsorptive sites for hydrophobic organic contaminants, Environ. Sci. Technol. 50 (2016) 3798–3808.
[15] M. Rani, U. Shanker, V. Jassal, Recent strategies for removal and degradation of persistent & toxic organochlorine pesticides using nanoparticles: a review, J. Environ. Manage. 190 (2017) 208–222.
[16] L. Mohammed, H.G. Gomaa, D. Ragab, J. Zhu, Magnetic nanoparticles for environmental and biomedical applications: a review, Particuology 30 (2017) 1–14.
[17] C. Su, Environmental implications and applications of engineered nanoscale magnetite and its hybrid nanocomposites: a review of recent literature, J. Hazard. Mater. 322 (2017) 48–84.
[18] L. Zhang, M. Fang, Nanomaterials in pollution trace detection and environmental improvement, Nano Today 5 (2010) 128–142.
[19] A.Z.M. Badruddoza, B. Bhattarai, R.P.S. Suri, Environmentally friendly β-cyclodextrin–ionic liquid polyurethane-modified magnetic sorbent for the removal of PFOA, PFOS, and Cr(VI) from water, ACS Sustain. Chem. Eng. 5 (2017) 9223–9232.
[20] N. Ezzatahmadi, G.A. Ayoko, G.J. Millar, R. Speight, C. Yan, J. Li, et al., Clay-supported nanoscale zero-valent iron composite materials for the remediation of contaminated aqueous solutions: a review, Chem. Eng. J. 312 (2017) 336–350.
[21] X. Zhao, W. Liu, Z. Cai, B. Han, T. Qian, D. Zhao, An overview of preparation and applications of stabilized zero-valent iron nanoparticles for soil and groundwater remediation, Water Res. 100 (2016) 245–266.
[22] Z.Y. Zhang, M. Lu, Z.Z. Zhang, M. Xiao, M. Zhang, Dechlorination of short chain chlorinated paraffins by nanoscale zero-valent iron, J. Hazard. Mater. 243 (2012) 105–111.
[23] H.I. Gomes, C. Dias-Ferreira, L.M. Ottosen, A.B. Ribeiro, Electroremediation of PCB contaminated soil combined with iron nanoparticles: effect of the soil type, Chemosphere 131 (2015) 157–163.
[24] Y.-M. Kim, K. Murugesan, Y.-Y. Chang, E.-J. Kim, Y.-S. Chang, Degradation of polybrominated diphenyl ethers by a sequential treatment with nanoscale zero valent iron and aerobic biodegradation, J. Chem. Technol. Biotechnol. 87 (2012) 216–224.

[25] Y. Zhuang, S. Ahn, A.L. Seyfferth, Y. Masue-Slowey, S. Fendorf, R.G. Luthy, Dehalogenation of polybrominated diphenyl ethers and polychlorinated biphenyl by bimetallic, impregnated, and nanoscale zerovalent iron, Environ. Sci. Technol. 45 (2011) 4896–4903.
[26] Z. Fang, X. Qiu, J. Chen, X. Qiu, Debromination of polybrominated diphenyl ethers by Ni/Fe bimetallic nanoparticles: influencing factors, kinetics, and mechanism, J. Hazard. Mater. 185 (2–3) (2011) 958–969.
[27] O.S. Arvaniti, Y. Hwang, H.R. Andersen, A.S. Stasinakis, N.S. Thomaidis, M. Aloupi, Reductive degradation of perfluorinated compounds in water using Mg-aminoclay coated nanoscale zero valent iron, Chem. Eng. J. 262 (2015) 133–139.
[28] X. Guan, Y. Sun, H. Qin, J. Li, I.M. Lo, D. He, et al., The limitations of applying zero-valent iron technology in contaminants sequestration and the corresponding countermeasures: the development in zero-valent iron technology in the last two decades (1994-2014), Water Res. 75 (2015) 224–248.
[29] M. Stefaniuk, P. Oleszczuk, Y.S. Ok, Review on nano zerovalent iron (nZVI): from synthesis to environmental applications, Chem. Eng. J. 287 (2016) 618–632.
[30] A.V.B. Reddy, Z. Yusop, J. Jaafar, Y.V.M. Reddy, A.B. Aris, Z.A. Majid, et al., Recent progress on Fe-based nanoparticles: synthesis, properties, characterization and environmental applications, J. Environ. Chem. Eng. 4 (2016) 3537–3553.
[31] Y.C. Lee, Y.P. Chen, M.J. Chen, J. Kuo, S.L. Lo, Reductive defluorination of perfluorooctanoic acid by titanium(III) citrate with vitamin B12 and copper nanoparticles, J. Hazard. Mater. 340 (2017) 336–343.
[32] Q. Li, G. Su, M. Zheng, M. Wang, Y. Liu, F. Luo, et al., Thermal oxidation degradation of 2,2′,4,4′-tetrabromodiphenyl ether over Lialpha TiO_x micro/nanostructures with dozens of oxidative product analyses and reaction mechanisms, Environ. Sci. Technol. 51 (2017) 10059–10071.
[33] Chemicals proposed for listing under the Convention. <http://chm.pops.int/TheConvention/ThePOPs/ChemicalsProposedforListing/tabid/2510/Default.aspx>, 2017 (accessed 18.10.31).
[34] Report of the Persistent Organic Pollutants Review Committee on the Work of Its Thirteenth Meeting: Risk Management Evaluation on Pentadecafluorooctanoic Acid (CAS No: 335-67-1, PFOA, Perfluorooctanoic Acid), Its Salts and PFOA-Related Compounds. UNEP/POPS/POPRC.13/7/Add.2. <http://chm.pops.int/TheConvention/POPsReviewCommittee/Meetings/POPRC13/Overview/tabid/5965/Default.aspx>, 2017.
[35] Y. Cao, K.H. Harada, T. Hitomi, T. Niisoe, P. Wang, Y. Shi, et al., Lactational exposure to short-chain chlorinated paraffins in China, Korea, and Japan, Chemosphere 173 (2017) 43–48.
[36] R. Sun, X. Luo, B. Tang, L. Chen, Y. Liu, B. Mai, Bioaccumulation of short chain chlorinated paraffins in a typical freshwater food web contaminated by e-waste in south china: bioaccumulation factors, tissue distribution, and trophic transfer, Environ. Pollut. 222 (2017) 165–174.
[37] S. Xin, W. Gao, Y. Wang, G. Jiang, Thermochemical emission and transformation of chlorinated paraffins in inert and oxidizing atmospheres, Chemosphere 185 (2017) 899–906.
[38] C. Gallistl, J. Sprengel, W. Vetter, High levels of medium-chain chlorinated paraffins and polybrominated diphenyl ethers on the inside of several household baking oven doors, Sci. Total Environ. 615 (2018) 1019–1027.
[39] T.J. McGrath, A.S. Ball, B.O. Clarke, Critical review of soil contamination by polybrominated diphenyl ethers (PBDEs) and novel brominated flame

retardants (NBFRs); concentrations, sources and congener profiles, Environ. Pollut. 230 (2017) 741–757.
[40] Y. Liu, T. Ruan, Y. Lin, A. Liu, M. Yu, R. Liu, et al., Chlorinated polyfluoroalkyl ether sulfonic acids in marine organisms from Bohai Sea, China: occurrence, temporal variations, and trophic transfer behavior, Environ. Sci. Technol., 51, 2017, pp. 4407–4414.
[41] S.J. Varjani, E. Gnansounou, A. Pandey, Comprehensive review on toxicity of persistent organic pollutants from petroleum refinery waste and their degradation by microorganisms, Chemosphere 188 (2017) 280–291.
[42] Z. Wang, J.M. Boucher, M. Scheringer, I.T. Cousins, K. Hungerbuhler, Toward a comprehensive global emission inventory of C4-C10 perfluoroalkanesulfonic acids (PFSAs) and related precursors: focus on the life cycle of C8-based products and ongoing industrial transition, Environ. Sci. Technol. 51 (2017) 4482–4493.
[43] J. Meng, S. Hong, T. Wang, Q. Li, S.J. Yoon, Y. Lu, et al., Traditional and new POPs in environments along the Bohai and Yellow Seas: an overview of China and South Korea, Chemosphere 169 (2017) 503–515.
[44] O. Aznar-Alemany, Y. Aminot, J. Vila-Cano, M. Kock-Schulmeyer, J.W. Readman, A. Marques, et al., Halogenated and organophosphorus flame retardants in European aquaculture samples, Sci. Total Environ. 612 (2018) 492–500.
[45] S. Corsolini, G. Sara, The trophic transfer of persistent pollutants (HCB, DDTs, PCBs) within polar marine food webs, Chemosphere 177 (2017) 189–199.
[46] H. Li, J. Fu, W. Pan, P. Wang, Y. Li, Q. Zhang, et al., Environmental behaviour of short-chain chlorinated paraffins in aquatic and terrestrial ecosystems of Ny-Alesund and London Island, Svalbard, in the Arctic, Sci. Total Environ. 590-591 (2017) 163–170.
[47] A. Speltini, A. Scalabrini, F. Maraschi, M. Sturini, A. Profumo, Newest applications of molecularly imprinted polymers for extraction of contaminants from environmental and food matrices: a review, Anal. Chim. Acta 974 (2017) 1–26.
[48] L. Chen, X. Wang, W. Lu, X. Wu, J. Li, Molecular imprinting: perspectives and applications, Chem. Soc. Rev. 45 (2016) 2137–2211.
[49] E. Morillo, J. Villaverde, Advanced technologies for the remediation of pesticide-contaminated soils, Sci. Total Environ. 586 (2017) 576–597.
[50] D. Jiang, G. Zeng, D. Huang, M. Chen, C. Zhang, C. Huang, et al., Remediation of contaminated soils by enhanced nanoscale zero valent iron, Environ. Res. 163 (2018) 217–227.
[51] R. Wang, T. Tang, G. Lu, K. Huang, H. Yin, Z. Lin, et al., Rapid debromination of polybrominated diphenyl ethers (PBDEs) by zero valent metal and bimetals: mechanisms and pathways assisted by density function theory calculation, Environ. Pollut. 240 (2018) 745–753.
[52] T. Tang, G. Lu, R. Wang, H. Chen, Y. Fang, K. Huang, et al., Debromination of polybrominated diphenyl ethers (PBDEs) by zero valent zinc: mechanisms and predicting descriptors, J. Hazard. Mater. 352 (2018) 165–171.
[53] M. Usman, K. Hanna, S. Haderlein, Fenton oxidation to remediate PAHs in contaminated soils: a critical review of major limitations and counter-strategies, Sci. Total Environ. 569-570 (2016) 179–190.
[54] S. Gligorovski, R. Strekowski, S. Barbati, D. Vione, Environmental implications of hydroxyl radicals ((*)OH), Chem. Rev. 115 (2015) 13051–13092.

[55] K.O. Badmus, J.O. Tijani, E. Massima, L. Petrik, Treatment of persistent organic pollutants in wastewater using hydrodynamic cavitation in synergy with advanced oxidation process, Environ. Sci. Pollut. Res. Int. 25 (2018) 7299–7314.
[56] H. Roth, Y. Gendel, P. Buzatu, O. David, M. Wessling, Tubular carbon nanotube-based gas diffusion electrode removes persistent organic pollutants by a cyclic adsorption—electro-Fenton process, J. Hazard. Mater. 307 (2016) 1–6.
[57] L. Santos-Juanes, F.S. García Einschlag, A.M. Amat, A. Arques, Combining ZVI reduction with photo-Fenton process for the removal of persistent pollutants, Chem. Eng. J. 310 (2017) 484–490.
[58] N.K. Cheruiyot, L.-C. Wang, S.-L. Lin, H.-H. Yang, Y.-T. Chen, Effects of selective catalytic reduction on the emissions of persistent organic pollutants from a heavy-duty diesel engine, Aerosol Air Qual. Res. 17 (2017) 1658–1665.
[59] A. Santos, S. Rodriguez, F. Pardo, A. Romero, Use of Fenton reagent combined with humic acids for the removal of PFOA from contaminated water, Sci. Total Environ. 563-564 (2016) 657–663.
[60] M. Trojanowicz, A. Bojanowska-Czajka, I. Bartosiewicz, K. Kulisa, Advanced oxidation/reduction processes treatment for aqueous perfluorooctanoate (PFOA) and perfluorooctanesulfonate (PFOS)—a review of recent advances, Chem. Eng. J. 336 (2018) 170–199.
[61] N. Merino, Y. Qu, R.A. Deeb, E.L. Hawley, M.R. Hoffmann, S. Mahendra, Degradation and removal methods for perfluoroalkyl and polyfluoroalkyl substances in water, Environ. Eng. Sci. 33 (2016) 615–649.
[62] H.V. Lutze, J. Brekenfeld, S. Naumov, C. von Sonntag, T.C. Schmidt, Degradation of perfluorinated compounds by sulfate radicals—new mechanistic aspects and economical considerations, Water Res. 129 (2018) 509–519.
[63] D. Huang, L. Yin, X. Lu, S. Lin, Z. Niu, J. Niu, Directional electron transfer mechanisms with graphene quantum dots as the electron donor for photodecomposition of perfluorooctane sulfonate, Chem. Eng. J. 323 (2017) 406–414.
[64] H. Tian, C. Gu, Effects of different factors on photodefluorination of perfluorinated compounds by hydrated electrons in organo-montmorillonite system, Chemosphere 191 (2018) 280–287.
[65] M.-J. Chen, A.-C. Yang, N.-H. Wang, H.-C. Chiu, Y.-L. Li, D.-Y. Kang, et al., Influence of crystal topology and interior surface functionality of metal-organic frameworks on PFOA sorption performance, Microporous Mesoporous Mater. 236 (2016) 202–210.
[66] Q. Zhuo, Q. Xiang, H. Yi, Z. Zhang, B. Yang, K. Cui, et al., Electrochemical oxidation of PFOA in aqueous solution using highly hydrophobic modified PbO_2 electrodes, J. Electroanal. Chem. 801 (2017) 235–243.
[67] Z. Song, X. Dong, N. Wang, L. Zhu, Z. Luo, J. Fang, et al., Efficient photocatalytic defluorination of perfluorooctanoic acid over BiOCl nanosheets via a hole direct oxidation mechanism, Chem. Eng. J. 317 (2017) 925–934.
[68] J. Chakraborty, S. Das, Molecular perspectives and recent advances in microbial remediation of persistent organic pollutants, Environ. Sci. Pollut. Res. Int. 23 (2016) 16883–16903.
[69] H. Sui, Y. Rong, J. Song, D. Zhang, H. Li, P. Wu, et al., Mechanochemical destruction of DDTs with Fe-Zn bimetal in a high-energy planetary ball mill, J. Hazard. Mater. 342 (2018) 201–209.

[70] G. Cagnetta, J. Robertson, J. Huang, K. Zhang, G. Yu, Mechanochemical destruction of halogenated organic pollutants: a critical review, J. Hazard. Mater. 313 (2016) 85–102.
[71] S. Deng, S. Kang, N. Feng, J. Zhu, B. Yu, X. Xie, et al., Mechanochemical mechanism of rapid dechlorination of hexachlorobenzene, J. Hazard. Mater. 333 (2017) 116–127.
[72] G. Cagnetta, J. Huang, M. Lu, B. Wang, Y. Wang, S. Deng, et al., Defect engineered oxides for enhanced mechanochemical destruction of halogenated organic pollutants, Chemosphere 184 (2017) 879–883.
[73] V.A. Arias Espana, M. Mallavarapu, R. Naidu, Treatment technologies for aqueous perfluorooctanesulfonate (PFOS) and perfluorooctanoate (PFOA): a critical review with an emphasis on field testing, Environ. Technol. Innovat. 4 (2015) 168–181.
[74] E.B. Miller, E.M. Zahran, M.R. Knecht, L.G. Bachas, Metal oxide semiconductor nanomaterial for reductive debromination: visible light degradation of polybrominated diphenyl ethers by Cu_2O@Pd nanostructures, Appl. Catal. B Environ. 213 (2017) 147–154.
[75] M. Arslan, A. Imran, Q.M. Khan, M. Afzal, Plant-bacteria partnerships for the remediation of persistent organic pollutants, Environ. Sci. Pollut. Res. Int. 24 (2017) 4322–4336.
[76] Y. Song, Y. Li, W. Zhang, F. Wang, Y. Bian, L.A. Boughner, et al., Novel biochar-plant tandem approach for remediating hexachlorobenzene contaminated soils: proof-of-concept and new insight into the rhizosphere, J. Agric. Food. Chem. 64 (2016) 5464–5471.
[77] W. Zhang, N. Liu, Y. Cao, X. Lin, Y. Liu, L. Feng, Superwetting porous materials for wastewater treatment: from immiscible oil/water mixture to emulsion separation, Adv. Mater. Interf. 4 (2017).
[78] H.I. Gomes, C. Dias-Ferreira, A.B. Ribeiro, Overview of in situ and ex situ remediation technologies for PCB-contaminated soils and sediments and obstacles for full-scale application, Sci. Total Environ. 445-446 (2013) 237–260.
[79] S.D. Richardson, S.Y. Kimura, Emerging environmental contaminants: challenges facing our next generation and potential engineering solutions, Environ. Technol. Innovat. 8 (2017) 40–56.
[80] A. Pariatamby, Y.L. Kee, Persistent organic pollutants management and remediation, Proc. Environ. Sci. 31 (2016) 842–848.
[81] J. Trivedi, D. Majumdar, Memory effect driven emissions of persistent organic pollutants from industrial thermal processes, their implications and management: a review, J. Environ. Manage. 119 (2013) 111–120.
[82] M. Vallejo, M. Fresnedo San Roman, I. Ortiz, A. Irabien, Overview of the PCDD/Fs degradation potential and formation risk in the application of advanced oxidation processes (AOPs) to wastewater treatment, Chemosphere 118 (2015) 44–56.
[83] R. Weber, A. Watson, M. Forter, F. Oliaei, Review article: persistent organic pollutants and landfills—a review of past experiences and future challenges, Waste. Manag. Res. 29 (2011) 107–121.
[84] X. Wang, P. Gong, C. Wang, J. Ren, T. Yao, A review of current knowledge and future prospects regarding persistent organic pollutants over the Tibetan Plateau, Sci. Total Environ. 573 (2016) 139–154.
[85] J.L. Luek, R.M. Dickhut, M.A. Cochran, R.L. Falconer, H. Kylin, Persistent organic pollutants in the Atlantic and southern oceans and oceanic atmosphere, Sci. Total Environ. 583 (2017) 64–71.
[86] E.T. Martin, C.M. McGuire, M.S. Mubarak, D.G. Peters, Electroreductive remediation of halogenated environmental pollutants, Chem. Rev. 116 (2016) 15198–15234.

[87] H.I. Gomes, C. Dias-Ferreira, L.M. Ottosen, A.B. Ribeiro, Electrodialytic remediation of polychlorinated biphenyls contaminated soil with iron nanoparticles and two different surfactants, J. Colloid. Interface. Sci. 433 (2014) 189–195.
[88] K.B. Pedersen, T. Lejon, P.E. Jensen, L.M. Ottosen, Simultaneous electrodialytic removal of PAH, PCB, TBT and heavy metals from sediments, J. Environ. Manage. 198 (2017) 192–202.
[89] P.P. Falciglia, D. Malarbì, V. Greco, F.G.A. Vagliasindi, Surfactant and MGDA enhanced—electrokinetic treatment for the simultaneous removal of mercury and PAHs from marine sediments, Sep. Purif. Technol. 175 (2017) 330–339.
[90] I. Sires, E. Brillas, M.A. Oturan, M.A. Rodrigo, M. Panizza, Electrochemical advanced oxidation processes: today and tomorrow. A review, Environ. Sci. Pollut. Res. Int. 21 (2014) 8336–8367.
[91] S.O. Ganiyu, M. Zhou, C.A. Martínez-Huitle, Heterogeneous electro-Fenton and photoelectro-Fenton processes: a critical review of fundamental principles and application for water/wastewater treatment, Appl. Catal. B Environ. 235 (2018) 103–129.
[92] W. Ren, D. Tang, M. Huang, J. Sun, K. Lv, Remarkable improved electro-Fenton efficiency by electric-field-induced catalysis of CeO_2, J. Hazard. Mater. 350 (2018) 88–97.
[93] M. Popescu, C. Sandu, E. Rosales, M. Pazos, G. Lazar, M.Á. Sanromán, Evaluation of different cathodes and reaction parameters on the enhancement of the electro-Fenton process, J. Electroanal. Chem. 808 (2018) 455–463.
[94] C. Trellu, N. Oturan, F.K. Keita, C. Fourdrin, Y. Pechaud, M.A. Oturan, Regeneration of activated carbon fiber by the electro-Fenton process, Environ. Sci. Technol. 52 (2018) 7450–7457.
[95] F. Gozzi, I. Sirés, A. Thiam, S.C. de Oliveira, A.M. Junior, E. Brillas, Treatment of single and mixed pesticide formulations by solar photoelectro-Fenton using a flow plant, Chem. Eng. J. 310 (2017) 503–513.
[96] E. Brillas, I. Sirés, M.A. Oturan, Electro-Fenton process and related electrochemical technologies based on Fenton's reaction chemistry, Chem. Rev. 109 (2009) 6570–6631.
[97] X. Liu, Y. Zhou, J. Zhang, L. Luo, Y. Yang, H. Huang, et al., Insight into electro-Fenton and photo-Fenton for the degradation of antibiotics: mechanism study and research gaps, Chem. Eng. J. 347 (2018) 379–397.
[98] M. Kahoush, N. Behary, A. Cayla, V. Nierstrasz, Bio-Fenton and Bio-electro-Fenton as sustainable methods for degrading organic pollutants in wastewater, Process Biochem. 64 (2018) 237–247.
[99] C. Trellu, B.P. Chaplin, C. Coetsier, R. Esmilaire, S. Cerneaux, C. Causserand, et al., Electro-oxidation of organic pollutants by reactive electrochemical membranes, Chemosphere 208 (2018) 159–175.
[100] P. Mandal, B.K. Dubey, A.K. Gupta, Review on landfill leachate treatment by electrochemical oxidation: drawbacks, challenges and future scope, Waste. Manag. 69 (2017) 250–273.
[101] W. Wu, Z.-H. Huang, T.-T. Lim, Recent development of mixed metal oxide anodes for electrochemical oxidation of organic pollutants in water, Appl. Catal. A General 480 (2014) 58–78.
[102] C.M. Dominguez, N. Oturan, A. Romero, A. Santos, M.A. Oturan, Lindane degradation by electrooxidation process: effect of electrode materials on oxidation and mineralization kinetics, Water Res. 135 (2018) 220–230.
[103] C.M. Dominguez, N. Oturan, A. Romero, A. Santos, M.A. Oturan, Removal of lindane wastes by advanced electrochemical oxidation, Chemosphere 202 (2018) 400–409.

[104] C. Fang, M. Megharaj, R. Naidu, Electrochemical Advanced Oxidation Processes (EAOP) to degrade per- and polyfluoroalkyl substances (PFASs), J. Adv. Oxid. Technol. 20 (2017) 20170014.
[105] V. Poza-Nogueiras, E. Rosales, M. Pazos, M.A. Sanroman, Current advances and trends in electro-Fenton process using heterogeneous catalysts—a review, Chemosphere 201 (2018) 399–416.
[106] S. Malato, J. Blanco, D.C. Alarcón, M.I. Maldonado, P. Fernández-Ibáñez, W. Gernjak, Photocatalytic decontamination and disinfection of water with solar collectors, Catal. Today 122 (2007) 137–149.
[107] S. Khan, X. He, J.A. Khan, H.M. Khan, D.L. Boccelli, D.D. Dionysiou, Kinetics and mechanism of sulfate radical- and hydroxyl radical-induced degradation of highly chlorinated pesticide lindane in UV/peroxymonosulfate system, Chem. Eng. J. 318 (2017) 135–142.
[108] M. Canle, M.I. Fernández Pérez, J.A. Santaballa, Photocatalyzed degradation/abatement of endocrine disruptors, Curr. Opin. Green Sustain. Chem. 6 (2017) 101–138.
[109] X. Chen, Q.D. Zhao, X.Y. Li, D. Wang, Enhanced photocatalytic activity of degrading short chain chlorinated paraffins over reduced graphene oxide/$CoFe_2O_4$/Ag nanocomposite, J. Colloid. Interface. Sci. 479 (2016) 89–97.
[110] W. Xiong, X. Li, Q. Zhao, Y. Shi, C. Hao, Insight into the photocatalytic mineralization of short chain chlorinated paraffins boosted by polydopamine and Ag nanoparticles, J. Hazard. Mater. 359 (2018) 186–193.
[111] C. Sun, W. Chang, W. Ma, C. Chen, J. Zhao, Photoreductive debromination of decabromodiphenyl ethers in the presence of carboxylates under visible light irradiation, Environ. Sci. Technol. 47 (2013) 2370–2377.
[112] R. Wang, H. Li, X. Tao, T. Tang, H. Lin, K. Huang, et al., Photodebromination behaviors of polybrominated diphenyl ethers in methanol/water systems: mechanisms and predicting descriptors, Sci. Total Environ. 595 (2017) 666–672.
[113] M. Lei, N. Wang, S. Guo, L. Zhu, Y. Ding, H. Tang, A one-pot consecutive photocatalytic reduction and oxidation system for complete debromination of tetrabromodiphenyl ether, Chem. Eng. J. 345 (2018) 586–593.
[114] W. Chen, X. Niu, J. Wang, A photocatalyst of graphene oxide (GO)/Ag_3PO_4 with excellent photocatalytic activity over decabromodiphenyl ether (BDE-209) under visible light irradiation, J. Photochem. Photobiol. A Chem. 356 (2018) 304–311.
[115] Y.-Y. Shao, W.-D. Ye, C.-Y. Sun, C.-L. Liu, Q. Wang, C.-C. Chen, et al., Enhanced photoreduction degradation of polybromodiphenyl ethers with Fe_3O_4-g-C_3N_4 under visible light irradiation, RSC Adv. 8 (2018) 10914–10921.
[116] S. Wang, Q. Yang, F. Chen, J. Sun, K. Luo, F. Yao, et al., Photocatalytic degradation of perfluorooctanoic acid and perfluorooctane sulfonate in water: a critical review, Chem. Eng. J. 328 (2017) 927–942.
[117] B. Xu, M.B. Ahmed, J.L. Zhou, A. Altaee, M. Wu, G. Xu, Photocatalytic removal of perfluoroalkyl substances from water and wastewater: mechanism, kinetics and controlling factors, Chemosphere 189 (2017) 717–729.
[118] K. Park, I. Ali, J.O. Kim, Photodegradation of perfluorooctanoic acid by graphene oxide-deposited TiO_2 nanotube arrays in aqueous phase, J. Environ. Manage. 218 (2018) 333–339.

[119] Y. Pi, X. Li, Q. Xia, J. Wu, Y. Li, J. Xiao, et al., Adsorptive and photocatalytic removal of persistent organic pollutants (POPs) in water by metal-organic frameworks (MOFs), Chem. Eng. J. 337 (2018) 351–371.
[120] D. Panda, S. Manickam, Recent advancements in the sonophotocatalysis (SPC) and doped-sonophotocatalysis (DSPC) for the treatment of recalcitrant hazardous organic water pollutants, Ultrason. Sonochem. 36 (2017) 481–496.
[121] E. Brillas, C.A. Martínez-Huitle, Decontamination of wastewaters containing synthetic organic dyes by electrochemical methods. An updated review, Appl. Catal. B Environ. 166-167 (2015) 603–643.
[122] S. Garcia-Segura, E. Brillas, Applied photoelectrocatalysis on the degradation of organic pollutants in wastewaters, J. Photochem. Photobiol. C Photochem. Rev. 31 (2017) 1–35.
[123] D. Cao, Y. Wang, X. Zhao, Combination of photocatalytic and electrochemical degradation of organic pollutants from water, Curr. Opin. Green Sustain. Chem. 6 (2017) 78–84.
[124] X. Meng, Z. Zhang, X. Li, Synergetic photoelectrocatalytic reactors for environmental remediation: a review, J. Photochem. Photobiol. C Photochem. Rev. 24 (2015) 83–101.
[125] G.G. Bessegato, T.T. Guaraldo, J.F. de Brito, M.F. Brugnera, M.V.B. Zanoni, Achievements and trends in photoelectrocatalysis: from environmental to energy applications, Electrocatalysis 6 (2015) 415–441.
[126] M.J. Martín de Vidales, M. Millán, C. Sáez, P. Cañizares, M.A. Rodrigo, Irradiated-assisted electrochemical processes for the removal of persistent pollutants from real wastewater, Sep. Purif. Technol. 175 (2017) 428–434.
[127] R. Daghrir, P. Drogui, D. Robert, Photoelectrocatalytic technologies for environmental applications, J. Photochem. Photobiol. A Chem. 238 (2012) 41–52.
[128] L.Y. Ozer, C. Garlisi, H. Oladipo, M. Pagliaro, S.A. Sharief, A. Yusuf, et al., Inorganic semiconductors-graphene composites in photo(electro) catalysis: synthetic strategies, interaction mechanisms and applications, J. Photochem. Photobiol. C Photochem. Rev. 33 (2017) 132–164.
[129] A. Eftekhari, V.J. Babu, S. Ramakrishna, Photoelectrode nanomaterials for photoelectrochemical water splitting, Int. J. Hydrogen. Energy 42 (2017) 11078–11109.
[130] K. Sivula, R. van de Krol, Semiconducting materials for photoelectrochemical energy conversion, Nat. Rev. Mater. 1 (2016) 15010.
[131] E. Zarei, R. Ojani, Fundamentals and some applications of photoelectrocatalysis and effective factors on its efficiency: a review, J. Solid State Electrochem. 21 (2016) 305–336.
[132] X. Quan, X. Ruan, H. Zhao, S. Chen, Y. Zhao, Photoelectrocatalytic degradation of pentachlorophenol in aqueous solution using a TiO_2 nanotube film electrode, Environ. Pollut. 147 (2007) 409–414.
[133] S. Shen, J. Chen, M. Wang, X. Sheng, X. Chen, X. Feng, et al., Titanium dioxide nanostructures for photoelectrochemical applications, Prog. Mater. Sci. 98 (2018) 299–385.
[134] D.S. Nair, M. Kurian, Heterogeneous catalytic oxidation of persistent chlorinated organics over cobalt substituted zinc ferrite nanoparticles at mild conditions: reaction kinetics and catalyst reusability studies, J. Environ. Chem. Eng. 5 (2017) 964–974.
[135] X. Meng, Z. Zhang, Two dimensional graphitic materials for photoelectrocatalysis: a short review, Catal. Today 315 (2018) 2–8.

[136] S. Kumar, K. Ojha, A.K. Ganguli, Interfacial charge transfer in photoelectrochemical processes, Adv. Mater. Interf. 4 (2017).
[137] P. Lianos, Review of recent trends in photoelectrocatalytic conversion of solar energy to electricity and hydrogen, Appl. Catal. B Environ. 210 (2017) 235–254.
[138] M. Hesari, X. Mao, P. Chen, Charge carrier activity on single-particle photo (electro)catalysts: toward function in solar energy conversion, J. Am. Chem. Soc. 140 (2018) 6729–6740.
[139] J. Chang, L. Zhang, P. Wang, Intelligent environmental nanomaterials, Environ. Sci. Nano 5 (2018) 811–836.
[140] J.P. da Costa, P.S.M. Santos, A.C. Duarte, T. Rocha-Santos, (Nano)plastics in the environment—sources, fates and effects, Sci. Total Environ. 566-567 (2016) 15–26.
[141] S.S. Patil, U.U. Shedbalkar, A. Truskewycz, B.A. Chopade, A.S. Ball, Nanoparticles for environmental clean-up: a review of potential risks and emerging solutions, Environ. Technol. Innovat. 5 (2016) 10–21.
[142] P. Schon, G. Ctistis, W. Bakker, G. Luthe, Nanoparticular surface-bound PCBs, PCDDs, and PCDFs—a novel class of potentially higher toxic POPs, Environ. Sci. Pollut. Res. Int. 24 (2017) 12758–12766.
[143] J. Gómez-Pastora, S. Dominguez, E. Bringas, M.J. Rivero, I. Ortiz, D.D. Dionysiou, Review and perspectives on the use of magnetic nanophotocatalysts (MNPCs) in water treatment, Chem. Eng. J. 310 (2017) 407–427.
[144] R. Dewil, D. Mantzavinos, I. Poulios, M.A. Rodrigo, New perspectives for Advanced Oxidation Processes, J. Environ. Manage. 195 (2017) 93–99.
[145] S. Bai, Y. Xiong, Some recent developments in surface and interface design for photocatalytic and electrocatalytic hybrid structures, Chem. Commun. 51 (2015) 10261–10271.
[146] P. Lanzafame, S. Perathoner, G. Centi, S. Gross, E.J.M. Hensen, Grand challenges for catalysis in the Science and Technology Roadmap on Catalysis for Europe: moving ahead for a sustainable future, Catal. Sci. Technol. 7 (2017) 5182–5194.
[147] S. Bai, L. Wang, Z. Li, Y. Xiong, Facet-engineered surface and interface design of photocatalytic materials, Adv. Sci. 4 (2017) 1600216.
[148] Y. Wang, H. Guo, Q. Zheng, W.A. Saidi, J. Zhao, Tuning solvated electrons by polar-nonpolar oxide heterostructure, J. Phys. Chem. Lett. 9 (2018) 3049–3056.
[149] L. Wen, R. Xu, C. Cui, W. Tang, Y. Mi, X. Lu, et al., Template-guided programmable janus heteronanostructure arrays for efficient plasmonic photocatalysis, Nano. Lett. 18 (2018) 4914–4921.

Further Reading

F.C. Moreira, R.A.R. Boaventura, E. Brillas, V.J.P. Vilar, Electrochemical advanced oxidation processes: a review on their application to synthetic and real wastewaters, Appl. Catal. B Environ. 202 (2017) 217–261.
B. Gomez-Ruiz, S. Gómez-Lavín, N. Diban, V. Boiteux, A. Colin, X. Dauchy, et al., Efficient electrochemical degradation of poly- and perfluoroalkyl substances (PFASs) from the effluents of an industrial wastewater treatment plant, Chem. Eng. J. 322 (2017) 196–204.

A. Soriano, D. Gorri, A. Urtiaga, Efficient treatment of perfluorohexanoic acid by nanofiltration followed by electrochemical degradation of the NF concentrate, Water Res. 112 (2017) 147–156.
Z. Xu, Y. Yu, H. Liu, J. Niu, Highly efficient and stable Zr-doped nanocrystalline PbO_2 electrode for mineralization of perfluorooctanoic acid in a sequential treatment system, Sci. Total Environ. 579 (2017) 1600–1607.
A. Farhat, J. Keller, S. Tait, J. Radjenovic, Removal of persistent organic contaminants by electrochemically activated sulfate, Environ. Sci. Technol. 49 (2015) 14326–14333.

8

POWER READY FOR DRIVING CATALYSIS AND SENSING: NANOMATERIALS DESIGNED FOR RENEWABLE ENERGY STORAGE

Anmin Liu[1] and Xuefeng Ren[2]

[1]State Key Laboratory of Fine Chemicals, School of Petroleum and Chemical Engineering, Dalian University of Technology, Panjin, P.R. China [2]School of Food and Environment, Dalian University of Technology, Panjin, P.R. China

8.1 Introduction to the Basics of Renewable Energy Storage and Opportunities From Nanomaterials

Environmental sensors represent a new way to sense and understand the environment and have huge potential in many areas of environmental sciences [1–4]. With excellent insensitivity to pressure and light, mechanical and chemical robustness, and the possibility of vertical or horizontal orientation, the all-solid-state potentiometric sensor is a promising platform for the determination of nitrate, nitrite, dihydrogen phosphate, and carbonate in situ [5]. The potential for water pollution outbreaks requires the development of rapid methods for water quality monitoring. For example, plasmonic nanostructures or nanoparticles are compelling candidates for the development of highly sensitive biosensors due to their unique localized surface plasmon resonances [6]. Based on environmental analyses by environmental sensors, pollution abatement and prevention should be paid more attention.

Environmental catalytic technologies play a major role in both pollution abatement and prevention [7,8]. Atmospheric pollutants such as nitrogen oxides (NOx), carbon monoxide, and volatile organic compounds (VOCs), chiefly emanating

Advanced Nanomaterials for Pollutant Sensing and Environmental Catalysis. DOI: https://doi.org/10.1016/B978-0-12-814796-2.00008-3

from industrial activities and transportation vehicles, are harmful to human health, while catalysis is one of the most effective and economic technologies to control serious air pollution problems [9]. Porous clay heterostructures (PCH) from natural pillared clays (bentonite with a high proportion of montmorillonite) can be used as supports for iron oxide for two reactions, namely the elimination of toluene and the selective oxidation of H_2S into elemental sulfur [10]. Gold catalysts exhibit a very high low temperature activity for the oxidation of carbon monoxide [11,12].

Wastewater pollution with heavy metals is an issue of great environmental concern. The future development of clean technologies for the treatment of wastewater loaded with heavy metals entails environment-friendly and sustainable processes that may allow for the simultaneous recovery of the metals and their reutilization as value-added catalysts in environmental applications [13].

The application of environmental catalysts and sensors in pollutant sensing and environmental catalysis provides an opportunity for renewable energies to drive catalysis and sensing. With the requirement of capacity and efficiency, nanometric materials are playing more and more important roles in renewable energy storage.

Batteries are currently being developed to power an increasingly diverse range of applications, from cars to microchips. How can the performance of batteries meet the demands of each application with the requirements of high energy and power density? And how can batteries become a sustainable technology for the future [14]? Renewable energy sources such as lithium-ion batteries (LIBs), sodium-ion batteries (NIBs), and fuel cells are important components for the reduction of dependence on fossil fuels and greenhouse gas emissions.

The choice of a battery for environmental catalysts and sensors can be approached from many perspectives. The most important factor affecting the choice are the application requirements. Batteries typically serve not just to supply the system with energy, but also to efficiently store energy harvested from the environment. In this way, energy can be stored for times when it cannot be directly extracted from the surroundings. Batteries can be primary or secondary. Primary batteries are non-rechargeable and can be one of the choices for environmental catalysts and sensors. They have many advantages including high capacity and temperature stability. Their main disadvantage is the need for periodic maintenance and replacement at the end of their life cycle. Secondary batteries are rechargeable. However, traditional rechargeable batteries generally have lower energy storage capability than primary batteries. The additional requirements for rechargeability and long operation make the choice of chemical systems and constructions be those more robust batteries [15,16]. However, rechargeable lithium batteries can store more than twice as much energy per unit weight and volume as other rechargeable batteries [17,18].

8.2 Nanomaterials for Lithium-Ion Batteries

With superexcellent high energy and power density, LIBs are widely used in power tools, hybrid/full electric vehicles, and portable electronics. For the purpose of energy storage from wind, solar, geothermal, and other renewable energy sources, LIBs with high energy efficiency are the best choice [19].

Each cell of a LIB stores electrical energy as chemical energy in two electrodes, a negative electrode and a positive electrode, separated by an electrolyte that transfers the ionic component of the chemical reaction inside the cell and forces the electronic component outside the battery, as displayed in Fig. 8.1. The electrolyte conducts the ionic component of the chemical reaction between the negative electrode and the positive electrode, but it forces the electronic component to traverse an external

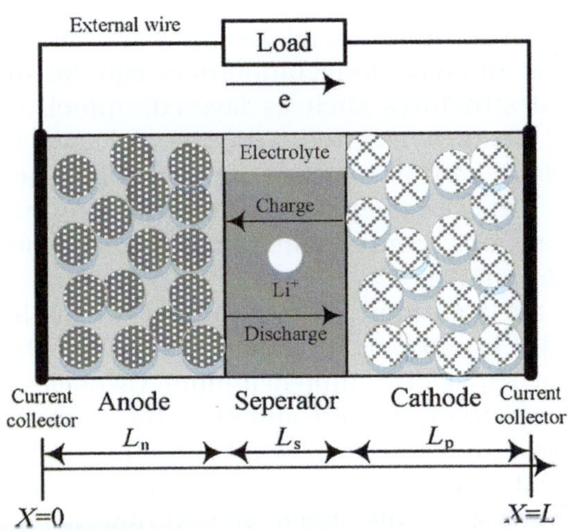

Figure 8.1 Schematic illustration of Li-ion battery [20].

circuit where it works; as the ionic mobility in the electrolyte is much smaller than the electronic conductivity in a metal, a cell has large-area electrodes separated by a thin electrolyte, and metallic current collectors deliver electronic current from/to the redox centers of the electrodes to/from posts that connect to the external circuit. A rechargeable battery has a reversible chemical reaction at the two electrodes, the same as renewable energy storage of LIB [21].

Renewable energy storage is playing more and more important roles; future generations of rechargeable LIBs will be required to power portable electronic devices, store electricity from renewable sources, and as a vital component in new hybrid electric vehicles. To achieve the increase in energy and power density essential to meet the future challenges of energy storage, new materials chemistry, and especially new nanomaterials chemistry, is essential, especially regarding new nanomaterials with new properties or combinations of properties, for use as electrodes and electrolytes in LIBs [22]. The electrochemical and energy storage performance of electrodes depend on the structural, physical, and chemical properties of electrode materials [23].

8.2.1 Positive Electrode Nanomaterials for Lithium-Ion Batteries

An intercalation positive electrode is a solid host network, which can store guest ions. The guest ions can be inserted into and be removed from the host network reversibly. In a LIB, Li^+ is the guest ion and the host network compounds are metal chalcogenides, transition metal oxides, and polyanion compounds. These intercalation compounds can be divided into several crystal structures such as layered, spinel, olivine, and tavorite [19]. Positive electrodes of intercalation compounds can retain their crystal structure upon lithium insertion, as shown in Fig. 8.2; even the host structure remains intact in spite of slight expansions, contractions, or distortions of the lattice upon Li^+ insertion [24].

Although the transition metals present in all intercalation compounds are capable of multiple electron transfer and thus present higher capacities, limited lithium vacancies inhibit the incorporation of more lithium and thus the charge transfer of more electrons to the structure. In addition, the covalency of the transition-metal dichalcogenide bond reduces the voltage of the reaction. Approaches have been carried out to improve the performance of positive electrode nanomaterials for LIBs [25].

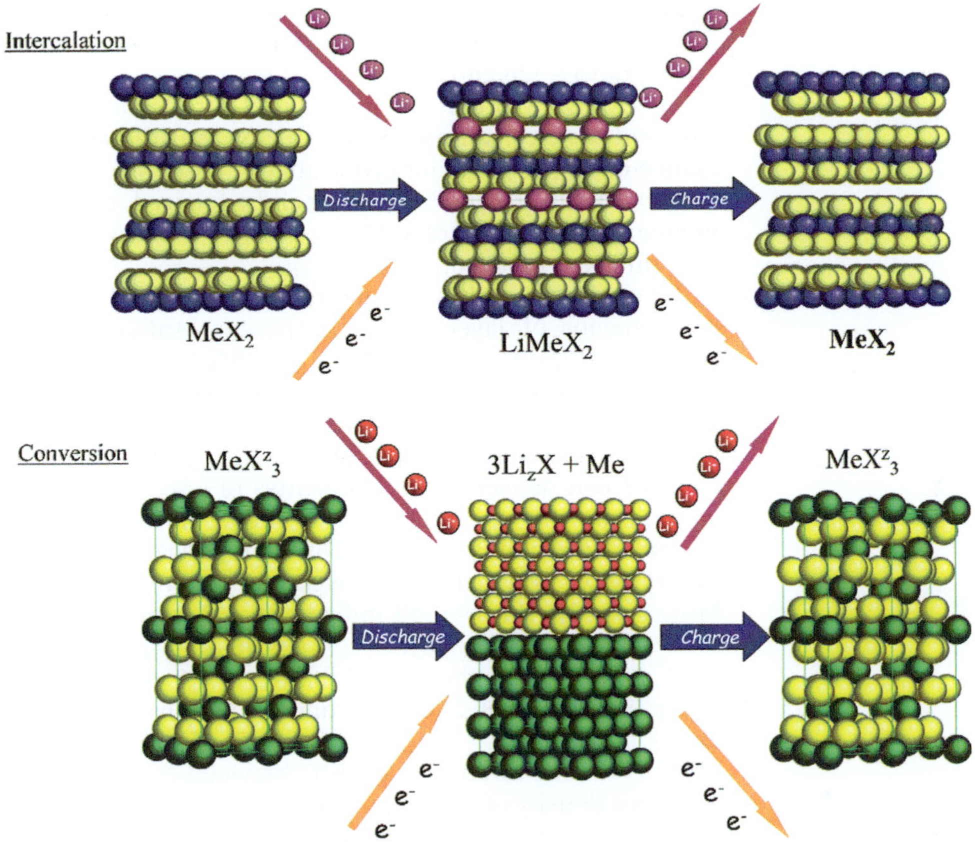

Figure 8.2 Schematic representation showing contrasting crystallographic reaction mechanisms occurring during discharge of positive electrodes with reaction mechanisms based on intercalation and reversible conversion, respectively [24].

As introduced by Goodenough, $LiCoO_2$ is the most commercially successful form of layered transition metal oxide positive electrodes till now, based on its relatively high theoretical specific capacity of 274 mA h g^{-1}, high theoretical volumetric capacity of 1363 mA h cm^{-3}, low self-discharge, high discharge voltage, and excellent cycling performance [26,27]. However, the high cost of Co, low thermal stability, and fast capacity fade at high current rates or during deep cycling of $LiCoO_2$ seriously limit the application of $LiCoO_2$. Doping with metals (Mn, Al, Fe, Cr) and coating with metal oxides (Al_2O_3, B_2O_3, TiO_2, ZrO_2) are the most commonly used effective ways to improve the performance of $LiCoO_2$ [19].

With similar structure and theoretical specific capacity (275 mA h g^{-1}) to $LiCoO_2$, but with a lower cost, $LiNiO_2$ is a promising positive electrode; however, the substitution of Li^+ sites by Ni^{2+} ions and its thermal instability limit the application of $LiNiO_2$ [28,29].

$LiMnO_2$ is a promising positive electrode due to its lower cost and toxicity compared to Co or Ni, but the cycling performance of $LiMnO_2$ is still beyond that of $LiCoO_2$ [30].

Subsequent researches are mainly focused on the performance improvement of $LiMnO_2$ positive electrodes by doping with different metals.

Lu et al. reported the structure, synthesis, and electrochemical behavior of layered $Li[Ni_xLi_{(1/3-2x/3)}Mn_{(2/3-x/3)}]O_2$ where $x = 1/3$, $5/12$, or $1/2$; $Li[Ni_xLi_{(1/3-2x/3)}Mn_{(2/3-x/3)}]O_2$ is derived from Li_2MnO_3 or $Li[Li_{1/3}Mn_{2/3}]O_2$ by the substitution of Li^+ and Mn^{4+} with Ni^{2+} while maintaining all the remaining Mn atoms in the 4^+ oxidation state. $Li[Ni_xLi_{(1/3-2x/3)}Mn_{(2/3-x/3)}]O_2$ with $x = 5/12$ can deliver steady capacities of 150 and 160 mA h g^{-1} at 30°C and 55°C, respectively, between 3.0 and 4.4 V using a current density of 30 mA g^{-1}. Differential scanning calorimetry experiments on charged electrodes of $Li[Ni_xLi_{(1/3-2x/3)}Mn_{(2/3-x/3)}]O_2$ for $x = 5/12$ indicate that this material should be safer than $LiCoO_2$. $Li[Ni_xLi_{(1/3-2x/3)}Mn_{(2/3-x/3)}]O_2$ with $x = 1/3$, $5/12$, and $1/2$ can be cycled between 2.0 and 4.6 V to give capacities of about 200, 180, and 160 mA h g^{-1}, respectively, at 30°C. $Li[Ni_xLi_{(1/3-2x/3)}Mn_{(2/3-x/3)}]O_2$ with $x = 1/3$ gives a capacity of 220 mA h g^{-1} at 55°C between 2.0 and 4.6 V using a current density of 30 mA g^{-1} [31].

Gerbrand Ceder et al. used ab initio computational modeling to infer that the combined use of low-valent transition-metal cations and low strain in the activated state are key strategies for increasing the rate capability of layered positive electrode materials, and they have successfully synthesized $Li(Ni_{0.5}Mn_{0.5})O_2$ with very little intralayer disordering to optimize those factors. In agreement with theoretical predictions it was found that positive materials retain their capacity at high rates. The substitution of Co for Ni and Mn can also be used to reduce Li–Ni exchange and improve rate performance, although the use of Co increases the cost and reduces the safety of materials. Although $Li(Ni_{0.5}Mn_{0.5})O_2$ displays an exciting combination of high rate and high capacity, several other factors such as thermal stability, cycle life, and the extra cost from the ion-exchange process, will need to be further investigated before its application in commercial products can be considered. If the outcome of such development studies is positive, $Li(Ni_{0.5}Mn_{0.5})O_2$ could be a potentially positive material for high rate applications [32].

Thackeray et al. studied the design of high capacity (> 200 mA h g^{-1}) Li_2MnO_3-stabilized $LiMO_2$ (M = Mn, Ni, or Co) electrodes for Li-ion batteries, they found that the structurally integrated xLi_2MnO_3 (1−x) $LiMO_2$ (M = Mn, Ni, or Co) compounds, comprised of two-layered components, provide exceptionally high capacities (200 mA h g^{-1}) when used as electrodes in high voltage (4.5–3.0 V) LIBs [33]. $LiNi_xCo_yMn_zO_2$ (NCM, $0 < x, y, z < 1$) is one of the most important positive electrode materials for LIBs due to its high capacity and cost effectiveness compared with $LiCoO_2$ [34]. However, the high-voltage operation of NCM (> 4.3 V) required for high capacity is inevitably accompanied by a more rapid capacity fade over numerous cycles. Kang et al. investigated the degradation mechanisms of $LiNi_{0.5}Co_{0.2}Mn_{0.3}O_2$ during cycling under various cutoff voltage conditions. The surface lattice structures of $LiNi_{0.5}Co_{0.2}Mn_{0.3}O_2$ are observed to suffer from an irreversible transformation; the type of transformation depends on the cutoff voltage conditions. The surface of the pristine rhombohedral phase tends to transform into a mixture of spinel and rock salt phases. Moreover, the formation of the rock salt phase is more dominant under a higher voltage operation (approximate to 4.8 V), which is attributable to the highly oxidative environment that triggers oxygen loss from the surface of the material. The presence of the ionically insulating rock salt phase may result in sluggish kinetics, thus deteriorating the capacity retention. This implies that the prevention of surface structural degradation can provide the means to produce and retain a high capacity, as well as stabilize the cycle life of $LiNi_{0.5}Co_{0.2}Mn_{0.3}O_2$ during high-voltage operations [35]. The investigations of $LiNi_xCo_yMn_zO_2$ give effective guidance for the performance improvement of $LiMO_2$ positive electrodes.

Yabuuchi et al. reported a novel lithium insertion material of $LiCo_{1/3}Ni_{1/3}Mn_{1/3}O_2$ for advanced Li-ion batteries. $LiCo_{1/3}Ni_{1/3}Mn_{1/3}O_2$ showed about 200 mA h g^{-1} of rechargeable capacity with rate capabilities as high as 1.6 A g^{-1} between voltages of 2.5 and 4.6 V. Preliminary cycle tests of a $LiCo_{1/3}Ni_{1/3}Mn_{1/3}O_2$ cell were also carried out for 30 cycles, which showed that the loss of rechargeable capacity was negligible. The thermal behavior of fully charged $LiCo_{1/3}Ni_{1/3}Mn_{1/3}O_2$ was examined by differential scanning calorimetry and it was shown that the exothermic reaction of $\Upsilon_{0.88}Li_{0.12}Co_{1/3}Ni_{1/3}Mn_{1/3}O_2$ with the electrolyte is much milder than that of NiO_2 or CoO_2. The potential application of this material for advanced lithium-ion batteries can be discussed from these results [36].

Shaju et al. prepared layered $Li(Ni_{1/3}Co_{1/3}Mn_{1/3})O_2$ through a mixed hydroxide method and characterized it by means of

X-ray diffraction, X-ray photoelectron spectroscopy (XPS), cyclic voltammetry, and charge–discharge cycling. The hexagonal lattice parameters obtained for the compound are: $a = 2.864$ and $c = 14.233$ angstrom. XPS studies showed that the predominant oxidation states of Ni, Co, and Mn in the compound are 2 +, 3 + and 4 +, respectively with small contents of Ni^{3+} and Mn^{3+} ions. An initial discharge capacity of 160 mA h g^{-1} was obtained in the range of 2.5–4.4 V and at a specific current of 30 mA g^{-1} of which 143 mA h g^{-1} was retained at the end of 40 charge–discharge cycles. At a lower current (10 mA g^{-1}) and in the voltage range of 2.5–4.7 V, a discharge capacity of 215 mA h g^{-1} was obtainable. From the voltage profile and cyclic voltammetry, the redox processes occurring at similar to 3.8 and similar to 4.6 V are assigned to the $Ni^{2+/4+}$ and $Co^{3+/4+}$ couples, respectively [37].

Yang-Kook Sun et al. found that when coated with a uniform and thin $Ni_3(PO_4)_2$ layer with a thickness of 10–20 nm using a simple dry ball-milling method, $Ni_3(PO_4)_2$-coated $LiNi_{0.8}Co_{0.15}Al_{0.05}O_2$ exhibited greatly improved lithium intercalation stability, having a capacity retention of 73% after 100 cycles at 55°C, while that of the pristine material was only 53%. This improvement originated from the stable charge transfer resistance between the positive electrode and the electrolyte and the suppression of surface degradation and bulk structure formation due to the protection of the $LiNi_{0.8}Co_{0.15}Al_{0.05}O_2$ by the $Ni_3(PO_4)_2$ coating [38].

Polyanion compound materials have been employed as positive electrodes such as $LiFePO_4$ [39], $LiMnPO_4$ [40], $LiCoPO_4$ [41], and $LiFeSO_4F$ [42,43]. With relatively high thermal stabilities and power capabilities, these polyanion compound materials are promising positive electrodes for LIBs; however, their low electrical and ionic conductivities limit the application of these materials [44,45].

Olivine-structured $LiFePO_4$ has been the focus of research toward developing low cost, high-performance positive electrode materials for LIBs. Various processes have been developed to synthesize $LiFePO_4$ or C/$LiFePO_4$ (carbon coating on $LiFePO_4$), and some of them are being used to mass produce C/$LiFePO_4$ at the commercial or pilot scale. Due to the low intrinsic electronic and ionic conductivities of $LiFePO_4$, a decrease in particle size and a carbon coating nanolayer on $LiFePO_4$ particle surfaces are necessary to achieve a high electrochemical performance. Significant progress has been made in understanding and controlling the phase purity, particle size, and carbon coating of C/$LiFePO_4$ composite materials [46].

8.2.2 Negative Electrode Nanomaterials for Lithium-Ion Batteries

Negative electrodes are important for the stability and cycle life of LIBs. The general operation of a LIB during the charging and discharging cycle is shown in Fig. 8.3. During the charging process, lithium ions are inserted into the interstitial sites of a graphitic anode, while the electrons migrate toward graphite via an external circuit (Fig. 8.3A and B). Upon connection to a load, the lithium ions are extracted from the anode and diffuse

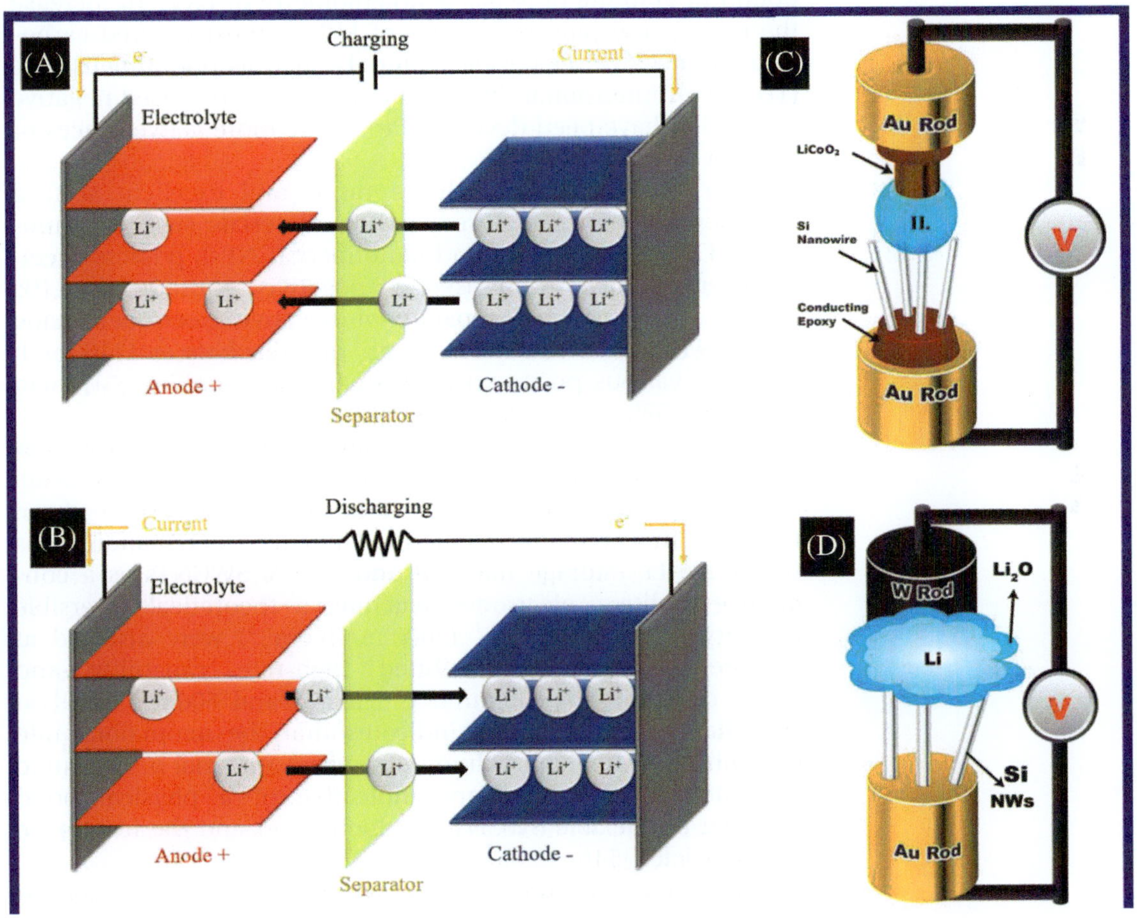

Figure 8.3 General operation of a LIB during the charging (A) and discharging (B) cycles. Lithium ions move from the cathode to the anode via the electrolyte during the charge cycle, while electrons travel through the external wire. The flow direction of charge/ions, both within the electrolyte and external circuit, is reversed during the discharge [47].

toward the cathode via the electrolyte, whereas the electrons transfer through the external wire [47]. Carbon negative electrodes have enabled LIBs to be commercially used and they represent the best choice for negative electrode materials. Carbonaceous materials, especially graphite, are the most used negative electrode materials for rechargeable LIBs [48]. They can avoid the problem of Li dendrite formation by the reversible intercalation of Li into the carbon host lattice, and this provides good cyclability and safety for the negative electrodes of LIBs. However, graphite has a limited theoretical capacity of 372 mA h g^{-1} since the most Li-enriched intercalation compound of graphite only has a stoichiometry of LiC_6. To increase the energy and power densities of LIBs, nanostructured carbonaceous negative electrode materials such as one-dimensional (1D), two-dimensional (2D), and porous carbon–based negative electrodes, have been developed to create more active spaces or sites for Li storage [48].

With high surface-to-volume ratios and excellent surface activities, 1D-nanostructured carbon materials such as nanotubes (CNTs), nanowires, and nanofibers (CNFs), possess excellent performance when used as negative electrodes in LIBs [49,50]. The reversible capacity of negative electrodes made from CNTs can exceed 460 mA h g^{-1} and reach up to 1116 mA h g^{-1} after various posttreatments such as ball-milling [51], acid oxidation [52], and metal-oxide cutting [53].

Ban Chunmei et al. reported a simple two-step process, that is, hydrothermal synthesis and vacuum filtration, to fabricate binder-free, high-rate capability Li-ion negative electrodes from nontoxic and abundant elements. Utilizing Fe_3O_4 nanorods as the active Li^+ storage material and 5 wt.% SWCNTs as a conductive additive, electrodes reaching a theoretical reversible capacity of 1000 mA h g^{-1} (2000 mA h cm^{-3}) were attained at C rate, which also exhibited high-rate capability and stable capacities of 800 mA h g^{-1} at 5C and 600 mA h g^{-1} at 10C. Raman spectroscopy indicates that the anomalous rate capability for this high-volume-expansion material is attributed to charge transfer, perhaps indicative of a physical bond between the flexible SWCNT conductive net and the iron-oxide nanoparticles [54].

Wang Wei et al. reported the synthesis of novel 1D heterostructures comprising vertically aligned multiwall CNTs (VACNTs) containing nanoscale amorphous/nanocrystalline Si droplets deposited directly onto the VACNTs with clearly defined spacing using a simple two-step liquid injection CVD process. The hallmark of these single reactor–derived heterostructures is an interfacial amorphous carbon layer anchoring the nanoscale Si clusters directly to the VACNTs. The defined spacing of the nanoscale Si combined with their tethered CNT architecture allow for the silicon to undergo reversible electrochemical alloying and dealloying with Li with minimal loss of contact with the underlying CNTs. These novel heterostructures thus exhibit impressive reversible stable capacities of approximately 2050 mA h g^{-1} with very good rate capabilities and an acceptable first cycle irreversible loss of approximately 20%, comparable to graphitic negative electrodes, indicating their promise as high capacity Li-ion negative electrodes [55].

Lee et al. studied the effect of pores in hollow CNFs on their negative electrode properties for a lithium rechargeable battery, and they found that the morphological features of the pHCNFs (porous hollow carbon nanofibers) are unique in that the pore size and volume increase while the total surface of the pores is reduced when compared to nonporous HCNFs (hollow carbon nanofibers). In addition, the spacing between the graphene layers in the pHCNFs increases due to the numerous layer-sequential mismatches, which occur in pore formation. These structural changes in HCNFs facilitate lithium-ion insertion and extraction, resulting in improved electrochemical performances. The initial capacity and reversible capacity rates were improved to 1003 mA h g^{-1} and 61.8%, respectively, which are 350 mA h g^{-1} and 7.9% higher than those of nonporous HCNFs. The coulombic efficiency showed a slight reduction, from 99.32% to 98.99% [56].

Graphene is an important 2D carbon material for application in electrochemical studies with its aromatic monolayer of honeycomb carbon lattice. When used as negative electrodes in LIBs, graphene-based materials delivered first-cycle charge and discharge capacities of 1233 and 672 mA h g^{-1}, respectively. Although the coulombic efficiency was low (55%), the reversible capacity can still be preserved to be about 502 mA h g^{-1} after 30 cycles [48].

Doped with metals and metal oxides, the capacity retention and cycle life of LIBs with graphene-based negative electrodes can be significantly improved. SnO_2-graphene nanocomposites prepared by Honma et al. were successfully used to prepare nanocrystalline Sn compounds (SnO_2, SnS_2, SnS) and GNS nanocomposites through hydrothermal methods. Charge–discharge studies on the nanocomposites indicate the high capacity of the nanocomposites for Li-ion storage during the initial cycles. The fast capacity fading in the subsequent cycles reveals the poor conductivity of the nanocomposites. The ratios of Sn compounds (SnO_2) and GNS were altered, among the examined SnO_2:GNS nanocomposite ratios (35:65, 50:50, and 80:20), the nanocomposite with a 50:50 wt.% ratio showed a high Li-ion storage capacity (400 mA h g^{-1} after 25 cycles) and good cyclability [57].

Cheng Hui-Ming et al. studied the oxygen bridges between NiO nanosheets and graphene for the improvement of lithium storage, and they found that NiO nanosheets (NiO NSs) were bonded strongly to the graphene through oxygen bridges. The oxygen bridges mainly originate from the pinning of the hydroxyl/epoxy groups from the graphene onto the Ni atoms of the NiO NSs. The calculated adsorption energies (137 and 1.84 eV for graphene with hydroxyl and epoxy, respectively) of a Ni adatom on oxygenated graphene by binding with oxygen are comparable with that on graphene (126 eV). However, the calculated diffusion barriers of the Ni adatom on the oxygenated graphene surface (223 and 1.69 eV for graphene with hydroxyl and epoxy, respectively) are much larger than that on the graphene (0.19 eV). Therefore the NiO NS is anchored strongly on the graphene through a C–O–Ni bridge, which allows for a high reversible capacity and excellent rate performance. The easy binding/difficult dissociating characteristic of Ni adatoms on the oxygenated graphene facilitates fast electron hopping from graphene to NiO and thus the reversible lithiation and delithiation of NiO [58].

Porous carbons, microporous (pore size <2 nm), mesoporous (2 nm$<$ pore size <50 nm), and macroporous (pore size >50 nm), with different pore sizes, high surface areas, and open pore structures are promising negative electrode materials for Li storage; based on their unique merits, porous carbons often show prominently increased capacities in comparison with traditional graphitic carbons [59].

Hu Yong-Sheng et al. studied the Li storage in hierarchically porous carbon monoliths with a relatively higher graphite-like ordered carbon structure. Macroscopic carbon monoliths with both mesopores and macropores were successfully prepared using meso/macroporous silica as a template and mesophase pitch as a precursor. Owing to the high porosity (providing ionic transport channels) and high electronic conductivity (approximately 0.1 S cm^{-1}), this porous carbon monolith with a mixed conducting 3D network shows a superior high-rate performance if used as a negative electrode material in electrochemical lithium batteries. A challenge for future research as to its applicability in batteries is the lowering of the irreversible capacity [59].

Besides carbon negative electrodes, metals [60], metal alloys [61], metal oxides [62] as well as metal-based compounds are important materials used as negative electrodes in LIBs.

Si-based LIB negative electrodes have recently received great attention as they offer specific capacity to an order of magnitude beyond that of conventional graphite. Magasinski reported a large-scale hierarchical bottom-up assembly route for the formation of Si on the nanoscale containing rigid and robust spheres with irregular channels for the rapid access of Li ions into the particle bulk. Large Si volume changes upon Li insertion and extraction are accommodated by the particles' internal porosity. Reversible capacities over five times higher than that of state-of-the-art negative electrodes (1950 mA h g^{-1}) and stable performances were attained. The synthesis process is simple, low-cost, safe, and broadly applicable, providing new avenues for the rational engineering of electrode materials with enhanced conductivity and power [63]. Zhen Zhou introduced core–double shell $Si/SiO_2/C$ nanocomposites as negative electrode materials for LIBs, the core–double shell $Si/SiO_2/C$ nanocomposites were prepared through a facile route with the combination of high-energy ball-milling, hydrothermal treatment, and annealing processes. SiO_2 and the carbon double shells effectively accommodated the volume swing of Si during repeated cycles and enhanced the electronic network between nanoparticles. Such composites present enhanced cyclic performances as negative electrode materials in LIBs. Si-based negative electrode materials with core–double shell nanostructure provide a strong structural buffer, high electrical conductivity, and low solid electrolyte interphase (SEI) film formation, which may be a feasible solution for their future practical applications [64].

Wang Chunsheng et al. studied the electrochemical performance of porous carbon(C)/tin(Sn) composite negative electrodes for NIBs and LIBs, and the electrochemical performance of mesoporous C/Sn negative electrodes in NIBs and LIBs was systematically investigated. The mesoporous C/Sn negative electrodes in a NIB showed a similar cycling stability but lower capacity and poorer rate capability than those observed in a LIB. The desodiation potentials of Sn negative electrodes are approximately 0.21 V lower than their delithiation potentials. The low capacity and poor rate capability of C/Sn negative electrodes in NIBs are mainly due to their large Na-ion size, resulting in slow Na-ion diffusion and large volume changes of porous C/Sn composite negative electrodes during alloy/dealloy reactions. A better understanding of the reaction mechanism between Sn and Na ions will provide useful insight toward exploring and designing new alloy-based negative electrode materials for NIBs [65].

Tsutomu Miyasaka et al. studied a high-capacity lithium-storage material in metal-oxide form which could replace carbon-based lithium intercalation materials in extensive use as the negative electrode of Li-ion rechargeable batteries. This tin-based amorphous composite oxide (TCO) contains Sn(II)-O as the active center for lithium insertion and other glass-forming elements, which make up an oxide network. The TCO negative electrode yields a specific capacity for reversible lithium adsorption that is more than 50% higher than those of the carbon families that persist after charge–discharge cycling when coupled with a lithium cobalt oxide positive electrode. Lithium-7 nuclear magnetic resonance measurements evidenced the high ionic state of lithium retained in the charged state, in which TCO accepted 8 moles of lithium ions per unit mole [66].

Yu-Guo Guo et al. bound SnO_2 nanocrystals in N-doped graphene sheets as negative electrode materials for LIBs. These hybrid negative electrode materials for LIBs were fabricated by binding SnO_2 nanocrystals (NCs) to N-doped reduced graphene oxide (N-rGO) sheets by means of an in situ hydrazine monohydrate vapor reduction method. The SnO_2NCs in the obtained SnO_2NC/N-rGO hybrid materials exhibited exceptionally high specific capacities and high rate capabilities. The bonds formed between the graphene and the SnO_2NCs limit the aggregation of in situ formed Sn nanoparticles, leading to stable hybrid negative electrode materials with long cycle lives [67].

Cheon et al. studied two-dimensional SnS_2 nanoplates with an extraordinarily high discharge capacity for LIBs; a facile synthesis of 2D layered-SnS_2 nanoplates, which had a nanoscale lateral size of 150 nm, was prepared. Their unique nanoscale characteristics including their finite lateral 2D morphology, cause electrodes fabricated from SnS_2 nanoplates to have remarkably high discharge capacities that are close to the theoretically limiting values. The laterally-confined 2D nanoplates constructed in this effort could be feasible alternative electrode materials for the next generation of LIBs [68].

With the development of LIB research and the promising developments in 2D layered materials, intercalation (nonfaradaic, into the lattice), pseudocapacitance (faradaic, charge transfer due to "intercalation" of Li within the Van der Waals gaps in some layered materials), and defect-mediated higher storage capacities may offer some routes toward better performances in battery and supercapacitor materials [69].

Their large surface-area-to-volume ratios and internal surface areas endow 2D materials with high mobility and high energy density; therefore, 2D materials could be promising candidates for LIBs with comprehensive investigations [70]. Following the discovery of graphene, a new family of 2D transition-metal carbides/nitrides, MXenes, derived from MAX phase precursors, have attracted extensive attention in recent years. The superior physical and chemical properties of MXenes include high mechanical strength, excellent electrical conductivity, multiple possible surface terminations, hydrophilic features, superior specific surface area, and the ability to accommodate intercalants. When applied as electrodes in lithium-based batteries, MXenes have demonstrated excellent performance [71,72].

Employing first-principles calculations, Ya-Meng Li et al. studied the electronic properties and electrochemical performance of double-metal MXenes, namely TiNbC and $TiNbCT_2$ (T = O, F, and OH) as negative electrode materials for LIBs. The results were addressed based on the adsorption stability and diffusion barrier of Li-ions, theoretical storage capacity, and average open circuit voltage (OCV). The calculations indicate that all the TiNbC and $TiNbCT_2$ monolayers exhibit metallic behaviors, ensuring their excellent electronic conductivity. It is found that monolayer $TiNbC(OH)_2$ is unstable for the adsorption of Li; however, $TiNbCF_2$ monolayer is unfavorable as anode material for LIBs because the F-groups and Li-ions could form ring-like structure. In the cases of TiNbC and $TiNbCO_2$ nanosheets, their moderate adsorption energies and low diffusion barriers toward Li display their high-rate performance for Li-ion transport. Moreover, the theoretical Li-ion storage capacity reaches up to 351 mA h g^{-1} (0.750 V for OCV) on the TiNbC monolayer and 290 mA h g^{-1} (1.497 V for OCV) on the $TiNbCO_2$ monolayer. Overall, this study suggests that double-metal TiNbC and $TiNbCO_2$ monolayers can be suitable negative electrode materials for LIBs [73]. Shuaikai Xu et al. combined Ti_2CT_x MXene with graphene oxide (GO) followed by a thermal treatment to fabricate flexible rGO/Ti_2CT_r films, in which the electrochemically active rGO and Ti_2CT_r NSs impede the stacking of layers and synergistically interact to produce ionically and electronically conducting electrodes. The effect of the thermal treatment on the electrochemical performance of Ti_2CT_x was evaluated. As a negative electrode for Li-ion storage, the thermally treated Ti_2CT_r possessed a higher capacity in comparison to the prepared Ti_2CT_x. The freestanding hybrid rGO/ Ti_2CT_r films exhibit excellent reversible capacity (700 mA h g^{-1} at 0.1 A g^{-}1), cycling stability, and rate performance. Additionally, flexible rGO/$Ti_3C_2T_r$ films were made using the same method and these also present improved capacity. Their study provided a simple yet effective approach to combine rGO with different MXenes, which can enhance their electrochemical properties for LIBs [74].

Changzhou Yuan et al. fabricated a novel SnO_x/Ti_3C_2 composite architecture material via a simple yet feasible hydrothermal strategy, in which ultrathin SnO_x NSs were uniformly anchored onto conductive Ti_3C_2. The smart integration of high theoretical capacity SnO_x and high-conductivity Ti_3C_2 combined the intrinsic advantages of the two and offered a pseudocapacitance-dominated Li^+-storage mechanism. Benefitting from its unique structural/componential merits, the fabricated SnO_x/Ti_3C_2 electrode exhibited large specific capacities and long-term cycling life. Impressively, the composite obtained a large reversible capacity of ~540 mA h g^{-1} at a current density of 500 mA h g^{-1} even after 1000 consecutive cycles. Kinetic investigation revealed the high participation of capacitive charges in the SnO_x/Ti_3C_2 negative electrode, which enormously benefitted the achievement of a long cycle life and a high-rate storage of Li ions. More promisingly, their contribution here provides a scalable methodology to synthesize advanced MXene-based hybrid anodes for advanced next-generation LIBs [75].

Sharona et al. measured the performance of MXenes (Ti_2CT_x) combined with electrolytic manganese dioxide (EMD) in three different weight ratios (MXene:EMD = 20:80; 50:50; 80:20) as negative electrodes for LIBs. A study of the structure, composition, and morphology of the synthesized materials was conducted. The materials were further investigated for their electrochemical properties in a half-cell configuration using impedance spectroscopy measurements, cyclic voltammetry, and galvanostatic charge–discharge cycling. The results showed that the combined MXene/EMD material had a greater cycling stability, capacity, and rate capability as compared to the EMD. The best MXene:EMD ratio was found to be 80:20. The capacity obtained for this material after 200 cycles was 460 mA h g^{-1} at a current density of 100 mA h g^{-1}. The Li-ion accessibility improved with cycling. This study provides the first insight into the viability of using one of the lightest known MXenes and EMD composites for improved LIB negative electrodes. As EMD is a low cost and abundant material, it provides great opportunities for improved capabilities for lightweight applications at an affordable cost [76].

Weibin Zhang et al. studied the phonon dispersion and electronic properties of Li^+ adsorbed on Mo_3N_2 with various surface group nanosheets via first principles calculations. All of the terminated-Mo_3N_2 showed excellent stability. This decreases the binding energy of Li atoms due to reduced charge transfer between Li and the Mo_3N_2 layer. The storage capacity and diffusion of Li^+ on the bare and T-terminated Mo_3N_2 (T = H, O, and

OH) was studied via ab initio density functional theory (DFT). The diffusion barriers and the corresponding path lengths of all three of the pathways illustrate that the Li diffusion follows the order: $Mo_3N_2O_2$ > Mo_3N_2 > $Mo_3N_2(OH)_2$ > $Mo_3N_2H_2$, in which the Li^+ migrates more easily and freely on the $Mo_3N_2O_2$ surface due to its lower energy barrier, exhibiting faster transport and higher charge–discharge rates for LIBs. Electronic structure analysis indicated that the excellent conductivity is suitable for use as an electrode material for LIBs [77].

8.2.3 Nanomaterials for Lithium–Sulfur Batteries

LIBs have become prominent over the past two decades, particularly for portable electronics, as they offer much higher energy density than other rechargeable systems. The current Li ion technology is based on insertion-compound negative electrode and positive electrode materials, which limit their charge-storage capacity and energy density. Li–S batteries (LSBs) have become attractive candidates for the next-generation high-energy rechargeable Li batteries because of their high theoretical energy density and cost effectiveness. Sulfur, one of the most abundant elements in the Earth's crust, offers a high theoretical capacity of 1672 mA h g^{-1}, which is an order of magnitude higher than those of transition-metal oxide positive electrodes. The high capacity is based on the conversion reaction of sulfur to form lithium sulfide (Li_2S) by reversibly incorporating two electrons per sulfur atom compared to the one or less than one electron per transition-metal ion in insertion-oxide positive electrodes. A Li–S cell is an electrochemical storage device through which electrical energy can be stored in sulfur electrodes. A schematic of the components in a single Li–S cell and its operation (charge and discharge) is shown in Fig. 8.4 [78]. A conventional Li–S cell consists of a lithium metal negative electrode, an organic electrolyte, and a sulfur composite positive electrode. Because sulfur is in a charged state, the cell operation starts with discharge. During the discharge reaction, lithium metal is oxidized at the negative electrode to produce lithium ions and electrons. The lithium ions produced move to the positive electrode through the electrolyte internally, while the electrons travel to the positive electrode through the external electrical circuit, and thereby an electrical current is generated. Sulfur is reduced to produce lithium sulfide by accepting the lithium ions and electrons at the positive electrode. The reactions occurring during discharge are given here, and backward reactions will occur during charge [79,80].

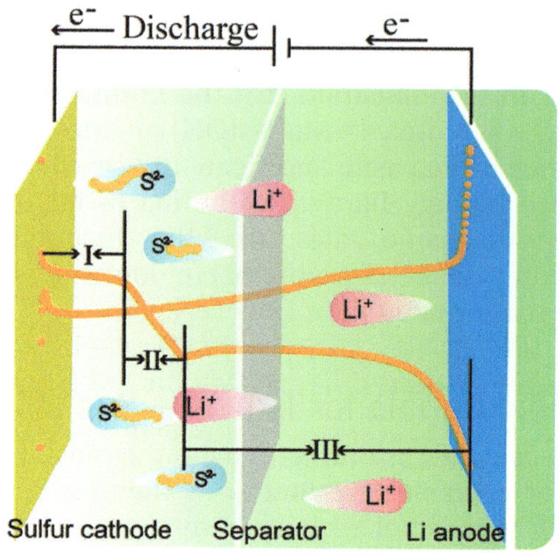

Figure 8.4 The electrochemical mechanism of Li—S batteries [78].

Rechargeable LSBs have attracted significant attention lately due to their high specific energy and low cost. They are promising candidates for several applications including portable electronics, electric vehicles, and grid-level energy storage. However, their poor cycle life and low power capability are major technical obstacles. Various nanostructured sulfur positive electrodes have been developed to address these issues as they provide greater resistance to pulverization, faster reaction kinetics, and better trapping of soluble polysulfides. Significant progress has been made with rechargeable LSBs in recent years by developing novel nanocomposites, efficient electrolytes, and novel cell configurations [81].

Ji Xiulei et al. reported a highly ordered interwoven composite for LSBs. The conductive mesoporous carbon framework precisely constrains sulfur nanofiller growth within its channels and generates essential electrical contact to the insulating sulfur. The structure provides access to Li^+ ingress/egress for reactivity with the sulfur, and it is speculated that the kinetic inhibition toward diffusion within the framework and the sorption properties of the carbon aid in trapping the polysulfides formed during redox. Polymer modification of the carbon surface further provides a chemical gradient that retards the diffusion of these large anions out of the electrode, thus facilitating a more-complete reaction. Reversible capacities of up to 1320 mA h g^{-1} are attained. The assembly process is simple and broadly applicable, conceptually providing new opportunities for materials scientists for tailored design that can be extended into many different electrode materials [82].

Dai Hongjie et al. reported the synthesis of a graphene–sulfur composite material by wrapping poly(ethylene glycol) (PEG)-coated submicrometer sulfur particles with mildly-oxidized graphene oxide sheets decorated by carbon black nanoparticles. The PEG and graphene coating layers are important to accommodating the volume expansion of the coated sulfur particles during discharge, trapping soluble polysulfide intermediates, and rendering the sulfur particles electrically conductive. The resulting graphene–sulfur composite showed high and stable specific capacities of up to approximately 600 mA h g^{-1} over more than 100 cycles, representing a promising positive electrode material for rechargeable lithium batteries with high energy density [83].

Mikhaylik et al. reported a quantitative analysis of the shuttle phenomenon in Li–S rechargeable batteries. Their work encompasses theoretical models of the charge process, charge and discharge capacity, overcharge protection, thermal effects, self-discharge, and a comparison of simulated and experimental data. The work focused on the features of polysulfide chemistry and polysulfide interaction with the Li negative electrode, a quantitative description of these phenomena, and their application in the development of a high-energy rechargeable battery. The objective was to present experimental evidence that self-discharge, charge–discharge efficiency, charge profile, and overcharge protection are all facets of the same phenomenon [84].

Jayaprakash et al. studied a facile, scalable procedure for synthesizing C–S nanocomposites based on mesoporous, hollow carbon capsules. The method uses a template-based approach for synthesizing hollow carbon particles with desirable features and vapor phase infusion of elemental sulfur into the carbon framework to produce a fast, efficient uptake of elemental sulfur. When evaluated as the positive electrode material in a Li–S secondary battery, the prepared C–S nanocomposite displayed outstanding electrochemical features at both low and high current densities. To the best of our knowledge the materials reported herein are among the first to offer extended cycle life and high charge rate capability in a secondary LSB. These observations are attributed to the sequestration of elemental sulfur in the carbon capsules and to its favorable effect in limiting polysulfide shuttling, as well as to the enhanced electron transport of the poorly conducting sulfur made possible by its close contact with the carbon framework [85].

Ji Liwen et al. employed a chemical approach to immobilize sulfur and lithium polysulfides (LiPSs) via the reactive functional groups on graphene oxide. This approach enabled the authors to obtain a uniform and thin (around tens of nanometers) sulfur coating on graphene oxide sheets by a simple chemical reaction–deposition strategy and a subsequent low-temperature thermal treatment process. Strong interaction between graphene oxide and sulfur or polysulfides enabled us to demonstrate LSBs with a high reversible capacity of 950–1400 mA h g^{-1} and stable cycling for more than 50 deep cycles at 0.1C (1C = 1675 mA g^{-1}) [86].

Based on first-principles calculations, Xiaobiao Liu et al. proposed a promising sulfur host material, S-terminated Ti_2C MXene, for LSBs. They demonstrated from that strong Ti–S interactions lead to stable S-terminated Ti_2C MXene (Ti_2CS_2). Compared with other surface-functionalized Ti_2C MXenes such as Ti_2CO_2 and Ti_2CF_2, Ti_2CS_2 has the highest affinity to polysulfides and thus the highest efficiency to suppress the shuttle phenomenon of polysulfides. Moreover, the metallic features of Ti_2CS_2 facilitate electrochemical activity during the charge and discharge processes. The low energy barrier of the Li diffusion on the Ti_2CS_2 surface is expected to assist the electrochemical process. These advantages of Ti_2CS_2 shed light on the design of sulfur host materials for high-performance LSBs [87].

Hong Pan et al. synthesized an accordion-like titanium oxide-Ti_3C_2 material as a high-efficiency sulfur host for a LSBs, which contains the features of high conductivity, high surface area, and rich LiPSs adsorption sites simultaneously. As a result, the S/$Ti_3C_2O_x$ cathode delivered a better rate performance and higher capacity than those of the S/Ti_3C_2 cathode. Their study provides an effective approach to improve the conductivity, suppress the shuttling effect, and remit volume expansion for LSBs. And it is a promising strategy to design advanced LSBs with high capacity and energy density [88].

Dewei Rao et al. systematically studied the behaviors of Li_2S_m on bare MXene materials, Ti_nX_{n-1} (X = C, N; n = 2, 3, 4), and functionalized Ti_2C by intensive density functional theory (DFT) calculations. The long-chain sulfides (Li_2S_4, Li_2S_6, Li_2S_8) can be entrapped by − OH and bare MXenes, and other sulfides (Li_2S and Li_2S_2) can be confined by − O and bare MXenes, resulting in the improvement of the utilization of the active materials of sulfur. The Coulomb interactions between Li_2S_m and MXenes dominate the binding, especially with Ti-S and H-S bonds, as well as Li − O interactions. Additionally, the excellent electronic conductivity of bare and OH-terminated MXenes would lead to a high rate performance of LSBs. This work well explains the influence of MXenes on the performance of LSBs found in experiments, which will provide theoretical guidance and support for the utilization of MXene in LSBs or maybe in other batteries [89].

Yu Yao et al. rationally designed an advanced S/Mxe/PDA hybrid (by confinement of sulfur in polydopamine-coated MXene nanosheets) as a high-performance LSB cathode. This advanced composite cathode exhibits an inner Mxe matrix that could immobilize polysulfides via the formation of stable Ti–S binding and the outer conductive PDA sheath not only possesses high mechanical strength for accommodating volume expansion, but also offers strong chemical adsorption for confining polysulfides. The results of DFT calculations and the visual observation experiments confirm the strong adsorbability between S/Mxe/PDA and polysulfides. The S/Mxe/PDA cathode delivers outstanding electrochemical properties even with high areal sulfur loadings (556 mA h g^{-1} after 330 cycles at 0.5C with a low degradation rate of 0.09% per cycle with 4.4 mg cm^{-2} sulfur loading) [90]. Weizhai Bao et al. prepared $Ti_3C_2T_x$/rGO/S composites through a facile and effective liquid-phase impregnation method. Owing to the synergistic effects of the $Ti_3C_2T_x$ NSs and the rGO NSs and the unique 3D layer structure, the $Ti_3C_2T_x$/rGO/S composites achieved both high capacity and stable cyclability as cathodes for LSBs. The XPS measurements confirmed the formation of Ti–S and S–O bonds in the $Ti_3C_2T_x$/rGO/S composites, illustrating the strong chemisorption capability for the trapping of sulfur and polysulfides. The $Ti_3C_2T_x$/rGO/S composites demonstrate high specific capacity and good cycling performance for application in LSBs [91].

8.3 Nanomaterials for Sodium-Ion Batteries

With a similar structure and working mode compared with LIBs, NIBs represent a promising renewable energy storage device based on the relative lower cost of Na than Li. In contrast to Li, Na resources are unlimited, and Na is one of the most abundant elements in the Earth's crust. Infinite Na resources are also found in the ocean. Additionally, Na is the second-lightest and -smallest alkali metal next to Li; on the basis of material abundance and standard electrode potential, rechargeable sodium batteries are an ideal alternative to LIBs [92–94].

8.3.1 Nanomaterials for Sodium-Ion Batteries

NIBs receive significant attention for electrochemical energy storage and conversion owing to their wide availability and the low cost of Na resources. However, NIBs face challenges of low specific energy, short cycling life, and insufficient specific power, owing to the heavy mass and large radius of Na^+ ions. As an important component of NIBs, positive electrode materials have a significant effect on the electrochemical performance of NIBs. The most recent advances and prospects of inorganic and organic positive electrode materials are summarized here. Among current positive electrode materials, layered transition-metal oxides achieve high specific energies of around 600 mW h g^{-1} owing to their high specific capacities of 180–220 mA h g^{-1} and their moderate operating potentials of 2.7–3.2 V (vs Na^+/Na). Porous $Na_3V_2(PO_4)_3/C$ nanomaterials exhibit excellent cycling performance with almost 100% retention over 1000 cycles owing to their robust structural frameworks. Recent emerging positive electrode materials such as amorphous $NaFePO_4$ and pteridine derivatives show interesting electrochemical properties and attractive prospects for application in NIBs. Future work should focus on strategies to enhance the overall performance of positive electrode materials in terms of specific energy, cycling life, and rate capability with cationic doping, anionic substitution, morphology fabrication, and electrolyte matching [95].

Liming Dai reported high-performance NIBs based on a 3D negative electrode from N-doped graphene foams; they found that the 3D N-doped graphene foams (N-GF) can be used as negative electrodes to significantly improve the overall performance of NIBs. Specifically, it was found that the 3D N-GF delivered an unusually high initial reversible capacity of 852.6 mA h g^{-1} at a current density of 1C (1C = 500 mA g^{-1}) between 0.02 and 3 V. After 150 cycles, the N-GF could still maintain a charge capacity of 594 mA h g^{-1} with 69.7% retention of the initial charge capacity, significantly outperforming previously reported carbonaceous materials. The observed superb performance of the 3D N-GF negative electrode in NIBs was attributed to synergistic effects associated with the 3D mesoporous structure with a well-defined porosity, large surface area, and enlarged lattice spacing between graphene layers, coupled with the N-doping-induced defects, to facilitate the diffusion of the large-size Na ions, enhance the storage of Na ions, and minimize the effect of volume expansion during discharge–charge processes [96].

Mai Liqiang et al. studied the effect of carbon matrix dimensions on the electrochemical properties of $Na_3V_2(PO_4)_3$ nanograins for high-performance symmetric NIBs and found that $Na_3V_2(PO_4)_3$ nanograins dispersed in acetylene carbon (AC) show the best electrochemical properties, thus making them a promising candidate for NIBs. This material exhibits a discharge capacity of 117.5 mA h g^{-1} at 0.5C in the initial stage, approximately 100% of the theoretical capacity of $Na_3V_2(PO_4)_3$, and a very stable cycling performance of 96.4% capacity retention at a 5C rate over 200 cycles. This synthesized material can be assembled into symmetric NIBs, and $Na_3V_2(PO_4)_3$/AC in the symmetric batteries exhibited excellent performance of 80% capacity retention after 200 cycles at 1C, and can tolerate a very high current rate of 10C. This superior performance may be attributed to the 3D sodium intercalation/deintercalation pathways and electron transport, which provide a significantly easier sodium path and better conductivity compared with $Na_3V_2(PO_4)_3$/CNT and $Na_3V_2(PO_4)_3$/graphite. Moreover, the 3D carbon framework provides robustness for the structure during cycling, thus endows this material with an excellent cycling performance [97].

Shi Zheng-Tian et al. synthesized hierarchical nanotubes assembled from MoS_2-carbon monolayer–sandwiched superstructure nanosheets for high-performance NIBs, in their studies the superstructure nanotubes were demonstrated as a robust negative electrode material for sodium storage with superior electrochemical performance. They deliver a high rate-capability and maintain discharge capacities of 295 and 187 mA h g^{-1} at high current densities of 10.0 and 20.0 A g^{-1}, respectively. Furthermore, they show a durable cycling life (capacity retention of 101.3%, 108.2%, and 107.8% after 200 cycles at current densities of 0.2, 0.5, and 1.0 A g^{-1}, respectively, in comparison to those of the second cycles), and an initial Coulombic efficiency as high as 84%. The MoS_2:C superstructure nanotubes performed best among the current MoS_2-based electrode materials [98].

Wang Chunsheng et al. developed a simple solid-state reaction method, in which carbon-coated SnS_2 (SnS_2/C) negative electrode materials were synthesized by annealing metallic Sn, sulfur powder, and polyacrylonitrile in a sealed vacuum glass tube. The SnS_2/C nanospheres with unique layered structure exhibit a high reversible capacity of 660 mA h g^{-1} at a current density of 50 mA g^{-1} and maintain at 570 mA h g^{-1} for 100 cycles with a degradation rate of 0.14% per cycle; demonstrating one of the best cycling performances in all reported SnS_2/C negative electrodes for NIBs to date. The superior cycling stability of SnS_2/C electrodes is attributed to the stable nanosphere morphology and structural integrity during charge–discharge cycles as evidenced by ex situ characterization [99].

Ni Jiangfeng et al. studied the sodium storage in $Na_2Ti_3O_7$ nanotube arrays; the fabrication of the nanotube arrays involved hydrothermal growing of $Na_2Ti_3O_7$ nanotubes, surface deposition of a thin layer of TiO_2, and subsequent sulfidation. The resulting nanoarrays exhibit a high electrochemical Na-storage activity that outperforms other $Na_2Ti_3O_7$-based materials. They deliver high reversible capacities of 221 mA h g^{-1} and exhibit a superior cycling efficiency and rate capability, retaining 78 mA h g^{-1} at 10C (1770 mA g^{-1}) over 10,000 continuous cycles. In addition, the full cell consisting of an $Na_2Ti_3O_7$ nanotube negative electrode and an $Na_{2/3}(Ni_{1/3}Mn_{2/3})O_2$ positive electrode is capable of delivering a specific energy of approximately 110 Wh kg^{-1} (based on the mass of both electrodes). The surface engineering provided useful tools for the development of high-performance negative electrode materials with robust power and cyclability [100].

Ji Xiaobo et al. reported a graphene-rich wrapped petal-like rutile TiO_2 tuned by carbon dots for high-performance sodium storage; they observed the function of CDs as “designer additives,” which induce rutile TiO_2 from nanoparticles to nanoneedles, which further self-organize to a 3D petal-like structure. In addition, each of the well-defined TiO_2 nanoneedles is uniformly and conformably capped with curved and layered graphene structures. Together with its nanoscale size, 3D petal-like structure, and graphene-rich wrapping, G/P–$RTiO_2$ exhibits outstanding sodium-storage properties. A high capacity of 245.3 mA h g^{-1} at a rate of 0.25C (83.75 mA g^{-1}) after 300 cycles is achieved; even at a high current density of 12.5C (4187.5 mA g^{-1}), a considerable capacity of 59.8 mA h g^{-1} can still be maintained. Notably, the reversible capacity up to 1100 cycles at a current density of 2.5C (837.5 mA g^{-1}) can still reach 144.4 mA h g^{-1}; even after 4000 cycles at 10C (3350 mA g^{-1}), a capacity retention of as high as 94.4% is obtained, demonstrating a marvelous, durable, long-term cycle life [101].

Zhou Haoshen et al. investigated monodispersed hierarchical Co_3O_4 spheres intertwined with carbon nanotubes for use as negative electrode materials in NIBs. The assemblies of the Co_3O_4 nanoparticles (about 5 nm), Co_3O_4/CNTs, have a uniform particle size of around 200 nm; used as a negative electrode material for NIBs its reversible capacity can reach 487 mA h g^{-1}, which is considerably higher than that of hard carbon. In an electrolyte of 1 M $NaPF_6$/EC + DEC with 2% FEC, the Co_3O_4/CNTs composite shows a good cycling performance. The composite also shows a high rate performance. Even at a high current of 3200 mA h g^{-1}, it still shows a high capacity of 184 mA h g^{-1}. The excellent electrochemical performance can be attributed to the formation of mixed electron and ion conducting networks and the monodispersed hierarchical spheres in the Co_3O_4/CNTs composite [102].

Xu Yang et al. studied the benefits of oxygen vacancies (OVs) in NIBs performance. A series of measurements show that OVs increase the electric conductivity and Na-ion diffusion coefficient, and the promotion from ultrathin coatings lie in the effective reduction of the cycling-induced solid–electrolyte interphase. The coated NSs exhibited a high reversible capacity and great rate capability with capacities of 283.9 (50 mA g^{-1}) and 179.3 mA h g^{-1} (1 A g^{-1}) after 100 cycles [103].

Zhang Shilin et al. reported S-doped mesoporous carbon from a surfactant-intercalated layered double hydroxide precursor as a high-performance negative electrode nanomaterial for both LIBs and NIBs. They prepared an S-doped mesoporous amorphous carbon (SMAC) from a commercially available alkyl surfactant sulfonate anion-intercalated NiAl-layered double hydroxide precursor via thermal decomposition and subsequent acid leaching. The resultant amorphous carbon was endowed with the integrated advantage of featuring a high reversible capacity and a long cycling stability, intrinsic doping of sulfur, large specific area, and broad mesopore size distribution. Electrochemical evaluation showed that the SMAC electrode exhibits highly enhanced electrochemical performances compared with electrodes of nondoped mesoporous and amorphous carbon prepared using a different surfactant (sodium laurate). A high reversible capacity of 958 mA h g^{-1} was achieved for the SMAC electrode after 110 cycles at 200 mA g^{-1}, and especially a super long cycle life with a reversible capacity of 579 mA h g^{-1} after 970 cycles at 500 mA g^{-1}. Moreover, the SMAC electrode can facilitate the reversible insertion–extraction of Na ions, owing to the proper specific area and mesopore size distribution, as well as the improved electronic conductivity resulting from the doping of sulfur [104].

Qian Yitai et al. studied $MoSe_2$-covered N,P-doped carbon NSs as a long-life and high-rate negative electrode material for NIBs. $MoSe_2$ grown on N,P-codoped carbon NSs was synthesized by a solvothermal reaction followed with a high-temperature calcination. This composite has an interlayer spacing of $MoSe_2$ expanded to facilitate Na-ion diffusion, $MoSe_2$

immobilized on carbon NSs to improve the charge-transfer kinetics, and N and P incorporated into carbon to enhance its interaction with active species upon cycling. These features greatly improve the electrochemical performance of this composite as compared to all the controls. It presents a specific capacity of 378 mA h g^{-1} after 1000 cycles at 0.5 A g^{-1}, corresponding to 87% of the capacity at the second cycle. Ex situ Raman spectra and high-resolution transmission electron microscopy images confirm that it is the element Se, rather than $MoSe_2$, that is, formed after the charging process. The interaction of the active species with modified carbon is simulated using density functional theory to explain this excellent stability. The superior rate capability, where the capacity at 15 A g^{-1} equals to approximately 55% of that at 0.5 A g^{-1}, could be associated with the significant contribution of pseudocapacitance. By pairing with homemade $Na_3V_2(PO_4)_3/C$, this composite also exhibits excellent performances in full batteries [105].

Satoshi Kajiyama et al. demonstrated that an MXene $Ti_3C_2T_x$ electrode in a nonaqueous Na^+ electrolyte exhibited reversible desolvated Na^+ intercalation/deintercalation into the stacked $Ti_3C_2T_x$ layers on sodiation/desodiation. The interlayer distance was expanded by the first sodiation process because of desolvated Na^+ intercalation and solvent molecule penetration. While the trapped Na^+ behaves as a pillar, the penetrated solvent molecules swell the interlayer space, both of which contribute to keeping the interlayer distance constant during the sodiation/desodiation processes. Therefore the electrochemical reaction is not accompanied by any substantial change in the interlayer distance, leading to a high cycle stability in addition to fast Na^+ diffusion in the expanded interlayer space. Such features make MXene-based electrodes promising potential candidates for advanced energy storage devices [106].

Yang et al. massively synthesized hierarchical layered $NaTi_2(PO_4)_3/Ti_3C_2$ (NTP/TC) nanocomposites as promising negative electrodes for NIBs via a simple yet scalable solvothermal strategy. Benefiting from the expanded interplanar distance, the high-capacity of the nano-NTP, and the high electrical conductivity of the Ti_3C_2, the NTP/TC electrode was endowed with more Na^+-storage sites, and reduced barriers of Na^+ ion mobility. More significantly, this flexible strategy is highly promising for constructing other MXene-based composites for versatile energy-related applications [107].

Youquan Zhang et al. synthesized SnS nanoparticle-modified $Ti_3C_2T_x$ MXene composites as negative electrodes for NIBs via hydrothermal and annealing methods. The SnS nanoparticles were anchored onto $Ti_3C_2T_x$ composites; the SnS/$Ti_3C_2T_x$ electrode exhibits a high reversible capacity and good rate performance. The superior electrochemical performance is due to the synergistic effects between the $Ti_3C_2T_x$ sheets and the SnS nanoparticles. The results demonstrate that incorporating a low dimensional material with a high specific capacity into the layers space of $Ti_3C_2T_x$ produces a promising material as a negative electrode for rechargeable batteries [108].

Mengli Tao et al. successfully synthesized a $CoNiO_2/Ti_3C_2T_x$ composite by a combination of hydrothermal and annealing treatments. The prepared interconnected multilayered-structure of the $CoNiO_2/Ti_3C_2T_x$ composite ensures fast ion diffusion and reduces volume expansion during the charging and discharging processes, thus leading to excellent electrochemical performance. This work indicates that the $CoNiO_2/Ti_3C_2T_x$ composite is a promising candidate for NIBs [109].

Cheng Zeng et al. employed the confined transformation of MXene/rGO hybrid films to fabricate integrated films of titanate/rGO with a sandwiched structure. Benefiting from a combination of the unique sandwiched structure, the reduced diffusion distance originating from the ultrathin sodium titanate (NTO)/potassium titanate (KTO) layers, the enhanced electronic conductivity of graphene, and the good compatibility between the layered NSs, the flexible layer titanate/rGO films exhibited excellent performance and long cycling stability for NIBs and potassium-ion batteries (PIBs) [110].

8.3.2 Nanomaterials for Sodium–Sulfur Batteries

Traditional Na–S batteries hold notable advantages including high energy density (theoretical value: 760 W h kg^{-1}) and efficiency (approaching 100%), low material cost (due to the rich abundances of Na and S in nature), and a long life, all these benefits make them especially promising for stationary storage applications [111].

It is noticeable that one of the most significant advantages of the present commercial tubular designed sodium–sulfur (NAS) battery is its higher energy density in comparison with LIBs and the vanadium redox flow batteries under development. The enhancement of its power density is very important in widening the application area of the NAS battery. Different designs of planar type new electrode systems workable at low temperature (100°C–200°C) in ambient-temperature sodium (ion) battery are attractive candidates. Further researches are still needed to improve the performance of NAS batteries; in order to improve their electrochemical performance, particular attention should be paid to the interface behaviors along the inner circuit, while the optimization of the ceramic electrolyte/electrode interfaces is among the best strategies [112].

Ambient- or room-temperature sodium–sulfur (RT Na–S) batteries are gaining much attention as a low-cost option for large-scale electrical energy storage applications. However, their adoption is hampered by severe challenges. The main challenges for RT Na–S batteries are similar to those for LSBs including the poor electronic conductivity of fully charged (elemental sulfur) and fully discharged (Na_2S_2, Na_2S) products, structural changes of the positive electrodes during cycling, migration of the intermediate charge/discharge products (sodium polysulfides) to the negative electrode, and the formation and growth of Na dendrites at the negative electrode. One promising strategy to overcome these critical issues is the integration of a solid electrolyte separator with Na–S chemistry. Compared to high-temperature molten electrode Na–S technology, the RT Na–S battery system is viable for a broad range of applications. From an economical point of view, RT Na–S chemistry provides an extremely low-cost energy-storage system. However, concerted efforts need to be made and many problems need to be overcome before a low-cost RT Na–S system can be implemented as a viable energy-storage technology [113].

Wang Jiulin et al. studied room temperature NAS batteries with sulfur composite positive electrode materials. In this study, the NAS batteries were assembled with a Na metal negative electrode, liquid electrolyte, and an S composite positive electrode. The charge/discharge curves of the NAS batteries indicated that Na could reversibly react with S embedded in the sulfurized polymer matrix at room temperature and the reactions of the NAS battery could be written as: $2Na + xS \rightarrow Na_2S_x$. During the first discharge process, the specific capacity was 654.8 mA h g^{-1} based on the composite positive electrode materials and the calculated discharge capacity of pure S was 1455 mA h g^{-1}. The reversible specific capacity of the S composite positive electrode was about 500 mA h g^{-1} and it kept stable in the subsequent cycles with close to 100% charge/discharge efficiencies [114].

Cheol-Wan Park et al. studied the discharge properties of an all-solid sodium–sulfur battery (ASNSB) using a poly(ethylene oxide) (PEO) electrolyte. The ASNSB using a PEO polymer electrolyte gave a high initial discharge capacity of 505 mA h g^{-1}

sulfur at 90°C with plateau potential regions at 2.28 and 1.73 V. From thermodynamic considerations, the lower plateau region should originate from the formation of Na_2S, Na_2S_2, and Na_2S_3 and the upper plateau region from Na_2S_4 and Na_2S_5. The discharge capacity decreased continuously during repeated charge–discharge cycling, but remained at 166 mA h g^{-1} S after 10 cycles, which is higher than was observed for a Na/PVDF/S battery at room temperature. The PEO electrolyte with $NaCF_3SO_3$ salt has a Na ion conductivity of 3.38×10^{-4} S cm^{-1} at 90°C. It is possible to fabricate an ASNSB at 90°C, but its cycling property has to be improved for practical applications [115].

Park et al. studied the room-temperature solid-state sodium/sulfur battery using polyvinylidene-fluoride-hexafluoropropene (PVDF) polymer electrolyte. Solid Na–S batteries may be composed of solid composite–type sulfur electrodes, sodium metal electrodes, and a PVDF gel polymer electrolyte. The PVDF gel polymer electrolyte with a tetraglyme plasticizer and $NaCF_3SO_3$ salt had a high Na-ion conductivity of 5.1×10^{-4} S cm^{-1} at 25°C. During the first discharge, the NAS battery showed two plateau potentials of 2.27 and 1.73 V, respectively. The first discharge capacity was 489 mA h g^{-1} S at room temperature, which was similar to the high-temperature battery. The discharge capacity drastically decreased by repeated charge–discharge cycling, and remained at 40 mA h g^{-1} after 20 cycles [116].

Douglas et al. demonstrated that ultrafine sizes (similar to 4.5 nm, average) of iron pyrite, or FeS_2, nanoparticles are advantageous to sustain reversible conversion reactions in NIBs and LIBs. This is attributed to the nanoparticle size comparable to or smaller than the diffusion length of Fe during cation exchange, yielding thermodynamically reversible nanodomains of converted Fe metal and Na_xS or Li_xS conversion products. This is comparable to bulk-like electrode materials, where kinetic and thermodynamic limitations of surface-nucleated conversion products inhibit successive conversion cycles. Reversible capacities over 500 and 600 mA h g^{-1} for Na and Li storage were observed for ultrafine nanoparticle with improved cycling and rate capability. Unlike alloying or intercalation processes, where SEI effects limit the performance of ultrafine nanoparticles, this work highlights the benefit of quantum dot length-scale nanocrystal electrodes for nanoscale metal sulfide compounds that store energy through chemical conversion reactions [117].

Wei Shuya et al. reported a room-temperature NAS battery that uses a microporous C–S composite positive electrode, and a liquid carbonate electrolyte containing the ionic liquid, 1-methyl-3-propylimidazolium-chlorate, tethered to SiO_2 nanoparticles. It is shown that these batteries can cycle stably at a rate of 0.5C (1C = 1675 mA h g^{-1}) with 600 mA h g^{-1} reversible capacity and nearly 100% Coulombic efficiency. By means of spectroscopic and electrochemical analysis, the particles were found to form an Na-ion conductive film on the negative electrode, which stabilizes the deposition of Na. They also found that S remains interred in the C pores and undergoes solid-state electrochemical reactions with the Na-ions [118].

Ryu Hosuk et al. studied the discharge reaction mechanism of an RT Na–S battery with a tetra ethylene glycol dimethyl ether (TEGDME) liquid electrolyte and $NaCF_3SO_3$ salt with a sodium ionic conductivity of 3.9×10^{-3} S cm^{-1} at 25°C. The discharge curve has two regions, that is, a sloping region between 2.23 and 1.66 V and a plateau region of 1.66 V. The Na–S cell using liquid electrolyte shows a first discharge capacity of 538 mA h g^{-1} at room temperature, which is lower than that of high-temperature NAS batteries. The discharge capacity remains at about 250 mA h g^{-1} until the 10th cycle after sharply decreasing in the second cycle. From XRD, DSC, and EDS results, the final discharge products are Na_2S_3 and Na_2S_2. The decrease in discharge capacity is due to a decrease in active material by the dissolution of S or Na polysulfides into the electrolyte and the irreversible reduction from Na sulfides to elemental S at full charge [119].

Wenzel Sebastian et al. studied the thermodynamics and cell chemistry of RT Na–S batteries with liquid and liquid–solid electrolyte, and it was shown by means of X-ray photoelectron spectroscopy that the cell reaction is incomplete, but it proved that the end members of the cell reaction (S and Na_2S) form among the expected polysulfide species Na_2S_x. The utilization of S can be improved by employing a solid electrolyte membrane (beta-alumina) that prevents the diffusion of the soluble polysulfide species toward the Na side. As an important finding, the Na^+ conduction within the solid electrolyte phase and across the two liquid–solid interfaces results in only small overpotentials. Nevertheless, the utilization of S in the present RT Na–S (475 mA h g^{-1}) batteries is lower than the theoretical value (1675 mA h g^{-1}). One probable reason is the chemical instability of the widely used PVDF binder. Also, the thermodynamic properties of RT Na–S batteries operating at room temperature are discussed and compared with the currently much more studied RT Li–S batteries [120].

8.4 Rechargeable Batteries for Driving Catalysis and Sensing

Accordingly, monitoring systems are seen as central tools for ecosystem-based environmental management, helping on one hand to accurately describe the water column and substrate biophysical properties, and on the other hand to correctly steer sustainability policies by providing timely and useful information to decision makers [121]. Power supplies that provide energy to run environmental sensors and drive environmental catalysis in environmental applications are important for the system reliability of environmental sensors. Collecting raw data from a wireless sensor network for environmental monitoring applications can be a difficult task due to the high energy consumption involved. This is especially difficult when the application requires specialized sensors that require very high energy consumption [122].

Directed energy research and development is generating more powerful portable devices designed to support operational, environmental, clinical, point detection, and remote-sensing applications [123]. Batteries, especially rechargeable batteries, typically serve not just to supply the system with energy, but also to efficiently store energy harvested from the environment. In this way, energy can be stored for times when it cannot be directly extracted from the surroundings [16].

Zhong Lin Wang et al. successfully fabricated a stand-alone, nanogenerator (NG)-powered mercury sensor utilizing ZnO nanowire (NW)-based generators and a SWCNT-based Hg^{2+} ion detector. Here, harvested energy from high-performance NGs was able to power all detection operations and light up an LED indicator consequently. The NGs were composed of 10 layers of a ZnO NW arrays and Au films both on flexible substrates which increased the number of effective contacts. They could utilize a single-walled carbon nanotube field effect transistor (SWNTFET) as a Hg^{2+} sensor owing to the difference in standard potential between SWCNTs and Hg^{2+} ions [124]. They also turned waste a milk-carton into a fully packaged arc-shaped triboelectric nanogenerator (AS-TENG) which could scavenge both wind and water flow energy for long-term wireless environmental monitoring. In general, milk cartons are laminated (from the inside out) with a layer of polyethylene (PE), a layer of Al foil, and a pulp layer. Such a laminated material is suitable as an electrical unit in AS-TENGs. The arc-shaped design of the AS-TENG was implemented to disrupt the flow, leading to vortex induced vibration, and thus the contact electrification of the AS-TENG results in power generation. By optimizing the structural parameters systematically, the AS-TENG achieved an instantaneous peak voltage of 600 V, and a current of 40 μA. Moreover, a self-powered wireless environmental monitoring system using an AS-TENG as a power supply was developed for in-situ real time water quality (pH value) monitoring and landslide early warning systems in natural environments. This study provides solid progress toward the practical application of TENGs in environmental monitoring. Considering the environment-friendly, portable, low cost, and easily scalable monitoring systems achieved, this work makes a significant step toward the practical application of TENGs for harvesting fluid flow energy to power wireless sensor nodes in long-term environmental monitoring [125].

For long-term applications of sensors and catalysis, in which power needs to be delivered for multiple years, larger batteries will need to be applied or a rechargeable battery in combination with an energy harvester. It is, however, expected that in future the performance of several types of batteries will be improved, and therefore the applicability of batteries will increase [126].

A key limitation to the practical incorporation of nanostructured materials into emerging applications is the challenge of achieving low-cost, high-throughput, and highly replicable scalable nanomanufacturing techniques to produce functional materials, envisioning the bottleneck for many commercial applications of nanomaterials to be low-cost, reliable, and scalable processing routes that build the foundation for product development and design, this nanomanufacturing approach, which seems unlimited in its versatility in the choice of materials, brings potential for many applications extending beyond energy storage into areas of energy conversion, sensing, catalysis, optoelectronics, protective coatings, and others [127].

8.5 Challenges and Future Perspectives

Environmental sensors represent a new way to sense and understand the environment, owning a huge potential in many areas of environmental sciences. Environmental catalytic technologies play a major role in both pollution abatement and prevention. Power supplies that provide energy to run environmental sensors and driving environmental catalysis in environmental applications are important for the system reliability of environmental sensors. The application of environmental catalysts and sensors in pollutant sensing and environmental catalysis brings opportunities for batteries to drive catalysis and sensing. The choice of battery for environmental catalysis and sensors can be approached from many perspectives, supplying the system with energy and storing energy harvested from the environment. In this way, rechargeable batteries are needed.

LIBs (Li–S) and NIBs (Na–S) have clear fundamental advantages based on their energy density, cycle life, and efficiency in renewable energy storage after systematic researches. However, continuous researches are urgently needed on the investigation of new electrodes with lower cost, high energy density, high-power density, as well as excellent cycle life and safety for their large-scale commercial application. Therefore enormous efforts have been made to explore nanomaterials for LIBs (Li–S) and NIBs (Na–S) to realize the requirements for their large-scale commercial application.

With the development and maturing of science for producing high-energy and high-power batteries, the preparation of nanostructured electrode materials with lower cost, high energy density, high-power density, as well as excellent cycle life and safety via more sustainable and greener strategies or from renewable resources can be addressed to limit environmental pollution and to secure a bright future for human beings and the world as a whole. LIBs (Li–S) and NIBs (Na–S) will no doubt have an ever greater impact on our lives in the years to come

Acknowledgments

Supports of the Natural Science Foundation of Liaoning Province (20180510020), the Fundamental Research Funds for the Central Universities (DUT18LK15 and DUT18LK21), and Supercomputing Center of Dalian University of Technology for this work are gratefully acknowledged.

References

[1] J.K. Hart, K. Martinez, Environmental sensor networks: a revolution in the earth system science? Earth Sci. Rev. 78 (2006) 177–191.

[2] H. Wang, X. Yuan, G. Zeng, Y. Wu, Y. Liu, Q. Jiang, et al., Three dimensional graphene based materials: synthesis and applications from energy storage and conversion to electrochemical sensor and environmental remediation, Adv. Colloid. Interface. Sci. 221 (2015) 41–59.

[3] A. Umar, R. Kumar, M.S. Akhtar, G. Kumar, S.H. Kim, Growth and properties of well-crystalline cerium oxide (CeO_2) nanoflakes for environmental and sensor applications, J. Colloid Interf. Sci. 454 (2015) 61–68.

[4] M. Srbinovska, C. Gavrovski, V. Dimcev, A. Krkoleva, V. Borozan, Environmental parameters monitoring in precision agriculture using wireless sensor networks, J. Clean. Prod. 88 (2015) 297–307.
[5] D. Yuan, A.H.C. Anthis, M.G. Afshar, N. Pankratova, M. Cuartero, G.A. Crespo, et al., All-solid-state potentiometric sensors with a multiwalled carbon nanotube inner transducing layer for anion detection in environmental samples, Anal. Chem. 87 (2015) 8640–8645.
[6] H. Wei, S.M.H. Abtahi, P.J. Vikesland, Plasmonic colorimetric and SERS sensors for environmental analysis, Environ. Sci. Nano 2 (2015) 120–135.
[7] R.J. Farrauto, R.M. Heck, Environmental catalysis into the 21st century, Catal Today 55 (2000) 179–187.
[8] S. Waclawek, V.V.T. Padil, M. Cernik, Major advances and challenges in heterogeneous catalysis for environmental applications: a review, Ecol. Chem. Eng. S Chem. I Inzynieria Ekologiczna S 25 (2018) 9–34.
[9] H. Dai, Environmental catalysis: a solution for the removal of atmospheric pollutants, Sci. Bull. 60 (2015) 1708–1710.
[10] R. Sanchis, J.A. Cecilia, M.D. Soriano, M.I. Vazquez, A. Dejoz, J.M. Lopez Nieto, et al., Porous clays heterostructures as supports of iron oxide for environmental catalysis, Chem. Eng. J. 334 (2018) 1159–1168.
[11] M.S. Scurrell, Thoughts on the use of gold-based catalysts in environmental protection catalysis, Gold Bull. 50 (2017) 77–84.
[12] O.A. Kirichenko, N.A. Dayshan, E.A. Redina, G.I. Kapustin, I.V. Mishin, O.P. Tkachenko, et al., Gold nanoparticles in environmental catalysis: influence of the Fe-modified alumina supports on the catalytic behavior of supported gold nanoparticles in CO oxidation in the presence of ammonia, Chem. Eng. J. 292 (2016) 62–71.
[13] B. Silva, I.C. Neves, T. Tavares, A sustained approach to environmental catalysis: reutilization of chromium from wastewater, Crit. Rev. Environ. Sci. Technol. 46 (2016) 1622–1657.
[14] M. Armand, J.M. Tarascon, Building better batteries, Nature 451 (2008) 652.
[15] M. Winter, R.J. Brodd, What are batteries, fuel cells, and supercapacitors? Chem. Rev. 104 (2004) 4245–4270.
[16] M. Prauzek, J. Konecny, M. Borova, K. Janosova, J. Hlavica, P. Musilek, Energy harvesting sources, storage devices and system topologies for environmental wireless sensor networks: a review, Sensors 18 (2018).
[17] B. Scrosati, Battery technology - challenge of portable power, Nature 373 (1995) 557–558.
[18] N. Oyama, T. Tatsuma, T. Sato, T. Sotomura, Dimercaptan-polyaniline composite electrodes for lithium batteries with high-energy density, Nature 373 (1995) 598–600.
[19] N. Nitta, F. Wu, J.T. Lee, G. Yushin, Li-ion battery materials: present and future, Mater. Today 18 (2015) 252–264.
[20] M. Jafari, K. Khan, L. Gauchia, Deterministic models of Li-ion battery aging: it is a matter of scale, J. Energy Storage 20 (2018) 67–77.
[21] J.B. Goodenough, K.-S. Park, The Li-ion rechargeable battery: a perspective, J. Am. Chem. Soc. 135 (2013) 1167–1176.
[22] P.G. Bruce, B. Scrosati, J.-M. Tarascon, Nanomaterials for rechargeable lithium batteries, Angew. Chem. Int. Edit. 47 (2008) 2930–2946.
[23] C. Julien, A. Mauger, K. Zaghib, H. Groult, Optimization of layered cathode materials for lithium-ion batteries, Materials 9 (2016).
[24] G.G. Amatucci, N. Pereira, Fluoride based electrode materials for advanced energy storage devices, J. Fluorine Chem. 128 (2007) 243–262.

[25] J.M. Tarascon, M. Armand, Issues and challenges facing rechargeable lithium batteries, Nature 414 (2001) 359–367.
[26] K. Mizushima, P.C. Jones, P.J. Wiseman, J.B. Goodenough, Li_xCoO_2 (0 < x < -1): a new cathode material for batteries of high energy density, Mater. Res. Bull. 15 (1980) 783–789.
[27] A. Du Pasquier, I. Plitz, S. Menocal, G. Amatucci, A comparative study of Li-ion battery, supercapacitor and nonaqueous asymmetric hybrid devices for automotive applications, J. Power Sour. 115 (2003) 171–178.
[28] H. Arai, S. Okada, Y. Sakurai, J.-i Yamaki, Thermal behavior of $Li_{1-y}NiO_2$ and the decomposition mechanism, Solid State Ionics 109 (1998) 295–302.
[29] A. Rougier, P. Gravereau, C. Delmas, Optimization of the composition of the $Li_{1-z}Ni_{1+z}O_2$ electrode materials: structural, magnetic, and electrochemical studies, J. Electrochem. Soc. 143 (1996) 1168–1175.
[30] A.R. Armstrong, P.G. Bruce, Synthesis of layered $LiMnO_2$ as an electrode for rechargeable lithium batteries, Nature 381 (1996) 499.
[31] Z.H. Lu, D.D. MacNeil, J.R. Dahn, Layered cathode materials Li $Ni_xLi_{(1/3-2x/3)}Mn_{(2/3-x/3)}$ O_2 for lithium-ion batteries, Electrochem. Solid State Lett. 4 (2001) A191–A194.
[32] K.S. Kang, Y.S. Meng, J. Breger, C.P. Grey, G. Ceder, Electrodes with high power and high capacity for rechargeable lithium batteries, Science 311 (2006) 977–980.
[33] M.M. Thackeray, S.-H. Kang, C.S. Johnson, J.T. Vaughey, R. Benedek, S.A. Hackney, Li_2MnO_3-stabilized $LiMO_2$ (M = Mn, Ni, Co) electrodes for lithium-ion batteries, J. Mater. Chem. 17 (2007) 3112–3125.
[34] C.C. Pan, C.E. Banks, W.X. Song, C.W. Wang, Q.Y. Chen, X.B. Ji, Recent development of $LiNi_xCo_yMn_zO_2$: impact of micro/nano structures for imparting improvements in lithium batteries, T Nonferr. Metal Soc. 23 (2013) 108–119.
[35] S.K. Jung, H. Gwon, J. Hong, K.Y. Park, D.H. Seo, H. Kim, et al., Understanding the degradation mechanisms of $LiNi_{0.5}Co_{0.2}Mn_{0.3}O_2$ cathode material in lithium ion batteries, Adv. Energy Mater. 4 (2014).
[36] N. Yabuuchi, T. Ohzuku, Novel lithium insertion material of $LiCo_{1/3}Ni_{1/3}Mn_{1/3}O_2$ for advanced lithium-ion batteries, J. Power Sour. 119 (2003) 171–174.
[37] K.M. Shaju, G.V.S. Rao, B.V.R. Chowdari, Performance of layered $Li(Ni_{1/3}Co_{1/3}Mn_{1/3})O_2$ as cathode for Li-ion batteries, Electrochim. Acta 48 (2002) 145–151.
[38] D.J. Lee, B. Scrosati, Y.K. Sun, $Ni_3(PO_4)_2$-coated Li $Ni_{0.8}Co_{0.15}Al_{0.05}O_2$ lithium battery electrode with improved cycling performance at 55°C, J. Power Sour. 196 (2011) 7742–7746.
[39] A.K. Padhi, K.S. Nanjundaswamy, J.B. Goodenough, Phospho-olivines as positive-electrode materials for rechargeable lithium batteries, J. Electrochem. Soc. 144 (1997) 1188–1194.
[40] G.H. Li, H. Azuma, M. Tohda, $LiMnPO_4$ as the cathode for lithium batteries, Electrochem. Solid State Lett. 5 (2002) A135–A137.
[41] K. Amine, H. Yasuda, M. Yamachi, Olivine $LiCoPO_4$ as 4.8 V electrode material for lithium batteries, Electrochem. Solid State Lett. 3 (2000) 178–179.
[42] P. Barpanda, M. Ati, B.C. Melot, G. Rousse, J.N. Chotard, M.L. Doublet, et al., A 3.90 V iron-based fluorosulphate material for lithium-ion batteries crystallizing in the triplite structure, Nat. Mater. 10 (2011) 772–779.
[43] N. Recham, J.N. Chotard, L. Dupont, C. Delacourt, W. Walker, M. Armand, et al., A 3.6 V lithium-based fluorosulphate insertion positive electrode for lithium-ion batteries, Nat. Mater. 9 (2010) 68–74.

[44] C. Delacourt, L. Laffont, R. Bouchet, C. Wurm, J.B. Leriche, M. Morcrette, et al., Toward understanding of electrical limitations (electronic, ionic) in $LiMPO_4$ (M = Fe, Mn) electrode materials, J. Electrochem. Soc. 152 (2005) A913–A921.
[45] D. Morgan, A. Van der Ven, G. Ceder, Li conductivity in Li_xMPO_4 (M = Mn, Fe, Co, Ni) olivine materials, Electrochem. Solid State Lett. 7 (2004) A30–A32.
[46] J.J. Wang, X.L. Sun, Understanding and recent development of carbon coating on $LiFePO_4$ cathode materials for lithium-ion batteries, Energy Environ. Sci. 5 (2012) 5163–5185.
[47] J. Woods, N. Bhattarai, P. Chapagain, Y. Yang, S. Neupane, In situ transmission electron microscopy observations of rechargeable lithium ion batteries, Nano Energy (2018).
[48] L.W. Ji, Z. Lin, M. Alcoutlabi, X.W. Zhang, Recent developments in nanostructured anode materials for rechargeable lithium-ion batteries, Energy Environ. Sci. 4 (2011) 2682–2699.
[49] A.C. Dillon, Carbon nanotubes for photoconversion and electrical energy storage, Chem. Rev. 110 (2010) 6856–6872.
[50] D. Deng, M.G. Kim, J.Y. Lee, J. Cho, Green energy storage materials: nanostructured TiO_2 and Sn-based anodes for lithium-ion batteries, Energy Environ. Sci. 2 (2009) 818–837.
[51] N. Dimov, S. Kugino, M. Yoshio, Carbon-coated silicon as anode material for lithium ion batteries: advantages and limitations, Electrochim. Acta 48 (2003) 1579–1587.
[52] N. Du, H. Zhang, B. Chen, J. Wu, X. Ma, Z. Liu, et al., Porous Co_3O_4 nanotubes derived from $Co_4(CO)_{12}$ clusters on carbon nanotube templates: a highly efficient material for Li-battery applications, Adv. Mater. 19 (2007) 4505.
[53] K.K.D. Ehinon, S. Naille, R. Dedryvere, P.E. Lippens, J.C. Jumas, D. Gonbeau, Ni_3Sn_4 electrodes for Li-ion batteries: Li-Sn alloying process and electrode/electrolyte interface phenomena, Chem. Mater. 20 (2008) 5388–5398.
[54] C.M. Ban, Z.C. Wu, D.T. Gillaspie, L. Chen, Y.F. Yan, J.L. Blackburn, et al., Nanostructured Fe_3O_4/SWNT electrode: binder-free and high-rate Li-Ion anode, Adv. Mater. 22 (2010) E145.
[55] W. Wang, P.N. Kumta, Nanostructured hybrid silicon/carbon nanotube heterostructures: reversible high-capacity lithium-ion anodes, ACS Nano 4 (2010) 2233–2241.
[56] B.S. Lee, S.B. Son, K.M. Park, G. Lee, K.H. Oh, S.H. Lee, et al., Effect of pores in hollow carbon nanofibers on their negative electrode properties for a lithium rechargeable battery, ACS Appl. Mater. Inter. 4 (2012) 6701–6709.
[57] M. Sathish, S. Mitani, T. Tomai, A. Unemoto, I. Honma, Nanocrystalline tin compounds/graphene nanocomposite electrodes as anode for lithium-ion battery, J. Solid State Electr. 16 (2012) 1767–1774.
[58] G.M. Zhou, D.W. Wang, L.C. Yin, N. Li, F. Li, H.M. Cheng, Oxygen bridges between NiO nanosheets and graphene for improvement of lithium storage, ACS Nano 6 (2012) 3214–3223.
[59] Y.-S. Hu, P. Adelhelm, B.M. Smarsly, S. Hore, M. Antonietti, J. Maier, Synthesis of hierarchically porous carbon monoliths with highly ordered microstructure and their application in rechargeable lithium batteries with high-rate capability, Adv. Funct. Mater. 17 (2007) 1873–1878.
[60] D. Larcher, S. Beattie, M. Morcrette, K. Edstroem, J.-C. Jumas, J.-M. Tarascon, Recent findings and prospects in the field of pure metals as

negative electrodes for Li-ion batteries, J. Mater. Chem. 17 (2007) 3759–3772.
[61] N. Tuan Poi, J.H. Kim, I.T. Kim, Electrochemical performance of Sn/SnO/Ni_3Sn composite anodes for lithium-ion batteries, J. Nanosci. Nanotech. 19 (2019) 1001–1005.
[62] P. Poizot, S. Laruelle, S. Grugeon, L. Dupont, J.M. Tarascon, Nano-sized transition-metaloxides as negative-electrode materials for lithium-ion batteries, Nature 407 (2000) 496–499.
[63] A. Magasinski, P. Dixon, B. Hertzberg, A. Kvit, J. Ayala, G. Yushin, High-performance lithium-ion anodes using a hierarchical bottom-up approach, Nat. Mater. 9 (2010) 353–358.
[64] L.W. Su, Z. Zhou, M.M. Ren, Core double-shell Si@SiO_2@C nanocomposites as anode materials for Li-ion batteries, Chem. Commun. 46 (2010) 2590–2592.
[65] Y.H. Xu, Y.J. Zhu, Y.H. Liu, C.S. Wang, Electrochemical performance of porous carbon/tin composite anodes for sodium-ion and lithium-ion batteries, Adv. Energy Mater. 3 (2013) 128–133.
[66] Y. Idota, T. Kubota, A. Matsufuji, Y. Maekawa, T. Miyasaka, Tin-based amorphous oxide: a high-capacity lithium-ion-storage material, Science 276 (1997) 1395–1397.
[67] X. Zhou, L.J. Wan, Y.G. Guo, Binding SnO_2 nanocrystals in nitrogen-doped graphene sheets as anode materials for lithium-ion batteries, Adv. Mater. 25 (2013) 2152–2157.
[68] J.W. Seo, J.T. Jang, S.W. Park, C.J. Kim, B.W. Park, J.W. Cheon, Two-dimensional SnS_2 nanoplates with extraordinary high discharge capacity for lithium ion batteries, Adv. Mater. 20 (2008) 4269–4273.
[69] D. McNulty, D.N. Buckley, C. O'Dwyer, Synthesis and electrochemical properties of vanadium oxide materials and structures as Li-ion battery positive electrodes, J. Power Sourc. 267 (2014) 831–873.
[70] X. Zhang, Z. Zhang, Z. Zhou, MXene-based materials for electrochemical energy storage, J. Energy Chem. 27 (2018) 73–85.
[71] X. Tang, X. Guo, W. Wu, G. Wang, 2D metal carbides and nitrides (MXenes) as high-performance electrode materials for lithium-based batteries, Adv. Energy Mater. 8 (2018).
[72] D. Xiong, X. Li, Z. Bai, S. Lu, Recent advances in layered Ti3C2Tx MXene for electrochemical energy storage, Small. 14 (2018).
[73] Y.-M. Li, W.-G. Chen, Y.-L. Guo, Z.-Y. Jiao, Theoretical investigations of TiNbC MXenes as anode materials for Li-ion batteries, J. Alloy Compd. 778 (2019) 53–60.
[74] S. Xu, Y. Dall'Agnese, J. Li, Y. Gogotsi, W. Han, Thermally reduced graphene/MXene film for enhanced Li-ion storage, Chem. Eur. J. 24 (2018) 18556–18563.
[75] X. Sun, Y. Liu, J. Zhang, L. Hou, J. Sun, C. Yuan, Facile construction of ultrathin SnO_x nanosheets decorated MXene (Ti_3C_2) nanocomposite towards Li-ion batteries as high performance anode materials, Electrochim. Acta 295 (2019) 237–245.
[76] S.A. Melchior, N. Palaniyandy, I. Sigalas, S.E. Iyuke, K.I. Ozoemena, Probing the electrochemistry of MXene (Ti2CTx)/electrolytic manganese dioxide (EMD) composites as anode materials for lithium-ion batteries, Electrochim. Acta 297 (2019) 961–973.
[77] W. Zhang, Y. Wang, J. Wang, Y. Xiao, Theoretical prediction of the structural, electronic, mobility, and dynamic properties of Mo3N2T2 (T = H, O and OH) with Li adsorption feature, Solid State Ionics 325 (2018) 238–246.

[78] L. Zhang, Y. Wang, Z. Niu, J. Chen, Advanced nanostructured carbon-based materials for rechargeable lithium-sulfur batteries, Carbon. N. Y. 141 (2019) 400–416.
[79] A. Manthiram, Y. Fu, S.H. Chung, C. Zu, Y.S. Su, Rechargeable lithium-sulfur batteries, Chem. Rev. 114 (2014) 11751–11787.
[80] Y.X. Yin, S. Xin, Y.G. Guo, L.J. Wan, Lithium-sulfur batteries: electrochemistry, materials, and prospects, Angew. Chem. Int. Edit. 52 (2013) 13186–13200.
[81] Y. Yang, G.Y. Zheng, Y. Cui, Nanostructured sulfur cathodes, Chem. Soc. Rev. 42 (2013) 3018–3032.
[82] X.L. Ji, K.T. Lee, L.F. Nazar, A highly ordered nanostructured carbon-sulphur cathode for lithium-sulphur batteries, Nat. Mater. 8 (2009) 500–506.
[83] H.L. Wang, Y. Yang, Y.Y. Liang, J.T. Robinson, Y.G. Li, A. Jackson, et al., Graphene-wrapped sulfur particles as a rechargeable lithium-sulfur battery cathode material with high capacity and cycling stability, Nano. Lett. 11 (2011) 2644–2647.
[84] Y.V. Mikhaylik, J.R. Akridge, Polysulfide shuttle study in the Li/S battery system, J. Electrochem. Soc. 151 (2004) A1969–A1976.
[85] N. Jayaprakash, J. Shen, S.S. Moganty, A. Corona, L.A. Archer, Porous hollow carbon@sulfur composites for high-power lithium-sulfur batteries, Angew. Chem. Int. Edit. 50 (2011) 5904–5908.
[86] L.W. Ji, M.M. Rao, H.M. Zheng, L. Zhang, Y.C. Li, W.H. Duan, et al., Graphene oxide as a sulfur immobilizer in high performance lithium/sulfur cells, J. Am. Chem. Soc. 133 (2011) 18522–18525.
[87] X. Liu, X. Shao, F. Li, M. Zhao, Anchoring effects of S-terminated Ti_2C MXene for lithium-sulfur batteries: a first-principles study, Appl. Surf. Sci. 455 (2018) 522–526.
[88] H. Pan, X. Huang, R. Zhang, D. Wang, Y. Chen, X. Duan, et al., Titanium oxide-Ti_3C_2 hybrids as sulfur hosts in lithium-sulfur battery: fast oxidation treatment and enhanced polysulfide adsorption ability, Chem. Eng. J. 358 (2019) 1253–1261.
[89] D. Rao, L. Zhang, Y. Wang, Z. Meng, X. Qian, J. Liu, et al., Mechanism on the improved performance of lithium sulfur batteries with MXene-based additives, J. Phys. Chem. C 121 (2017) 11047–11054.
[90] Y. Yao, W. Feng, M. Chen, X. Zhong, X. Wu, H. Zhang, et al., Boosting the electrochemical performance of Li-S batteries with a dual polysulfides confinement strategy, Small. 14 (2018).
[91] W. Bao, X. Xie, J. Xu, X. Guo, J. Song, W. Wu, et al., Confined sulfur in 3D MXene/reduced graphene oxide hybrid nanosheets for lithium-sulfur battery, Chem. Eur. J. 23 (2017) 12613–12619.
[92] N. Yabuuchi, K. Kubota, M. Dahbi, S. Komaba, Research development on sodium-ion batteries, Chem. Rev. 114 (2014) 11636–11682.
[93] M.D. Slater, D. Kim, E. Lee, C.S. Johnson, Sodium-ion batteries, Adv. Funct. Mater. 23 (2013) 947–958.
[94] V. Palomares, P. Serras, I. Villaluenga, K.B. Hueso, J. Carretero-Gonzalez, T. Rojo, Na-ion batteries, recent advances and present challenges to become low cost energy storage systems, Energy Environ. Sci. 5 (2012) 5884–5901.
[95] X.D. Xiang, K. Zhang, J. Chen, Recent advances and prospects of cathode materials for sodium-ion batteries, Adv. Mater. 27 (2015) 5343–5364.
[96] J.T. Xu, M. Wang, N.P. Wickramaratne, M. Jaroniec, S.X. Dou, L.M. Dai, High-performance sodium ion batteries based on a 3D anode from nitrogen-doped graphene foams, Adv. Mater. 27 (2015) 2042–2048.

[97] S. Li, Y.F. Dong, L. Xu, X. Xu, L. He, L.Q. Mai, Effect of carbon matrix dimensions on the electrochemical properties of $Na_3V_2(PO_4)_3$ nanograins for high-performance symmetric sodium-ion batteries, Adv. Mater. 26 (2014) 3545–3553.
[98] Z.T. Shi, W.P. Kang, J. Xu, Y.W. Sun, M. Jiang, T.W. Ng, et al., Hierarchical nanotubes assembled from MoS_2-carbon monolayer sandwiched superstructure nanosheets for high-performance sodium ion batteries, Nano Energy 22 (2016) 27–37.
[99] J.J. Wang, C. Luo, J.F. Mao, Y.J. Zhu, X.L. Fan, T. Gao, et al., Solid-state fabrication of SnS_2/C nanospheres for high-performance sodium ion battery anode, ACS Appl. Mater. Inter. 7 (2015) 11476–11481.
[100] J.F. Ni, S.D. Fu, C. Wu, Y. Zhao, J. Maier, Y. Yu, et al., Superior sodium storage in $Na_2Ti_3O_7$ nanotube arrays through surface engineering, Adv. Energy Mater. 6 (2016).
[101] Y. Zhang, C.W. Foster, C.E. Banks, L.D. Shao, H.S. Hou, G.Q. Zou, et al., Graphene-rich wrapped petal-like rutile TiO_2 tuned by carbon dots for high-performance sodium storage, Adv. Mater. 28 (2016) 9391.
[102] Z.L. Jian, P. Liu, F.J. Li, M.W. Chen, H.S. Zhou, Monodispersed hierarchical Co_3O_4 spheres intertwined with carbon nanotubes for use as anode materials in sodium-ion batteries, J. Mater. Chem. A 2 (2014) 13805–13809.
[103] Y. Xu, M. Zhou, X. Wang, C.L. Wang, L.Y. Liang, F.B. Grote, et al., Enhancement of sodium ion battery performance enabled by oxygen vacancies, Angew. Chem. Int. Edit. 54 (2015) 8768–8771.
[104] S.L. Zhang, F. Yao, L. Yang, F.Z. Zhang, S.L. Xu, Sulfur-doped mesoporous carbon from surfactant-intercalated layered double hydroxide precursor as high-performance anode nanomaterials for both Li-ion and Na-ion batteries, Carbon. N. Y. 93 (2015) 143–150.
[105] F.E. Niu, J. Yang, N.N. Wang, D.P. Zhang, W.L. Fan, J. Yang, et al., $MoSe_2$-covered N,P-doped carbon nanosheets as a long-life and high-rate anode material for sodium-ion batteries, Adv. Funct. Mater. 27 (2017).
[106] S. Kajiyama, L. Szabova, K. Sodeyama, H. Iinuma, R. Morita, K. Gotoh, et al., Sodium-ion intercalation mechanism in MXene nanosheets, ACS Nano 10 (2016) 3334–3341.
[107] C. Yang, X. Sun, Y.R. Zhang, Y. Liu, Q.A. Zhang, C.Z. Yuan, Facile synthesis of hierarchical $NaTi_2(PO_4)_3/Ti_3C_2$ nanocomposites with superior sodium storage performance, Mater. Lett. 236 (2019) 408–411.
[108] Y. Zhang, B. Guo, L. Hu, Q. Xu, Y. Li, D. Liu, et al., Synthesis of SnS nanoparticle-modified MXene ($Ti_3C_2T_x$) composites for enhanced sodium storage, J. Alloy Compd. 732 (2018) 448–453.
[109] M. Tao, Y. Zhang, R. Zhan, B. Guo, Q. Xu, M. Xu, A chemically bonded $CoNiO_2$ nanoparticles/MXene composite as anode for sodium-ion batteries, Mater. Lett. 230 (2018) 173–176.
[110] C. Zeng, F. Xie, X. Yang, M. Jaroniec, L. Zhang, S.-Z. Qiao, Ultrathin titanate nanosheets/graphene films derived from confined transformation for excellent Na/K ion storage, Angew. Chem. Int. Edit. 57 (2018) 8540–8544.
[111] S. Xin, Y.X. Yin, Y.G. Guo, L.J. Wan, A high-energy room-temperature sodium-sulfur battery, Adv. Mater. 26 (2014) 1261–1265.
[112] Z.Y. Wen, Y.Y. Hu, X.W. Wu, J.D. Han, Z.H. Gu, Main challenges for high performance NAS battery: materials and interfaces, Adv. Funct. Mater. 23 (2013) 1005–1018.
[113] A. Manthiram, X.W. Yu, Ambient temperature sodium-sulfur batteries, Small 11 (2015) 2108–2114.

[114] J.L. Wang, J. Yang, Y. Nuli, R. Holze, Room temperature Na/S batteries with sulfur composite cathode materials, Electrochem. Commun. 9 (2007) 31–34.
[115] C.W. Park, H.S. Ryu, K.W. Kim, J.H. Ahn, J.Y. Lee, H.J. Ahn, Discharge properties of all-solid sodium-sulfur battery using poly (ethylene oxide) electrolyte, J. Power Sour. 165 (2007) 450–454.
[116] C.W. Park, J.H. Ahn, H.S. Ryu, K.W. Kim, H.J. Ahn, Room-temperature solid-state sodium/sulfur battery, Electrochem. Solid State Lett. 9 (2006) A123–A125.
[117] A. Douglas, R. Carter, L. Oakes, K. Share, A.P. Cohn, C.L. Pint, Ultrafine iron pyrite (FeS_2) nanocrystals improve sodium-sulfur and lithium-sulfur conversion reactions for efficient batteries, ACS Nano 9 (2015) 11156–11165.
[118] S.Y. Wei, S.M. Xu, A. Agrawral, S. Choudhury, Y.Y. Lu, Z.Y. Tu, et al., A stable room-temperature sodium-sulfur battery, Nat. Commun. 7 (2016).
[119] H. Ryu, T. Kim, K. Kim, J.H. Ahn, T. Nam, G. Wang, et al., Discharge reaction mechanism of room-temperature sodium-sulfur battery with tetra ethylene glycol dimethyl ether liquid electrolyte, J. Power Sour. 196 (2011) 5186–5190.
[120] S. Wenzel, H. Metelmann, C. Raiss, A.K. Durr, J. Janek, P. Adelhelm, Thermodynamics and cell chemistry of room temperature sodium/sulfur cells with liquid and liquid/solid electrolyte, J. Power Sour. 243 (2013) 758–765.
[121] F. Grimaccia, R. Johnstone, M. Mussetta, A. Pirisi, R.E. Zich, Evolutionary techniques for Sensor Networks Energy Optimization in Marine Environmental Monitoring, in: B. Huang, A.J. Plaza (Eds.). High-Performance Computing in Remote Sensing Ii, 2012.
[122] S. Chatterjea, P. Havinga, Improving temporal coverage of an energy-efficient data extraction algorithm for environmental monitoring using wireless sensor networks, Sensors 9 (2009) 4941–4954.
[123] M.D. Colvard, M. Naiman, L. Danziger, L. Hanley, Handheld directed energy sensor for environmental monitoring and clinician safety, Aviat. Space Environ. Med. 81 (2010) 602–604.
[124] M. Lee, J. Bae, J. Lee, C.-S. Lee, S. Hong, Z.L. Wang, Self-powered environmental sensor system driven by nanogenerators, Energy Environ. Sci. 4 (2011) 3359–3363.
[125] Z. Zhou, X. Li, Y. Wu, H. Zhang, Z. Lin, K. Meng, et al., Wireless self-powered sensor networks driven by triboelectric nanogenerator for in-situ real time survey of environmental monitoring, Nano Energy 53 (2018) 501–507.
[126] J.F.M. Oudenhoven, R.J.M. Vullers, R. Schaijk, A review of the present situation and future developments of micro-batteries for wireless autonomous sensor systems, Int. J. Energy Res. 36 (2012) 1139–1150.
[127] L. Oakes, T. Hanken, R. Carter, W. Yates, C.L. Pint, Roll-to-roll nanomanufacturing of hybrid nanostructures for energy storage device design, ACS Appl. Mater. Inter. 7 (2015) 14201–14210.

9

COLLOIDAL SEMICONDUCTOR QUANTUM DOT–BASED MULTICOMPONENT ARTIFICIAL SYSTEM FOR HYDROGEN PHOTOGENERATION

Ping Wang
State Key Laboratory of Electroanalytical Chemistry, Changchun Institute of Applied Chemistry, Chinese Academy of Sciences, Changchun, P.R. China

9.1 Introduction to the Basics of Nanomaterials for Hydrogen Photogeneration

Rapid economic development greatly increases the demand for global energy consumption. In 2008, the worldwide energy consumption had already reached 15 TW, and this value is estimated to double by the middle of this century [1]. Controversially, the traditional, limited, nonrenewable fossil fuels such as coal, crude oil, and natural gas, which are now the primary global energy resources, will apparently not meet this great demand in the near future. In addition, accompanying the combustion of these traditional fossil fuels, air pollution and global warming are becoming increasingly serious problems, which start to influence our daily life and cannot be ignored anymore. Therefore scientists all over the world are trying to find out an alternative, which is expected to be environment-friendly, renewable, easily obtained, and low cost, to replace them. Among the different choices, solar energy is considered as an ideal substitute and seems to well meet all of the mentioned requirements.

Benefitting from the pioneering work of Honda and Fujishima in 1972 [2], where they first reported the phenomenon of photoelectrochemical water splitting by utilizing TiO_2

Advanced Nanomaterials for Pollutant Sensing and Environmental Catalysis. DOI: https://doi.org/10.1016/B978-0-12-814796-2.00009-5

and Pt as the photoanode and photocathode, respectively, under the illumination of UV light, tremendous interests have been devoted to this promising research area and various types of semiconductors that are composed of similar photoelectrochemical systems are expected to push its efficiency into a higher value. Meanwhile, the following breakthrough of technology for the synthesis of nanomaterials has triggered an attempt to use semiconductor nanomaterials as photocatalysts to realize solar light water splitting. The photocatalytic water splitting reaction (as indicated in Eq. (9.1)), which needs the standard Gibbs free energy change of 237 kJ mol^{-1}, mainly involves three processes, namely (1) the photogeneration of electron–hole pairs in the semiconductor nanomaterial after the absorption of proper incident light; (2) the separation of the electron–holes pairs and the migration of the electrons and holes to the surface of the nanomaterial separately; and (3) charge (electrons and holes) transfer to the corresponding cocatalysts to reduce H^+ to H_2 or oxidize OH^- to O_2, respectively.

$$H_2O \xrightarrow{\text{Photocatalyst}} H_2\uparrow + \frac{1}{2}O_2\uparrow \tag{9.1}$$

However, due to the complexity of building the complete system and its relatively low efficiency, most research now focuses on the half reaction of proton reduction ($2H^+ + 2e^- \rightarrow H_2\uparrow$). The H_2 photogeneration system is normally composed of three components, namely (1) a light absorber, which is used to absorb incident light and photogenerated electron–hole pairs; (2) a cocatalyst, which acts as the active site to accept photogenerated electrons and further use them to reduce H^+ to produce H_2; (3) a sacrificial electron donor, which is used to neutralize the holes left to keep the entire system stable.

Many efforts have been put into developing an efficient light absorber, and these have been systematically reviewed by others [3–5]; therefore this will not be addressed again in this chapter. Among the different types of candidates, colloidal semiconductor quantum dots (QDs) are considered as a good option because of several reasons. The first and most important reason, is that their bandgap can be consecutively tuned due to the quantum confinement effect, and consequently, their conduction band can be easily adjusted to the proper energy level to meet the requirements for the reduction of protons [6]. Second, their absorbance can also be tuned to cover the most solar spectral region together with a large absorption cross-section [7,8]. Fig. 9.1 presents an example of band energy level and PL emission variations of colloidal CdTe of different sizes.

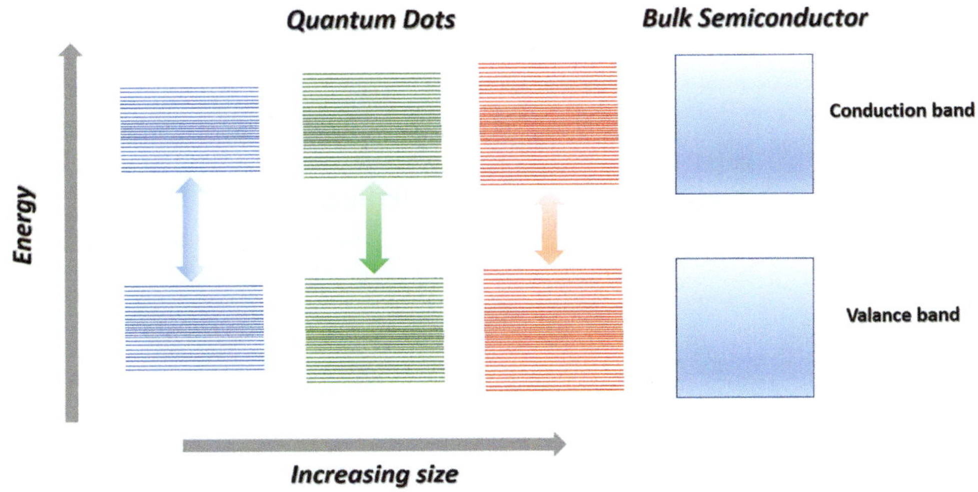

Figure 9.1 Simplified energy level diagrams of semiconductor nanocrystals of decreasing size (from right to left) alongside a bulk semiconductor material.

Third, they possess superior photostability during H_2 photogeneration measurements. Although it is considered that QDs would be unstable due to the oxidation reaction by the holes due to the photocatalytic process, the existence of a sacrificial electron donor, which can quickly consume the holes left in the QDs, eliminates this concern, and the QDs in the system usually display relatively long-term stability [9].

Noble metals such as Pt [10–15], Pd [16], Rh [17], and Au [18] are commonly used as cocatalysts to reach a high efficiency for photocatalytic water splitting, and Pt has been confirmed to be the most effective due to its low overpotential for proton reduction to H_2 [19]. However, their scarcity means that they are not desirable choices for further potentially large-scale photocatalytic H_2 production. Therefore some scientists have turned their interests to exploring low cost, but still highly efficient, substitutes to replace them. Today, several noble metal–free cocatalysts such as metal sulfides (NiS [20,21], WS_2, and MoS_2 [22,23]), Ni [24,25], hydrogenases (including their mimics, [FeFe] [26–28] and [NiFeSe] [29]), molecular nickel [9,30–34], and cobalt complexes [35–38], have been discovered to possess the probable ability to realize an efficient water splitting. Though, there is still a long way to realize that goal. Compared with the two components mentioned, studies on the sacrificial electron donor seem to gain much less attention. However, hole scavenging sometimes also plays an important role in the rate-limiting step [10].

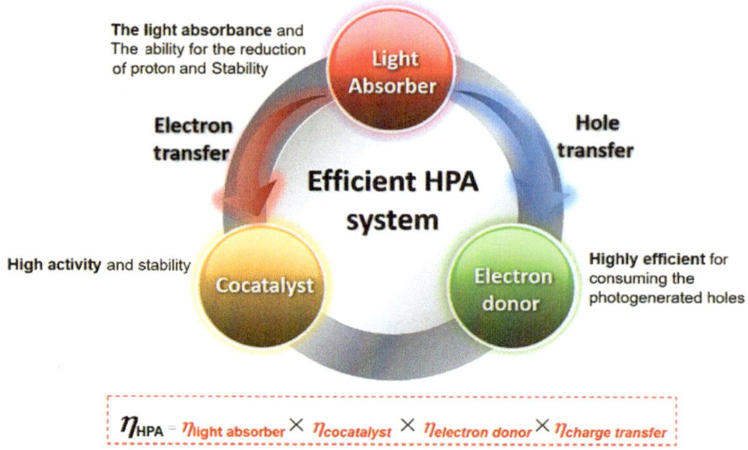

Figure 9.2 Schematic illustration of the factors and their relationships that affect the efficiency of the H_2-producing artificial system (η_{HPA}).

Due to the characteristics of this multicomponent H_2-producing artificial (HPA) system, there are many factors that affect the overall efficiency of the system. As indicated in Fig. 9.2, the efficiency of the HPA system (η_{HPA}) is determined by four items multiplied by each other; which are $\eta_{light\ absorber}$, $\eta_{cocatalyst}$, $\eta_{electron\ donor}$, and $\eta_{charge\ transfer}$, respectively, where $\eta_{light\ absorber}$ is correlated with the light absorbance, the ability for the reduction of protons, and the stability of the light absorber; $\eta_{cocatalyst}$ is dependent on the activity and stability of the cocatalyst; $\eta_{electron\ donor}$ is decided by the rate of the consumption of photogenerated holes; while, $\eta_{charge\ transfer}$ is related to the charge transfer processes between the light absorber and the cocatalyst or electron donor. It can be seen that the HPA system is of high complexity and, therefore, there is still a lot of work to do toward realizing its practical application. This chapter will review recent works mainly relating to three main aspects including the design of colloidal semiconductor QDs, surface treatments, and the selection of cocatalyst, which have significant effects on the efficiency of the HPA system. Confined by the knowledge of the author, the chapter may not include all the related works.

9.2 Structure Design of Colloidal Semiconductor Quantum Dots

Colloidal semiconductor QDs, as the light absorber in the HPA system, are responsible for the absorption of the incident light and its further conversion into photogenerated electrons and holes for the following proton reduction reaction. The processes occurring in QDs during photocatalytic water splitting include the photogeneration and separation of electron–hole pairs and the further migration of the electrons and holes to the surface of the QDs. To realize solar light water splitting, the absorption of QDs should match the solar spectrum as much as possible and the photogenerated electrons should possess enough ability to reduce protons to H_2. In addition, it is highly desirable that the electrons and holes in the QDs be separated efficiently and that charge carriers migrate to the surface of the QDs to contribute to the reaction. Therefore to meet the mentioned requirements, the electronic states (energy levels and band structure) of the QDs, which greatly affect the proton reduction ability of electrons and charge carrier (electrons and holes) distribution, are the primary concern when scientists design the structure of QDs to achieve a higher catalytic efficiency in HPA systems.

9.2.1 Single Component Quantum Dots

Besides the mentioned merits of colloidal semiconductor QDs as a light absorber, single component QDs are commonly investigated for H_2 photogeneration because their synthesis method is relatively easy. Furthermore, their redox potentials and absorption band can be readily tuned by simply changing their size. Several types of single component II–VI semiconductor QDs such as CdS [28,39–41], CdSe [9,26,30], CdTe [27,38,42,43] have been used for photocatalytic water splitting.

CdS nanorods are used as a light absorber due to their greater photostability as well as larger absorption cross-section (molar absorptivities $\sim 10^7\,M^{-1}$ at 405 nm) and surface area [28]. More importantly, no matter whether this material is in the bulk or nanoscale dimension, it possesses a suitable energy level for both proton reduction and water oxidation [3]. King et al., combined CdS nanorods with *Clostridium acetobutylicum* [FeFe]-hydrogenase I (CaI) [28], where the CdS nanorods and the CaI were used as the light absorber and cocatalyst, respectively (Fig. 9.3). In the presence of ascorbic acid (AA) as the sacrificial electron donor, the quantum yield (QY) of the HPA system could reach 20% with a turnover frequency (TOF) value (mol H_2 (mol CaI)$^{-1}$ s^{-1}) of 380 under the illumination of 405 nm light; and more interestingly, under the illumination of high intensity white light (30,000 $\mu Em^{-2}\,s^{-1}$), the TOF is even higher than that of photosystem I hybrid systems.

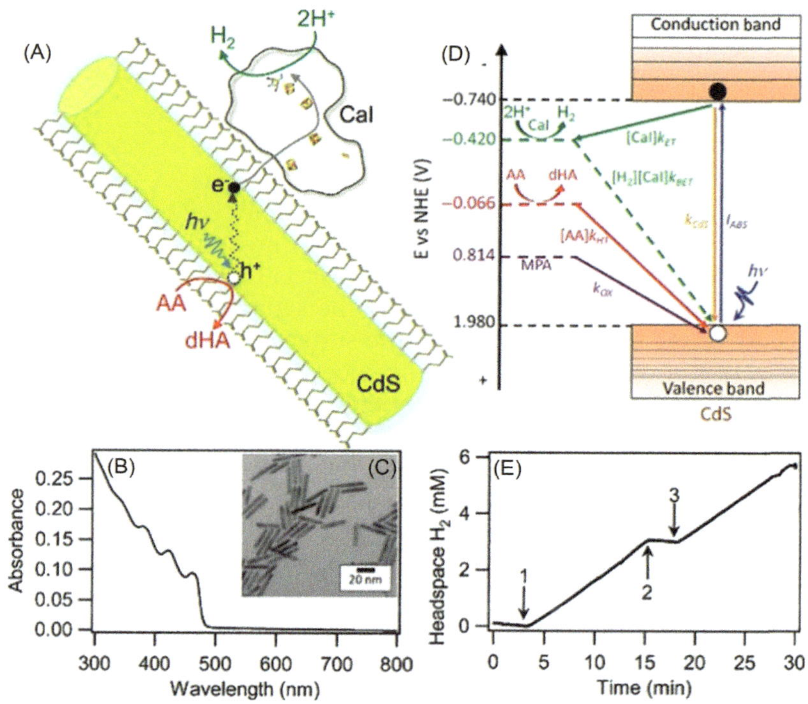

Figure 9.3 Properties of CdS nanorods and CdS/CaI complexes. (A) The proposed scheme for photocatalytic H_2 production by CdS/CaI complexes. Electron transfer [FeS]–clusters and the catalytic H-cluster are shown in yellow (sulfur) and orange (iron). CdS and CaI are drawn to scale, while MPA molecules are enlarged ~5x. (B) UV – vis absorption spectrum of CdS nanorods in water. (C) TEM of typical CdS nanorods with an average size of 30 × 4.5 nm based on the measurement of at least 200 nanorods. (D) Energy level diagram showing processes in photoexcited CdS that are relevant to H_2 production. (Abbreviations: AA, ascorbic acid; dHA, dehydroascorbate; CaI, [FeFe]-hydrogenase; Eg, bandgap energy; kET, rate constant for ET from CdS to CaI; kHT, rate constant for HT from CdS to AA; MPA, mercaptopropionic acid; kOX, rate of photooxidation of surface-bound MPA ligands; kBET, rate constant for ET from CaI to CdS (H_2 uptake); IABS, flux of absorbed photons; kCdS, rate of excited state decay in CdS, including both radiative and nonradiative pathways) (e.g., electron – hole recombination, carrier trapping). Potentials are shown versus NHE (pH 7, 1 atm H_2). (E) Photocatalytic H_2 production by CdS/CaI complexes in 50 mM Tris–HCl, 275 nM CdS, 360 nM CaI, and 100 mM AA illuminated with white light at 10 000 $\mu Em^{-2}\ s^{-1}$. Arrows show changes in illumination: (1) light on, (2) light off, (3) light on. Reprinted with permission from K.A. Brown, M.B. Wilker, M. Boehm, G. Dukovic, P.W. King, Characterization of photochemical processes for H2 production by CdS nanorod-[FeFe] hydrogenase complexes, J. Am. Chem. Soc. 134 (2012) 5627–5636. Credit: Copyright 2012 American Chemical Society.

Banni et al., investigated the catalytic activity for H_2 photogeneration of different types of CdS–PdX hybrids using sulfide and sulfite as sacrificial electron donors [39]. They found that CdS–Pd_4S with tip exchange exhibits the highest yield at 3.25% under the illumination of 405 nm light (Fig. 9.4), and they attributed this to parasitic absorbance by the Pd_4S region

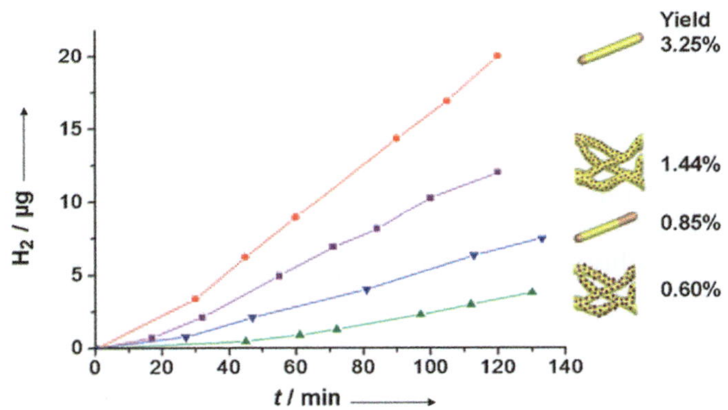

Figure 9.4 Photocatalytic H_2 production by CdS–PdX hybrids in an aqueous solution of sulfide and sulfite, excited at $\lambda = 405$ nm. Experiments were performed at the limit, whereby nearly all the photons are absorbed by the sample. Yields relate H_2 formation to the number of photons absorbed. ● CdS–Pd_4S tip exchange; ■ CdS–Pd0 small dots; ▼ CdS–Pd_4S body exchange; ▲ CdS–Pd0 large dots. Control experiments of CdS nanorods and Pd nanoparticles gave yields of 0.03% and 0.003%, respectively. Hydrogen evolution was not detected for Pd0 nanoparticles without CdS. Credit: Copyright 2010, John Wiley & Sons, Inc.

together with the possible raised Fermi level and the lower electron–hole recombination rate.

CdSe QDs as a light absorber also exhibit superior stability and good activity for photocatalytic water reduction. Eisenberg et al. [9] employed CdSe coupled with an Ni-based molecular cocatalyst which will be discussed here. With the use of AA as the sacrificial electron donor, the HPA system exhibits a QY of over 36% under illumination at 520 nm and excellent long-term stability for at least 360 h. They also investigated the size effect of CdSe QDs on the catalytic efficiency of the system, which showed significant size dependence. They attributed this to the size-related change of reduction ability of the photogenerated electrons. Also, by combining CdSe QDs with the artificial [FeFe]-hydrogenases, while using AA as the sacrificial electron door, the HPA system showed an exceptional turnover number (TON) and TOF with respect to Fe_2S_2 of up to 27,135 and 3.6 s^{-1} [26].

Through the assembly of CdTe QDs with hydrogenase using AA as the sacrificial electron donor, the HPA system exhibited a QY of 9% under the illumination of monochromatic light and an AM 1.5 efficiency of 1.8% [43]. Wu et al., composed an HPA system with CdTe QDs and Co^{2+} as well as AA, and this system showed exceptional activity (25 μmol h^{-1} mg^{-1}, 219,100 mol H_2

per mol QD or 59,600 mol H_2 per mol Co) and stability (at least 70 h) for photocatalytic water reduction [42].

In addition to these typical II–VI semiconductor QDs, scientists are also trying to find environment-friendly (cadmium-free) QDs to realize photocatalytic water splitting. For example, Cabot et al., reported the synthesis of Cu_2ZnSnS_4 (CZTS) with Pt and Au islands for H_2 photogeneration from water [44] (Fig. 9.5). Although no exact QY for this HPA system was given

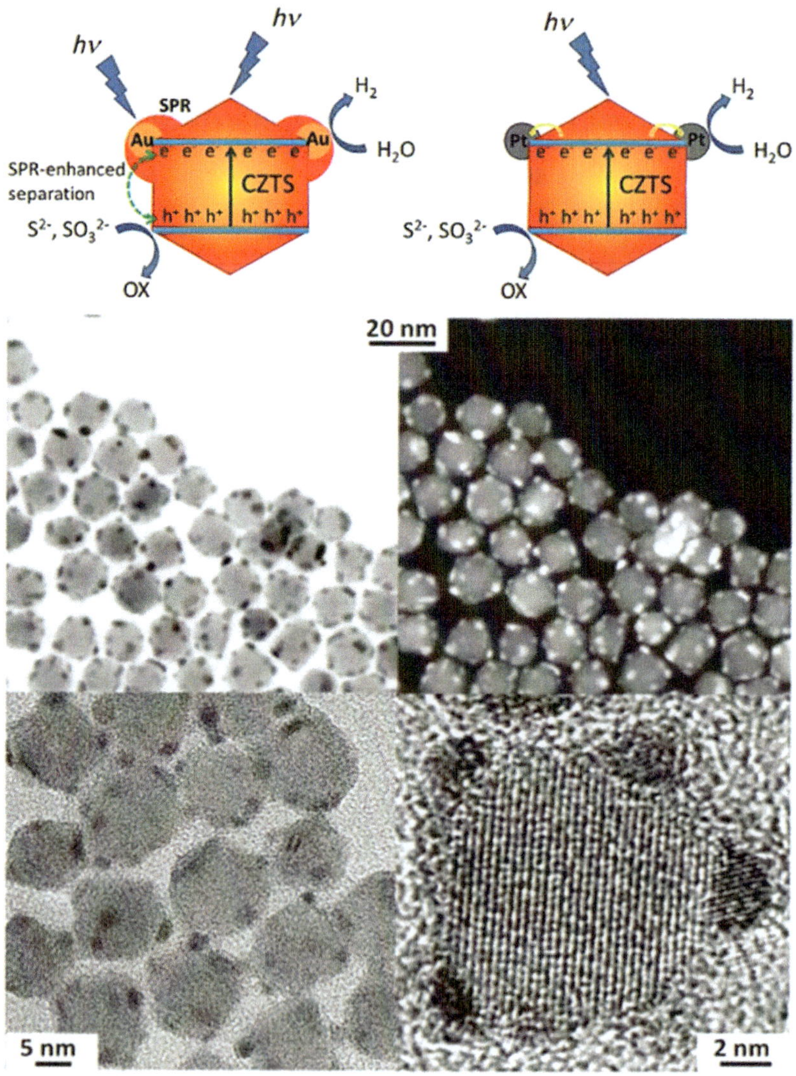

Figure 9.5 Illustration of the possible mechanism of enhancement of the H_2 evolution rate in CZTS–Au and CZTS–Pt heterostructured nanoparticles in the presence of S^{2-} and SO_3^{2-} hole scavengers (up). Bright-field and dark-field TEM micrographs and HRTEM image of CZTS–Pt heterostructured nanoparticles (down). Reprinted with permission from X. Yu, A. Shavel, X. An, Z. Luo, M. Ibáñez, A. Cabot, Cu2ZnSnS4-Pt and Cu2ZnSnS4-Au heterostructured nanoparticles for photocatalytic water splitting and pollutant degradation, J. Am. Chem. Soc. 136 (2014) 9236–9239. Credit: Copyright 2014 American Chemical Society.

in their work, an eightfold enhancement of activity was observed for CZTS–Pt compared with that of the bare samples.

9.2.2 Quasi-Type II CdSe/CdS Dot-in-Rod Nanorods

CdSe/CdS dot-in-rod nanorods are of type I band alignment; however, due to their unique structure, when being excited, the photogenerated holes are three-dimensionally confined to the CdSe region, while the electrons are delocalized throughout the whole nanorod [45], which actually could be considered as quasi-type II QDs. Since the parameters of this type of nanorod such as the length of the nanorods and the seed size can be well controlled, it is considered as an interesting model to systematically study the relationship between catalytic activity and charge transfer behavior, and many works have been carried out that relate to this topic [10,12–15,46–48].

Amirav and Alivisatos systematically investigated the catalytic activity of CdSe/CdS dot-in-rod nanorods with different lengths and CdSe seed sizes, where tipped Pt at one end of the nanorods was used as the cocatalyst [13]. They found that for the nanorod with the same CdSe seed size, its catalytic activity will be enhanced with the increasing of its length (Fig. 9.6). They attributed this to the probable improved charge separation by increasing the distance of the reaction sites for reduction and oxidation. Under the optimal conditions, the apparent QY of the HPA system could reach 20% at 450 nm, and photostability could also be improved compared to the CdS rods without a CdSe seed.

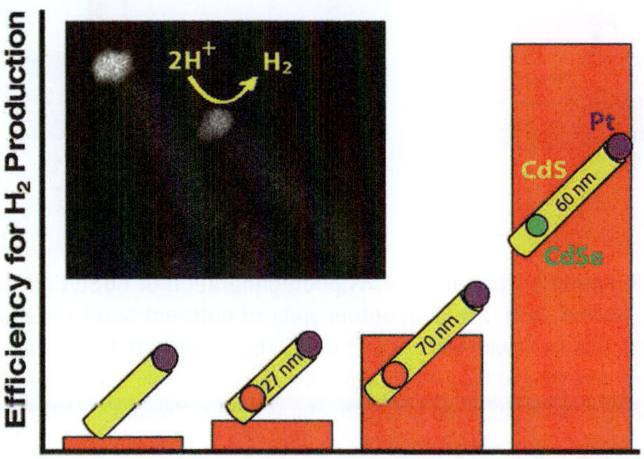

Figure 9.6 Dark-field TEM image of Pt-tipped CdSe/CdS dot-in-rod nanorods and the efficiency for H_2 photogeneration of CdSe/CdS dot-in-rod nanorods with different parameters. Reprinted with permission from L. Amirav, A. P. Alivisatos, Photocatalytic hydrogen production with tunable nanorod heterostructures, J. Phys. Chem. Lett. 1 (2010) 1051–1054. Credit: Copyright 2010 American Chemical Society.

Lian et al., introduced a common redox mediator, methylviologen (MV^{2+}), into a QD-based HPA system [12]. They found that compared to CdSe seeds, CdSe/CdS core–shell QDs, and CdS nanorods, CdSe/CdS dot-in-rod nanorods exhibit obviously higher QYs (Fig. 9.7), which was attributed to the ultrafast electron transfer of methylviologen, fast hole removal by the sacrificial electron donor, and the slow charge recombination of the dot-in-rod structure. This work demonstrates the importance of tailoring the wave function of electrons and holes on light harvesting and charge separation. Furthermore, by employing different additional sacrificial electron donors (methanol or sulfite) in CdSe/CdS dot-in-rod nanorod-based HPA systems, they further found that the hole transfer rate to the electron donor is a key efficiency-limiting step and should also be considered when designing a QD-based HPA system to achieve an efficient solar energy conversion [14].

Zamkov et al., reported a method to realize the isotropic etching of CdSe/CdS dot-in-rod structures [15]. As indicated in Fig. 9.8, through the etching process, both the reduction (Pt) and oxidation (CdSe domain) active sites are in direct contact with the external environment, which facilitates the transfer of photogenerated holes from the CdSe domain by direct contact with the sacrificial electron donor, thus pushing the whole reaction toward H_2 photogeneration. The results showed that the catalytic activity of etched CdSe/CdS dimers could be enhanced by about 3–4 times upon etching treatment.

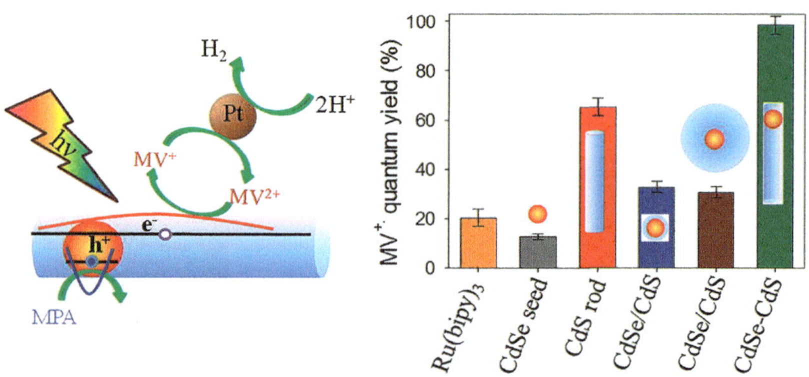

Figure 9.7 Schematic illustration of H_2 photogeneration of CdSe/CdS dot-in-rod based HPA system (left). MV^{2+} quantum yield of different types of QDs and Ru (bipy)$_3$ (right). Reprinted with permission from H. Zhu, N. Song, H. Lv, C.L. Hill, T. Lian, Near unity quantum yield of light-driven redox mediator reduction and efficient H2 generation using colloidal nanorod heterostructures, J. Am. Chem. Soc. 134 (2012) 11701–11708. Credit: Copyright 2012 American Chemical Society.

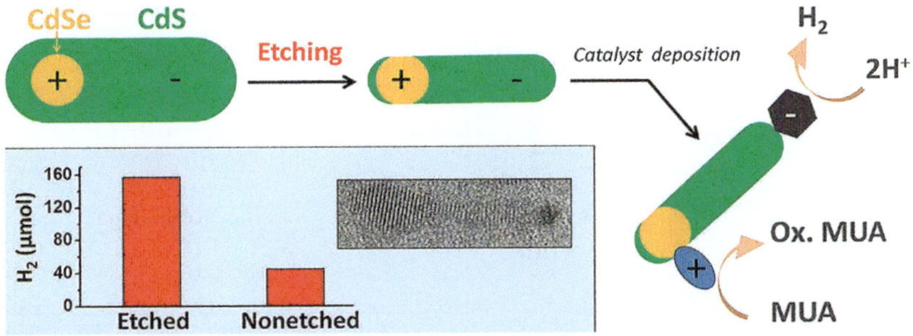

Figure 9.8 Schematic illustration of etching process and comparison of H_2 photogeneration of etched and nonetched CdSe/CdS dot-in-rod nanorods. Reprinted with permission from E. Khon, K. Lambright, R.S. Khnayzer, P. Moroz, D. Perera, E. Butaeva, et al., Improving the catalytic activity of semiconductor nanocrystals through selective domain etching, Nano. Lett. 13 (2013) 2016–2023. Credit: Copyright 2013 American Chemical Society.

9.2.3 Core–Shell Quantum Dots

Through the formation of a core–shell structure, photogenerated charge transfer behavior could be modulated or the stability of QDs could be improved, and these could be beneficial for further improving the efficiency of HPA systems. For instance, Larsen et al., discovered that by coating a CdSe core with a CdS shell, under the same testing conditions, an approximately 10-fold enhancement of visible light–induced H_2 photogeneration for CdSe/CdS core–shell QDs compared to that of a CdSe core alone could be observed (Fig. 9.9) [49]. The authors ascribed this increase to the passivation of the deep surface trap states which was identified by transient absorption and PL measurements. CdSe/ZnS core–shell QDs were also used as a light absorber in an HPA system because the ZnS shell could dramatically increase the photostability and fluorescence QYs of the QDs compared to those without the shell [35].

In addition, due to the type II band alignment of ZnSe/CdS core–shell QDs, the photogenerated electron–hole pairs can be effectively separated by transferring the electrons and holes to the shell and core domains, respectively, which is considered to be beneficial for photocatalytic application. However, there is a concern that the holes localized in the core domain cannot efficiently react with the external environment. While, Zamkov et al., found that in the presence of a sacrificial electron donor, the transfer of holes from the core to the shell is nearly an order of magnitude faster than the electron–hole recombination time [50]. Therefore this is not a problem for practical photocatalytic water splitting.

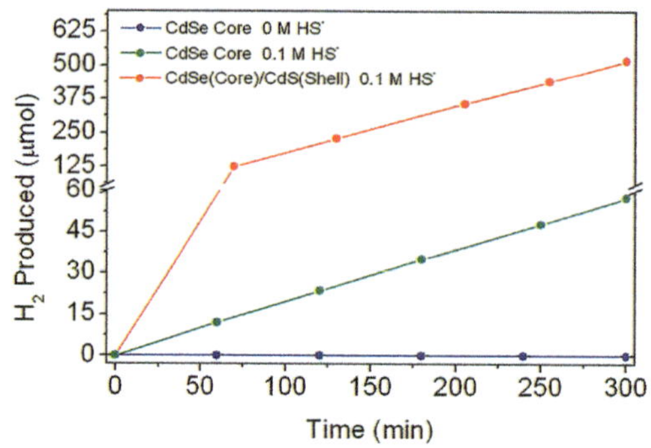

Figure 9.9 Photocatalytic hydrogen evolution for CdSe and CdSe/CdS QDs in pure H_2O and 0.1 M HS^- solution under visible light illumination from a 300 W Xe lamp equipped with a 400 nm long pass filter. Note the change in scale of the *y*-axis after the break. Reprinted with permission from A. Thibert, F.A. Frame, E. Busby, M.A. Holmes, F.E. Osterloh, D.S. Larsen, Sequestering high-energy electrons to facilitate photocatalytic hydrogen generation in CdSe/CdS nanocrystals, J. Phys. Chem. Lett. 2 (2011) 2688–2694. Credit: Copyright 2011 American Chemical Society.

It was recently found that CdS/CdSe core–shell QDs with typical reverse type I band alignment could also be used as a light absorber for an efficient photocatalytic H_2 photogeneration [31]. For this type of QD, both electrons and holes could be easily transferred to the surface of the QDs to contribute to the reactions of photocatalytic water splitting. The migration of charge carriers (electrons and holes) to the shell region may decrease the reduction ability of the electrons and increase the probability of charge carriers recombination, which is not beneficial for the proton reduction. However, under the same testing conditions, the catalytic activity of CdS/CdSe core–shell QDs is about five-fold higher than that of CdS/CdSe core–shell QDs with similar core and shell sizes (Fig. 9.10A). This demonstrates the importance of transferring the charge carriers to the surface of the QDs for improving the activity of H_2 photogeneration. The effect of CdSe shell thickness on the efficiency of photocatalytic water splitting was also further investigated [51]. It was shown that with increases in the CdSe shell thickness, the catalytic activity of the HPA system gradually increased, while its efficiency does not show a similar trend; there is an optimal CdSe shell thickness (1.1 monolayer shell thickness) for an HPA system to reach the highest QY (30.9%) under the illumination of 420 nm of light (Fig. 9.10B). This was ascribed to the two competing processes, the increase of the reaction probability of charge carriers for proton reduction–related reactions, and a decrease in the charge separation efficiency, which codetermine this phenomenon.

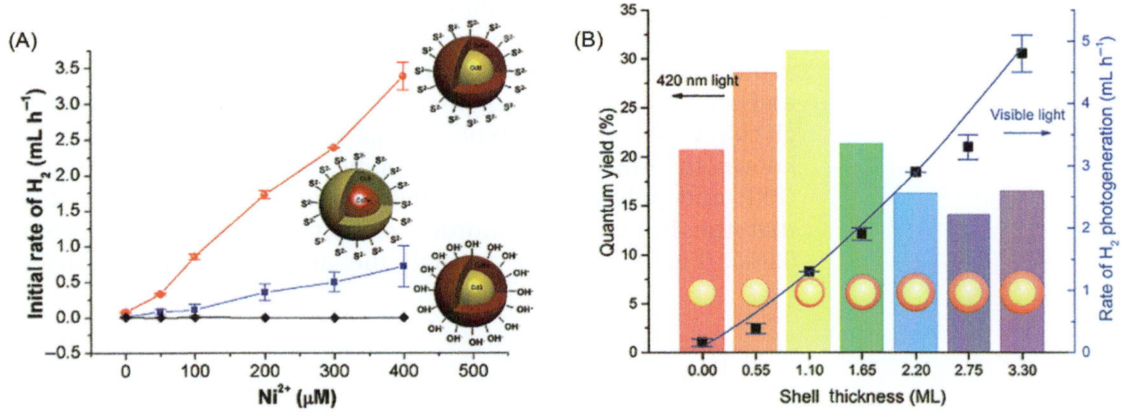

Figure 9.10 (A) The dependence of the initial rate of H_2 of the system containing the same concentration (0.075 μM) of QDs (S^{2-} capped CdS/CdSe and CdS/CdSe, OH^- capped CdS/CdSe core–shell QDs) with the conditions of 0.8 M ascorbic acid at pH 5.2, on the concentration of Ni^{2+}. (B) Quantum yield and H_2 photogeneration rate of the CdS/CdSe QD-based HPA system versus the shell thickness under the illumination of 420 nm and visible light, respectively. Reprinted with permission from P. Wang, M. Wang, J. Zhang, C. Li, X. Xu, Y. Jin, Shell thickness engineering significantly boosts the photocatalytic H2 evolution efficiency of CdS/CdSe core/shell quantum dots, ACS Appl. Mater. Interfaces 9 (2017) 35712–35720. Credit: (A) Reproduced with the permission of The Royal Society of Chemistry; (B) Copyright 2017 American Chemical Society.

9.3 Surface Treatments

Most colloidal semiconductor QDs, especially those with a uniform size or complicated structures are synthesized through the organic phase synthesis method, and the obtained QDs are normally capped with a long-chain organic ligand, which makes the synthesized QDs undispersible in water. Therefore post surface treatments to QDs are usually needed before they can be used as a light absorber in photocatalytic water splitting reactions. Besides, since both the electrons and holes need to cross the surface of QDs to reach the cocatalyst and the sacrificial electron donor, respectively, it can be imagined that the surface properties of QDs are also important factors that affect the catalytic efficiency of HPA systems.

Some works that are focused on the influence of ligands on the catalytic activity of QDs have been reported [52–56]. For instance, Uri Banin et al., employed Au-tipped CdS nanorods as a model system to study the effects of surface coatings on the catalytic activity of H_2 photogeneration [41]. In their work, they chose different types of thiolated alkyl ligands to modify the surface of Au-tipped CdS nanorods; these include (1) mercaptocarboxylic acids with different chain lengths, namely mercaptoundecanoic, mercaptohexanoic acid, and mercaptopropionic acid (MPA); (2) L-glutathione (GSH), which is a tripeptide with two carboxylate and one amine end groups per ligand; (3) mer-acptosulfonioc acid, which has a different charged end group; and (4) *O*-(2-carboxyethyl)-*O*′-(2-mercaptoethyl)heptaethylene glycol; while another type of ligand is polymers, which are known to provide excellent stability such as polyethylenimine (PEI) and poly(styrene-*co*-maleic anhydride). The results show that the PEI-coated nanorods exhibit the highest QY (up to 6.3%) (Fig. 9.11), and through the analysis of the results of

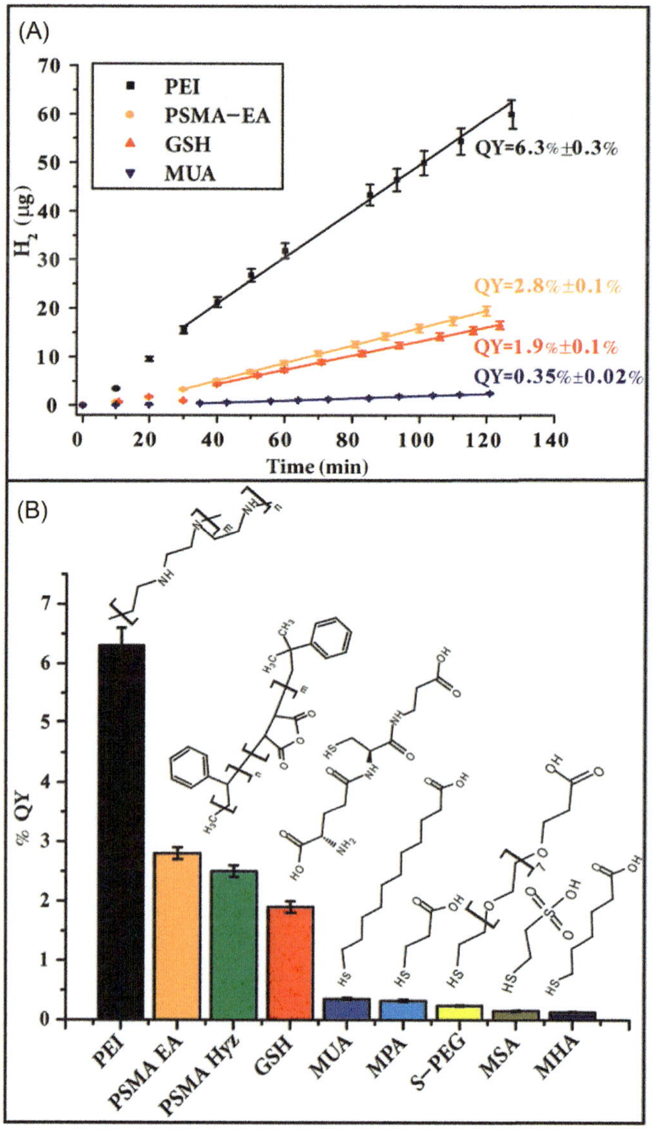

Figure 9.11 (A) Kinetic hydrogen evolution measurements by CdS–Au HNPs for different surface coatings. Straight lines represent the linear fits from which the percentage QY was extracted. (B) Apparent photocatalysis percentage QY values for a wide range of surface coatings including thiolated alkyl ligands, L-glutathione (GSH), and polymer coatings. PEI exhibits the highest QY. Credit: Copyright 2014, John Wiley & Sons, Inc.

transient absorption spectroscopy, fluorescence intensity, and lifetime measurements, the authors mainly ascribed this to the passivation of surface traps.

However, sometimes, surface traps may be beneficial for improving the catalytic activity of H_2 photogeneration. For example, three types of ligands were chosen with different numbers of thiol functional groups, namely poly(acrylic acid) (PAA, average $M_w = \sim 1800$), 3-mercaptopropionic acid (3-MPA), and meso-2,3-dimercaptosuccinic acid (DMSA), to coat CdSe/CdSe core–shell QDs [57]. PL intensity measurements indicate that with by increasing the number of thiol groups in the ligand, the PL QY is gradually decreased from 64.8% to 4.6%, and furthermore, time-resolved PL decay results show that the average fluorescence lifetime of QDs is decreased from 30.5 ns for the PAA coating to 27.7 ns for the 3-MPA coating and 22.5 ns for the DMSA coating (Fig. 9.12), which reflect the

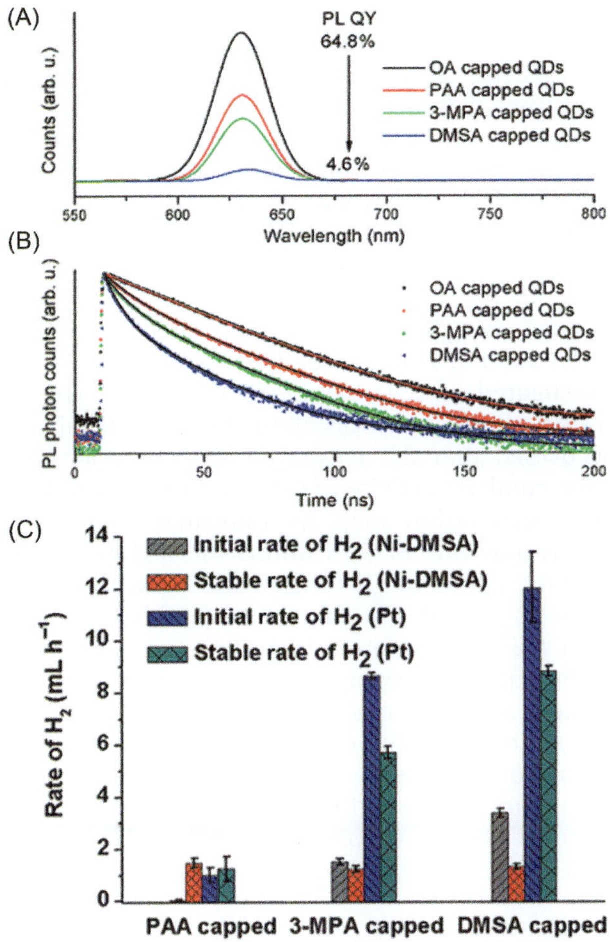

Figure 9.12 (A) PL spectra of the CdSe/CdS QDs capped with different ligands; (B) normalized time-resolved PL decay curves of the CdSe/CdS QDs capped with different ligands. The solid lines are the fitting results of the entire data series; (C) comparison of the initial and stable rate of H_2 photogeneration of the HPA systems with QDs capped with the three different ligands using Ni–DMSA and Pt as the catalyst, respectively. Credit: Reproduced with the permission of The Royal Society of Chemistry.

ligand-induced surface traps by the thiol groups [58]. Interestingly, with an increase of the thiol groups in the ligand, the catalytic activity of the QDs for H_2 photogeneration is gradually increased. This was supposed to be attributed to the different charge separation efficiencies determined by the ligand-induced surface traps.

Besides the mentioned organic ligands, inorganic ligands can also be used for coating QDs for photocatalytic water splitting since it has been demonstrated that electronically transparent inorganic ligands could enhance the electronic coupling between QDs by replacing the original insulating organic ligands [59,60]. For instance, S^{2-}-capped CdS/CdSe core–shell QDs were used as the light absorber of an HPA system, and it was shown that under optimal conditions, the QY of the system could reach as high as 20.6% under the illumination of 520 nm [31]. Furthermore, Reisner et al., reported that by removing the ligand from CdS QDs, their catalytic activity for H_2 photogeneration could be increased to 175 times higher than that of 3-MPA-capped QDs in the presence of a cobalt-based cocatalyst and Na_2SO_3 as the sacrificial electron donor [53]. They attributed this distinct difference to the role of 3-MPA as a physical barrier, therefore restricting the incorporation of the Co-based cocatalyst onto the particles and limiting substrate or product diffusion to and from the surface of the QDs.

9.3.1 Metal

Pt has been recognized as the most effective cocatalyst for the proton reduction reaction, and therefore it has been intensively investigated in QD-based HPA systems [10,12–15,44, 61,62]. Besides the many works about Pt as a cocatalyst, Amirav et al., discovered that the cocatalyst sites play a vital role in affecting the catalytic activity of CdSe/CdS dot-in-rod nanorods for H_2 photogeneration [62]. As indicated in Fig. 9.13, the nanorod with only one Pt tip exhibits the highest apparent QY at 27%, and the apparent QY of the nanorod with two Pt tips is decreased to 18%, while for the nanorod with multiple Pt tips, its apparent QY is even lower than 1%. The authors ascribed this to the fact that hydrogen evolution is a two-electron reaction, and there are two steps in hydrogen evolution. One is the bonding of hydrogen to the catalyst: $H++e-+*\rightarrow H^*$, ding of hydrogen to the catalysts to the parent QY QY at 27tion; while the following step involves the release of molecular hydrogen through one of these two processes: $2H^*\rightarrow H_2+2^*$ or $H++e-+H^*\rightarrow H^*_2+^*$ [63]. Therefore it is favorable if two

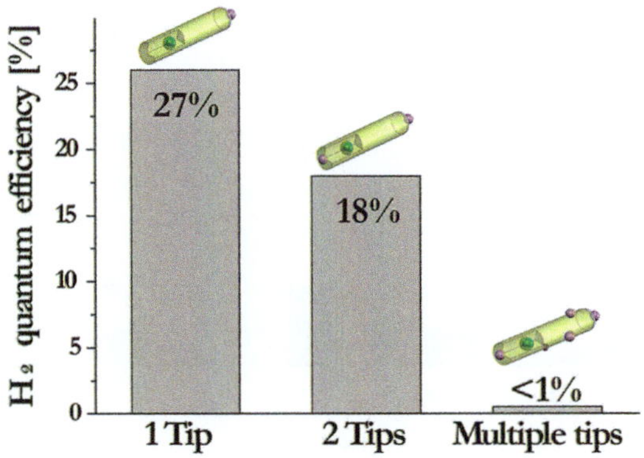

Figure 9.13 The photocatalytic quantum efficiency for the hydrogen reduction half reaction obtained with CdSe/CdS nanorod photocatalysts decorated with single, double, or multiple Pt reduction sites. Reprinted with permission from Y. Nakibli, P. Kalisman, L. Amirav, Less is more: the case of metal cocatalysts, J. Phys. Chem. Lett. 6 (2015) 2265–2268. Credit: Copyright 2015 American Chemical Society.

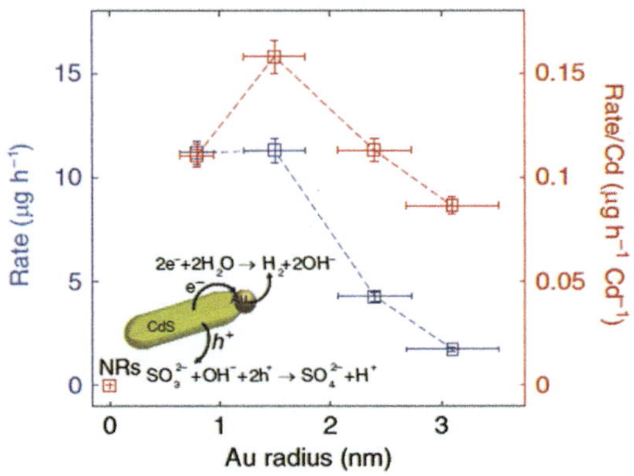

Figure 9.14 Hydrogen production rate (lower line) and Cd normalized rate (upper line) curves as a function of Au size domain in the hybrid nanoparticles. Negligible rates are measured for CdS nanorods. Credit: Copyright 2016, Nature Publishing Group.

photogenerated electrons can be transferred to the cocatalyst site and then the two-bonded hydrogen atoms would be close enough to form molecular hydrogen.

In addition, Au nanoparticles as a cocatalyst exhibit an interesting size-dependent HPA system efficiency [64]. Through varying the Au size in an Au-tipped CdS hybrid model system, Banin et al., found that there is an optimal Au domain size to obtain the highest catalytic activity for H_2 photogeneration (Fig. 9.14). Further transient absorption measurements, hydrogen evolution kinetics, and theoretical modeling indicate that optimum balancing between the positive charge separation efficiency and the negative back recombination determine this phenomenon.

Due to the scarcity of noble metals, other nonnoble metals such as Ni [24,25,47,48] have been investigated as a potential substitute for Pt. For instance, by employing isopropyl alcohol as the sacrificial electron donor, Ni-tipped CdSe/CdS dot-in-rod

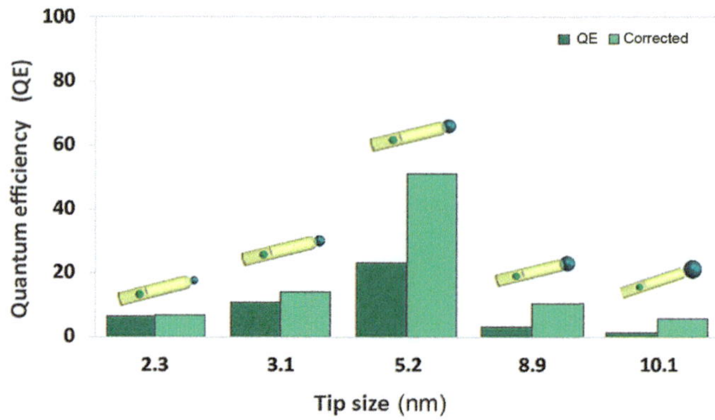

Figure 9.15 Photocatalytic quantum efficiency for the water reduction half reaction obtained with CdSe/CdS nanorod photocatalysts decorated with different-sized Ni tips. Experimental quantum efficiency in dark-green bars, and quantum efficiency corrected for metal absorption in light-green bars. Reprinted with permission from Y. Nakibli, L. Amirav, Selective growth of Ni tips on nanorod photocatalysts, Chem. Mater. 28 (2016) 4524–4527. Credit: Copyright 2018 American Chemical Society. Note: For interpretation of the references to color in this figure legend, the reader is referred to the web version of this chapter.

nanorods exhibited comparable activity for H_2 photogeneration to that of Pt-tipped nanorods [47]. Furthermore, Ni also shows a similar size-dependent activity to that of Au, as shown in Fig. 9.15; for an Ni-tipped CdSe/CdS dot-in-rod nanorod–based HPA system, the activity sharply increased from 7% for a 2.3 nm tip to the maximum of 50.9% for a 5.2 nm tip and then decreased to 6% for a 10.1 nm tip. The following fluorescence measurements show a reverse trend compared to that of the efficiency of H_2 photogeneration. After further detailed analysis of the transient absorption data, the authors ascribed the existence of the optimal Ni tip size for photocatalytic water splitting to the two competing processes, that is, the Coulomb blockade charging energy and the size-dependent Schottky barrier between the Ni tip and the semiconductor.

9.3.2 Metal Sulfides

Many metal sulfides such as MoS_3 [65], Pd_4S [39], WS_2 [40], NiS [66,67], and CoS [67] have been used as cocatalysts in colloidal semiconductor QD-based HPA systems. For example, Alivisatos et al. [65] reported a method for uniformly coating amorphous MoS3 onto CdSe/CdS dot-in-rod nanorods, as shown in Fig. 9.16. Under the optimal conditions, this hybrid shows the maximum activity of 100 mmol H_2 h^{-1} g^{-1} and an

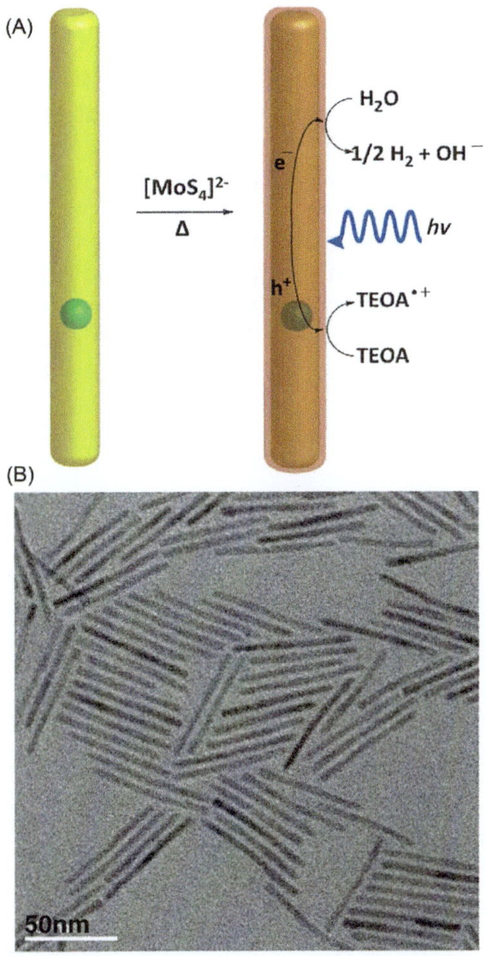

Figure 9.16 (A) MoS_3 deposition on a CdSe-seeded CdS nanorod with photocatalytic hydrogen production in the visible range using triethanolamine (TEOA) as a sacrificial reductant. (B) Bright-field TEM image of CdSe-seeded CdS nanorods (length 60 nm). The scale bar is 50 nm. Credit: Copyright 2011, John Wiley & Sons, Inc.

apparent QY of 10% under the illumination of 450 nm light. However, the obtained hybrid is not very stable. The H_2 photogeneration activity of the HPA system gradually decreased during the measurements, and the authors attributed this to the dissolution of amorphous MoS_3 from the surface of the nanorods, which was confirmed by ICP-OES measurements. As for the palladium sulfides, actually, they are usually considered as undesirable byproducts after the poisoning of metallic palladium by H_2S [68], and therefore these compounds have not been investigated as possible cocatalysts for H_2 photogeneration. However, Banin et al., developed an organic phase cation exchange method to synthesize Pd_4S–CdS hybrid nanorods, and these exhibit a catalytic efficiency of 3.25% under the proper conditions due to the possibly raised Fermi level of the metallic Pd_4S [39].

Differently from anchoring the cocatalyst onto the QDs in two steps (growing the metal sulfide cocatalyst onto the pre-synthesized QDs), Zhang et al., reported a facile one-pot method to synthesize WS_2–CdS or MoS_2–CdS nanohybrids, where the single layer WS_2 or MoS_2 with lateral sizes of 4–10 nm are selectively grown on the Cd-rich (0001) surface of wurtzite CdS QDs [40]. Further photocatalytic experiments indicated that the catalytic activity for H_2 evolution of the obtained nanohybrids was about 16 and 12 times higher than that of pure CdS QDs, respectively, under the illumination of visible light (>420 nm) (Fig. 9.17).

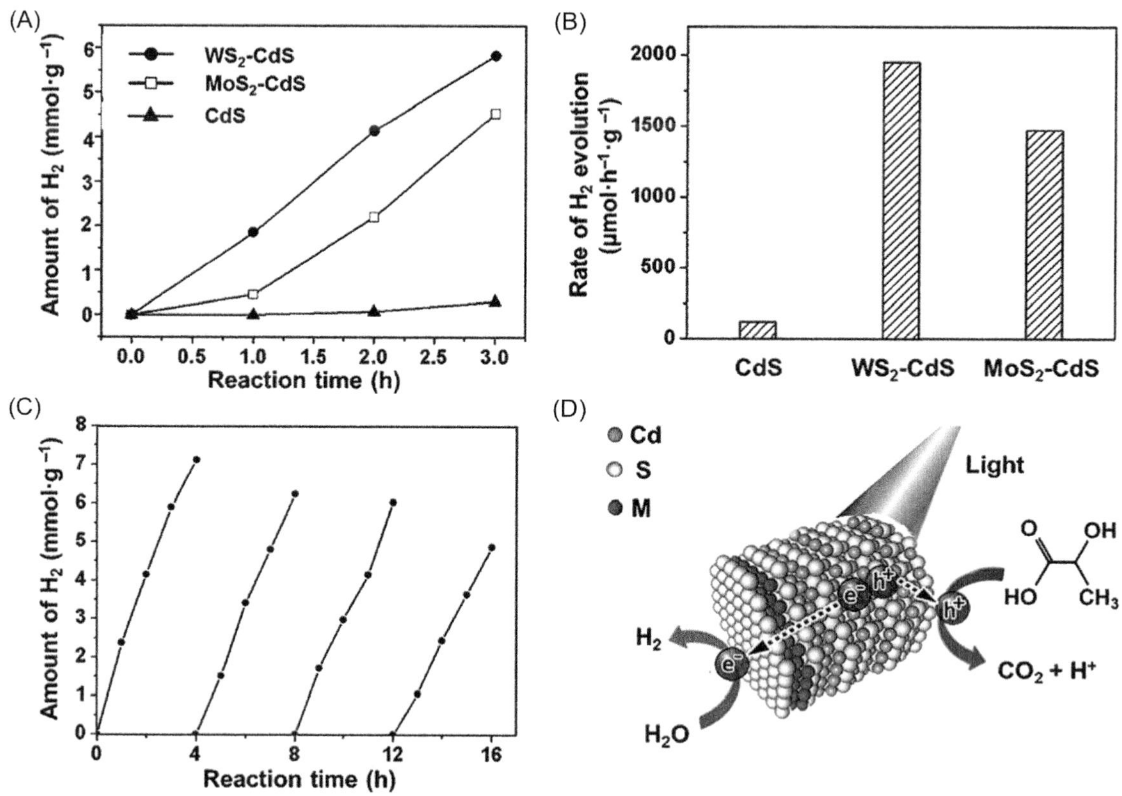

Figure 9.17 Photocatalytic activity of MS_2–CdS nanohybrids for H_2 evolution reaction. (A) Time-dependent photocatalytic H_2 evolution for WS_2–CdS, MoS_2–CdS, and pure CdS. (B) Comparison of the H_2 evolution rate under visible light irritation for WS_2–CdS, MoS_2–CdS, and pure CdS. (C) Cycling test of photocatalytic H_2 evolution for WS_2–CdS. (D) Schematic illustration of the photocatalytic process of MS_2–CdS nanohybrids in a lactic acid solution. Credit: Copyright 2014, John Wiley & Sons, Inc.

Besides, Wu et al., developed an in situ method to prepare an NiS–CdSe/CdS nanohybrid [66]. In their method, Ni^{2+} ions were introduced into 3-MPA-capped CdSe QDs, then, under the illumination of visible light (>400 nm), a CdS shell and an NiS core were formed. Under optimal conditions, the HPA system exhibited an H_2 constant evolution rate of ~153 h^{-1} mg^{-1} and a QY of 11.2% at 410 nm. Furthermore, Ye et al., reported that by using the protocol of controlled decarboxylation of ethylenediaminetetraacetate (EDTA) in an Ni (or Co)–EDTA complex and a CdS mixture precursor, the Ni (or Co) could be anchored onto the surface of CdS though strong bonding to the surface edge of sulfur atoms, and further it can be used as a cocatalyst for H_2 photogeneration [67] (Fig. 9.18). Through this method, the obtained Ni–CdS and Co–CdS nanohybrids exhibited high activities of 3.1 and 4.3 mmol h^{-1} and apparent QYs of 56.2% and 67.5%, respectively, under the illumination of 420 nm light.

9.3.3 Hydrogenases and Their Mimics

Hydrogenases are enzymes that can reversibly catalyze the conversion of protons and hydrogen molecules with high efficiency. As hydrogen activating catalysts, they possess some superior characteristics such as low activation energy, a wide range of oxygen sensitivities, and active sites composed of earth-abundant elements [69]. Hence, scientists have tried to employ hydrogenases and their mimics as cocatalysts to interact with colloidal semiconductor QDs for photocatalytic water splitting [26–28,43,70].

For instance, King et al., studied the self-assembly, charge transfer kinetics, and catalytic activity of a hybrid complex consisting of CaI- and 3-MPA-capped CdTe QDs [43]. As indicated in Fig. 9.19, through self-assembly, the obtained hybrid could realize H_2 photogeneration under the illumination of solar light in the presence of AA as the sacrificial electron donor, where the CdTe QDs and CaI act as the light absorber and cocatalyst, respectively. The authors found that both the concentration of sacrificial electron donor (AA) and the ratio of QDs and CaI could be rate-limiting steps for H_2 evolution. Under the optimal conditions, the QY under the illumination of monochromatic light could reach 9%, and the obtained TON of H_2 photogeneration under illumination was 25 mol H_2 mol^{-1} CaI s^{-1}. They further used CaI as a cocatalyst to complex with the 3-MPA-capped CdS nanorods for the photocatalytic reduction of protons to H_2, and interestingly, the CdS/CaI complex–based HPA system exhibited a QY of up to 20% under the illumination of 405 nm light, and a

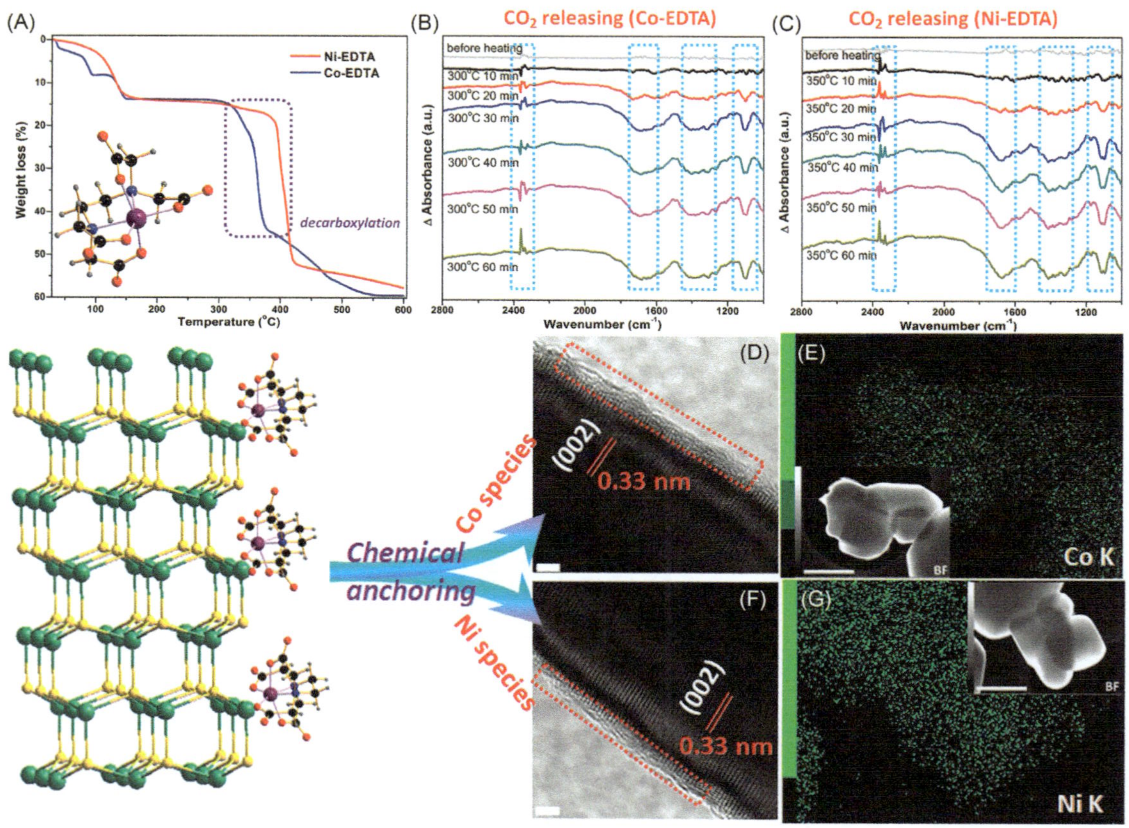

Figure 9.18 (A) The thermogravimetric analysis (TGA) curve of Co—EDTA and Ni—EDTA in argon; (B, C) in situ FI-IR spectra measured during the calcination of Co—EDTA and Ni—EDTA at 300°C and 350°C, respectively. At the bottom: Schematic illustration of the chemical anchoring of Co/Ni species on the CdS surface through the controlled decomposition of an M—EDTA precursor. (D, F) HRTEM of the resulting CdS—CoE—300 and CdS—NiE—350 (scale bar: 2 nm); (E, G) EDS elemental mapping images of CdS—CoE—300 and CdS—NiE—350; inset in (E, G) STEM image of CdS—CoE—300 and CdS—NiE—350 (scale bar: 1 μm). Credit: Copyright 2017, John Wiley & Sons, Inc.

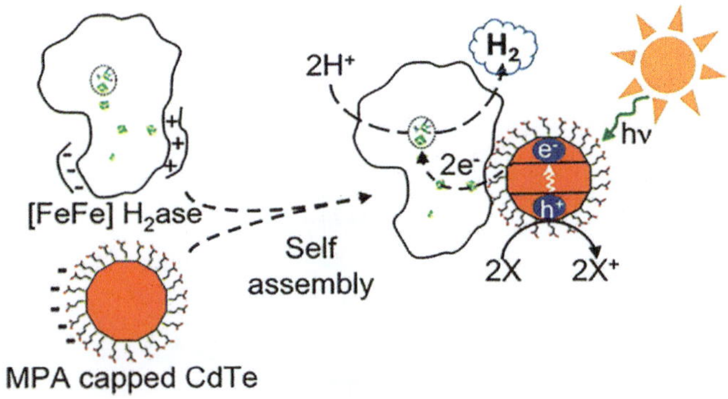

Figure 9.19 Schematic illustration of H_2 photogeneration of the self-assembled 3-MPA-capped CdTe QDs and *Clostridium acetobutylicum* [FeFe]-hydrogenase I complex under the illumination of solar light. Reprinted with permission from K.A. Brown, S. Dayal, X. Ai, G. Rumbles, P.W. King, Controlled assembly of hydrogenase-CdTe nanocrystal hybrids for solar hydrogen production, J. Am. Chem. Soc. 132 (2010) 9672–9680. Credit: Copyright 2010 American Chemical Society.

TOF of 380–900 s^{-1} [28]. Although this system exhibits a high efficiency for H_2 photogeneration, its activity only lasts for up to 4 h, and the authors ascribed this to the photooxidation of CaI by the CdS nanorods. Similarly, Dyer et al., coupled CdSe/CdS dot-in-rod nanorods with a [NiFe] soluble hydrogenase from *Pyrococcus furiosus* for photocatalytic H_2 production [70]. In their system, the redox mediator, propyl-bridged 2-2′-biyridinium, which is used to shuttle photogenerated electrons to the catalyst, was introduced, and under the optimal conditions, the H_2 production QY of the HPA system could reach as high as 77% (the wavelength of the illumination light was not exactly indicated in their work), and more importantly, the stability of the HPA system was greatly increased over 385 h under the illumination of 450 nm light, which is significantly longer than that of the system reported by King et al. [28].

Inspired by these attractive results, scientists are trying to develop artificial hydrogenase mimics with similar active centers for further improving the activity of H_2 production. For example, Wu et al., reported an artificial water-soluble [FeFe] hydrogenase mimic 1 as a cocatalyst to realize H_2 photogeneration in water in the presence AA- and 3-MPA-capped CdTe QDs (Fig. 9.20) [27]. Under the optimal conditions, this HPA system exhibited a TON and TOF of up to 505 and 50 h^{-1}, respectively, and 17.6 mL of H_2 could be obtained after 10 h of visible light illumination (>400 nm). Based on this work, they anchored Fe_2S_2 active sites on the side chain of PAA, and then coupled

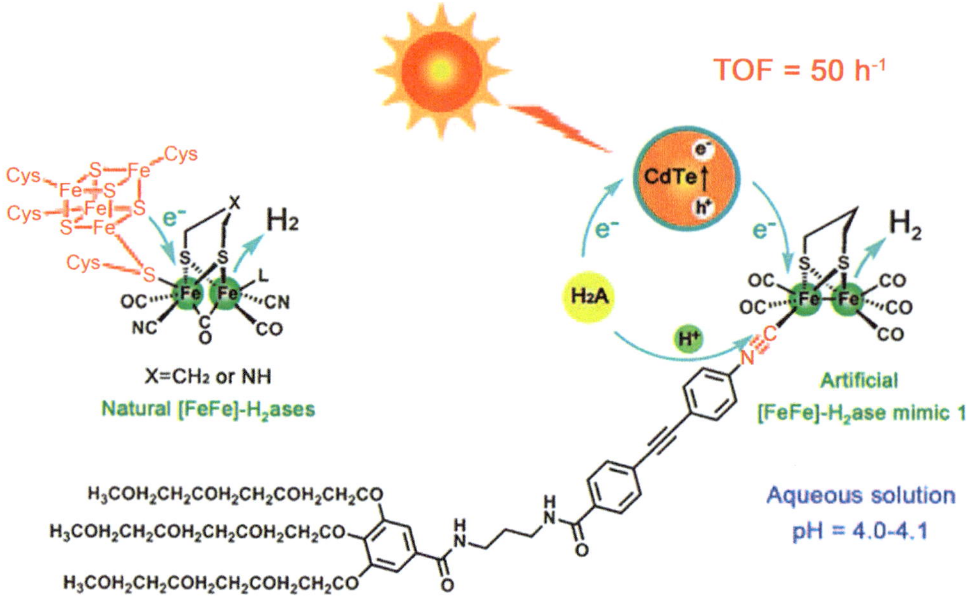

Figure 9.20 The structure of natural [FeFe] hydrogenase and schematic illustration of H_2 photogeneration the HPA system composed of artificial [FeFe] hydrogenase mimic 1, CdTe QDs, and AA under the illumination of solar light. Credit: Copyright 2011, John Wiley & Sons, Inc.

this with CdSe QDs to form a complex. In the presence of AA as the sacrificial electron donor, the HPA system shows a much higher TON (based on Fe_2S_2 active sites) and an initial TOF of up to 27,135 and 3.6 s^{-1}, respectively [26]. However, it should be noted that the stability of these hydrogenase mimics is not yet good enough and they will usually be degraded within several hours of H_2 photogeneration. This may be the primary goal of future work on exploring new artificial hydrogenase mimics, while retaining an efficient catalytic activity.

9.3.4 Molecule

Some small molecules, which usually contain earth-abundant metals such as Ni [9,29,31,32,34,51,57] and Co [35–37], have recently been found to possess efficient catalytic activity and good stability for photocatalytic H_2 evolution. These molecules are usually easy to get through the coordination of metal ions with organic molecules at low cost, and therefore, they have gained tremendous interest. Until now, some Ni- or Co-based molecules have been used as cocatalysts in colloidal semiconductor QD-based HPA systems.

For example, Li et al., coupled cobaloximes (CoIII complexes) with CdS nanocrystals to form a complex, and in the presence of triethanolamine as the sacrificial electron donor, the HPA system exhibited a TON of up to 171 (CoIII(dmgH) 2pycl 1) under the illumination of visible light and an apparent QY of 9.1 % under the illumination of 420 nm light [36]. Chen et al., investigated the catalytic activity for H_2 photogeneration of the cobaloxime hybrid and CdSe/ZnS core–shell QDs (Fig. 9.21) [35]. In the presence of triethanolamine as the sacrificial electron donor, the HPA system showed a TON of over 10,000 of H_2 per QD in 10 h. The authors confirmed the electron transfer from the QDs to the cobaloxime catalyst with an average time constant of 105 ps, which is much faster than that of the charge recombination (»3 ns), by transient absorption measurements, and they attributed this to the long-lived catalytically active sites.

Despite the attractive results of these works, the stability of molecule catalysts is still a problem; however, Eisenberg et al., reported a Ni-based molecule catalyst with exceptional longevity and no decrease of activity for over two weeks under the illumination of light in water [9]. Besides, this Ni-based molecule catalyst can be easily obtained by introducing Ni^{2+} ions into a system of dihydrolipoic acid (DHLA)-capped CdSe QDs, then

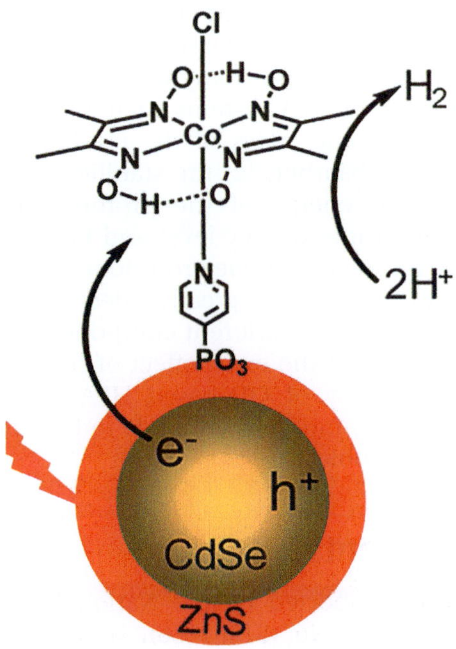

Figure 9.21 Schematic illustration of H_2 photogeneration of the HPA system with CdSe/ZnS core–shell QDs and cobaloxime as the light absorber and the cocatalyst, respectively. Reprinted with permission from J. Huang, K.L. Mulfort, P. Du, L.X. Chen, Photodriven charge separation dynamics in CdSe/ZnS core/shell quantum dot/cobaloxime hybrid for efficient hydrogen production, J. Am. Chem. Soc. 134 (2012) 16472–16475. Credit: Copyright 2012 American Chemical Society.

the Ni–DHLA catalyst, which could be used as a cocatalyst for H_2 photogeneration, is formed. In addition to the excellent stability of the Ni–DHLA catalyst under illumination, it also shows high activity for photocatalytic H_2 evolution. Under the optimal conditions, the corresponding HPA system gives a TON of over 600,000 and a QY of over 36% under the illumination of 520 nm light by employing AA as the sacrificial electron donor. The authors found that the Ni–DHLA molecule catalyst was formed by the coordination of the Ni^{2+} ions with the S donors of the DHLA. Based on this, they further developed a series of nickel bis(chelate) complexes as cocatalysts for photocatalytic H_2 evolution [33], and the results show that some of these are found to be highly active for photocatalytic proton reduction in water. It was recently also found that besides DHLA, other small organic molecules with the thiol functional group such as, 3-MPA or meso-2,3-dimercaptosuccinic acid can also be used to complex with Ni^{2+} ions to form an active cocatalyst for H_2 photogeneration [31,57]. These findings may provide a hint for the future design of efficient molecule catalysts for photocatalytic H_2 photogeneration.

9.4 Conclusion

In summary, there is still a long journey ahead toward the goal of achieving a highly efficient colloidal semiconductor QD-based HPA system. All the discussions in this chapter clearly demonstrate that there are so many factors that can determine the overall efficiency of HPA systems such as (1) the absorption properties, the electronic states of the photogenerated charge carriers of the light absorber; (2) the stability and the ability to lower the activation energy of the proton reduction of the cocatalyst; (3) the proper energy level and fast consumption of the holes left of the sacrificial electron donors; and (4) the surface properties of QDs, which greatly affect the charge carrier transfer behaviors between different components. A further and deeper understanding of the exact effect of these factors on the photocatalytic efficiency of QD-based HPA systems is highly desired and, of course, much work needs to be done in order to fully address these problems.

Acknowledgments

This work was supported by the National Natural Science Foundation of China (No. 51502286), and Prof. Ping Wang thanks the Prof. Yongdong Jin from the Changchun Institute of Applied Chemistry, Chinese Academy of Sciences for the help with writing the chapter.

References

[1] P. Du, R. Eisenberg, Catalysts made of earth-abundant elements (Co, Ni, Fe) for water splitting: recent progress and future challenges, Energy Environ. Sci. 5 (2012) 6012–6021.

[2] A. Fujishima, K. Honda, Electrochemical photolysis of water at a semiconductor electrode, Nature 238 (1972) 37–38.

[3] X. Chen, S. Shen, L. Guo, S.S. Mao, Semiconductor-based photocatalytic hydrogen generation, Chem. Rev. 110 (2010) 6503–6570.

[4] A. Kudo, Y. Miseki, Heterogeneous photocatalyst materials for water splitting, Chem. Soc. Rev. 38 (2009) 253–278.

[5] J. Ran, J. Zhang, J. Yu, M. Jaroniec, S.Z. Qiao, Earth-abundant cocatalysts for semiconductor-based photocatalytic water splitting, Chem. Soc. Rev. 43 (2014) 7787–7812.

[6] A. Vaneski, J. Schneider, A.S. Susha, A.L. Rogach, Colloidal hybrid heterostructures based on II–VI semiconductor nanocrystals for photocatalytic hydrogen generation, J. Photochem. Photobiol. C: Photochem. Rev. 19 (2014) 52–61.

[7] M.B. Wilker, K.J. Schnitzenbaumer, G. Dukovic, Recent progress in photocatalysis mediated by colloidal II-VI nanocrystals, Isr. J. Chem. 52 (2012) 1002–1015.

[8] A. Rogach, Semiconductor nanocrystal quantum dots: synthesis, assembly, Spectroscopy and Applications, Springer, 2008.

[9] Z. Han, F. Qiu, R. Eisenberg, P.L. Holland, T.D. Krauss, Robust photogeneration of H_2 in water using semiconductor nanocrystals and a nickel catalyst, Science 338 (2012) 1321–1324.

[10] P. Kalisman, Y. Nakibli, L. Amirav, Perfect photon-to-hydrogen conversion efficiency, Nano. Lett. 16 (2016) 1776–1781.

[11] Y.-S. Chen, P.V. Kamat, Glutathione-capped gold nanoclusters as photosensitizers. visible light-induced hydrogen generation in neutral water, J. Am. Chem. Soc. 136 (2014) 6075–6082.

[12] H. Zhu, N. Song, H. Lv, C.L. Hill, T. Lian, Near unity quantum yield of light-driven redox mediator reduction and efficient H_2 generation using colloidal nanorod heterostructures, J. Am. Chem. Soc. 134 (2012) 11701–11708.

[13] L. Amirav, A.P. Alivisatos, Photocatalytic hydrogen production with tunable nanorod heterostructures, J. Phys. Chem. Lett. 1 (2010) 1051–1054.

[14] K. Wu, Z. Chen, H. Lv, H. Zhu, C.L. Hill, T. Lian, Hole removal rate limits photodriven H_2 generation efficiency in CdS-Pt and CdSe/CdS-Pt semiconductor nanorod–metal tip heterostructures, J. Am. Chem. Soc. 136 (2014) 7708–7716.

[15] E. Khon, K. Lambright, R.S. Khnayzer, P. Moroz, D. Perera, E. Butaeva, et al., Improving the catalytic activity of semiconductor nanocrystals through selective domain etching, Nano. Lett. 13 (2013) 2016–2023.

[16] R. Su, R. Tiruvalam, A.J. Logsdail, Q. He, C.A. Downing, M.T. Jensen, et al., Designer Titania-supported Au–Pd nanoparticles for efficient photocatalytic hydrogen production, ACS Nano 8 (2014) 3490–3497.

[17] M. Kazuhiko, S. Naoyuki, I. Takahiro, O. Hajime, X. Anke, L. Daling, et al., Preparation of core–shell-structured nanoparticles (with a noble-Metal or

metal oxide core and a chromia shell) and their application in water splitting by means of visible light, Chem. Euro. J. 16 (2010) 7750–7759.
[18] M. Murdoch, G.I.N. Waterhouse, M.A. Nadeem, J.B. Metson, M.A. Keane, R. F. Howe, et al., The effect of gold loading and particle size on photocatalytic hydrogen production from ethanol over Au/TiO_2 nanoparticles, Nat. Chem. 3 (2011) 489–492.
[19] J. Yang, D. Wang, H. Han, C. Li, Roles of cocatalysts in photocatalysis and photoelectrocatalysis, Acc. Chem. Res. 46 (2013) 1900–1909.
[20] C. Kong, S. Min, G. Lu, Dye-sensitized NiSx catalyst decorated on graphene for highly efficient reduction of water to hydrogen under visible light irradiation, ACS Catal. 4 (2014) 2763–2769.
[21] W. Zhang, Y. Wang, Z. Wang, Z. Zhong, R. Xu, Highly efficient and noble metal-free NiS/CdS photocatalysts for H_2 evolution from lactic acid sacrificial solution under visible light, Chem. Commun. 46 (2010) 7631–7633.
[22] X. Zong, H. Yan, G. Wu, G. Ma, F. Wen, L. Wang, et al., Enhancement of photocatalytic H_2 evolution on CdS by loading MoS_2 as cocatalyst under visible light irradiation, J. Am. Chem. Soc. 130 (2008) 7176–7177.
[23] J. Chen, X.-J. Wu, L. Yin, B. Li, X. Hong, Z. Fan, et al., One-pot synthesis of CdS nanocrystals hybridized with single-layer transition-metal dichalcogenide nanosheets for efficient photocatalytic hydrogen evolution, Angew. Chem. Int. Ed. 54 (2015) 1210–1214.
[24] M. Zhukovskyi, P. Tongying, H. Yashan, Y. Wang, M. Kuno, Efficient photocatalytic hydrogen generation from Ni nanoparticle decorated CdS nanosheets, ACS Catal. 5 (2015) 6615–6623.
[25] G. Han, Y.-H. Jin, R.A. Burgess, N.E. Dickenson, X.-M. Cao, Y. Sun, Visible-light-driven valorization of biomass intermediates integrated with H_2 production catalyzed by ultrathin Ni/CdS nanosheets, J. Am. Chem. Soc. 139 (2017) 15584–15587.
[26] F. Wang, W.J. Liang, J.X. Jian, C.B. Li, B. Chen, C.H. Tung, et al., Exceptional poly(acrylic acid)-based artificial [FeFe]-hydrogenases for photocatalytic H_2 production in water, Angew. Chem. Int. Ed. 52 (2013) 8134–8138.
[27] F. Wang, W.G. Wang, X.J. Wang, H.Y. Wang, C.H. Tung, L.Z. Wu, A highly efficient photocatalytic system for hydrogen production by a robust hydrogenase mimic in an aqueous solution, Angew. Chem. Int. Ed. 50 (2011) 3193–3197.
[28] K.A. Brown, M.B. Wilker, M. Boehm, G. Dukovic, P.W. King, Characterization of photochemical processes for H_2 production by CdS nanorod-[FeFe] hydrogenase complexes, J. Am. Chem. Soc. 134 (2012) 5627–5636.
[29] C.A. Caputo, M.A. Gross, V.W. Lau, C. Cavazza, B.V. Lotsch, E. Reisner, Photocatalytic hydrogen production using polymeric carbon nitride with a hydrogenase and a bioinspired synthetic Ni catalyst, Angew. Chem. Int. Ed. 53 (2014) 11538–11542.
[30] A. Das, Z. Han, M.G. Haghighi, R. Eisenberg, Photogeneration of hydrogen from water using CdSe nanocrystals demonstrating the importance of surface exchange, Proc. Natl. Acad. Sci. USA. 110 (2013) 16716–16723.
[31] P. Wang, J. Zhang, H. He, X. Xu, Y. Jin, Efficient visible light-driven H_2 production in water by CdS/CdSe core/shell nanocrystals and an ordinary nickel-sulfur complex, Nanoscale 6 (2014) 13470–13475.
[32] Z. Han, W.R. McNamara, M.S. Eum, P.L. Holland, R. Eisenberg, A nickel thiolate catalyst for the long-lived photocatalytic production of hydrogen in a noble-metal-free system, Angew. Chem. Int. Ed. 51 (2012) 1667–1670.

[33] A. Das, Z. Han, W.W. Brennessel, P.L. Holland, R. Eisenberg, Nickel complexes for robust light-driven and electrocatalytic hydrogen production from water, ACS Catal. (2015) 1397–1406.
[34] B.C.M. Martindale, G.A.M. Hutton, C.A. Caputo, E. Reisner, Solar hydrogen production using carbon quantum dots and a molecular nickel catalyst, J. Am. Chem. Soc. 137 (2015) 6018–6025.
[35] J. Huang, K.L. Mulfort, P. Du, L.X. Chen, Photodriven charge separation dynamics in CdSe/ZnS core/shell quantum dot/cobaloxime hybrid for efficient hydrogen production, J. Am. Chem. Soc. 134 (2012) 16472–16475.
[36] F. Wen, J. Yang, X. Zong, B. Ma, D. Wang, C. Li, Photocatalytic H_2 production on hybrid catalyst system composed of inorganic semiconductor and cobaloximes catalysts, J. Catal. 281 (2011) 318–324.
[37] Y. Sun, J. Sun, J.R. Long, P. Yang, C.J. Chang, Photocatalytic generation of hydrogen from water using a cobalt pentapyridine complex in combination with molecular and semiconductor nanowire photosensitizers, Chem. Sci. 4 (2013) 118–124.
[38] C. Gimbert-Suriñach, J. Albero, T. Stoll, J. Fortage, M.-N. Collomb, A. Deronzier, et al., Efficient and limiting reactions in aqueous light-induced hydrogen evolution systems using molecular catalysts and quantum dots, J. Am. Chem. Soc. 136 (2014) 7655–7661.
[39] Y. Shemesh, J.E. Macdonald, G. Menagen, U. Banin, Synthesis and photocatalytic properties of a family of CdS-PdX hybrid nanoparticles, Angew. Chem. Int. Ed. 50 (2011) 1185–1189.
[40] J. Chen, X.-J. Wu, L. Yin, B. Li, X. Hong, Z. Fan, et al., One-pot synthesis of CdS nanocrystals hybridized with single-layer transition-metal dichalcogenide nanosheets for efficient photocatalytic hydrogen evolution, Angew. Chem. Int. Ed. 54 (2014) 1210–1214.
[41] Y. Ben-Shahar, F. Scotognella, N. Waiskopf, I. Kriegel, S. Dal Conte, G. Cerullo, et al., Effect of surface coating on the photocatalytic function of hybrid CdS–Au Nanorods, Small. 11 (2014) 462–471.
[42] Z.-J. Li, X.-B. Li, J.-J. Wang, S. Yu, C.-B. Li, C.-H. Tung, et al., A robust "artificial catalyst" in situ formed from CdTe QDs and inorganic cobalt salts for photocatalytic hydrogen evolution, Energy Environ. Sci. 6 (2013) 465–469.
[43] K.A. Brown, S. Dayal, X. Ai, G. Rumbles, P.W. King, Controlled assembly of hydrogenase-CdTe nanocrystal hybrids for solar hydrogen production, J. Am. Chem. Soc. 132 (2010) 9672–9680.
[44] X. Yu, A. Shavel, X. An, Z. Luo, M. Ibáñez, A. Cabot, Cu_2ZnSnS_4-Pt and Cu_2ZnSnS_4-Au heterostructured nanoparticles for photocatalytic water splitting and pollutant degradation, J. Am. Chem. Soc. 136 (2014) 9236–9239.
[45] D.V. Talapin, R. Koeppe, S. Götzinger, A. Kornowski, J.M. Lupton, A.L. Rogach, et al., Highly emissive colloidal CdSe/CdS heterostructures of mixed dimensionality, Nano. Lett. 3 (2003) 1677–1681.
[46] K. Wu, H. Zhu, T. Lian, Ultrafast exciton dynamics and light-driven H_2 evolution in colloidal semiconductor nanorods and Pt-Tipped nanorods, Acc. Chem. Res. 48 (2015) 851–859.
[47] Y. Nakibli, L. Amirav, Selective growth of Ni tips on nanorod photocatalysts, Chem. Mater. 28 (2016) 4524–4527.
[48] Y. Nakibli, Y. Mazal, Y. Dubi, M. Wächtler, L. Amirav, Size matters: cocatalyst size effect on charge transfer and photocatalytic activity, Nano. Lett. 18 (2018) 357–364.

[49] A. Thibert, F.A. Frame, E. Busby, M.A. Holmes, F.E. Osterloh, D.S. Larsen, Sequestering high-energy electrons to facilitate photocatalytic hydrogen generation in CdSe/CdS nanocrystals, J. Phys. Chem. Lett. 2 (2011) 2688–2694.
[50] D. Perera, R. Lorek, R.S. Khnayzer, P. Moroz, T. O'Connor, D. Khon, et al., Photocatalytic activity of core/shell semiconductor nanocrystals featuring spatial separation of charges, J. Phys. Chem. C 116 (2012) 22786–22793.
[51] P. Wang, M. Wang, J. Zhang, C. Li, X. Xu, Y. Jin, Shell thickness engineering significantly boosts the photocatalytic H_2 evolution efficiency of CdS/CdSe core/shell quantum dots, ACS Appl. Mater. Interfaces 9 (2017) 35712–35720.
[52] A. Jana, K.N. Lawrence, M.B. Teunis, M. Mandal, A. Kumbhar, R. Sardar, Investigating the control by quantum confinement and surface ligand coating of photocatalytic efficiency in chalcopyrite copper indium diselenide nanocrystals, Chem. Mater. 28 (2016) 1107–1120.
[53] C.M. Chang, K.L. Orchard, B.C.M. Martindale, E. Reisner, Ligand removal from CdS quantum dots for enhanced photocatalytic H_2 generation in pH neutral water, J. Mater. Chem. A 4 (2016) 2856–2862.
[54] V. Rajendran, M. Lehnig, C.M. Niemeyer, Photocatalytic activity of colloidal CdS nanoparticles with different capping ligands, J. Mater. Chem. 19 (2009) 6348–6353.
[55] E.A. Weiss, Designing the surfaces of semiconductor quantum dots for colloidal photocatalysis, ACS Energy Lett. 2 (2017) 1005–1013.
[56] Z. Zhang, K. Edme, S. Lian, E.A. Weiss, Enhancing the rate of quantum-dot-photocatalyzed carbon–carbon coupling by tuning thecomposition of the dot's ligand shell, J. Am. Chem. Soc. 139 (2017) 4246–4249.
[57] P. Wang, J. Zhang, H. He, X. Xu, Y. Jin, The important role of surface ligand on CdSe/CdS core/shell nanocrystals in affecting the efficiency of H_2 photogeneration from water, Nanoscale 7 (2015) 5767–5775.
[58] E. Zillner, S. Fengler, P. Niyamakom, F. Rauscher, K. Köhler, T. Dittrich, Role of ligand exchange at CdSe quantum dot layers for charge separation, J. Phys. Chem. C 116 (2012) 16747–16754.
[59] M.V. Kovalenko, M. Scheele, D.V. Talapin, Colloidal nanocrystals with molecular metal chalcogenide surface ligands, Science 324 (2009) 1417–1420.
[60] D.V. Talapin, J.-S. Lee, M.V. Kovalenko, E.V. Shevchenko, Prospects of colloidal nanocrystals for electronic and optoelectronic applications, Chem. Rev. 110 (2010) 389–458.
[61] K.P. Acharya, R.S. Khnayzer, T. O'Connor, G. Diederich, M. Kirsanova, A. Klinkova, et al., The role of hole localization in sacrificial hydrogen production by semiconductor–metal heterostructured nanocrystals, Nano. Lett. 11 (2011) 2919–2926.
[62] Y. Nakibli, P. Kalisman, L. Amirav, Less is more: the case of metal cocatalysts, J. Phys. Chem. Lett. 6 (2015) 2265–2268.
[63] B. Hinnemann, P.G. Moses, J. Bonde, K.P. Jørgensen, J.H. Nielsen, S. Horch, et al., Biomimetic hydrogen evolution: MoS_2 nanoparticles as catalyst for hydrogen evolution, J. Am. Chem. Soc. 127 (2005) 5308–5309.
[64] Y. Ben-Shahar, F. Scotognella, I. Kriegel, L. Moretti, G. Cerullo, E. Rabani, et al., Optimal metal domain size for photocatalysis with hybrid semiconductor-metal nanorods, Nat. Commun. 7 (2016) 10413.
[65] M.L. Tang, D.C. Grauer, B. Lassalle-Kaiser, V.K. Yachandra, L. Amirav, J.R. Long, et al., Structural and electronic study of an amorphous MoS_3

hydrogen-generation catalyst on a quantum-controlled photosensitizer, Angew. Chem. Int. Ed. 50 (2011) 10203–10207.
[66] Z.J. Li, J.J. Wang, X.B. Li, X.B. Fan, Q.Y. Meng, K. Feng, et al., An exceptional artificial photocatalyst, Ni-CdSe/CdS core/shell hybrid, made in situ from CdSe quantum dots and nickel salts for efficient hydrogen evolution, Adv. Mater. 25 (2013) 6613–6618.
[67] G. Zhao, Y. Sun, W. Zhou, X. Wang, K. Chang, G. Liu, et al., Superior photocatalytic H_2 production with cocatalytic Co/Ni species anchored on sulfide semiconductor, Adv. Mater. 29 (2017) 1703258.
[68] C.P. O'Brien, A.J. Gellman, B.D. Morreale, J.B. Miller, The hydrogen permeability of Pd_4S, J. Membr. Sci. 371 (2011) 263–267.
[69] W. Lubitz, H. Ogata, O. Rüdiger, E. Reijerse, Hydrogenases, Chem. Rev. 114 (2014) 4081–4148.
[70] B. Chica, C.-H. Wu, Y. Liu, M.W.W. Adams, T. Lian, R.B. Dyer, Balancing electron transfer rate and driving force for efficient photocatalytic hydrogen production in CdSe/CdS nanorod-[NiFe] hydrogenase assemblies, Energy Environ. Sci. 10 (2017) 2245–2255.

10

NANOCARBON-BASED HYBRIDS AS ELECTROCATALYSTS FOR HYDROGEN AND OXYGEN EVOLUTION FROM WATER SPLITTING

Muhammad Adnan Younis and Yang Hou
Key Laboratory of Biomass Chemical Engineering of Ministry of Education, College of Chemical and Biological Engineering, Zhejiang University, Hangzhou, P.R. China

10.1 Introduction to the Principles of Water Splitting Through Electrocatalysis

Impending energy crises along with the resulting environmental issues have prompted scientists to develop high performance, environment-friendly, and cost-effective renewable energy resources. Among renewable resources, sunlight, wind-powered, and water resources are found to be appealing ways to produce hydrogen energy by water splitting in a sustainable way. Hydrogen energy plays an important role in the energy sector and chemical industry. It is an essential feedstock to produce NH_3 and various hydrocarbons.

In recent years, electrochemical water splitting has gained widespread attention due to its negligible environmental pollution and wide range of applications [1]. Electrochemical water electrolysis is a promising way to produce hydrogen evolution reaction (HER) and oxygen evolution reaction (OER), as shown

Advanced Nanomaterials for Pollutant Sensing and Environmental Catalysis. DOI: https://doi.org/10.1016/B978-0-12-814796-2.00010-1

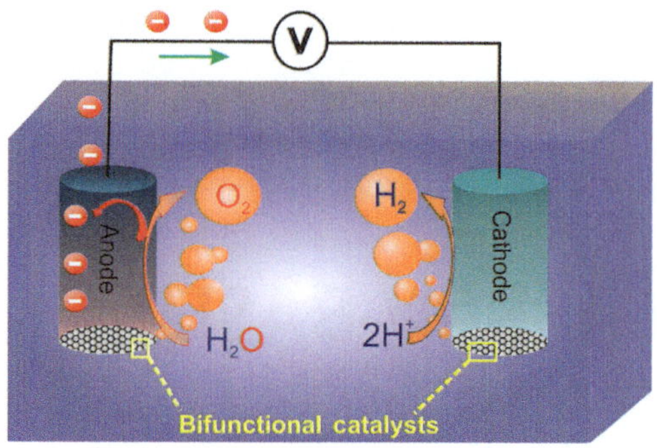

Figure 10.1 Schematic illustration of bifunctional catalyst for electrocatalytic water splitting.

in Fig. 10.1. The anode is involved in water oxidation, which splits water molecules into oxygen gas with accompanying protons and electrons, while hydrogen gas evolves at the cathode by the recombination of electrons and protons. Considering that the water splitting process has sluggish reaction kinetics, the development of highly-efficient electrocatalysts is the key to making water electrolysis economically viable.

In this regard, water electrolysis requires effective catalysts to facilitate both the HER and OER processes. Noble metals such as Pt and their hybrid materials are considered to be the most effective catalysts for HER, while the best electrocatalysts for OER are Ir/Ru-based compounds. However, the high cost and high scarcity of precious metals limit their widespread use. Therefore intensive research efforts have been dedicated to the development of nonprecious electrocatalysts such as transition-metal carbides and sulfides, which can act as effective electrocatalyst for HER, while transition-metal oxides (Ni_2P/NiO_x, CoO_x), hydroxides [NiOOH, $CoNi(OH)_x$], phosphides (CoP_x), and nitrides (CoN_x) are studied for OER in alkaline solutions [2–8]. Along with the enormous efforts in the development of nonprevious metal-based catalysts to replace noble metals, a new class of nanocarbon-based materials have been introduced which dramatically increase the efficiency of water electrolyzers for HER and OER. Additionally, the developed electrocatalysts for both HER and OER should operate in a wide pH range to achieve overall water splitting. Therefore this condition remains challenging for most electrocatalysts because catalysts with highly active basic electrolytes might have low activity or bad stability in strong acidic conditions. Keeping this in mind, a bifunctional catalyst should deliver integrated benefits for HER and OER, which can be obtained by exhibiting a binding force toward hydrogenated oxygen intermediates. Additionally, a bifunctional electrocatalyst with good catalytic activity might further simplify the system architecture and reduce the cost as well. Recently, the application of a water electrolyzer has made it possible to use a single bifunctional catalyst to achieve good HER and OER performance in a single system. In this regard, bifunctional heterogeneous electrocatalysts have been introduced with excellent catalytic activity and unique structure design. Transition metal–based compounds such as Co-based, Fe-based, and Ni-based compounds have been considered for use as bifunctional electrocatalysts due to their multiple advantages [9]. In this chapter, a detailed discussion will be provided on nanocarbons and their hybrids, which are used as electrocatalysts for HER and OER.

$$\text{Totalreaction}\quad H_2O \rightarrow {}^{\bullet}H_2 + 1/2O_2 \tag{10.1}$$

In acidic solution:

$$\text{Cathode}\quad 2H^+ + 2e^- \rightarrow {}^{\bullet}H_2 \tag{10.2}$$

$$\text{Anode}\quad H_2O \rightarrow {}^{\bullet}2H^+ + 1/2O_2 + 2e^- \tag{10.3}$$

In neutral or alkaline solution:

$$\text{Cathode}\quad 2H_2O + 2e^- \rightarrow {}^{\bullet}H_2 + 2OH^- \tag{10.4}$$

$$\text{Anode}\quad 2OH^- \rightarrow {}^{\bullet}H_2O + 1/2O_2 + 2e^- \tag{10.5}$$

10.2 Nanocarbons

10.2.1 1D Nanocarbon-Based Hybrids

One-dimensional nanocarbons such as carbon nanofibers (CNFs) and carbon nanotubes (CNTs) have been applied for many electrochemical related applications because of their attractive properties including high electrical conductivity, strong mechanical strength, and high surface area. CNFs can be prepared using the chemical vapor decomposition (CVD) of hydrocarbons. During the CVD process, graphitic layers are produced that slant at any angle along with the nanofibers' axis, which produces graphitic layers in the CNFs [10,11]. In contrast, CNTs can be synthesized as single-walled carbon nanotubes (SWCNTs) or multiwalled carbon nanotubes (MWCNTs) having concentric cylinders with diameters of 1.4 nm and 10–20 nm, respectively [12,13]. Among 1D nanocarbons, CNTs have been widely applied in HER and OER electrocatalysis. For instance, Zou et al., developed a highly-efficient cobalt-embedded N-rich CNT (Co-NRCNT) electrocatalyst using a simple pyrolysis method for HER. The composite precursors used in the synthesis methods were cheap and easily accessible, which makes this Co-NRCNT catalyst cost effective. The Co-NRCNTs hybrid showed excellent electrochemical performance for HER, and it can be applicable in a wide range of pH values (0–14). The Co-NRCNTs hybrid was found to be a good HER electrocatalyst due to several reasons. First, it has a high HER activity close to that of commercial Pt/C. Second, it works well under various electrolytes including acidic, neutral, and basic media. Such properties make this Co-NRCNT hybrid to be considered as an effective electrocatalyst for OER and overall water splitting [14].

10.2.2 2D Nanocarbon-Based Hybrids

Two-dimensional nanocarbon structures include carbon nanosheets, graphene nanosheets (GNSs), and graphene-based composites. Among these, graphene-based composites have attracted attention as metal-free electrocatalysts, and also as supports for the growth of metals or metal oxides in energy conversion systems. GNSs have exceptional properties with high electrical conductivity of 103–104 $S\,m^{-1}$, large surface area of 2630 $m^2\,g^{-1}$, good thermal conductivity of 5000 $Wm^{-1}\,K^{-1}$, and excellent mechanical strength [15]. Furthermore, GNSs with rich surface defects can be synthesized by the chemical exfoliation treatment of graphite [16]. Due to their defects, unique electronic structures, and special edge sites, 2D GNSs can act as electrocatalysts by themselves. Moreover, 2D-structured porous nanocarbons such as graphene platelets, carbon nanosheets, and carbon nanoplates have porous structures and large aspect ratios, which could reduce the ion transport distances. Notice that among 2D-structured nanocarbons, porous graphene and porous carbon nanosheets have been largely employed for energy conversion and storage technologies [17–21].

Recently, porous carbon nanosheets such as hierarchically porous carbon nanosheets and N-doped hierarchically porous carbon nanosheets (N-HPCNs) have attracted great attention because of their intrinsically high porosity and large specific surface area [22]. Although HPC has great potential to be used as an electrocatalyst, it has been noticed that doping with heteroatoms such as N, S, and P atoms may boost the catalytic properties of carbon-based materials. Among the heteroatoms, N-doping (e.g., N-doped carbon nanosheets) is well known as an effective strategy to enhance their electrocatalytic properties because N-doping can increase their catalytic exposed active sites [23–25].

10.2.3 3D Nanocarbon-Based Hybrids

Besides the 1D and 2D nanostructured carbons, 3D nanocarbons such as carbon aerogels, 1D CNTs/2D graphene hybrids, and carbon frameworks possess excellent properties including porous structure and large specific surface area, which makes them suitable candidates to be used as supports for the growth of 3D nanocarbon hybrid electrocatalysts [26,27].

Amongst 3D nanocarbons, 3D nanostructured hybrids consisting of 1D CNTs and 2D graphene have been intensively investigated due to their multifunctional material structures with attractive catalytic properties such as excellent conductivity and good durability [28,29]. The incorporated CNTs may increase basal spacing and bridge the defects for electron transfer. Recently, 3D nanocarbon hybrids have been successfully fabricated by employing different synthesis strategies, while the multiple disadvantages such as the low yields, complex processes, high costs, and special instrumentations in the preparation process largely limit their practical applications. It is the need of the hour to discover and explore efficient strategies to fabricate 3D nanocarbon hybrid materials. Different kinds of 3D nanocarbons including N-doped CNT/graphene nanoribbon (N-csCNT-GNR) hybrids [30] and 2D N-doped porous nanosheet (PCN)/1D CNT hybrids [31] have been synthesized and applied in electrochemical water splitting [32]. These hybrids consist of satisfactory active sites and desirable porosities, which are necessary to enhance the catalytic performance toward HER and OER. Rationally, 3D graphene-based materials assist as a support for the growth of active catalysts and produce a high specific surface area to enhance electrocatalytic activities [30–32].

10.3 Synthesis, Structural Characteristics, and Electrochemical Performance of Nanocarbon-Based Hybrids

10.3.1 Noble Metal/Nanocarbon Electrocatalysts

In recent years, electrochemical water splitting has attracted great attention for the large-scale production of hydrogen energy. In this regard, noble metals were found to be potential candidates for electrochemical overall water splitting. Among noble metals, Pt-based and Ir-based electrocatalysts were found to be the most efficient electrocatalysts for HER and OER, respectively. Furthermore, the high HER and OER activities of noble-metal nanoparticles were related to their particle sizes and shapes [33–36]. For instance, the use of Pt nanoparticles in high concentrations may result in an increase in particle size and uncontrollable growth of Pt nanoparticles. In general, electrodeposition with a minimum concentration of Pt precursors could produce small-size and highly-efficient Pt nanoparticles [37]. Moreover, in a few studies, it has been introduced that a lesser amount of Pt contamination can effectively enhance the electrocatalytic activities of catalysts [38–40]. For instance, Ji et al., [33] reported a Pt/N-doped carbon-based electrocatalyst using the anodic dissolution of a Pt electrode to conduct HER. The Pt/N-doped carbon material was developed by a simple pyrolysis treatment of poly(2,6 diaminopridine) in an NH_3 atmosphere. The loading amount of Pt was only 1.5 wt.%, but its catalytic performance was much better than commercial Pt/C (20 wt.%).

Furthermore, it was observed that well-dispersed N-doping, which acted as Pt anchor sites, is the most important factor contributing to highly active Pt nanoparticles. Du et al., [41] introduced a novel synthesis strategy of CNF-supported noble-metal Pt–Co alloy nanoparticles with core–shell structures. The fabricated noble-metal nanocrystal/CNF system consisting of Pt–Co alloy and CNFs was found to be an efficient bifunctional electrocatalyst with excellent catalytic activity for overall water splitting [42]. The complex hierarchical structures may facilitate the diffusion of active species due to their large surface area, which speeds up surface reactions. Meanwhile, 1D CNF nanostructures offer more contact surfaces to support electrocatalysts, while the 3D architecture of CNFs provide fast electron transport and gas diffusion due to the large interstices and interfaces of the electrolytes and the nanocrystal surfaces. In this regard, the N-doped CNFs and Pt–Co alloy (Pt–Co/NCNFs) nanoparticles with a core–shell structure were synthesized using an electrospinning method. It was observed that the nanocarbons of Pt–Co alloy were homogenously dispersed over the entirety of the CNFs [43]. When serving as bifunctional electrode materials for HER and OER, the novel 1D Pt–Co/NCNFs hybrid material exhibited excellent catalytic activity with an onset potential of 20 mV for HER. Meanwhile, the prepared 1D Pt–Co/NCNFs hybrid showed excellent catalytic activity for OER with a small Tafel slope of 76 mV dec^{-1}, an overpotential of 400 mV at a current density of 10 mA cm^{-2}, and an onset potential of 310 mV.

In an earlier study, Pt nanoparticles were fabricated and loaded onto Ni nanofibers/N-doped carbon (Ni-NCNFs-Pt) and applied as an electrocatalyst for HER [44]. The prepared Ni-NCNFs-Pt showed excellent electrochemical performance with a small overpotential of 47 mV in 0.5 M H_2SO_4. The use of NCNFs as a substrate facilitates a fast electron transport, which ultimately leads to a remarkable HER performance. Similarly, Au nanoparticles supported on porous nanocarbons were fabricated by the direct pyrolysis treatment of metal–organic frameworks (MOFs). The MOFs served as both a template and a precursor for the preparation of metal incorporated porous nanocarbons with a large surface area, controlled morphology, and uniform porosity. For instance, a Zn/Fe-embedded carbon shell wrapped by Au nanoparticles was prepared using Zn/Fe-based MOFs as a precursor. The Zn/Fe-based hybrid catalyst showed a high HER performance with a low onset potential of 80 mV in 0.5 M H_2SO_4 [45]. Additionally, multi-metal nanoparticles, either homogeneous or heterogeneous alloys, have the ability to tailor the geometries and enhance the electrochemical activities of nanoparticles. Similarly, an Au–Cu/CNF membrane was synthesized through an environment-friendly approach and employed as an electrode in HER. The synthesized Au–Cu/CNFs showed outstanding electrocatalytic performance with a Tafel slope of 70 mV dec^{-1} and an overpotential of 83 mV with remarkable stability. The bimetallic hybrid structure paved the way to explore more unique nanocrystals for electrochemical water splitting.

Furthermore, a core–shell structured nanocomposite based on Au nanoparticles/Zn–Fe-embedded porous nanocarbons (Au/Zn–Fe–C) was obtained from Zn/Fe-based MOFs for HER [45] (Fig. 10.2). The Zn–Fe compound was embedded into porous nanocarbons, which were encapsulated with Au nanoparticles having sizes of between 50 and

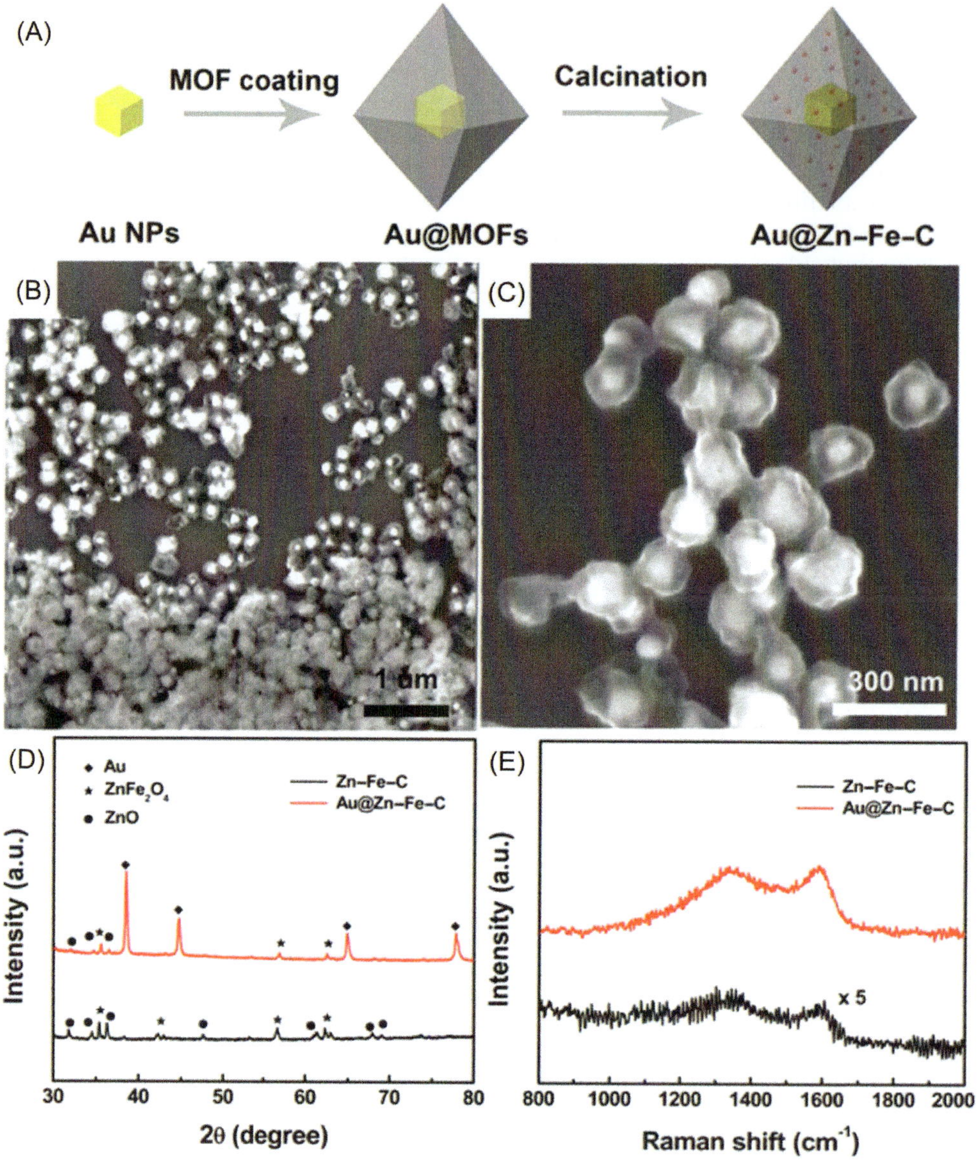

Figure 10.2 (A) Schematic illustration of the synthetic process of Au/Zn–Fe–C core–shell nanostructures, (B–C) FESEM images of Au/Zn–Fe–C, (D) XRD patterns, and (E) Raman spectra of Zn–Fe–C and Au/Zn–Fe–C. Reprinted with permission from J. Lu, W. Zhou, L. Wang, J. Jia, Y. Ke, L. Yang, et al., Core–shell nanocomposites based on gold nanoparticle@zinc–iron-embedded porous carbons derived from metal–organic frameworks as efficient dual catalysts for oxygen reduction and hydrogen evolution reactions, ACS Catal. 6 (2016) 1045–1053. Copyright 2016, American Chemical Society.

100 nm. The prepared Au/Zn–Fe–C hybrid showed good HER performance and achieved a high current density 10 mA cm^{-2} at −0.12 V with an onset potential of −0.08 V in 0.5 M H_2SO_4. It is believed that the HER performance is related to the incorporation of the Au nanoparticles, which are responsible for modulating and enhancing the electron flows of the Zn–Fe compound in the porous nanocarbon layers. Thus metal-embedded porous nanocarbons synthesized from the MOFs were introduced as an effective electrocatalyst to replace other counterparts for HER.

In addition to the Au/Zn–Fe–C hybrid, another precious metal–based bimetallic catalyst with CNFs (Au–Cu/CNFs) was synthesized using Au–Cu alloy for electrochemical HER performance. The adjustment of the precursors to the CNFs was optimized to control the physical properties such as structure, morphology, and composition of the Au–Cu bimetallic alloy. Moreover, the bimetallic alloy was altered from a $AuCu_3$ phase to a Au_3Cu phase when the content of Cu was increased [46]. Furthermore, the prepared bimetallic hybrid was found to be an efficient catalyst to test electrochemical performance and it showed excellent catalytic performance toward HER. The HER performance of the bimetallic hybrid is mainly due to the synergic effects which come from the interaction between uncoordinated high-density atoms and the interfacial effects of the bimetallic alloy. Additionally, the 3D structure of the Au–Cu/CNFs network gave a rapid flow of charges and fast movements toward the desorption of gas [46]. The novel 3D integrated Au–Cu/CNFs structure showed the lowest overpotential of 83 mV, a Tafel slope of 70 mV dec^{-1}, and the highest current density of 0.790 mA cm^{-2} at −0.13 V among all control samples. It was noticed that the charge transfer played an effective role in electrochemical activities.

Due to the presence of a lone pair, the N atom is considered to be a suitable candidate to prepare N-doped carbon hybrid catalysts. The presence of electrons in p-orbital in carbon frameworks together with the lone pair of pyridinic N may enhance the catalytic activity of N-doped carbon-based materials. N-doped carbon-based hybrid catalysts can be synthesized using different nitrogen precursors, and a few kinds of N-doped carbon-based materials have been reported for HER. In this regard, a novel structure of Au/NCNRs/CNFs was developed through the integration of Au nanoparticles on N-doped carbon nanorods (NCNRs) [47]. Typically, the synthesis method was based on the growth of polyaniline nanorods (PNRs) over the surface of CNFs, and then the Au nanoparticles were decorated on the PNRs/CNF hybrid. Different characterization techniques were employed to measure the chemical state, morphology, and structure of the Au/NCNRs/CNF hybrid. The novel Au/NCNR/CNF hybrid as a promising HER catalyst showed excellent durability, a Tafel slope of 93 $mV\,dec^{-1}$, and an onset potential of 126 mV in 0.5 M H_2SO_4. The excellent HER performance of the Au/NCNR/CNF hybrid could be due to the synergistic interaction of the Au nanoparticles and the NCNRs/CNFs. It was noticed that the incorporation of N atoms in nanocarbon materials enhanced the electron donor ability of the materials, which is also one of the reasons the further synthesis of N-doped nanocarbon–based hybrid materials attracts attention. Conclusively, the doping of metal atoms into N-doped nanocarbons might be beneficial due to the fine particle and size construability, which results in the improvement of electrochemical water splitting.

10.3.1.1 Nonnoble Metal/Nanocarbon Electrocatalysts

Nanocarbons are considered as inert for multiple electrochemical reactions such as HER and OER. However, the surface modification of nanocarbon matrices with nonnoble-metal nanoparticle– or compound–based dopants can make it electrochemically active for HER. Dopants can be single atom or multi-atoms, and dopants usually differ in electron density from carbon atoms. Therefore dopants attack the valance energy orbitals of adjacent carbon atoms. It was noticed that dopants can alter the electronic properties of carbon materials, which in turn provides active sites for carbon substances to give high electrochemical activity toward HER. These carbon materials are doped by precious metals, metal phosphides, and metal sulfides, and are reported to be efficient electrocatalysts toward HER [48]. Among various carbon supports, CNTs and porous carbon–based materials have excellent properties including good pore structure and high conductivity to prevent chemical attacks. These properties of CNTs and porous carbon make them suitable matrix supports for both HER and OER. The fabrication of nanocarbon matrices together with nonnoble-metal atoms may enhance the conductivity and dispersion of the active phase of materials. Furthermore, instead of the alteration of the active metal phase, the incorporation of transition-metal atoms may also enhance electrocatalytic activity. For instance, Wu et al., [49] synthesized a novel hybrid based on the incorporation of Fe atoms into a carbon matrix for HER electrocatalysis using a simple pyrolysis method. This work highlighted that the incorporation of Fe atoms into N-doped carbon can facilitate the adsorption–desorption process, which ultimately effects the kinetics involved during H_2 evolution and enhanced HER performance. The prepared hybrid catalyst showed excellent HER performance and achieved a current density of 10 mA cm^{-2} at a low overpotential of 143 mV with a small Tafel slope of 40 mV dec^{-1} in 0.5 M H_2SO_4.

Among nonnoble metal–based catalysts, transition metal phosphides are found to be good electrocatalysts for HER. The distance between metal atoms was increased by the introduction of P atoms, which shrink the d bandgap of the metal. This shrinkage of the d bandgap enhanced the density of states that was placed at near-fermi levels. This property of transition-metal phosphides enables them to replace noble metals for water splitting. Furthermore, the active sites in hydrogenases are found to be mimics in the transition-metal phosphides, while transition-metal phosphides possess less-active sites, and obtain poor conductivities which affect their catalytic performance. The drawbacks related to the conductivity of transition-metal phosphides can be solved by the introduction of carbon supports. Carbon supports can alter the textural properties and enhance the conductivity of transition-metal compounds. Among nanocarbon supports, CNTs acted as suitable carbon supports and were used for the growth of transition-metal phosphides. Additionally, in transition-metal phosphides, the metal sites acted as active sites, whereas the P atoms served as proton acceptors. Until now, only a few carbon-based transition-metal phosphides have been synthesized such as CoP/C [50], FeP/C [51], NiP/C [52]. Among transition-metal phosphide–based materials, Co-based phosphides are considered as highly suitable candidates for HER. In this regard, Sun et al., synthesized a CoP/CNT [53] catalyst by phosphidation treatment of Co_3O_4/CNTs with a P precursor at a low temperature, and the obtained CoP/CNT hybrid exhibited a higher HER catalytic performance with a lower onset potential of 40 mV than that of Co_3O_4/CNTs. Moreover, the Faradaic efficiency of the CoP/CNT hybrid catalyst was nearly 100%. A synergistic effect was observed, which could be due to the coupling interaction between the CoP and the CNTs. The presence of CNTs provided a higher conductivity which makes this CoP/CNTs hybrid a suitable candidate for HER. Instead of CoP, other transition-metal phosphides with nanocarbons were also synthesized such as FeP and Ni_2P, and this Ni-based phosphide material required an overpotential of around 170 mV to achieve a current density of 10 mA cm^{-2}. Additionally, You et al., [54] investigated a novel metal phosphide/nanocarbon hybrid for electrochemical overall water splitting. Moreover, the Co–P/NC–based bifunctional catalyst was synthesized using a MOFs derived route. The CoP was presented in nanopolyhedron and the Co_2P nanoparticles were incorporated into N-doped carbon supports. The obtained Co–P/NC catalyst possessed a large porosity and a high specific surface area of 183 $m^2\,g^{-1}$, which is beneficial to perform electrochemical overall water splitting. The prepared Co–P/NC hybrid achieved the highest current density of 165 mA cm^{-2} at 2.0 V with an excellent durability. Furthermore, the Co–P/NC hybrid exhibited excellent HER and OER performances with an overpotential of 154 and 319 mV in 0.1 M KOH, respectively.

Similarly, it was observed that bimetallic oxides have higher HER and OER activities than that of single metal oxides. In this regard, Zheng et al. [55] synthesized a CoMnO/CN hybrid bifunctional electrocatalyst by coating of CoMn oxide nanoparticles on highly-ordered nanocarbons for HER and OER. During this synthesis strategy, the carbon specie, N dopant, and bimetallic effect present in this CoMnO/CN composite provided a synergic effect to achieve high catalytic activity. The CoMnO nanoparticles had a strong ability for the adsorption of OH^-, while the adsorption of O_2 was weaker than that of CoO and MnO. Therefore the CoMnO nanoparticles appear to be efficient electrocatalysts for OER. On the other hand, the active sites in the CoMnO/CN hybrid for HER were N-doped carbon frameworks. The N-doped carbon frameworks allowed for a fast transfer of ions and electrons toward the CoMnO. Additionally, the CoMnO/CN hybrid possessed an ordered superlattice structure. In this structure, the presence of carbon layers between adjacent CoMnO nanoparticles was effectively found to enhance the active sites. The carbon layers also prevented the dissolution or aggregation of CoMnO nanoparticles. One can conclude that the combination of N-doped carbon and metal-oxide nanoparticles may provide a strong synergic effect, which results in the enhanced performance and catalytic activity of materials. Keeping this in mind, Hou et al., reported a nanostructured Co_3O_4/N-PC dodecahedron by the thermal treatment of a ZIF-67 precursor for OER (Fig. 10.3). This hybrid structure was composed of Co_3O_4 nanoparticles that were embedded into N-doped porous carbon (N-PC). The OER performance of the Co_3O_4/N-PC hybrid was impressive with a slightly higher onset potential of 1.52 V than commercial Ir/C (1.45 V). However, the prepared Co_3O_4/N-PC hybrid showed much a smaller onset potential than the N-PC and Co_3O_4 nanoparticles. Meanwhile, the current density achieved by the Co_3O_4/N-PC hybrid was higher than that of Co_3O_4 and N-PC. It was observed that the

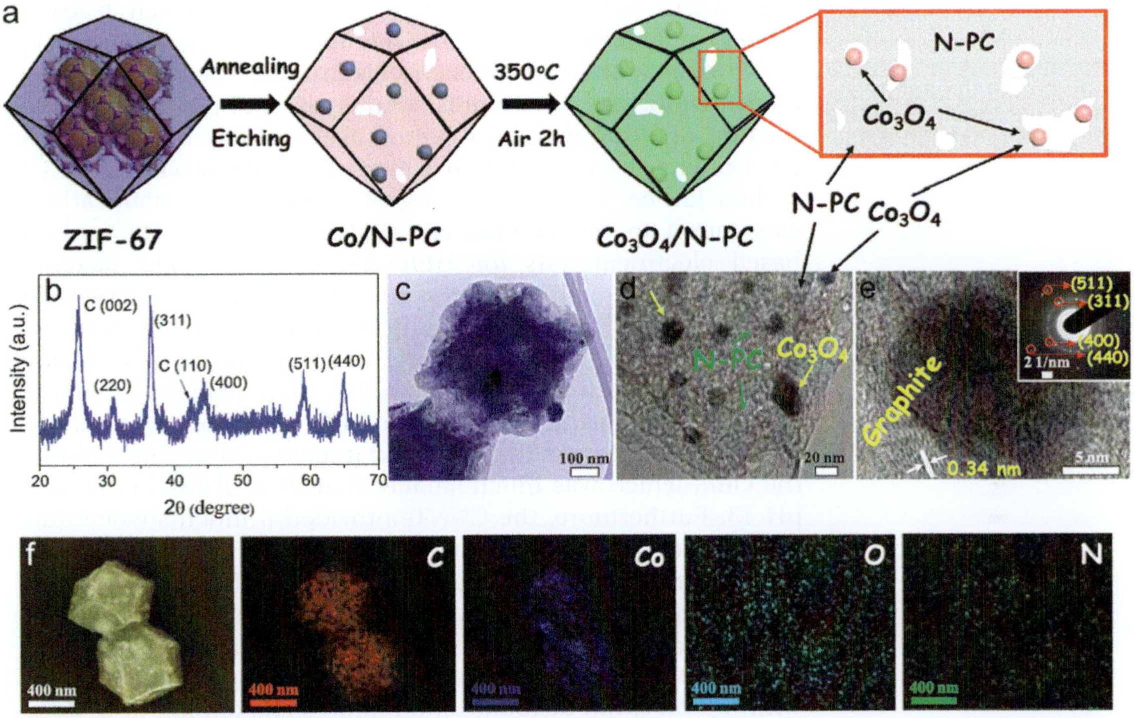

Figure 10.3 (A) Synthesis procedure and schematic diagram, (B) XRD pattern of Co_3O_4/N-PC, (C–D) TEM images, (E) HRTEM image, and (F) FESEM and EDX mapping images of Co_3O_4/N-PC. Inset (E) shows corresponding SAED pattern. Reprinted with permission from Y. Hou, J. Li, Z. Wen, S. Cui, C. Yuan, J. Chen, Co_3O_4 nanoparticles embedded in nitrogen-doped porous carbon dodecahedrons with enhanced electrochemical properties for lithium storage and water splitting, Nano Energy 12 (2015) 1–8, Copyright 2015, Elsevier.

OER catalytic activity of Co_3O_4 was remarkably increased due to the presence of the N-PC [56].

10.3.2 Metal-Free/Nanocarbon Electrocatalysts for OER and HER

Along with precious-metal and nonprecious metal–based catalysts, metal-free carbon-based materials have been introduced as more favorable electrocatalysts due to their tunable structure, high electronic conductivity, and strong tolerance to alkaline/acid environments. In this regard, various heteroatoms such as P, N, B, and O, among others, are used as dopants. The doping of nanocarbons with other heteroatoms may enhance their electrochemical properties. Since heteroatoms have different electronegativity and size, their presence in nanocarbon frameworks might cause charge distribution, which leads to higher catalytic activity toward OER and HER [57].

Nitrogen atoms have different electronic structures to carbon atoms, while they have almost similar sizes. Therefore the introduction of N atoms into a carbon skeleton might change their electronic structure, while it causes small disturbances on the carbon lattice. Thus the doping of N atoms into nanocarbons was found to be an effective strategy to synthesize carbon-based electrocatalysts for HER and OER [58]. For instance, Yadav et al., [59] synthesized carbon nitrogen nanotubes (CNNTs) using the CVD method. To synthesize the CNNTs, chitin was used as an N precursor, which is the most abundant source of nitrogen atoms on Earth. The prepared CNNTs showed a high efficiency and durability for OER. The potential of CNNTs to achieve a current density of 10 mA cm^{-2} was 1.68 V during the OER, which was much smaller than that of Pt/C (1.78 V) at pH 13. Furthermore, the CNNTs provided a much smaller Tafel slope (38 mV dec^{-1}) than undoped CNNT (658 mV dec^{-1}). This huge difference in Tafel slopes between CNNTs and undoped CNTs suggested that the catalytic activities of CNTs might be increased by varying the doping levels of N [60].

In addition to doping with N atoms, CNTs can also be doped with B atoms, but at room temperature. The doping of B atoms might reduce the resistivity and enhance the graphitization of CNTs. Meanwhile, the doping of B atoms into CNTs might shift their Fermi levels into the valence band, which would ultimately enhance their electrochemical performances for HER and OER. In this regard, B-doped MWCNTs were synthesized by the thermal treatment of MWCNTs with boric acid. The prepared B-doped MWCNT catalyst showed excellent OER activity in 0.1 M KOH, which was impressively increased by increasing the B dopant content.

In addition to doping with N and B atoms, doping with O atoms has been found to be impressive for HER activity in acidic electrolytes. O atoms can be doped by the introduction of carboxyl groups into the structure of CNTs. The doping of O atoms makes CNTs highly efficient toward electrochemical activities. Furthermore, an electrocatalyst based on O-doped CNTs achieved a current density of 10 mA cm^{-2} at an overpotential of 220 mV, while the onset potential reported was about 100 mV. In another study, a carbon cloth–based electrode, after acidic oxidation treatment, was employed for OER. The resultant O-doped carbon showed good conductivity and high catalytic performance with a Faradic efficiency of 100%, a Tafel slope of 82 mV dec^{-1}, an onset potential of 328 mV, and an overpotential of 477 mV at a current density of 10 mA cm^{-2}. These outstanding electrochemical results paved the way to investigate and explore more O-doped nanocarbons for HER and OER [61].

As is well known, single atom–doped nanocarbons have great potential for HER and OER, while their catalytic performance still needs to be enhanced. In this regard, it was noticed that codoping with heteroatoms such as P or S can play a crucial role in enhancing their catalytic activity [62]. Recently, S/N and P/N codoped nanocarbon-based materials have been reported for both HER and OER [63]. Additionally, carbon nanosheets codoped with S and N atoms were also reported and evaluated for electrochemical HER properties. The catalyst was synthesized by the thermal treatment of peanut root nodules, and the resultant S,N-doped carbon nanosheets showed impressive HER performance with a Tafel slope of 80 mV dec^{-1}, a low overpotential of only 27 mV, and good stability in 0.5 M H_2SO_4. The N-doping might increase the electron density on C atoms, while S-doping can extend this electron density to adjacent C and S atoms, which leads to a better HER performance as compared with other catalysts doped with only N atoms. In another example, Liu et al., [64] reported S and N codoped carbon using a scalable method. The codoped carbon material showed a good durability and high catalytic activity with a Tafel slope of 57.4 mV dec^{-1}, an onset potential of 12 mV, and an overpotential of 97 mV at a current density of 10 mA cm^{-2} for HER. It was suggested that the synergic effect between C–S–C moieties and N-dopants resulted in the excellent HER performance.

Additionally, experimental studies confirmed that the presence of intrinsic defects in nanocarbon networks such as edge defects and topological defects also played crucial roles in electrochemical performances. For instance, Dai et al., [65] reported that the edge of graphite is much more active than the basal plane, which was confirmed using an in situ self-designed microdroplet system. Additionally, N and B atoms were used as dopants to prepare a highly-efficient electrocatalyst based on defective nanocarbon materials for OER. The N,B-doped carbon catalyst was synthesized by the pyrolysis of ethyl cellulose and benzeneboronic acid under a continuous NH_3 flow. The resultant N,B-doped carbon hybrid provided an exceptional OER performance with good durability and high stability. The prepared N,B-doped carbon catalyst was much more efficient than precious metal–based Pt/C and RuO_2 catalysts for water splitting. High activity and exceptional performances of heteroatom codoped nanocarbons were observed due to the presence of abundant carbon defects and heteroatom codopants [66].

Additionally, P atoms possess the same number of electrons as N atoms; however, the lower electronegativity and larger atomic size make P atoms a suitable candidate for the adsorption of O*. Additionally, P atoms have a lone pair of electrons that create a local charge density toward O*, which is effective for electrocatalysis during the OER. The doping of P atoms with other heteroatoms can also further enhance the OER performance of nanocarbons, which ultimately initiates an enhanced OER performance. Keeping this in mind, an N,P-codoped carbon catalyst was prepared using phosphoric acid, urea, and glucose as precursors. This novel bottom-up strategy provided cross-linked carbon nanosheets with high porosity and dopant concentration. In another example, a P,N-codoped nanocarbon was prepared using a bacterium strain. The prepared P,N-codoped nanocarbon showed excellent HER performance with a low onset potential of 76 mV, and a small Tafel slope of 58.4 mV dec^{-1} with high stability and durability (Table 10.1) [83]. These electrochemical results of the P,N-codoped carbon catalyst were comparable to some of the best metal-free catalysts for HER. Subsequently, Zhang et al., [81] synthesized P,N-codoped mesoporous nanocarbons (NPMCs) using a template-free method. The prepared NPMCs with a large surface area of 1663 $m^2\,g^{-1}$ showed remarkable OER performance, which was reflected by their low onset potential of 0.94 V in alkaline media. Thus codoping can be considered as a useful approach to synthesize metal-free carbon-based electrocatalysts for electrochemical water splitting.

10.4 Electrochemical Properties of Electrocatalysts Supported on Graphene/Modified-Graphene

Amongst 2D materials, graphene has gained remarkable research attention for a wide range of electrochemical applications. However, graphene exhibits quite a poor catalytic performance owing to its zero-bandgap properties. The formation of graphene hybrids with other materials, and the heteroatom-doping of graphene are considered to be the most effective ways to boost the catalytic activity of graphene and graphene-based hybrid materials for various electrochemical reactions. When graphene is used as a support, it offers commendable

Table 10.1 Summary of Carbon-Based Materials for Electrocatalytic Water Splitting

Material	Application	Synthesis Method	Electrolyte	Onset Potential Versus RHE (mV)	Overpotential Versus RHE (mV)	Tafel Slope (mV dec^{-1})	Ref.
CNT/$MoSe_2$	HER	Solvothermal	0.5 M H_2SO_4	70	178	58	[67]
CoP/CNTs	HER	Hydrothermal	0.5 M H_2SO_4	40	70–122	54	[53]
Co-NRCNTs	HER	Carbonization	0.5 M H_2SO_4	–	260	–	[14]
MoS_2/CNTs	HER	Solvothermal	0.5 M H_2SO_4	90	–	107.4	[68]
FeCo/NCNTs	HER	Hydrothermal	0.1 M H_2SO_4	70	110	72	[69]
$Fe_{0.9}Co_{0.1}S_2$/CNTs	HER	Solvothermal	0.5 M H_2SO_4	90	120	46	[70]
FeCo/N CNTs	HER	CVD	0.1 M H_2SO_4	70	270	63	[69]
CoNi/NC	HER	Hydrothermal	0.1 M H_2SO_4	50	142	105	[71]
Co_3O_4/GCN	HER/OER	Pyrolyzation	0.5 M H_2SO_4	30	90	–	[72]
CoP/carbon	HER	Solvothermal	0.5 M H_2SO_4	77.74	112	56.7	[73]
Co/N-doped carbon	HER	Hydrothermal	0.1 M KOH	49	200	79.3	[74]
Co/NGF	HER	Solvothermal	0.5 M H_2SO_4	14	125	93.9	[75]
Co/N-doped carbon/Ti mesh	HER	Hydrothermal	0.5 M H_2SO_4	56	106	78.2	[76]
Fe_3C-graphene	HER	CVD	0.5 M H_2SO_4	32	166.6	46	[77]
Mo_2C/NC	HER	Pyrolysis	0.5 M H_2SO_4	–	156	60	[78]
Co_3O_4/NPC	OER	Pyrolysis	0.1 M KOH	147	–	70	[79]
N,P/Graphene	HER	Pyrolysis	0.5 M H_2SO_4	420	–	90	[80]
Porous C_3N_4/N-doped graphene	HER	Pyrolysis	0.5 M H_2SO_4	8	270	49.1	[81]
N,S/carbon	HER	CVD	0.5 M H_2SO_4	130	142	80.5	[64]
N,S/carbon 800	HER	Pyrolysis	0.5 M H_2SO_4	12	–	57	[64]
CNNTs	OER	CVD	0.1 M KOH	69	110	330	[59]
C_3N_4/NG	HER	Pyrolysis	0.5 M H_2SO_4	240	–	51	[82]
N,B-doped carbon	OER	Pyrolyzation	1.0 M KOH	157	–	84	[66]

benefits such as (1) accelerating charge-transfer kinetics, (2) enhancing electrical conductivity, (3) increasing the solubility and stability of loaded catalysts, (4) increasing the lifetime of catalysts, and (5) providing more active sites for catalytic reactions. The synergistic interactions between graphene and transition-metal compounds can tune the electronic structure of active centers, which ultimately results in an enhanced catalytic performance [84].

It is noteworthy to mention that the electrochemical inertness of graphene makes it an inactive catalyst for HER and OER. However, it has been reported that the HER and OER catalytic properties of graphene can be boosted via chemical doping treatments. Based on the chemical doping method, graphene can be distinguished into two types, namely (1) substitutionally doped graphene and (2) adsorptive doped graphene. In the substitution doping, carbon atoms in the graphene are substituted by foreign atoms, while the latter doping approach is obtained by the adsorption of metal, gas, or organic molecules onto the surface of graphene [84]. Both doping approaches can modulate the electronic structure of graphene, which makes it an electrocatalytically active material for water splitting.

10.4.1 Graphene/Modified-Graphene Supported Precious-Metal Electrocatalysts

Recently, precious-metal catalysts have gained much popularity due to their excellent catalytic performance for OER and HER. However, upon interaction with carbon materials, precious-metal catalysts show dramatically enhanced electrocatalytic activities. Apart from the serious issues associated with precious metals such as durability and high price, noble metals remain the best catalysts with high efficiencies. Several efforts have been made to maximize the catalytic performance of precious metals by either coupling with graphene or by engineering their morphology or composition. In this regard, Kayan et al., deposited highly dispersed Au nanoparticles on reduced graphene oxide (rGO) nanosheets and investigated the resulting composite's electrocatalytic HER activity. The prepared rGO/Au composite exhibited significant improvements in electron transfer kinetics. Also, the Au nanoparticles loaded onto the rGO nanosheets increased the electron density, which promoted an efficient adsorption of hydrogen atoms on the surface of the rGO nanosheets, thus resulting in an enhanced HER activity [85]. Besides single precious metal–based catalysts, dual precious metal-based catalysts (such as, Au, Ag, Pt, and Pd) have also been investigated toward achieving high OER and HER activities. Among such binary metal compounds, Pt–Au alloy–based electrocatalysts have been found to be interesting because alloy-based materials can enhance the electrochemically active surface area and modify the electronic band structure of metallic Pt. Thus an N-doped graphene–based noble-metal alloy (Pt–Au/N–G) hybrid was fabricated to evaluate its electrochemical performance. The good electrochemical activities of the Pt–Au/N–G hybrid were attributed to the synergistic effect between the Pt–Au alloy and the N-doped graphene [86].

10.4.2 Graphene/Modified-Graphene Supported Nonprecious-Metal Electrocatalysts

Undoubtedly, graphene-supported noble-metal catalysts have made great progress for electrochemical water splitting, but achieving a scalable synthesis, their high cost, and the scarcity of precious metals still remain challenging for researchers. To overcome these issues, several nonnoble metal-based compounds such as transition-metal particles and transition-metal sulfides, oxides, phosphides, and carbides have been introduced in the past few decades [87,88]. Furthermore, carbon-based electrocatalysts decorated with transition metal–N complex structures specially M–N–C, M–Fe, Co, Ni have gained attention due to their tunable electronic structures and fast electron transfer capacities. The difference in electronegativity between transition metals and carbon enable M–N–C structured electrocatalysts to reduce the kinetics barriers involved in HER/OER. Therefore Lei et al., [89] designed a unique 3D-structured electrocatalyst based on an $Fe-N_4$ complex coordinated with CNFs supported by an electrochemically exfoliated graphene (EG). The FeN_4/CNF/EG hybrid was synthesized using the simple pyrolysis of polyaniline with electrodeposited EG. The prepared hybrid exhibited impressive OER performance with a Tafel slope of 129 mV dec^{-1} and achieved a current density of 10 mA cm^{-2} with an overpotential of 294 mV, much lower than the overpotential (320 mV) of commercial Ir/C in 0.5 M H_2SO_4. The significantly enhanced OER performance of the catalyst was attributed to the presence of highly active sites on the $Fe-N_4$ complex which were responsible for boosting the OER activity by reducing the potential barriers between O*, OOH*, and O_2 generations. Besides the Fe–N complex, other metal–N complex–based catalysts have also been explored for HER and

OER. For instance, Ni nanoparticles were embedded into porous carbon anchored with an Ni–N_x complex for HER electrocatalysis [90]. The novel Ni–NP/Ni–N–C/EG hybrid was synthesized using hydrothermal and pyrolysis methods. The prepared catalysts outperformed previously reported transition metal–based electrocatalysts for HER in a basic solution. The novel hybrid was supported with EG to provide conductive support and it achieved a current density of 10 mA cm^{-2} with an overpotential of 147 mV and a Tafel slope of 114 mV dec^{-1}. The excellent HER performance of the catalyst was associated with the synergistic effect between the Ni–NP and the Ni–N–C. The presence of Ni–NP and Ni–N–C in the Ni–NP/Ni–N–C/EG hybrid could effectively enhance the adsorption of OH^- and reduce the kinetic barriers involved in water dissociation. Similarly, Xu et al., investigated a Ni/N complex–doped graphene hybrid (Ni–NC) prepared by annealing at high temperature [91]. The prepared Ni–NC catalyst showed excellent catalytic performance for both HER and OER in 1.0 M KOH and achieved a current density of 10 mA cm^{-2} and overpotentials of 205 and 280 mV, respectively. Hou et al., synthesized an N-doped graphene/cobalt embedded porous carbon polyhedron (N/Co-doped PCP//NRGO) hybrid electrocatalyst using a pyrolysis method [92]. The N/Co-doped PCP//NRGO hybrid showed excellent HER performance with a low onset potential of 58 mV, while the overpotential was 1.66 V at a current density of 10 mA cm^{-2} for OER in an acid electrolyte. The high catalytic efficiency of the N/Co-doped PCP//NRGO hybrid can be attributed to the presence of NRGO, the impressive features of porous carbon, the doping effect of the N/Co, and the good connection between the porous carbon and the NRGO. A synergic effect between the NRGO and the N/Co-doped PCP originated from a dual active-site mechanism, which makes this catalyst suitable for use in large scale water splitting technologies.

Besides metal particle–loaded nanocarbons, transition-metal oxides have also been extensively reported with excellent OER catalytic activity because the incorporation of metal oxides with graphene supports can increase electron transfer at the interfaces. Additionally, it was noticed that the introduction of such nonnoble metal–based compounds can significantly enhance the stability and catalytic activity of graphene. Furthermore, nanostructures suffer from aggregation, dissolution, and sintering during electrochemical operations, which can be significantly reduced by the introduction of GNSs. Additionally, graphene can impressively enhance the electrical conductivity of such hybrids, which is an essential property for electrochemical water splitting.

Keeping this in mind, extensive research has been conducted to synthesize nonnoble metal/graphene–based hybrids for OER and HER. The reported graphene-based hybrids showed much better durability and higher HER and OER catalytic activity than those of electrocatalysts prepared without graphene supports. Apart from good intrinsic catalytic activity, an excellent OER catalyst also requires abundant surface-active sites. It was observed that layered double hydroxides (LDHs) have multiple active sites and they have shown interesting OER intrinsic catalytic activity at each active site [93]. Aiming to get high OER activities, a number of investigations have reported on graphene/LDHs–based hybrids. In this regard, an N-doped graphene framework (NGF) hybridized with NiFe LDHs was prepared, which showed excellent OER activity with a current density of 10 mA cm^{-2} at a low overpotential of 337 mV and a Tafel slope of 45 mV dec^{-1} with a stable current density of up to 12,000 s in 0.1 M KOH. This result showed that the introduction of the 3D NGF into the NiFe LDHs/NGF hybrid provided a highly electrochemically active surface area [93,94]. Also, the hybridization of NiFe LDHs and NGF developed a strong charge transfer from the Ni centers to the graphene. Furthermore, the charge transfer imparted more positive charge to the Ni atoms, which is beneficial for the adsorption of OH^- during the OER.

Subsequently, Hou et al., synthesized a ternary hybrid of NiFe–LDH/$Co_{0.85}Se$ nanosheets vertically oriented on electrochemically exfoliated graphene (EG) and applied for electrochemical water splitting (Fig. 10.4). The 3D hybrid showed remarkable electrocatalytic performance toward overall water splitting and attained the highest current density of 20 mA cm^{-2} at 1.71 V after 10 h. Moreover, the 3D hybrid delivered a current density of 10 mA cm^{-2} at an overpotential of 260 mV in 1.0 M KOH for HER. However, the OER performance of the 3D hybrid was found to be excellent and it achieved current densities of 150 and 250 mA cm^{-2} at overpotentials of 1.50 and 1.51 V, respectively. The higher catalytic activity of the graphene-based 3D hybrid was corresponded to a large surface area of 164 m^2 g^{-1} and the synergistic effect between EG and $Co_{0.85}Se$, which facilitated a minimization of the charge transfer resistance [32].

Apart from metal oxides and metal hydroxides, transition-metal sulfides are also considered to be effective electrocatalysts for HER and OER. Amongst metal sulfides, molybdenum disulfides (MoS_2)–based catalysts have attracted much attention due to their abundance and intrinsic catalytic activities toward HER.

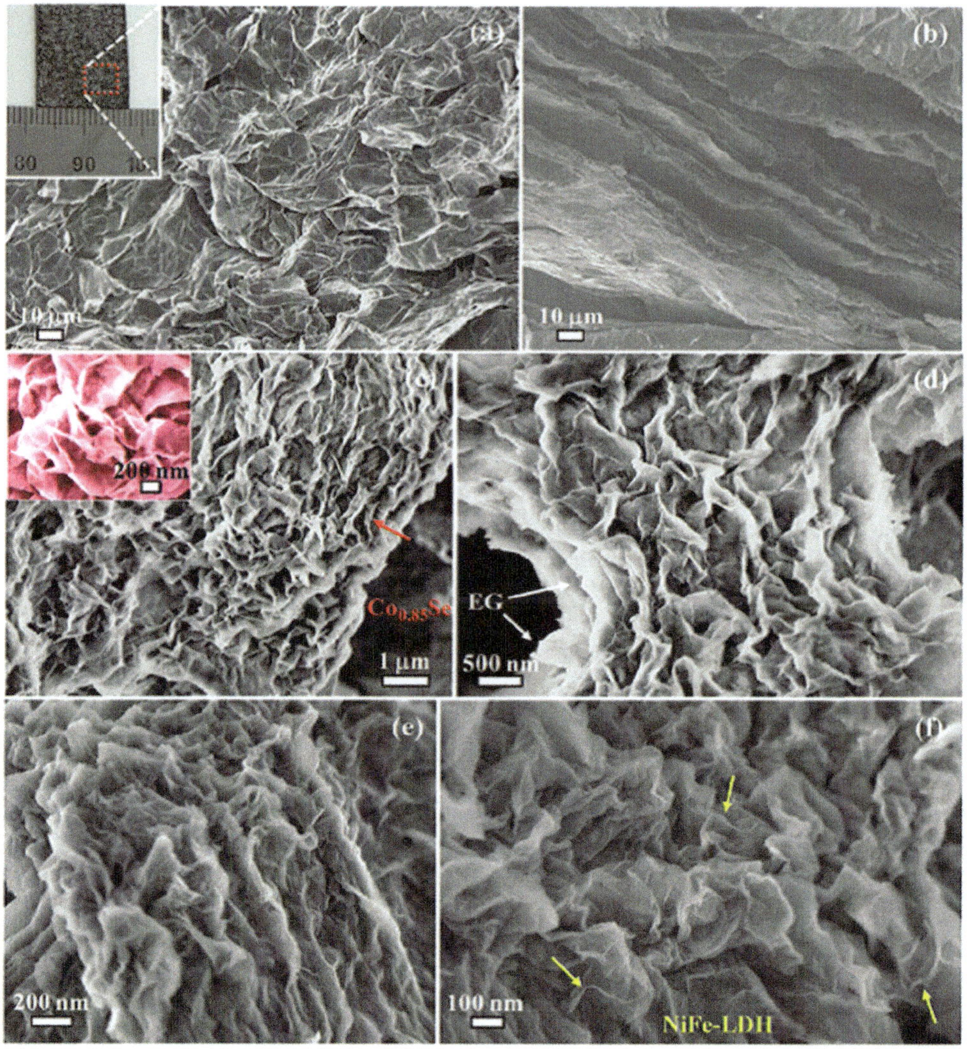

Figure 10.4 (A–B) FESEM images of EG, (C–D) EG/$Co_{0.85}$Se, and (E–F) EG/$Co_{0.85}$Se/NiFe–LDH. Inset: (A) the corresponding EG photograph and (C) enlarged FESEM image of EG/$Co_{0.85}$Se. Reprinted with permission from Y. Hou, M. R. Lohe, J. Zhang, S. Liu, X. Zhuang, X. Feng, Vertically oriented cobalt selenide/NiFe layered-double-hydroxide nanosheets supported on exfoliated graphene foil: an efficient 3D electrode for overall water splitting, Energy Environ. Sci. 9 (2016) 478–483. Copyright 2016, Royal Society of Chemistry.

Bulk MoS_2 is electrochemically inert for HER, whereas artificially designed nanostructured MoS_2 showed impressive catalytic performances for HER due to their more-exposed edge sites. Therefore using MoS_2 with rich exposed edge sites might be a suitable strategy to obtain a high HER performance.

Accordingly, a MoS_2/rGO hybrid with abundant exposed active sites was synthesized by a solvothermal method, which was found to be a suitable method to prevent the aggregation and dissolution of MoS_2 nanoparticles [95]. The MoS_2/rGO hybrid showed impressive catalytic activity for HER with a comparatively low Tafel slope from 94 to 41 mV dec^{-1} and it retained its stability for more than 1000 cycles in acidic conditions.

Besides MoS_2, the synergic effect of rGO has also been applied to others sulfides such as WS_2, MoS_2–$CoMo_2S_4$, and MoS_x, and all of them displayed excellent HER performances [96,97]. In this regard, a self-supported 3D CoS_2/graphene–CNT hybrid was synthesized and exhibited impressive HER performance compared to electrocatalysts without graphene [98,99]. The presence of graphene/CNT conductive networks provided a high conductivity and charge transfer resistance to the hybrid, which was exceptionally higher than that of bare CoS_2 which was 225 Ω compared to 70 Ω, respectively. Hou et al., reported a 3D hybrid electrocatalyst in which Co_9S_8 nanosheets were anchored on Ni_3Se_2 nanosheets grown on the surface of electrochemical exfoliation of graphene foil (Fig. 10.5). The existence of graphene supports provided good conductivity and high surface area to the 3D hybrid, and this synergistic effect led to a great electrochemical performance [100]. Similarly, Cao et al., [101] fabricated a unique 3D-structured electrocatalyst which consisted of N-doped carbon embedded with core–shell CoS_2 and an EG conductive support. The novel 3D CoS_2/CoS_2–NC/EG was synthesized using hydrothermal and pyrolysis methods. The prepared hybrid catalyst achieved a current density of 10 mA cm^{-2} at an overpotential of 210 mV and a Tafel slope of 61.9 mV dec^{-1}. The unique 3D hybrid outperformed the commercial precious Ir/C (84.5 mV dec^{-1}) and well-established state-of-the-art CoS_2-based electrocatalysts for OER. Additionally, the 3D structured CoS_2/CoS_2–NC/EG hybrid exhibited a faster charge transfer ability for OER than EG/CoS_2 and EG/CoS_2–NC. The impressive OER performance of the CoS_2/CoS_2–NC/EG hybrid was attributed to the unique 3D core–shell structure and the synergistic effect between the CoS_2 and the CoS_2–NC. The core part of the hybrid was composed of CoS_2 (20 nm) and the shell part was composed of CoS_2–NC (10 nm) derived from $Co(OH)_2$ nanosheets and ZiF-67, respectively.

Besides metal sulfides, transition-metal phosphides have attracted attention because they have a similar catalytic mechanism to hydrogenase. Therefore transition-metal phosphides were hybridized with graphene to increase their catalytic

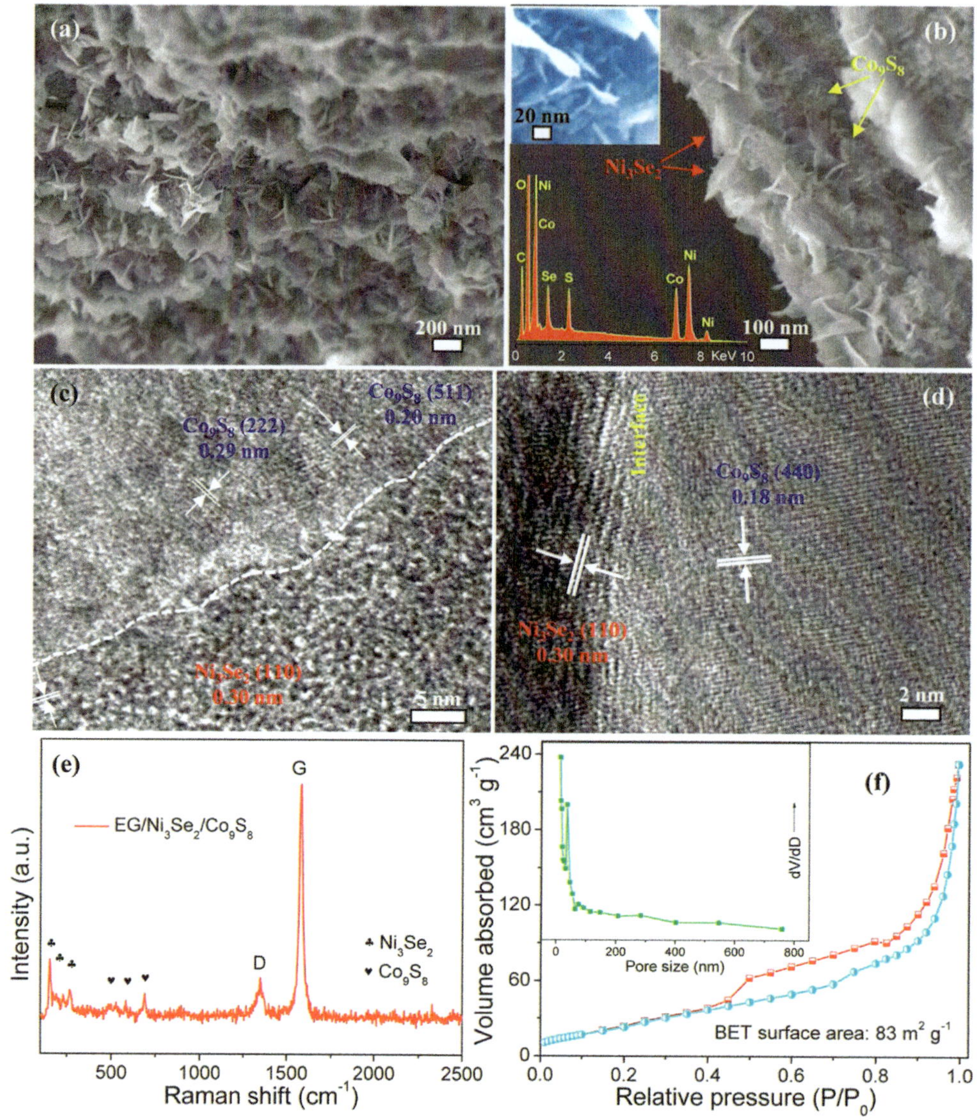

Figure 10.5 (A–B) FESEM images, (C–D) HRTEM images, (E) Raman spectrum, (F) N_2 adsorption isotherm, and corresponding pore size distribution of EG/Ni_3Se_2/Co_9S_8. Inset: corresponding EDX spectrum and enlarged FESEM image. Reprinted with permission from Y. Hou, M. Qiu, G. Nam, M.G. Kim, T. Zhang, K. Liu, et al., Integrated hierarchical cobalt sulfide/nickel selenide hybrid nanosheets as an efficient three-dimensional electrode for electrochemical and photoelectrochemical water splitting, Nano. Lett. 17 (2017) 4202–4209. Copyright 2017, American Chemical Society.

properties toward OER and HER. In this regard, a CoP/graphene hybrid was prepared by a combined pyrolysis and phosphating process. The CoP/graphene hybrid showed impressive electrocatalytic activity for HER in 1.0 M KOH electrolyte with a low

Tafel slope of 38 mV dec^{-1}, which was lower than that of bare CoPt (60 mV dec^{-1}). The Tafel slope of the CoP/graphene hybrid was marginally close to Pt/C (36 mV dec^{-1}). Moreover, the CoP/graphene hybrid achieved a high durability and stability with a marginal decay in current density after 20 h. It was also observed that the charge transfer resistance was reduced, which allowed for a rapid flow of charge during HER [102].

An iron-based metal phosphide (FeP) was loaded onto a graphene support and this FeP/graphene hybrid was applied to catalyze HER. The well-structured FeP/graphene hybrid showed excellent HER performance with a Tafel slope of 50 mV dec^{-1}, a marginally small onset potential of 30 mV, and an overpotential of 123 mV at a current density of 10 mA cm^{-2}. Also, the prepared FeP/graphene hybrid showed a high durability with an undecayed current density of up to 19 h. The synergic effects between the FeP nanoparticles with their rich active sites and the conductive graphene supports led to the enhanced electrochemical activity of the FeP/graphene hybrid. Apart from the FeP, other metal phosphides such as MoP, Fe_2P, and Ni_2P can also be hybridized with graphene to explore their electroactivity for water splitting.

Other than metal sulfides and metal phosphides, metal carbides also have great potential for HER, and various Mo_2C-based hybrids are reported using graphene/CNTs as carbon supports [103]. The HER activity of the graphene/CNT-supported Mo_2C hybrid was enhanced due to multiple reasons including (1) the graphene/CNT supports stabilized the Mo_2C nanoparticles, (2) the high specific surface area of the graphene/CNT support allowed for the Mo_2C nanoparticles to form intimate contacts with the electrolyte, and (3) the graphene/CNT supports provided efficient pathways to enhance electron transfer between the electrocatalysts.

Keeping this in mind, a Mo_2C/CNT–graphene hybrid was introduced for HER. The Mo_2C/CNT–graphene hybrid showed impressive catalytic activity and stability with a Tafel slope of 58 mV and a small onset potential of 62 mV under acidic conditions. It was supposed that the high conductivity and low hydrogen binding of the Mo_2C/CNT–graphene hybrid led to the impressive catalytic activity. Furthermore, the graphene/CNT hybrid prevented Mo_2C nanoparticles from aggregating and provided a large surface area for contact with the electrolyte, thus resulting in the enhanced catalytic activity toward HER [103]. A typical bifunctional catalyst–based 3D crumpled graphene (CG)–cobalt oxide (CoO) hybrid was prepared using aerosolization and a high-temperature nanocrystal growth method. The CG–CoO hybrid showed impressive OER activity with good durability. The 3D hollow graphene structure provided a stable anchor site and a large surface area for the growth of CoO nanocrystals. This architecture allowed the catalyst to use its full surface by preventing the agglomeration of the GNSs. Furthermore, the presence of N-doped graphene impressively enhanced the catalytic activity toward OER with a small overpotential of 0.34 V [104]. Subsequently, Cao et al., [105] designed a novel 3D-structured hybrid consisting of vertically aligned $Co(OH)_2$ nanosheets on an EG substrate with the in situ coating of ZIF-67 ultrathin nanosheets through a vapor phase hydrothermal strategy. The ultrathin ZIF-67 nanosheets served as the shell and $Co(OH)_2$ served as the core to provide a core–shell structure. The unique 3D-structured EG/$Co(OH)_2$/ZIF-67 hybrid demonstrated an excellent OER performance in an alkaline solution, featured by a low Tafel slope of 63 mV dec^{-1} and an overpotential of 280 mV to achieve a current density of 10 mA cm^{-2}. Interestingly, the Tafel slope of the EG/$Co(OH)_2$/ZIF-67 hybrid was far lower than that of the previously reported ZIF-67-based catalysts and much smaller than the Tafel slope of commercial Ir/C (80 mV dec^{-1}). The significantly enhanced OER performance of the hybrid was attributed to the excellent conductive support of EG and the strong synergistic effect between the ultrathin ZIF-67 and the $Co(OH)_2$ nanosheet array, which provided minimum resistance toward the charge transfer process. In situ Raman results revealed the conversion of Co-O into Co-OOH bonds, which might be the active sites of EG/$Co(OH)_2$/ZIF-67 hybrid to boosting OER. Furthermore, a strong electron coupling effect was also observed between the ZIF-67 and the $Co(OH)_2$ nanosheets in the EG/$Co(OH)_2$/ZIF-67 hybrid, which can also be regarded as an important factor in the enhancement of OER activity.

10.4.3 Graphene/Modified-Graphene Supported Metal-Free Electrocatalysts

Recently, graphene has emerged as a state of the art nanocarbon with extraordinary properties. As a bandgap is necessary for many catalytic applications, the zero bandgap of single-layer graphene is a major obstacle that restricts its electrocatalytic applications [106,107]. Current studies have suggested that bandgap engineering of graphene by either surface modification or heteroatom doping can remarkably boost the catalytic properties of graphene [108–110]. It has been noticed that heteroatom-doped graphene has great potential to be used as an electrocatalyst for various electrochemical applications such as water splitting [111], fuel cells [112], lithium-ion batteries [113], supercapacitors [112], and orthers. For instance, Sathe et al., prepared B-substituted graphene (B-SuG) through a wet electrochemical method [114]. The prepared B-SuG showed remarkable electrochemical performances for HER, including faster HER kinetics than that of undoped graphene. Also, this B-SuG had a much higher HER performance than that of defective graphene (DeG), suggesting that the existence of B-dopants affected the catalytic performance of graphene. Chen et al., synthesized S- and N-codoped graphene using a CVD method [115]. The prepared S- and N-codoped graphene contained high crystallinities, geometric defects, and high concentrations of dopants. The N- and S-dopants were homogeneously dispersed on the surface of porous graphene and occupied the structural defects of the graphene lattice. The best S,N-doped graphene catalyst was prepared at 800°C, which showed a great HER performance in acid electrolyte with a Tafel slope of 80 mV dec^{-1} and a comparable onset potential of 130 mV [116]. Additionally, Dai et al., developed bifunctional 2D N,S-doped graphene nanosheets (SHG) using a simple pyrolysis method [117]. The prepared N,S-doped graphene achieved a current density of 10 mA cm^{-2} at overpotentials of 300 and 230 mV for HER and OER in 0.1 M KOH, respectively. The excellent bifunctional catalytic performance was attributed to the good electron transport and abundant active sites of SHG. Furthermore, Zheng et al., [80] prepared a heteroatom-codoped graphene-based catalyst, and mentioned that the codoping of P and N atoms effectively catalyzed HER reactions. To confirm this prediction, the N,P-codoped graphene was prepared by the thermal treatment of graphene. The resultant codoped graphene consisted of N and P atoms at about 4.6and 1.63 at.%. The carbon atoms in the carbon networks could be activated by the introduction of the P,N-dopant, which ultimately gave a dramatic increase in the performance of the codoped nanocarbons. The prepared N,P-codoped graphene achieved an overpotential of 420 mV at a current density of 10 mA cm^{-2}, and a Tafel slope of 91 mV dec^{-1}. These results combined with theoretical calculations suggested that the codoped carbon materials have a great potential to replace state-of-the-art precious-metal catalysts. Subsequently, Xu et al., introduced an N,P,O-tridoped porous graphite carbon on an oxidized carbon cloth (ONPPGC/OCC), which was synthesized by a simple pyrolysis method for HER and OER [118]. The prepared ONPPGC/OCC catalyst achieved a current density of 10 mA cm^{-2} at overpotentials of 446 and 410 mV for HER and OER in 1.0 M KOH, respectively. The excellent bifunctional performance of the ONPPGC/OCC hybrid was attributed to its 3D porous structure, good electrical conductivity, and rich active sites.

Graphitic carbon nitride (g-C_3N_4) is a newly developed catalyst, which has been rarely explored as an electrocatalyst due to its low electrical conductivity, which leads to a low performance and impedes further advancements. It was found that the incorporation of graphene into g-C_3N_4 could improve the catalytic activity and promote the electrical conductivity of g-C_3N_4, which can be beneficial for a wide range of electronic and photoelectronic applications [119]. For instance, a g-C_3N_4/graphene hybrid showed great promise as an excellent electrocatalyst for HER and OER. Importantly, several research groups have reported enhanced HER activities with the use of g-C_3N_4/graphene composites, which has opened new promising avenues toward metal-free electrocatalysts with appreciable catalytic activity. Further, metal-free g-C_3N_4/graphene catalysts could also provide an opportunity to replace precious noble metals for HER applications [87,120]. The proposed catalytic mechanism of g-C_3N_4/graphene hybrids for enhanced HER activity is as follows: g-C_3N_4 contains carbon and nitrogen atoms that increase OH adsorption and the hydrogen concentration on the surface (Hads), respectively, due to the presence of a positive charge and free electrons in N atoms. Then, the hydrogen from a vicinal OH and reduced hydrogen (Hads) are recombined to form H_2 gas. For instance, Qiao et al., [121] designed a hybrid catalyst by coupling g-C_3N_4 with N-doped graphene for electrocatalytic HER. Theoretical calculations were performed to study the electronic interaction between N-doped graphene and g-C_3N_4. The g-C_3N_4/NG hybrid demonstrated a Tafel slope of 51 mV dec^{-1}, much lower than individual g-C_3N_4, N-doped graphene, and mixture of g-C_3N_4/NG, indicating superior HER activity (Fig. 10.6) [82]. Impressively, the g-C_3N_4/NG hybrid showed great stability in both basic and acidic media, indicating strong covalent bonding between the N-doped graphene and the g-C_3N_4. These findings suggested that the synergistic interaction between the conductive N-graphene and the g-C_3N_4 promoted the HER.

In addition to the HER, g-C_3N_4-based hybrids have also been investigated for OER applications. For instance, Sun et al., [122] explored the potential of ultrathin g-C_3N_4/graphene hybrids as efficient OER electrocatalysts in an alkaline solution. Both g-C_3N_4 nanosheets and the g-C_3N_4/graphene hybrid produced a

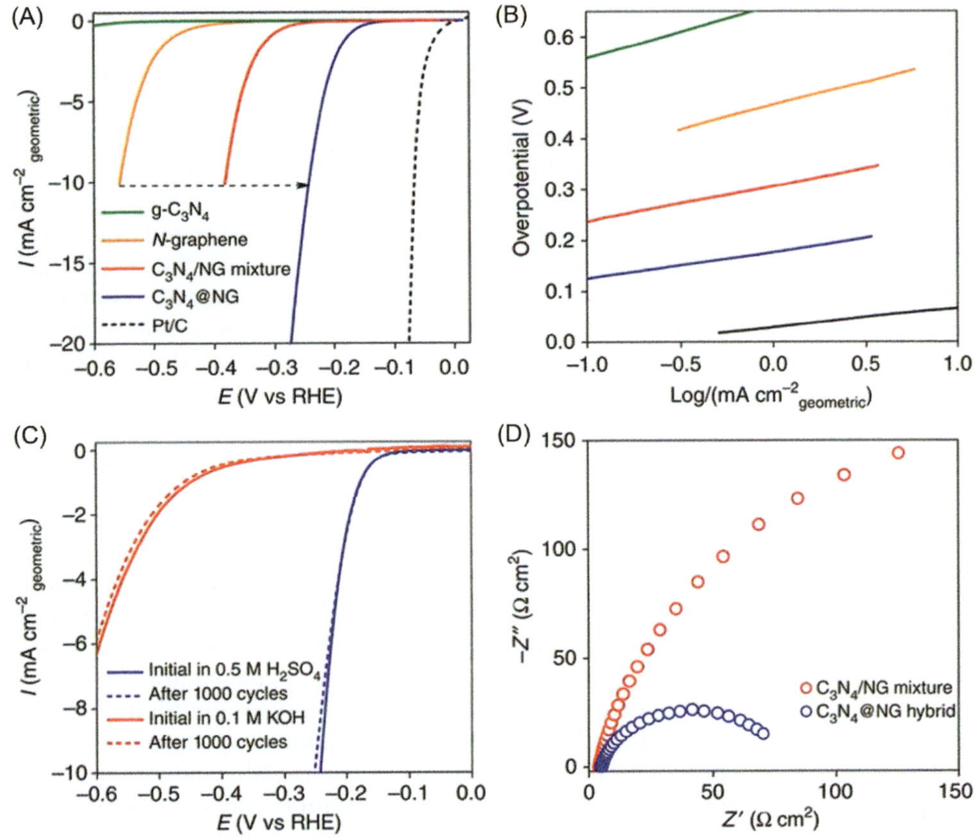

Figure 10.6 (A–B) The HER polarization curves and Tafel plots for four metal-free electrocatalysts and 20% Pt/C (electrolyte: 0.5 M H_2SO_4). The curve referring to C_3N_4/NG was recorded for a sample with 33 wt.% of g-C_3N_4 in the hybrid. (C) Polarization curves recorded for C_3N_4/NG hybrid before and after 1000 potential sweeps under acidic and basic conditions. (D) Electrochemical impedance spectroscopy data for C_3N_4/NG hybrid and C_3N_4/NG mixture in H_2SO_4. Reprinted with permission from Y. Zheng, Y. Jiao, Y. Zhu, L.H. Li, Y. Han, Y. Chen, et al., Hydrogen evolution by a metal-free electrocatalyst, Nat. Commun. 5 (2014) 3783. Copyright 2014, Macmillan Publishers Limited.

number of oxygen bubbles adsorbed on the electrode surfaces. Further, the g-C_3N_4/graphene hybrid showed a more positive onset potential of 0.58 V for OER with a larger anodic current density than that of individual g-C_3N_4 nanosheets and graphene. The potential applied by the g-C_3N_4/graphene hybrid to achieve a current density of 10 mA cm^{-2} was about 805 mV, which was much lower than that applied by the g-C_3N_4 nanosheets (975 mV) in basic electrolyte. Additionally, the smaller Tafel slope of 68.5 mV dec^{-1} for the C_3N_4/graphene hybrid compared to that of the g-C_3N_4 nanosheets (120.9 mV dec^{-1}) clearly corresponded to their more satisfactory OER activity than that of the g-C_3N_4/graphene hybrid. The excellent OER performance of the g-C_3N_4/graphene hybrid might be due to several reasons, namely (1) the π-π interaction of the g-C_3N_4 and the graphene leads to effective electronic contact, (2) the ultrathin g-C_3N_4 nanosheets allow for efficient electron transfer from the graphene to the g-C_3N_4, and (3) the pyridinic N of the g-C_3N_4 nanosheets is considered as an active site for OER. It is noteworthy that few studies have investigated the potential of g-C_3N_4/graphene hybrids for OER. Finally, the g-C_3N_4/graphene hybrid retained 100% OER catalytic performance after 1000 cycles with excellent durability, which makes it a potentially low-cost substitute for noble or nonnoble metal–based catalysts for OER.

10.5 Conclusions and Future Perspectives

Nanocarbon-based electrocatalysts such as 1D, 2D, and 3D nanocarbons possess great potential for electrocatalytic water splitting in terms of HER and OER due to their unique properties such as excellent electrical conductivity, good thermal conductivity, and large surface area. Increasingly, precious metals– and nonprecious metals–based nanocarbons provide excellent electrocatalytic activity due their high durability and the synergistic effect between metals and nanocarbon supports, but their large-scale production is still challenging. Metal-free (S, N, O, P atoms, etc.) doped nanocarbon materials with abundant active sites and controllable molecular structures exhibit great potential for electrochemical water splitting. Although an increasing number of metal-free carbon-based catalysts are emerging for both the HER and OER, the electrocatalytic performance of carbon-based catalysts in comparison with their metal-based counterparts still remains elusive. Hence, most metal-free carbon-based HER and OER electrocatalysts are inferior to their metal counterparts. For instance, MoS_2 for HER and RuO_2 for OER still provide better performances than those of metal-free carbon-based electrocatalysts.

Furthermore, intrinsic graphene is incapable of catalyzing HER and OER in the water-splitting reaction, but the functionalization of graphene with noble metals, nonnoble metals, and metal-free materials provides promising and highly efficient catalytic activities because graphene supports provide large surface areas and more exposed active sites for HER and OER catalysts. The functionalization of graphene by chemical doping with heteroatoms can effectively activate its electrocatalytic capability and modulate its electronic structure. Graphene as a support can provide good stability and can suppress aggregation, catalyst bleaching, and sintering under acidic and basic conditions. These outstanding benefits of graphene for electrocatalytic water splitting have encouraged scientists, and significant research progress has been achieved in the development of chemically-doped graphene and graphene hybrids as effective catalysts for HER and OER and bifunctional catalysts for overall water splitting.

Additionally, the use of EG as a conductive support significantly enhances the HER and OER electrocatalytic performance of graphene-based hybrids which can be rationalized as follows: (1) the hybridization of metal sulfides and metal/N–C catalysts on EG supports can effectively provide abundant exposed $M-N_x$ active sites and improve the dispersity of loaded catalysts. (2) The coupling of EG with $M-N_x$ structured catalysts demonstrates a strong electron coupling effect, which can efficiently boost the charge-transfer and mass transportation process from transition metal to EG, resulting in more positively-charged metal centers, which facilitates the HER and OER processes. (3) EG offers a good conductive network between catalysts and reduces the kinetics barriers present between HER and OER intermediate species and thereby accelerates the charge transfer kinetics in overall water splitting. (4) EG-based hybrids have good chemical stability and strong durability, which is beneficial to conduct water splitting in basic and acidic solutions.

Undoubtedly, nanocarbon-, modified graphene–, and graphene-based hybrids have shown remarkable electrocatalytic performance in terms of HER and OER. However, the performance of graphene hybrids cannot match that of well-established noble metal–based electrocatalysts. Therefore a number of great challenges still need to be overcome in order to achieve their practical applications. For example, it is the need of the hour to move toward the development of efficient bifunctional electrocatalysts that promote both high HER and OER activity simultaneously rather than the extensive exploration of ways to improve the electrocatalytic activity of HER and OER individually. Currently, most HER catalysts perform efficiently in acid electrolytes, while OER catalysts have shown their remarkable activities in basic and neutral solutions. This poses a potential problem when the HER is coupled with the OER in the overall water splitting. Therefore the development of electrocatalysts to conduct HER in alkaline solutions and finding OER and HER catalysts that can work efficiently over a wide range of pH values with high performance are highly desirable.

Increasingly, the reaction mechanisms, roles of doping substances, and identification of real active sites of carbon-based catalysts still need to be further explored. Theoretical and experimental studies combined with in situ microscopic techniques such as in situ Raman might be useful to provide valuable guidance toward the design and fabrication of effective bifunctional electrocatalysts. Additionally, increasing the number of exposed active sites might enhance the electrocatalytic performance of carbon-based bifunctional materials is still answerable. The fabrication of electrocatalysts with different carbon nanostructures such as CNTs, CNFs, and CNRs, may increase their exposed catalytic active sites. Lastly, the stability and activity of carbon-based materials as electrocatalysts must be assessed by their performance in a real water electrolyzer. Until now, all the evaluation data belong to liquid testing which does not give a clear understanding of performance in an electrolyzer.

Acknowledgments

Y. Hou is grateful for the support of the National Natural Science Foundation of China (51702284, 21878270), the Zhejiang Provincial Natural Science Foundation of China (LR19B060002), and the Startup Foundation for Hundred-Talent Program of Zhejiang University (112100-193820101/001/022).

References

[1] X. Liu, L. Dai, Carbon-based metal-free catalysts, Nat. Rev. Mater. 1 (2016) 16064.

[2] W.-F. Chen, J.T. Muckerman, E. Fujita, Recent developments in transition metal carbides and nitrides as hydrogen evolution electrocatalysts, Chem. Commun. 49 (2013) 8896–8909.

[3] P. Xiao, W. Chen, X. Wang, A review of phosphide-based materials for electrocatalytic hydrogen evolution, Adv. Energy Mater. 5 (2015) 1500985.

[4] Z. Chen, Y. Ha, Y. Liu, H. Wang, H. Yang, H. Xu, et al., In situ formation of cobalt nitrides/graphitic carbon composites as efficient bifunctional electrocatalysts for overall water splitting, ACS Appl. Mater. Interfaces 10 (2018) 7134–7144.

[5] X. Chen, P. Gao, H. Liu, J. Xu, B. Zhang, Y. Zhang, et al., In situ growth of iron-nickel nitrides on carbon nanotubes with enhanced stability and activity for oxygen evolution reaction, Electrochim. Acta 267 (2018) 8–14.

[6] L. Marc, C. Guylhaine, A. Markus, S. Menny, Highly porous materials as tunable electrocatalysts for the hydrogen and oxygen evolution reaction, Adv. Funct. Mater. 25 (2015) 393–399.

[7] W. Zhou, X.-J. Wu, X. Cao, X. Huang, C. Tan, J. Tian, et al., Ni3S2 nanorods/ Ni foam composite electrode with low overpotential for electrocatalytic oxygen evolution, Energy Environ. Sci. 6 (2013) 2921–2924.
[8] L. Siwen, W. Yongcheng, P. Sijia, Z. Lijuan, M.A.-E. Abdullah, Z. Hui, et al., Co–Ni-based nanotubes/nanosheets as efficient water splitting electrocatalysts, Adv. Energy Mater. 6 (2016) 1501661.
[9] Y. Yan, B. Xia, Z. Xu, X. Wang, Recent development of molybdenum sulfides as advanced electrocatalysts for hydrogen evolution reaction, ACS Catal. 4 (2014) 1693–1705.
[10] M. Endo, Grow carbon fibers in the vapor phase, Chemtech (1988).
[11] H. Hui, Z. Wenkui, G. Yongping, Z. Xiaobin, T. Jiangping, Carbon nanotubes as a secondary support of catalyst layer in oxygen diffusion electrode and their electrochemical properties, Chin. J. Chem. Phys. 18 (2005) 428.
[12] H. Huang, W. Zhang, M. Li, Y. Gan, J. Chen, Y. Kuang, Carbon nanotubes as a secondary support of a catalyst layer in a gas diffusion electrode for metal air batteries, J. Colloid Interface Sci. 284 (2005) 593–599.
[13] V.N. Popov, Carbon nanotubes: properties and application, Mater. Sci. Eng. R. Rep. 43 (2004) 61–102.
[14] X. Zou, X. Huang, A. Goswami, R. Silva, B.R. Sathe, E. Mikmekova, et al., Cobalt-embedded nitrogen-rich carbon nanotubes efficiently catalyze hydrogen evolution reaction at all pH values, Angew. Chem. Int. Ed. 53 (2014) 4372–4376.
[15] A.A. Balandin, S. Ghosh, W. Bao, I. Calizo, D. Teweldebrhan, F. Miao, et al., Superior thermal conductivity of single-layer graphene, Nano. Lett. 8 (2008) 902–907.
[16] P.V. Kamat, Graphene-based nanoassemblies for energy conversion, J. Phys. Chem. Lett. 2 (2011) 242–251.
[17] H. Wang, Z. Xu, A. Kohandehghan, Z. Li, K. Cui, X. Tan, et al., Interconnected carbon nanosheets derived from hemp for ultrafast supercapacitors with high energy, ACS Nano 7 (2013) 5131–5141.
[18] Y. Fang, Y. Lv, R. Che, H. Wu, X. Zhang, D. Gu, et al., Two-dimensional mesoporous carbon nanosheets and their derived graphene nanosheets: synthesis and efficient lithium ion storage, J. Am. Chem. Soc. 135 (2013) 1524–1530.
[19] F. Xiaoming, Y. Chang, Y. Juan, L. Zheng, H. Chao, Z. Mengdi, et al., A layered-nanospace-confinement strategy for the synthesis of two-dimensional porous carbon nanosheets for high-rate performance supercapacitors, Adv. Energy Mater. 5 (2015) 1401761.
[20] Y. Lei, W. Qin, Z. Peixin, Z. Junmin, W. Dongrui, L. Yongliang, et al., Scalable 2D hierarchical porous carbon nanosheets for flexible supercapacitors with ultrahigh energy density, Adv. Mater. 30 (2018) 1706054.
[21] H. Fan, W. Shen, Carbon nanosheets: synthesis and application, ChemSusChem 8 (2015) 2004–2027.
[22] G. Wang, X. Hu, L. Liu, Y. Yu, H. Lv, A. Chen, Nitrogen-doping hierarchically porous carbon nanosheets for supercapacitor, J. Mater. Sci. Mater. Electron. 29 (2018) 5363–5372.
[23] X. Zheng, W. Lv, Y. Tao, J. Shao, C. Zhang, D. Liu, et al., Oriented and interlinked porous carbon nanosheets with an extraordinary capacitive performance, Chem. Mater. 26 (2014) 6896–6903.
[24] H. Guang-Ping, L. An-Hui, D. Wei, J. Zhen-Yu, Z. Xiang-Qian, Z. Jin-Tao, et al., Sandwich-type microporous carbon nanosheets for enhanced supercapacitor performance, Adv. Energy Mater. 3 (2013) 1421–1427.

[25] Z. Qi-Long, X. Wei, A. Tomoki, Z. Ruqiang, X. Qiang, Metal-organic framework-derived honeycomb-like open porous nanostructures as precious-metal-free catalysts for highly efficient oxygen electroreduction, Adv. Mater. 28 (2016) 6391–6398.
[26] Y. Zhao, C. Hu, Y. Hu, H. Cheng, G. Shi, L. Qu, A versatile, ultralight, nitrogen-doped graphene framework, Angew. Chem. Int. Ed. 51 (2012) 11371–11375.
[27] Z.-S. Wu, S. Yang, Y. Sun, K. Parvez, X. Feng, K. Müllen, 3D nitrogen-doped graphene aerogel-supported Fe_3O_4 nanoparticles as efficient electrocatalysts for the oxygen reduction reaction, J. Am. Chem. Soc. 134 (2012) 9082–9085.
[28] Y. Liu, D.C. Higgins, J. Wu, M. Fowler, Z. Chen, Cubic spinel cobalt oxide/ multi-walled carbon nanotube composites as an efficient bifunctionalelectrocatalyst for oxygen reaction, Electrochem. Commun. 34 (2013) 125–129.
[29] X. Liu, M. Park, M.G. Kim, S. Gupta, G. Wu, J. Cho, Integrating NiCo alloys with their oxides as efficient bifunctional cathode catalysts for rechargeable zinc–air batteries, Angew. Chem. Int. Ed., 54, 2015, pp. 9654–9658.
[30] X. Lu, H.M. Chan, C.-L. Sun, C.-M. Tseng, C. Zhao, Interconnected core–shell carbon nanotube–graphene nanoribbon scaffolds for anchoring cobalt oxides as bifunctional electrocatalysts for oxygen evolution and reduction, J. Mater. Chem. A 3 (2015) 13371–13376.
[31] H. Yang, C. Shumao, W. Zhenhai, G. Xiaoru, F. Xinliang, C. Junhong, Strongly coupled 3D hybrids of N-doped porous carbon nanosheet/CoNi alloy-encapsulated carbon nanotubes for enhanced electrocatalysis, Small 11 (2015) 5940–5948.
[32] Y. Hou, M.R. Lohe, J. Zhang, S. Liu, X. Zhuang, X. Feng, Vertically oriented cobalt selenide/NiFe layered-double-hydroxide nanosheets supported on exfoliated graphene foil: an efficient 3D electrode for overall water splitting, Energy Environ. Sci. 9 (2016) 478–483.
[33] L. Ji, J. Wang, S. Zuo, Z. Chen, In situ preparation of Pt nanoparticles supported on N-doped carbon as highly efficient electrocatalysts for hydrogen production, J. Phys. Chem. C 121 (2017) 8923–8930.
[34] D.V. Esposito, S.T. Hunt, A.L. Stottlemyer, K.D. Dobson, B.E. McCandless, R. W. Birkmire, et al., Low-cost hydrogen-evolution catalysts based on monolayer platinum on tungsten monocarbide substrates, Angew. Chem. Int. Ed. 49 (2010) 9859–9862.
[35] X. Huang, Z. Zeng, S. Bao, M. Wang, X. Qi, Z. Fan, et al., Solution-phase epitaxial growth of noble metal nanostructures on dispersible single-layer molybdenum disulfide nanosheets, Nat. Commun. 4 (2013) 1444.
[36] H. Yin, S. Zhao, K. Zhao, A. Muqsit, H. Tang, L. Chang, et al., Ultrathin platinum nanowires grown on single-layered nickel hydroxide with high hydrogen evolution activity, Nat. Commun. 6 (2015) 6430.
[37] M.-C. Tsai, T.-K. Yeh, C.-H. Tsai, An improved electrodeposition technique for preparing platinum and platinum–ruthenium nanoparticles on carbon nanotubes directly grown on carbon cloth for methanol oxidation, Electrochem. Commun. 8 (2006) 1445–1452.
[38] G. Dong, M. Fang, H. Wang, S. Yip, H.-Y. Cheung, F. Wang, et al., Insight into the electrochemical activation of carbon-based cathodes for hydrogen evolution reaction, J. Mater. Chem. A. 3 (2015) 13080–13086.
[39] K.-I. Ota, S. Nishigori, N. Kamiya, Dissolution of platinum anodes in sulfuric acid solution, J. Electroanal. Chem. 257 (1988) 205–215.

[40] R. Liu, D. Wu, X. Feng, K. Müllen, Nitrogen-doped ordered mesoporous graphitic arrays with high electrocatalytic activity for oxygen reduction, Angew. Chem. Int. Ed. 122 (2010) 2619–2623.
[41] T. Yang, H. Zhu, M. Wan, L. Dong, M. Zhang, M. Du, Highly efficient and durable PtCo alloy nanoparticles encapsulated in carbon nanofibers for electrochemical hydrogen generation, Chem. Commun. 52 (2016) 990–993.
[42] A. Zhao, J. Masa, W. Xia, A. Maljusch, M.-G. Willinger, G. Clavel, et al., Spinel Mn–Co oxide in N-doped carbon nanotubes as a bifunctional electrocatalyst synthesized by oxidative cutting, J. Am. Chem. Soc. 136 (2014) 7551–7554.
[43] X. Liu, M. Zhang, T. Yang, L. Wang, H. Zhu, S. Wang, et al., Carbon nanofibers as nanoreactors in the construction of PtCo alloy carbon core-shell structures for highly efficient and stable water splitting, Mater. Des. 109 (2016) 162–170.
[44] M. Li, Y. Zhu, N. Song, C. Wang, X. Lu, Fabrication of Pt nanoparticles on nitrogen-doped carbon/Ni nanofibers for improved hydrogen evolution activity, J. Colloid Interface Sci. 514 (2018) 199–207.
[45] J. Lu, W. Zhou, L. Wang, J. Jia, Y. Ke, L. Yang, et al., Core–shell nanocomposites based on gold nanoparticle@zinc–iron-embedded porous carbons derived from metal–organic frameworks as efficient dual catalysts for oxygen reduction and hydrogen evolution reactions, ACS Catal. 6 (2016) 1045–1053.
[46] J. Wang, H. Zhu, D. Yu, J. Chen, J. Chen, M. Zhang, et al., Engineering the composition and structure of bimetallic Au–Cu alloy nanoparticles in carbon nanofibers: self-supported electrode materials for electrocatalytic water splitting, ACS Appl. Mater. Interfaces 9 (2017) 19756–19765.
[47] M. Zhang, S. Wang, T. Li, J. Chen, H. Zhu, M. Du, Nitrogen and gold nanoparticles co-doped carbon nanofiber hierarchical structures for efficient hydrogen evolution reactions, Electrochim. Acta 208 (2016) 1–9.
[48] W. Jing, X. Fan, J. Haiyan, C. Yiqing, W. Yong, Non-noble metal-based carbon composites in hydrogen evolution reaction: fundamentals to applications, Adv. Mater. 29 (2017) 1605838.
[49] X. Wu, S. Han, D. He, C. Yu, C. Lei, W. Liu, et al., Metal organic framework derived fe-doped $CoSe_2$ incorporated in nitrogen-doped carbon hybrid for efficient hydrogen evolution, ACS Sustain. Chem. Eng. 6 (2018) 8672–8678.
[50] Z. Li, L. Zhang, X. Ge, C. Li, S. Dong, C. Wang, et al., Core-shell structured CoP/FeP porous microcubes interconnected by reduced graphene oxide as high performance anodes for sodium ion batteries, Nano Energy 32 (2017) 494–502.
[51] D. Xiong, X. Wang, W. Li, L. Liu, Facile synthesis of iron phosphide nanorods for efficient and durable electrochemical oxygen evolution, Chem. Commun. 52 (2016) 8711–8714.
[52] Y. Pan, N. Yang, Y. Chen, Y. Lin, Y. Li, Y. Liu, et al., Nickel phosphide nanoparticles-nitrogen-doped graphene hybrid as an efficient catalyst for enhanced hydrogen evolution activity, J. Power Sources 297 (2015) 45–52.
[53] Q. Liu, J. Tian, W. Cui, P. Jiang, N. Cheng, A.M. Asiri, et al., Carbon nanotubes decorated with CoP nanocrystals: a highly active non-noble-metal nanohybrid electrocatalyst for hydrogen evolution, Angew. Chem. Int. Ed. 53 (2014) 6710–6714.
[54] B. You, N. Jiang, M. Sheng, S. Gul, J. Yano, Y. Sun, High-performance overall water splitting electrocatalysts derived from cobalt-based metal–organic frameworks, Chem. Mater. 27 (2015) 7636–7642.

[55] J. Li, Y. Wang, T. Zhou, H. Zhang, X. Sun, J. Tang, et al., Nanoparticle superlattices as efficient bifunctional electrocatalysts for water splitting, J. Am. Chem. Soc. 137 (2015) 14305–14312.
[56] Y. Hou, J. Li, Z. Wen, S. Cui, C. Yuan, J. Chen, Co_3O_4 nanoparticles embedded in nitrogen-doped porous carbon dodecahedrons with enhanced electrochemical properties for lithium storage and water splitting, Nano Energy 12 (2015) 1–8.
[57] L. Dai, Carbon-based catalysts for metal-free electrocatalysis, Curr. Opin. Electrochem. 4 (2017) 18–25.
[58] W. Wei, L. Haiwei, P. Khaled, Z. Xiaodong, F. Xinliang, M. Klaus, Nitrogen-doped carbon nanosheets with size-defined mesopores as highly efficient metal-free catalyst for the oxygen reduction reaction, Angew. Chem. Int. Ed. 126 (2014) 1596–1600.
[59] R.M. Yadav, J. Wu, R. Kochandra, L. Ma, C.S. Tiwary, L. Ge, et al., Carbon nitrogen nanotubes as efficient bifunctional electrocatalysts for oxygen reduction and evolution reactions, ACS Appl. Mater. Interfaces 7 (2015) 11991–12000.
[60] L. Chen, Z. Li, Z. Chen, L. Yang, M. Zhou, B. He, et al., Three-dimensional nitrogen-doped carbon nanotubes/carbon nanofragments complexes for efficient metal-free electrocatalyst towards oxygen reduction reaction, Int. J. Hydrogen Energy 43 (2018) 6158–6166.
[61] H. Chuangang, D. Liming, Carbon-based metal-free catalysts for electrocatalysis beyond the ORR, Angew. Chem. Int. Ed. 55 (2016) 11736–11758.
[62] X. Gong, S. Liu, C. Ouyang, P. Strasser, R. Yang, Nitrogen- and phosphorus-doped biocarbon with enhanced electrocatalytic activity for oxygen reduction, ACS Catal. 5 (2015) 920–927.
[63] C.H. Choi, M.W. Chung, H.C. Kwon, S.H. Park, S.I. Woo, B, N- and P, N-doped graphene as highly active catalysts for oxygen reduction reactions in acidic media, J. Mater. Chem. A 1 (2013) 3694–3699.
[64] X. Liu, W. Zhou, L. Yang, L. Li, Z. Zhang, Y. Ke, et al., Nitrogen and sulfur co-doped porous carbon derived from human hair as highly efficient metal-free electrocatalysts for hydrogen evolution reactions, J. Mater. Chem. A 3 (2015) 8840–8846.
[65] A. Shen, Y. Zou, Q. Wang, R.A.W. Dryfe, X. Huang, S. Dou, et al., Oxygen reduction reaction in a droplet on graphite: direct evidence that the edge is more active than the basal plane, Angew. Chem. Int. Ed. 126 (2014) 10980–10984.
[66] T. Sun, J. Wang, C. Qiu, X. Ling, B. Tian, W. Chen, et al., B, N codoped and defect-rich nanocarbon material as a metal-free bifunctional electrocatalyst for oxygen reduction and evolution reactions, Adv. Sci. 5 (2018) 1800036.
[67] Y. Huang, H. Lu, H. Gu, J. Fu, S. Mo, C. Wei, et al., A CNT@$MoSe_2$ hybrid catalyst for efficient and stable hydrogen evolution, Nanoscale 7 (2015) 18595–18602.
[68] Y. Yan, X. Ge, Z. Liu, J.Y. Wang, J.M. Lee, X. Wang, Facile synthesis of low crystalline MoS_2 nanosheet-coated CNTs for enhanced hydrogen evolution reaction, Nanoscale 5 (2013) 7768–7771.
[69] J. Deng, P. Ren, D. Deng, L. Yu, F. Yang, X. Bao, Highly active and durable non-precious-metal catalysts encapsulated in carbon nanotubes for hydrogen evolution reaction, Energy Environ. Sci. 7 (2014) 1919–1923.
[70] D.Y. Wang, M. Gong, H.L. Chou, C.J. Pan, H.A. Chen, Y. Wu, et al., Highly active and stable hybrid catalyst of cobalt-doped FeS_2 nanosheets-carbon

nanotubes for hydrogen evolution reaction, J. Am. Chem. Soc. 137 (2015) 1587–1592.
[71] D. Jiao, R. Pengju, D. Dehui, B. Xinhe, Enhanced electron penetration through an ultrathin graphene layer for highly efficient catalysis of the hydrogen evolution reaction, Angw. Chem. Int. Ed., 54, 2015, pp. 2100–2104.
[72] M. Tahir, N. Mahmood, X. Zhang, T. Mahmood, F.K. Butt, I. Aslam, et al., Bifunctional catalysts of Co_3O_4@GCN tubular nanostructured (TNS) hybrids for oxygen and hydrogen evolution reactions, Nano Res. 8 (2015) 3725–3736.
[73] M. Li, X. Liu, Y. Xiong, X. Bo, Y. Zhang, C. Han, et al., Facile synthesis of various highly dispersive CoP nanocrystal embedded carbon matrices as efficient electrocatalysts for the hydrogen evolution reaction, J. Mater. Chem. A 3 (2015) 4255–4265.
[74] L. Yang, J. Yu, Z. Wei, G. Li, L. Cao, W. Zhou, et al., Co-N-doped MoO_2 nanowires as efficient electrocatalysts for the oxygen reduction reaction and hydrogen evolution reaction, Nano Energy 41 (2017) 772–779.
[75] D. Hou, W. Zhou, K. Zhou, Y. Zhou, J. Zhong, L. Yang, et al., Flexible and porous catalyst electrodes constructed by Co nanoparticles@nitrogen-doped graphene films for highly efficient hydrogen evolution, J. Mater. Chem. A 3 (2015) 15962–15968.
[76] W. Zhou, Y. Zhou, L. Yang, J. Huang, Y. Ke, K. Zhou, et al., N-doped carbon-coated cobalt nanorod arrays supported on a titanium mesh as highly active electrocatalysts for the hydrogen evolution reaction, J. Mater. Chem. A 3 (2015) 1915–1919.
[77] X. Fan, Z. Peng, R. Ye, H. Zhou, X. Guo, M3C (M: Fe, Co, Ni) nanocrystals encased in graphene nanoribbons: an active and stable bifunctional electrocatalyst for oxygen reduction and hydrogen evolution reactions, ACS Nano 9 (2015) 7407–7418.
[78] L. Yipu, Y. Guangtao, L. Guo-Dong, S. Yuanhui, A. Tewodros, C. Wei, et al., Coupling Mo_2C with nitrogen-rich nanocarbon leads to efficient hydrogen-evolution electrocatalytic sites, Angew. Chem. Int. Ed., 54, 2015, pp. 10752–10757.
[79] T.Y. Ma, S. Dai, M. Jaroniec, S.Z. Qiao, Metal–organic framework derived hybrid Co_3O_4-carbon porous nanowire arrays as reversible oxygen evolution electrodes, J. Am. Chem. Soc. 136 (2014) 13925–13931.
[80] Y. Zheng, Y. Jiao, L.H. Li, T. Xing, Y. Chen, M. Jaroniec, et al., Toward design of synergistically active carbon-based catalysts for electrocatalytic hydrogen evolution, ACS Nano 8 (2014) 5290–5296.
[81] J. Zhang, Z. Zhao, Z. Xia, L. Dai, A metal-free bifunctional electrocatalyst for oxygen reduction and oxygen evolution reactions, Nat. Nanotechnol. 10 (2015) 444.
[82] Y. Zheng, Y. Jiao, Y. Zhu, L.H. Li, Y. Han, Y. Chen, et al., Hydrogen evolution by a metal-free electrocatalyst, Nat. Commun. 5 (2014) 3783.
[83] H. Jiang, Y. Zhu, Y. Su, Y. Yao, Y. Liu, X. Yang, et al., Highly dual-doped multilayer nanoporous graphene: efficient metal-free electrocatalysts for the hydrogen evolution reaction, J. Mater. Chem. A 3 (2015) 12642–12645.
[84] L. Jiayuan, Z. Zeqiong, M. Yuanyuan, Q. Yongquan, Graphene and their hybrid electrocatalysts for water splitting, ChemCatChem 9 (2017) 1554–1568.
[85] D. Balun Kayan, D. Koçak, M. İlhan, Electrocatalytic hydrogen production on GCE/RGO/Au hybrid electrode, Int. J. Hydrogen. Energy 43 (2018) 10562–10568.

[86] G. Yang, Y. Li, R.K. Rana, J.-J. Zhu, Pt–Au/nitrogen-doped graphene nanocomposites for enhanced electrochemical activities, J. Mater. Chem. A 1 (2013) 1754–1762.
[87] Y. Jiao, Y. Zheng, M. Jaroniec, S.Z. Qiao, Design of electrocatalysts for oxygen- and hydrogen-involving energy conversion reactions, Chem. Soc. Rev. 44 (2015) 2060–2086.
[88] X. Zou, Y. Zhang, Noble metal-free hydrogen evolution catalysts for water splitting, Chem. Soc. Rev. 44 (2015) 5148–5180.
[89] C. Lei, H. Chen, J. Cao, J. Yang, M. Qiu, Y. Xia, et al., Fe-N_4 sites embedded into carbon nanofiber integrated with electrochemically exfoliated graphene for oxygen evolution in acidic medium, Adv. Energy Mater. 8 (2018) 1801912.
[90] C. Lei, Y. Wang, Y. Hou, P. Liu, J. Yang, T. Zhang, et al., Efficient alkaline hydrogen evolution on atomically dispersed Ni–Nx Species anchored porous carbon with embedded Ni nanoparticles by accelerating water dissociation kinetics, Energy Environ. Sci (2018).
[91] Y. Xu, W. Tu, B. Zhang, S. Yin, Y. Huang, M. Kraft, et al., Nickel nanoparticles encapsulated in few-layer nitrogen-doped graphene derived from metal–organic frameworks as efficient bifunctional electrocatalysts for overall water splitting, Adv. Mater. 29 (2017) 1605957.
[92] Y. Hou, Z. Wen, S. Cui, S. Ci, S. Mao, J. Chen, An advanced nitrogen-doped graphene/cobalt-embedded porous carbon polyhedron hybrid for efficient catalysis of oxygen reduction and water splitting, Adv. Funct. Mater. 25 (2015) 872–882.
[93] T. Cheng, W. Hao-Fan, Z. Xiao-Lin, L. Bo-Quan, Z. Qiang, Advances in hybrid electrocatalysts for oxygen evolution reactions: rational integration of NiFe layered double hydroxides and nanocarbon, Part Part Syst. Charact. 33 (2016) 473–486.
[94] C. Xuan, Z. Peng, K. Xia, J. Wang, W. Xiao, W. Lei, et al., Self-supported ternary Ni-Fe-P nanosheets derived from metal-organic frameworks as efficient overall water splitting electrocatalysts, Electrochim. Acta 258 (2017) 423–432.
[95] Y. Li, H. Wang, L. Xie, Y. Liang, G. Hong, H. Dai, MoS_2 nanoparticles grown on graphene: an advanced catalyst for the hydrogen evolution reaction, J. Am. Chem. Soc. 133 (2011) 7296–7299.
[96] Y. Jieun, V. Damien, A.S. Joon, K. Dongwoo, K.A. Young, C. Manish, et al., Two-dimensional hybrid nanosheets of tungsten disulfide and reduced graphene oxide as catalysts for enhanced hydrogen evolution, Angew. Chem. Int. Ed., 52, 2013, pp. 13751–13754.
[97] D. Voiry, J. Yang, J. Kupferberg, R. Fullon, C. Lee, H.Y. Jeong, et al., High-quality graphene via microwave reduction of solution-exfoliated graphene oxide, Science 353 (2016) 1413–1416.
[98] A. Khoshroo, M. Mazloum-Ardakani, M. Forat-Yazdi, Enhanced performance of label-free electrochemical immunosensor for carbohydrate antigen 15-3 based on catalytic activity of cobalt sulfide/graphene nanocomposite, Sens. Actuators B. Chem. 255 (2018) 580–587.
[99] P. Shengjie, L. Linlin, H. Xiaopeng, S. Wenping, S. Madhavi, G.M. Subodh, et al., Cobalt sulfide nanosheet/graphene/carbon nanotube nanocomposites as flexible electrodes for hydrogen evolution, Angew. Chem. Int. Ed. 126 (2014) 12802–12807.
[100] Y. Hou, M. Qiu, G. Nam, M.G. Kim, T. Zhang, K. Liu, et al., Integrated hierarchical cobalt sulfide/nickel selenide hybrid nanosheets as an

efficient three-dimensional electrode for electrochemical and photoelectrochemical water splitting, Nano Lett. 17 (2017) 4202–4209.
[101] J. Cao, C. Lei, B. Yang, Z. Li, L. Lei, Y. Hou, et al., Zeolitic imidazolate framework-derived core-shell-structured CoS_2/CoS_2-N-C supported on electrochemically exfoliated graphene foil for efficient oxygen evolution, Batteries Supercaps 0 (2018).
[102] L. Jiao, Y.-X. Zhou, H.-L. Jiang, Metal–organic framework-based CoP/reduced graphene oxide: high-performance bifunctional electrocatalyst for overall water splitting, Chem. Sci. 7 (2016) 1690–1695.
[103] D.H. Youn, S. Han, J.Y. Kim, J.Y. Kim, H. Park, S.H. Choi, et al., Highly active and stable hydrogen evolution electrocatalysts based on molybdenum compounds on carbon nanotube–graphene hybrid support, ACS Nano 8 (2014) 5164–5173.
[104] S. Mao, Z. Wen, T. Huang, Y. Hou, J. Chen, High-performance bi-functional electrocatalysts of 3D crumpled graphene–cobalt oxide nanohybrids for oxygen reduction and evolution reactions, Energy Environ. Sci. 7 (2014) 609–616.
[105] J. Cao, C. Lei, J. Yang, X. Cheng, Z. Li, B. Yang, et al., An ultrathin cobalt-based zeolitic imidazolate framework nanosheet array with a strong synergistic effect towards the efficient oxygen evolution reaction, J. Mater. Chem. A 6 (2018) 18877–18883.
[106] C.N.R. Rao, K. Gopalakrishnan, A. Govindaraj, Synthesis, properties and applications of graphene doped with boron, nitrogen and other elements, Nano Today 9 (2014) 324–343.
[107] C.N.R. Rao, K.S. Subrahmanyam, H.S.S.R. Matte, B. Abdulhakeem, A. Govindaraj, D. Barun, et al., A study of the synthetic methods and properties of graphenes, Sci. Technol. Adv. Mater. 11 (2010) 054502.
[108] S. Mukherjee, T. Kaloni, Electronic properties of boron-and nitrogen-doped graphene: a first principles study, J. Nanopart. Res. 14 (2012) 1059.
[109] L. Panchakarla, K. Subrahmanyam, S. Saha, A. Govindaraj, H. Krishnamurthy, U. Waghmare, et al., Synthesis, structure, and properties of boron-and nitrogen-doped graphene, Adv. Mater. 21 (2009) 4726–4730.
[110] C. Zhu, S. Dong, Recent progress in graphene-based nanomaterials as advanced electrocatalysts towards oxygen reduction reaction, Nanoscale 5 (2013) 1753–1767.
[111] U. Maitra, U. Gupta, M. De, R. Datta, A. Govindaraj, C.N.R. Rao, Highly effective visible-light-induced H_2 generation by single-layer 1T-MoS_2 and a nanocomposite of few-layer 2H-MoS_2 with heavily nitrogenated graphene, Angew. Chem. Int. Ed., 52, 2013, pp. 13057–13061.
[112] K. Gopalakrishnan, A. Govindaraj, C.N.R. Rao, Extraordinary supercapacitor performance of heavily nitrogenated graphene oxide obtained by microwave synthesis, J. Mater. Chem. A. 1 (2013) 7563–7565.
[113] Z.-S. Wu, W. Ren, L. Xu, F. Li, H.-M. Cheng, Doped graphene sheets as anode materials with superhigh rate and large capacity for lithium ion batteries, ACS Nano. 5 (2011) 5463–5471.
[114] B.R. Sathe, X. Zou, T. Asefa, Metal-free B-doped graphene with efficient electrocatalytic activity for hydrogen evolution reaction, Catal. Sci. Technol. 4 (2014) 2023–2030.
[115] Y. Ito, W. Cong, T. Fujita, Z. Tang, M. Chen, High catalytic activity of nitrogen and sulfur co-doped nanoporous graphene in the hydrogen evolution reaction, Angew. Chem. Int. Ed., 54, 2015, pp. 2131–2136.

[116] Y. Chen, R. Ren, Z. Wen, S. Ci, J. Chang, S. Mao, et al., Superior electrocatalysis for hydrogen evolution with crumpled graphene/tungsten disulfide/tungsten trioxide ternary nanohybrids, Nano Energy 47 (2018) 66–73.

[117] C. Hu, L. Dai, Multifunctional carbon-based metal-free electrocatalysts for simultaneous oxygen reduction, oxygen evolution, and hydrogen evolution, Adv. Mater. 29 (2017) 1604942.

[118] J. Lai, S. Li, F. Wu, M. Saqib, R. Luque, G. Xu, Unprecedented metal-free 3D porous carbonaceous electrodes for full water splitting, Energy Environ. Sci. 9 (2016) 1210–1214.

[119] Z. Lu, J. Wang, S. Huang, Y. Hou, Y. Li, Y. Zhao, et al., N,B-codoped defect-rich graphitic carbon nanocages as high performance multifunctional electrocatalysts, Nano Energy 42 (2017) 334–340.

[120] Z. Yang, Z. Jing, Q. Liangti, Graphitic carbon nitride/graphene hybrids as new active materials for energy conversion and storage, ChemNanoMat 1 (2015) 298–318.

[121] Z. Yao, J. Yan, J. Mietek, Q.S. Zhang, Advancing the electrochemistry of the hydrogen-evolution reaction through combining experiment and theory, Angew. Chem. Int. Ed., 54, 2015, pp. 52–65.

[122] T. Jingqi, L. Qian, M.A. Abdullah, A.A. Khalid, S. Xuping, Ultrathin graphitic C_3N_4 nanosheets/graphene composites: efficient organic electrocatalyst for oxygen evolution reaction, ChemSusChem 7 (2014) 2125–2130.

INDEX

Note: Page numbers followed by "*f*" and "*t*" refer to figures and tables, respectively.

A

Activated carbon, 251
Adsorption, 225
Advanced oxidation processes (AOPs), 205, 264–265
AgNPs. *See* Silver nanoparticles (AgNPs)
Air pollutants
  catalytic pathways of nanomaterials
    oxidative catalysis, 233–234
    reductive catalysis, 230–233
  nanomaterials for eliminating
    adsorption by nanomaterials, 225–227
    separation through nanostructured membranes, 227–229
  pollutants and challenges
    organic, 224–225
    typical inorganic, 224
  removal by nanomaterials, 203–208
Alkanethiols, 47
All-solid sodium–sulfur battery (ASNSB), 334–335
α-(1,4)-linked D-glucopyranoses, 253–254
Amorphous composite oxide, 319–320
Anodic oxidation (AO), 280
Anodic stripping voltammetry (ASV), 97–98
Anthropogenic-polluted water sources, 270
AO. *See* Anodic oxidation (AO)
AOPs. *See* Advanced oxidation processes (AOPs)
Aromatic hydrocarbons, 262
Artificial atom, 119–120
Ascorbic acid (AA), 351
ASNSB. *See* All-solid sodium–sulfur battery (ASNSB)
ASV. *See* Anodic stripping voltammetry (ASV)
AuNPs. *See* Gold nanoparticles (AuNPs)
AuNRs. *See* Gold nanorods (AuNRs)
Au-tipped CdS nanorods, 359–361, 360*f*

B

Bare Au nanoparticles, 58*f*
BEF process. *See* Bio-electro-Fenton (BEF) process
β-cyclodextrin (β-CD), 253–254
β-cyclodextrin–ionic liquid (β-CD-IL), 253–254
BF process. *See* Bio-Fenton (BF) process
Binary nanocrystals, 147–153
Biochar–plant tandem approach, 269
Bio-electro-Fenton (BEF) process, 277–278
Bio-Fenton (BF) process, 277–278
Bohr diameter, 142–144
Boron nitride (BN), 227
Bovine serum albumin (BSA), 81–82
Butanoic acid, 266

C

C_4–C_{10} perfluoroalkane sulfonic acids, 262
Carbon cycle, 224
Carbon nanofibers (CNFs), 252–253, 381–382
Carbon nanomaterials, 252–253
Carbon nanosheets, 382–383
Carbon nanotube (CNT), 1–3, 252–253, 381–382
Carbon nitrogen nanotubes (CNNTs), 392
Catalysis, 22–23
Catalyst, 119
Catalytic activity, 43
Catalytic pollutant removal, 147
CBs. *See* Conduction bands (CBs)
Charge-transfer process, 294
Chemical vapor decomposition (CVD), 381–382
Chemical vapor deposition, 169–170
Chitosan, 48
Chlorinated n-decane (CP10), 255
Chlorinated paraffins (CPs), 255, 259
Chlorinated polyfluoroalkyl ether sulfonic acids (Cl-FESAs), 262
Cl-FESAs. *See* Chlorinated polyfluoroalkyl ether sulfonic acids (Cl-FESAs)
CNFs. *See* Carbon nanofibers (CNFs)
CNNTs. *See* Carbon nitrogen nanotubes (CNNTs)
CNT. *See* Carbon nanotube (CNT)
Cobalt-based cocatalyst, 362
Cobalt-embedded N-rich CNT (Co NRCNT) electrocatalyst, 381–382

Cold trapping, 272
Colorimetrics, 87–90
Colorimetric sensing
aggregation of NMPs, 47–49
disassembly of aggregated NMPs, 49–50
sensitivity of NMPs, 50
surface plasmon resonance, 46–47
Conduction bands (CBs), 121–122
Co-NRCNT. *See* Cobalt-embedded N-rich CNT (Co-NRCNT) electrocatalyst
Contact mode, 58*f*
Copper NPs (CuNPs), 257–258
Coprecipitation, 169
CoS-NP SSNBs. *See* Single-shelled nanoboxes assembled by nanoparticles (CoS-NP SSNBs)
CoS-NS SSNBs. *See* Single-shelled nanoboxes assembled by nanosheets (CoS-NS SSNBs)
Covalent organic frameworks (COFs), 228–229
CP10. *See* Chlorinated n-decane (CP10)
CPs. *See* Chlorinated paraffins (CPs)
Cr(VI) ions, 253–254
Cr(VI)-perfluorinated compound (PFC), 253–254
Crumpled graphene–cobalt oxide (CG–CoO) hybrid, 403–404
Crystallographic reaction mechanisms, 310, 311*f*
Cu_2ZnSnS_4 (CZTS), 354–355, 354*f*
Cubic iron-based MOFs (Fe-soc-MOF), 3
CuNPs. *See* Copper NPs (CuNPs)
CVD. *See* Chemical vapor decomposition (CVD)
Cyclic adsorption–electro-Fenton process, 265

D

DDT. *See* Dichlorodipheny ltrichloroethane (DDT)
Decabromodiphenyl ether, 259–260
DEG. *See* Diethylene glycol (DEG)
Density functional theory (DFT), 282, 326–327
Deposition-precipitation, 169
DET. *See* Direct electron transfer (DET)
DFT. *See* Density functional theory (DFT)
DHSs. *See* Double-layer hollow silica shells (DHSs)
Dichlorodipheny ltrichloroethane (DDT), 249
Diethylene glycol (DEG), 18–19
Differential scanning calorimetry, 312
Dimethyl sulfoxide (DMSO), 10–11
Diphenyl ether–degrading bacteria, 256
Direct electron transfer (DET), 190–191, 191*f*
Direct emulsions, 16–17
DMSA. *See* meso-2,3-Dimercaptosuccinic acid (DMSA)
DMSO. *See* Dimethyl sulfoxide (DMSO)
Double-layer hollow silica shells (DHSs), 13
DOX. *See* Drug doxorubicin (DOX)
Drug doxorubicin (DOX), 14–15

E

EAOPs. *See* Electrochemical advanced oxidation processes (EAOPs)
EDTA. *See* Ethylene diamine tetraacetic acid (EDTA)
EG. *See* Exfoliated graphene (EG)
Electrocatalysis. *See also* Nanocarbons
acidic solution, 381
electrochemical water splitting, 379–380
hydrogen energy, 379
hydrogen evolution reaction (HER), 379–380, 380*f*
neutral/alkaline solution, 381
oxygen evolution reaction (OER), 379–380, 380*f*
water electrolyzers, 380–381
Electrochemical advanced oxidation processes (EAOPs), 275
Electrochemical sensors
for constructing electrochemical sensing, 66–68
electroactive labels, 65–66
electron transferring, 69–71
Electrokinetic (EK) remediation, 255–256
Electrolytic manganese dioxide (EMD), 322
Electron–hole separation efficiency, 170
Electrooxidative remediation, 274
Electrostatic layer-by-layer (LBL), 10–11
Eley-Rideal mechanism, 182
EMD. *See* Electrolytic manganese dioxide (EMD)
EMS. *See* Ethylmethanesulfonate (EMS)
Environmental Protection Agency (EPA), 45, 99–100
EPA. *See* Environmental Protection Agency (EPA)
Ethylene diamine tetraacetic acid (EDTA), 92–93
Ethylmethanesulfonate (EMS), 149–150
Exfoliated graphene (EG), 399, 400*f*
External functionalization methods, 252

F

Fabrication methodology, hollow structures
hard-templating methods
carbon-based templates, 13–14
ceramic templates, 14
on inorganic and complex salts, 14–15
natural materials as, 15
polymer-based hard templates, 10–12
silica-based hard templates, 12–13
self-template methods
Galvanic replacement, 21
Kirkendall effect, 20–21
Ostwald ripening, 19–20
surface-protected etching, 18–19
soft-templating methods
emulsion-based, 16–17
gas bubble-based, 17
micelle/vesicle-based, 17
template-free methods, 21–22
Fenton oxidation, 264
Fe-soc-MOF. *See* Cubic iron-based MOFs (Fe-soc-MOF)
Flame retardants, 261
Fluorescence, 43
Fluorescence-based sensing
assembly with quantum dots, 55
quenching by SPR, 53–55
ultrasmall gold nanoparticles, 51–53
Fluorescence mode, 90–93
Fluorescence resonance energy transfer (FRET), 87, 92–96
Forster resonance energy transfer, 43
Fullerene, 252–253

G

Galvanic replacement, 21
Gas diffusion electrode (GDE), 265
Gas hourly space velocity (GHSV), 231–233
GDE. *See* Gas diffusion electrode (GDE)
GO. *See* Graphene oxide (GO)
Gold nanoparticles (AuNPs), 40, 165–166
catalytic properties of, 165–167
environmental catalysis of
catalytic decomposition of volatile organic compounds, 183–184
catalytic oxidation of carbon monoxide, 177–183
ozone decomposition, 188
removal of nitrogen oxides, 185–188
synthesis and characterization of, 167–177
factors for catalytic activity, 175–177
supported gold nanoparticles preparation, 167–170
Gold nanorods (AuNRs), 41
Gold nanospheres (AuNSs), 41
GQDs. *See* Graphene quantum dots (GQDs)
Graphene (GR), 252–253, 287
Graphene/modified-graphene
benefits, 394–396
metal-free electrocatalysts, 404–408, 407*f*
nonprecious-metal electrocatalysts
CG–CoO hybrid, 403–404
charge transfer resistance, 399
EG, 399, 400*f*
electrochemical exfoliation, 401, 402*f*
FeP, 403
hydrothermal and pyrolysis methods, 401
metal particle–loaded nanocarbons, 398–399
Mo_2C/CNT–graphene hybrid, 403–404
molybdenum disulfides (MoS_2)-based catalysts, 399–401
N/Co-doped PCP//NRGO hybrid, 397–398
NGF, 399
Ni–NP/Ni–N–C/EG hybrid, 397–398
rGO, synergic effect, 401
3D-structured electrocatalyst, 397–398
transition-metal phosphides, 401–403
precious-metal electrocatalysts, 396–397
types, 396
zero-bandgap properties, 394–396
Graphene nanosheets (GNSs), 382
Graphene oxide (GO), 11–12, 321
Graphene quantum dots (GQDs), 68, 266–267
Guar gum, 271

H

HAADFSTEM. *See* High-angle annular dark-field scanning TEM (HAADFSTEM)
Halogenated flame retardants (HFRs), 261
Hard-templating methods, hollow structures
carbon-based templates, 13–14
ceramic templates, 14
inorganic and complex salts, 14–15
natural materials as, 15
polymer-based hard templates, 10–12
silica-based hard templates, 12–13
H-atom transfer mechanism, 256
Heavy metal (HM), 45

Hexachlorobenzene (HCB), 268
HFRs. *See* Halogenated flame retardants (HFRs)
High-angle annular dark-field scanning TEM (HAADFSTEM), 5–6
HM. *See* Heavy metal (HM)
Hollow silica shells (HSs), 13
Hollow structures
  catalysis, 22–23
  fabrication methodology of
    hard-templating methods, 9–15
    self-template methods, 18–21
    soft-templating methods, 15–17
  features of
    morphology, 8–9
    shell number, 8
    size, thickness, and intershell spacing, 8
  hybrid materials, 6–7
  multishelled hierarchy, 3–5
  open structure materials, 5–6
  sensors, 23–24
HSCs. *See* Hybrid supercapacitors (HSCs)
HSs. *See* Hollow silica shells (HSs)
Hybrid supercapacitors (HSCs), 26
Hydrogenases, 367–370, 369*f*, 370*f*
Hydrogen evolution reaction (HER), 230
Hydrogen photogeneration
  cocatalyst selection, 350
  colloidal CdTe, 348–349
  colloidal semiconductor quantum dots, 348–349. *See also* Quantum dots (QDs)
  energy level, 348–349, 349*f*
  factors, 350, 350*f*
  Gibbs free energy, 347–348
  global energy resources, 347
  H_2-producing artificial (HPA) system, 350
  photoelectrochemical water splitting, 347–348
  surface treatments
    Au-tipped CdS nanorods, 359–361, 360*f*
    catalytic efficiency, 359
    CdSe/CdSe core–shell QDs, 361–362, 361*f*
    charge separation efficiencies, 361–362
    cobalt-based cocatalyst, 362
    hydrogenases, 367–370, 369*f*, 370*f*
    metal, 362–364, 363*f*, 364*f*
    metal sulfides, 364–367, 365*f*, 366*f*, 368*f*
    molecule, 370–372, 371*f*
Hydroxide method, 313–314

I

ICPMS. *See* Inductively coupled plasma-mass spectrometry (ICPMS)
IFCT. *See* Interfacial charge transfer (IFCT)
Impregnation, 168–169
Indium phosphide (InP), 83–84
Inductively coupled plasma-mass spectrometry (ICPMS), 99–100
InP. *See* Indium phosphide (InP)
Interfacial charge transfer (IFCT), 290
Iron-based metal phosphide (FeP), 403

J

Janus heteronanostructure arrays, 294–295

K

Kirkendall effect, 20–21
KTO. *See* Potassium titanate (KTO)

L

Langmuir-Hinshelwood mechanism, 179–180
LED. *See* Light emitting diodes (LED)
LEMF. *See* Local electromagnetic field (LEMF)
LIB. *See* Lithium-ion batteries (LIB)
Light emitting diodes (LED), 79
Lindane, 281
Linear sweep voltammetry (LSV), 97–98
Lithium-ion batteries (LIB), 26–27
  chemical energy, 309
  components, 309–310, 309*f*
  energy and power density, 309
  lithium–sulfur (Li–S) batteries
    Coulomb interactions, 326–327
    DFT, 326–327
    electrochemical mechanism, 323, 324*f*
    interwoven composite, 324–325
    Mxe matrix, 327
    polymer modification, 324–325
    quantitative analysis, 325
    sulfur positive electrodes, 324
    template-based approach, 325
    transition-metal oxide positive electrodes, 323
  negative electrode nanomaterials
    amorphous composite oxide, 319–320
    carbon negative electrodes, 315–316
    charging and discharging cycle, 315–316, 315*f*
    EMD, 322
    GO, 321
    graphene, 317
    hydrothermal synthesis, 316

macroscopic carbon monoliths, 318
mesoporous C/Sn negative electrodes, 319
metal-based compounds, 318
MXenes, 322
NiO nanosheets, 318
phonon dispersion, 322–323
Si-based LIB negative electrodes, 319
surface-area-to-volume ratios, 320–321
surface-to-volume ratios, 316
VACNTs, 316–317
vacuum filtration, 316
positive electrode nanomaterials
crystallographic reaction mechanisms, 310, 311*f*
degradation mechanisms, 313
differential scanning calorimetry, 312
host network compounds, 310
hydroxide method, 313–314
$LiCoO_2$, 311
$LiMnO_2$, 312
$LiNiO_2$, 311–312
multiple electron transfer, 310
olivine-structured $LiFePO_4$, 314
polyanion compound materials, 314
rock salt phase, 313
thermal behavior, 313
transition-metal dichalcogenide bond, 310
XPS, 313–314
Local electromagnetic field (LEMF), 190–191, 191*f*
Localized surface plasmon resonance (LSPR), 96–97
Lotus root-like multichannel carbon (LRC), 4–5
LRC. *See* Lotus root-like multichannel carbon (LRC)
LSPR. *See* Localized surface plasmon resonance (LSPR)
LSV. *See* Linear sweep voltammetry (LSV)

M

MA. *See* Mercaptoacetic acid (MA)
Macroscopic carbon monoliths, 318
Magnetic-core NPs, 253
Magnetic cyclodextrin nanocomposites, 253–254
Magnetic nanoparticles (MNPs), 253
Magnetic nanophotocatalysts (MNPCs), 293
Mass spectroscopy (MS), 99–100
Matryoshkas-like hollow nanostructure, 3–4
MCCPs. *See* Medium-chain chlorinated paraffins (MCCPs)
Medium-chain chlorinated paraffins (MCCPs), 261
Melamineformaldehyde (MF), 10–11
Mercaptoacetic acid (MA), 90
Mercaptopropionic acid (MPA), 92–93
3-Mercaptopropionic acid (MPA), 124
Meso-2,3-dimercaptosuccinic acid (DMSA), 361–362
Mesoporous silica (MS), 12
Metal dichalcogenides (TMDs), 228–229
Metallic nanostructured catalyst, 209–210
Metal nanoparticles (MNPs), 97
Metal organic frameworks (MOFs), 3, 138, 139*f*, 226, 385
MF. *See* Melamineformaldehyde (MF)
Microparticles (MPs), 253
Mie theory, 39–41
MIPs. *See* Molecularly imprinted polymers (MIPs)
Mixed-matrix membranes (MMMs), 228–229
Mixed metal oxides (MMOs), 279
MMOs. *See* Mixed metal oxides (MMOs)
MNPCs. *See* Magnetic nanophotocatalysts (MNPCs)
MNPs. *See* Magnetic nanoparticles (MNPs); Metal nanoparticles (MNPs)
MOFs. *See* Metal organic frameworks (MOFs)
Molecularly imprinted polymers (MIPs), 262–263
Molybdenum disulfides (MoS_2)-based catalysts, 399–401
MPA. *See* 3-mercaptopropionic acid (MPA); Mercaptopropionic acid (MPA)
MS. *See* Mass spectroscopy (MS); Mesoporous silica (MS)
Multielectron reduction process, 283
Multiwalled carbon nanotubes (MWCNTs), 381–382
MWCNTs. *See* Multiwalled carbon nanotubes (MWCNTs)
MXenes, 322

N

"Naked eye" analytical method, 87–88
Nanocarbons
metal-free
atom–doped nanocarbons, 393
carbon defects, 393–394
CNNTs, 392
codoped carbon material, 393
electrocatalytic water splitting, 394, 395*t*

Nanocarbons (*Continued*)
electrochemical properties, 391–392
heteroatom codopants, 393–394
nitrogen atoms, 392
O-doped CNTs, 392–393
P,N-codoped mesoporous nanocarbons, 394
noble metal
Au/Zn–Fe–C core–shell nanostructures, 385–387, 386*f*
bifunctional electrocatalyst, 384–385
bimetallic alloy, 387
MOFs, 385
NCNRs, 387–388
Ni-NCNFs-Pt, 385
nonnoble, 388–391, 391*f*
PNRs, 387–388
Pt-based and Ir-based electrocatalysts, 384
Zn/Fe-based hybrid catalyst, 385
1D nanocarbon-based hybrids, 381–382
3D nanocarbon-based hybrids, 383
2D nanocarbon-based hybrids, 382–383
Nanocomposite catalysts, 219–222
Nanocrystals (NCs), 79, 105–107
Nanocrystals for catalysis
CO_2 conversion
metal chalcogenides nanocrystals, 142–145
metal oxide nanocrystals, 145–147
reduction, 141–142
compositions
metal chalcogenides binary nanocrystals, 126–130
Metal Oxide Nanocrystals, 131–132
metal oxide nanocrystals, 131–132
one-dimensional nanocrsytals-based catalysts, 134–136
ternary nanocrystals, 130–131
three-dimensional nanocrystals-based catalysts, 138–140
two-dimensional nanocrystal-based catalysts, 136–138
for pollutant removal
binary nanocrystals, 147–153
ternary/quaternary nanocrystals, 153–154
Nanofiber-based materials, 252
Nanofibrous media, 252
Nanorod (NR), 129–130
Nanoscale zero-valent iron (nZVI), 254–255, 257
Nanosensors, 85–87
design and applications principles of, 87–99
functional components in, 86*f*
III-V nanocrystals for, 104–105
II-VI nanocrystals for, 101–104
Nanostructure adsorbents, 208–209
Nanostructured photoanodes, 286–287
NAS batteries. *See* Sodium-sulfur (NAS) batteries
NBFRs. *See* Novel brominated flame retardants (NBFRs)
NCs. *See* Nanocrystals (NCs)
N-doped carbon nanorods (NCNRs), 387–388
N-doped graphene foams (N-GF), 328
N-doped graphene framework (NGF), 399
N-doped hierarchically porous carbon nanosheets (N-HPCNs), 382–383
N-GF. *See* N-doped graphene foams (N-GF)
NIBs. *See* Sodium-ion batteries (NIBs)
NiO nanosheets, 318
NMPs. *See* Noble-metal nanoparticles (NMPs)
N-nitrosonornicotine (NNN), 149–150
NNN. *See* *N*-nitrosonornicotine (NNN)
Noble-metal nanoparticles (NMPs), 39
aggregation of dispersed, 47–49
catalytic activity, 43
colorimetric sensing, 50
fluorescence, 43
physical properties, 40*f*
sensing platform, 44*t*
surface functionalization, 44
surface plasmon resonance, 39–43
Nonmetallic nanostructured catalysts, 210–218
Nonrenewable energy sources, 251
Novel brominated flame retardants (NBFRs), 261–262
NTO. *See* Sodium titanate (NTO)

O

1-Octadecene (ODE), 128
ODE. *See* 1-octadecene (ODE)
Oil-in-water (O/W) emulsion, 16–17
Oleic acid (OA), 128
Ostwald ripening, 19–20
OVs. *See* Oxygen vacancies (OVs)
Oxidative catalysis, 233–234
Oxygen vacancies (OVs), 331
Ozone decomposition, 188

P

PAA. *See* Poly(acrylic acid) (PAA)

PAH. *See* Poly allylamine hydrochloride (PAH)
PAHs. *See* Polycyclic aromatic hydrocarbons (PAHs)
PAN. *See* Polyacrylonitrile (PAN)
PBA. *See* Prussianblue-analog (PBA)
PCBs. *See* Polychlorinated biphenyls (PCBs)
PCDDs. *See* Polychlorinated dibenzo-*p*-dioxins (PCDDs)
PCDFs. *See* Polychlorinated dibenzofurans (PCDFs)
PCR. *See* Polymerase chain reaction (PCR)
PDADMAC. *See* Poly-(diallyldimethylammonium chloride) (PDADMAC)
PDDA. *See* Poly(dimethyl-diallyl ammonium chloride) (PDDA)
PEI. *See* Polyethylenimine (PEI)
Perfluorinated acetic acid, 266
Perfluorooctane sulfonate (PFOS), 253–254
Perfluorooctane sulfonyl fluoride (POSF), 259
Perfluorooctanoic acid (PFOA), 253–254
Persistent organic pollutants (POPs)
  anthropogenic-polluted water sources, 270
  antifouling, 270–271
  AOPs, 264–265
  applications, 249
  aromatic hydrocarbons, 262
  artificial microplastics, 292
  biochar–plant tandem approach, 269
  C_4–C_{10} perfluoroalkane sulfonic acids, 262
  catabolic genes, 267–268
  charge-transfer process, 294
  chemical coagulants, 271
  Cl-FESAs, 262
  “cold trapping”, 272
  control and degradation, 249, 250*f*
  cost-efficient photoelectrocatalytic reactor, 286
  CPs, 259
  cyclic adsorption–electro-Fenton process, 265
  DDT, 249
  decabromodiphenyl ether, 259–260
  degradation products, 270
  electrochemical techniques
    AO, 280
    application, 281
    BEF process, 277–278
    BF process, 277–278
    disadvantages, 276
    EAOPs, 275
    electrooxidative remediation, 274
    environmental-economic index, 277
    halogenated environmental pollutants, 273–274
    Hg/PAH removal, 275
    landfill leachate treatment, 278–279
    MMOs, 279
    PFASs, 279–280
    recalcitrant organochlorine pesticides, 279
    REM, 278
    SPEF process, 277
    suspended electrodialytic remediation, 274–275
    thermal decomposition, 273–274
    UCF, 276
    Zr-PbO_2 film electrodes, 280
  electronic waste (e-waste) recycling, 260–261
  environmental degradation, 249
  environmental microorganisms, 267–268
  environmental purification, 290–291, 291*f*
  Fenton oxidation, 264
  Fenton treatment, 270
  flame retardants, 261
  GDE, 265
  GQDs, 266–267
  GR, 287
  guar gum, 271
  halogenated organic contaminants, 258–259
  HCB, 268
  HFRs, 261
  IFCT, 290
  innate plant growth–promoting mechanisms, 269
  inorganic semiconductors, 287
  Janus heteronanostructure arrays, 294–295
  MCCPs, 261
  mechanochemical destruction, 268
  memory effect–driven emission, 271
  metal oxide semiconductor nanostructures, 269
  microbial degradation, 273
  minority-carrier diffusion, 287–288
  MIPs, 262–263
  MNPCs, 293
  MOFs, 267
  nanostructured photoanodes, 286–287
  NBFRs, 261–262
  nutrition and pollution stress, 267–268
  optical absorption, 287–288
  organic tissues and long-term toxicity, 249
  organo-modified montmorillonite, 267
  PCBs, 249
  PCDD/PCDFs, 271–272
  PCDDs, 259
  PCDFs, 259
  PEC, 285–286
  perfluorooctane sulfonic acid, 259
  pesticide pollution, 263
  petroleum refinery waste, 262

Persistent organic pollutants (POPs) (*Continued*)
PFASs, 266
PFOA, 260, 265–267
photocatalytic dechlorination process, 268–270
photocatalytic techniques
aerobic aqueous solutions, 282–283
cyclic adsorption/photodegradation, 284–285
debromination reactions, 282
density functional theory, 282
Fenton reagents, 281
lindane, 281
multielectron reduction process, 283
nanostructured photocatalysts, 281–282
photodebromination behaviors, 282
ROS, 284–285
SPC, 285
visible light–harvesting photocatalysts, 281–282
plastic antibodies, 262–263
plasticizers, 261
POSF, 259
properties, 250
removal
adsorption methods, 251–254
bimetallic Ni/Fe NPs, 256
borohydride reduction method, 255
clay minerals, 254
CP10, 255
CPs, 255
CuNPs, 257–258
dechlorination reactions, 255
dehalogenation kinetics, 256
electrokinetic remediation, 255–256
environmental purification and remediation, 250–251
groundwater remediation technologies, 254
H-atom transfer mechanism, 256
limitations, ZVI, 257
metal and metal oxide engineered nanomaterials, 254
nanobased techniques, 250–251
nanoscale zero-valent iron, 254–255, 257
PBDEs, 256, 258
PCBs, 255–256
PFCs, 256
recalcitrant pollutants, 250–251
short-chain chlorinated paraffins, 255
zero-valent iron materials, 254
selective catalytic reduction techniques, 265
self-healing membranes, 270–271
soil bioremediation process, 263–264
Stockholm Convention, 249, 258–259, 262, 271–272
sulfate radicals, 266
surface modification approach, 288–289
surface-modified semiconductor particles, 291–292
synthetic process, 287
target and nontarget chemical analyses, 270–271
three-dimensional (3D) network polymer binding sites, 262–263
TiO_2 nanostructures, 288–289
two-dimensional graphitic materials, 289–290
water management, 266
xanthan gum, 271
zeolitic imidazolate frameworks, 267
zero-valent metals, 264
zinc ferrite nanocomposite catalysts, 289
Petroleum refinery waste, 262
PFOA. *See* Perfluorooctanoic acid (PFOA)
PFOS. *See* Perfluorooctane sulfonate (PFOS)
Phonon dispersion, 322–323
Photocatalysis, 122–123, 125*f*, 130, 133–134, 136–137, 141
Photocatalytic dechlorination process, 268–270
Photocatalytic degradation, 24–25
Photocatalytic energy conversion, 25–26
Photocatalytic oxidation (PCO), 211
Photodebromination behaviors, 282
Photoelectrocatalysis (PEC), 285–286
Photothermal effect, 191–192
$\pi - \pi$ stacking action, 63–64
Plasmon-coupling effect, 42
Plasticizers, 261
PMMA. *See* Poly (methyl methacrylate) (PMMA)
PNRs. *See* Polyaniline nanorods (PNRs)
Poly(acrylic acid) (PAA), 12–13
Poly(dimethyl-diallyl ammonium chloride) (PDDA), 104
Poly (methyl methacrylate) (PMMA), 10
Polyacrylonitrile (PAN), 4–5
Poly allylamine hydrochloride (PAH), 11
Polyaniline nanorods (PNRs), 387–388
Polychlorinated biphenyls (PCBs), 63–64, 249
Polychlorinated dibenzofurans (PCDFs), 259

Polychlorinated dibenzo-*p*-dioxins (PCDDs), 259
Polycyclic aromatic hydrocarbons (PAHs), 60–61
Poly-(diallyldimethylammonium chloride) (PDADMAC), 10–11
Polyethylenimine (PEI), 11–12
Polymerase chain reaction (PCR), 66
Poly(sodium-*p*-styrenesulfonate) (PSS), 11
Polystyrene (PS), 10–12
Polyvinyl chloride (PVC), 228
Polyvinylidene-fluoride–hexafluoropropene (PVDF) polymer electrolyte, 335
POPs. *See* Persistent organic pollutants (POPs)
POSF. *See* Perfluorooctane sulfonyl fluoride (POSF)
Potassium titanate (KTO), 333
Propionic acid, 266
Prussianblue-analog (PBA), 5
PS. *See* Polystyrene (PS)
PSS. *See* Poly(sodium-*p*-styrenesulfonate) (PSS)
PVDF. *See* Polyvinylidene-fluoride–hexafluoropropene (PVDF) polymer electrolyte

Q
QD-metal organic framework (MOF)
  bottle around the ship, 138–139, 139*f*
  ship in the bottle, 138–139, 139*f*
QDs. *See* Quantum dots (QDs)
Quantum dots (QDs), 123–124, 348–349
  core–shell quantum dots, 357–358, 358*f*, 359*f*
  future of, 109–110
  proton reduction reaction, 350–351
  quasi-type II CdSe/CdS dot-in-rod nanorods, 355–356, 355*f*, 356*f*, 357*f*
  single component quantum dots
    ascorbic acid (AA), 351
    CdSe QDs, 353
    CdS nanorods, 351, 352*f*
    CdS–PdX hybrids, 352–353, 353*f*
    Cu_2ZnSnS_4 (CZTS), 354–355, 354*f*
    quantum yield (QY), 351
    turnover frequency (TOF), 351
  typical II–VI semiconductor QDs, 354–355
Quantum size effect, 79–80
Quantum yield (QY), 51–52
Quasi-type II CdSe/CdS dot-in-rod nanorods, 355–356, 355*f*, 356*f*, 357*f*
QY. *See* Quantum yield (QY)

R
Raman scattering, 55–56
Raman spectroscopy
  detection of environmental pollutants, 60–62
  detection of food residual pesticides, 62–63
  NMPs and assemblies as substrates, 57–60
  NMPs and films as substrates, 60
  spectrum-based sensing, 63–65
Rayleigh scattering, 55–56
Reactive electrochemical membrane (REM), 278
Reactive nanomaterials
  metallic nanostructured catalyst, 209–210
  nanocomposite catalysts, 219–222
  nanostructured adsorbents, 208–209
  nonmetallic nanostructured catalysts, 210–218
Reactive oxygen species (ROS), 205, 284–285
Recalcitrant organochlorine pesticides, 279
Receptors, 100
Rechargeable batteries, 337–338
Recombination, 213
Redox mechanism, 180
REM. *See* Reactive electrochemical membrane (REM)
Renewable energy storage
  atmospheric pollutants, 307–308
  environmental catalysts, 308
  environmental sensors, 307
  LIBs, 308. *See also* Lithium-ion batteries (LIB)
  NAS batteries, 333–336
  NIBs, 308
  PCH, 307–308
  rechargeable batteries, 337–338
  secondary batteries, 308–309
  sodium-ion batteries. *See* Sodium-ion batteries
  value-added catalysts, 308
  wastewater pollution, 308
  water quality monitoring, 307
ROS. *See* Reactive oxygen species (ROS)

S
Scanning tunneling microscopy/spectroscopy (STM/STS), 175
SCCPs. *See* Short-chain chlorinated paraffins (SCCPs)
SCR techniques. *See* Selective catalytic reduction (SCR) techniques
S-doped mesoporous amorphous carbon (SMAC), 331
Selective catalytic reduction (SCR) techniques, 265
Semiconductor nanocrystals
  electronic structure of, 79–81
  environmental catalysis
    catalysis and flexibility, 119–125

Semiconductor nanocrystals (*Continued*)
  photoluminescence of, 81–85
Semiconductor Quantum Dot
  composition of nanosensors, 85–87
  nanosensor design and applications
    colorimetrics, 87–90
    fluorescence mode, 90–93
    fluorescence resonance energy transfer (FRET), 93–96
    surface-enhanced Raman scattering, 96–97
    voltammetry, 97–99
Semiheterogeneous catalyst, 119–120
Semivolatile organic compounds (SVOCs), 225
Sensing of Metal Ions
  II-VI nanocrystals for nanosensors, 101–104
  III-V nanocrystals for nanosensors, 104–105
  ternary nanocrystals, 105–107
  in vivo and in vitro, 107–109
Sensors, 23–24
  for constructing electrochemical sensing, 66–68
  electroactive labels, 65–66
  electron transferring, 69–71
SERS effect, 57–59
SHINERS, 58*f*
Short-chain chlorinated paraffins (SCCPs), 255, 259
Silver nanoparticles (AgNPs), 40
Single-shelled hollow structures, 1–3
Single-shelled nanoboxes assembled by nanoparticles (CoS-NP SSNBs), 27–28
Single-shelled nanoboxes assembled by nanosheets (CoS-NS SSNBs), 27–28
Single-walled carbon nanotubes (SWCNTs), 381–382
SMAC. *See* S-doped mesoporous amorphous carbon (SMAC)
Sodium-ion batteries (NIBs), 308
  charge–discharge cycles, 329–330
  $CoNiO_2/Ti_3C_2T_x$ composite, 333
  electrochemical properties, 328
  hydrothermal and annealing, 332–333
  KTO, 333
  long-term cycle life, 330–331
  material abundance, 327
  MoS_2-carbon monolayer–sandwiched super-structure nanosheets, 329
  N-GF, 328
  NTO, 333
  OVs, 331
  SMAC, 331
  sodiation process, 332
  solvothermal reaction, 331–332
  3D sodium intercalation/deintercalation pathways, 329
Sodium-sulfur (NAS) batteries, 333–336
Sodium titanate (NTO), 333
Soil bioremediation process, 263–264
Solar photoelectro-Fenton (SPEF) process, 277
Solvothermal reaction, 331–332
Sonophotocatalysis (SPC), 285
Sorption isotherms, 253–254
SPC. *See* Sonophotocatalysis (SPC)
SPEF process. *See* Solar photoelectro-Fenton (SPEF) process
SPR. *See* Surface plasmon resonance (SPR)
SPs. *See* Superparticles (SPs)
Stern-Volmer constant, 91–92
Stern-Volmer equation, 91–92
STM/STS. *See* Scanning tunneling microscopy/spectroscopy (STM/STS)
Stokes' shift, 91–92
Supercapacitors, 27–28
Superparamagnetism, 253
Superparticles (SPs), 103–104
Surface-enhanced Raman scattering, 96–97
Surface-enhanced Raman spectrum, 42–43
Surface functionalization, 44
Surface-modified semiconductor particles, 291–292
Surface plasmon resonance (SPR), 39–43, 166–167, 191*f*
  control of
    properties, 46–47
  properties
    metal composition, 40
    morphology of materials, 41
    particle size, 41
    surrounding environmental media, 41
  SPR-enhanced photocatalysis, 170*f*, 171
    for air purification, 192–193
    design of, 189
    mechanism of, 189–192
    photocatalytic treatment of wastewater, 194–195
Suspended electrodialytic remediation, 274–275
SWCNTs. *See* Single-walled carbon nanotubes (SWCNTs)

T

TEGDME liquid electrolyte. *See* Tetra ethylene glycol dimethyl ether (TEGDME) liquid electrolyte
Ternary nanocrystals, 105–107, 130–131
TERS. *See* Tip-enhanced Raman scattering (TERS)

Tetra ethylene glycol dimethyl ether (TEGDME) liquid electrolyte, 336
Tetrahydrofuran (THF), 10–11
3,3',5,5'-Tetramethylbenzidine (TMB), 90
TGA. *See* Thioglycolic acid (TGA)
THF. *See* Tetrahydrofuran (THF)
Thioglycolic acid (TGA), 88–90, 150–151
Tip-enhanced Raman scattering (TERS), 57–59
Tip-enhanced Raman spectroscopy, 58*f*
Titania nanotube (TNT), 209
TNT. *See* Trinitrotoluene (TNT)
Transient differential absorption (TDA), 137–138
Transmission electron microscopy (TEM), 3
Trinitrotoluene (TNT), 61–62
Tri-n-octylphosphine (TOP), 126–128
Trioctylphosphine oxide (TOPO), 126–128
Two-dimensional graphitic materials, 289–290

U
UCF. *See* Unidirectional carbon fiber (UCF)
UCNPs. *See* Upconversion nanoparticle (UCNPs)
Unidirectional carbon fiber (UCF), 276
Upconversion nanoparticle (UCNPs), 107–109

V
VACNTs. *See* Vertically aligned multiwall CNTs (VACNTs)
Valence bands (VBs), 121–122
van der Waals' force, 63–64
Vertically aligned multiwall CNTs (VACNTs), 316–317
VOCs. *See* Volatile organic compounds (VOCs)
Volatile organic compounds (VOCs), 183, 204
Voltammetry, 97–99
Volumetric flow rate, 238

W
Water-based emulsion, 16–17
WHO. *See* World Health Organization (WHO)
World Health Organization (WHO), 45, 99–100

X
Xanthan gum, 271
XPS. *See* X-ray photoelectron spectroscopy (XPS)
X-ray fluorescence spectroscopy, 99–100
X-ray photoelectron spectroscopy (XPS), 175, 313–314

Z
Zeolitic imidazole frameworks (ZIFs), 226, 267
ZIFs. *See* Zeolitic imidazolate frameworks (ZIFs)
Zinc ferrite nanocomposite catalysts, 289
Zn/Fe-based hybrid catalyst, 385
Zr-doped nanocrystalline PbO_2 (Zr-PbO_2) film electrodes, 280

CPI Antony Rowe
Eastbourne, UK
November 20, 2019